CONTENTS

AUTHORS' PREFACE: HOW TO USE THIS BOOK

Important words or concepts are emphasised in the text in bold or italics. In many cases, more details of each of these words and concepts are given either in the same section, or elsewhere in the book. Don't forget to use the index if you want to find out more information!

The syllabus material in this book is arranged in *Topic Groups*. Each Topic Group has a particular *theme* which links the individual topics. Remember, you can revise in any order you like. For instance, you may want to revise them immediately after you have covered that part of the syllabus with your teacher. You can often understand topics better if you also revise related topics elsewhere in the book. We have provided "Links With Other Topics" tables after each topic, to help you find your way around the book.

Each Topic Group concludes with – *Review Questions,* to test your understanding of the preceding sections. In many cases the Review exercises are based on actual examination questions, so that you get used to dealing with them. Answers to Review Questions can be found at the end of the *chapter.* We have also included *Practice Examination Questions* at the end of most chapters. The answers to these, with an examiner's comments, can be found at the end of the *book*. You will also find a number of *actual student's answers*, with the examiner's comments on these answers to help you see what the examiner is looking for.

Martin Barker, Alan Jones and Chris Millican

ACKNOWLEDGEMENTS

We are grateful to the following Examination Groups for permission to reproduce questions which have appeared in their examination papers. The answers, or hints on any answers, are solely the responsiblility of the authors.

Midland Examining Group (MEG); Northern Examinations and Assessment Board (NEAB); Northern Ireland Schools Examinations and Assessment Council (NISEAC); Southern Examining Group (SEG); University of London Examinations and Assessment Council (ULEAC); Welsh Joint Education Committee (WJEC).

We are grateful to the following for permission to reproduce copyright material; Royal Mail.

Many thanks to Clive Hurford for his encouragement during the writing of this new edition, and for his help in preparing the drawings.

Martin Barker, Alan Jones and Chris Millican

GCSE/KEY STAGE 4

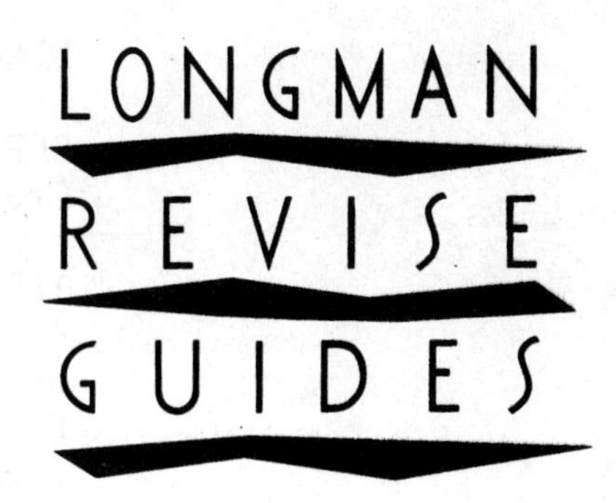

BIOLOGY

Martin Barker, Alan Jones and Chris Millican

Longman

LONGMAN REVISE GUIDES

SERIES EDITORS:
Geoff Black and Stuart Wall

TITLES AVAILABLE:
Art and Design
Biology*
Business Studies
Chemistry*
Economics
English*
English Literature*
French
Geography
German
Home Economics
Information Systems*
Mathematics*
Mathematics: Higher Level*
Music
Physics*
Religious Studies
Science*
Sociology
Spanish
Technology*
World History

* new editions for Key Stage 4

Longman Group Ltd,
Longman House, Burnt Mill, Harlow,
Essex CM20 2JE, England
and Associated Companies throughout the world.

First Published 1988
Second Edition 1994

ISBN 0 582 23772 6

British Library Cataloguing-in-Publication Data

A catalogue record for this book is
available from the British Library

Set by 19 QQ in 10/12pt Century Old Style
Printed in Great Britain by William Clowes Ltd.,
Beccles and London

ASSESSMENT IN BIOLOGY

GCSE AND KEY STAGE 4 BIOLOGY

THE SYLLABUSES

TARGET RANGES OF LEVELS

ACHIEVING THE TARGETS

TERMINAL EXAMINATION PAPERS

SPELLING, PUNCTUATION AND GRAMMAR

YOUR FINAL CERTIFICATE

GETTING STARTED

All pupils in state maintained schools are required, by law, to study National Curriculum Science (single award) or Science (double award) or the three separate sciences (Biology, Chemistry and Physics) at Key Stage 4.

The syllabuses for all science courses prepared by the various examination groups are based on the requirements of the National Curriculum for Science. However, the syllabuses do vary although all the Biology syllabuses will contain a common core, Attainment Target 2 : Life and Living Processes. All the syllabus content of the core is covered in this book. However, the extension material required for the different Biology syllabuses will vary from one examination group to another. It is important, therefore, that you check with your teacher exactly which syllabus you are studying, and to what level, so that you do not revise work at a higher level than that required for your examination.

At the end of each *topic group* in this book there will be a series of sample questions and outline answers which may be used to assess your knowledge and understanding of each area of the syllabus. There are also some typical student answers with comments from an experienced examiner. Many of the questions are based on sample KS4/GCSE questions provided by the examination groups. They will provide excellent opportunities for your preparation for the final examinations which you will sit at the end of your course.

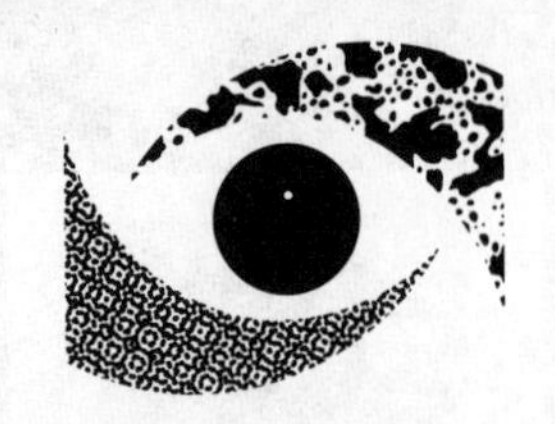

ESSENTIAL PRINCIPLES

The syllabuses designed for Biology by the various examination boards are each designed as one of a suite of three separate GCSE science syllabuses (Science: Biology, Science: Chemistry, and Science: Physics) through which National Curriculum Double Science can also be assessed and reported. The syllabuses can also be taken independently by candidates who are not subject to National Curriculum requirements, i.e. pupils educated at independent or non-state maintained schools.

Biology syllabuses must conform to the National Criteria for Biology, the National Criteria for Science, the National General Criteria and the requirements of the National Curriculum, Science Order.

The knowledge and understanding which is to be assessed is described in terms of a subject "core" and a subject "extension". The subject core is common for the syllabuses prepared by all the examination boards. This "core" comprises the National Curriculum Science Attainment Target 2 (Sc. 2) The "extension" however is represented by *additional* strands and these may differ for each of the examination boards.

ATTAINMENT TARGETS

All science syllabuses are based on one or more of the four *Attainment Targets* for Science.

Attainment Target 2: Life and Living Processes is the attainment target which represents the core of the Biology Key Stage 4/GCSE syllabus.

The five strands of Attainment Target 2 are:

Strand (i) Life and living processes
Strand (ii) Variation and the mechanism of inheritance and evolution
Strand (iii) Populations and human influences within ecosystems
Strand (iv) Energy flows and cycles of matter within ecosystems
Strand (v) This will contain the extension work specified by the examination board; e.g. "Use by humans of other organisms" and "Human health and disease" (ULEAC Biology Syll.A)

Each Attainment Target is weighted in the Scheme of Assessment. The assessment of Attainment Target 1 : Scientific Investigation (see later) is weighted at 25% and will be assessed by your teachers as Coursework during your course. The Subject Core (AT2) is weighted at 45–50% and the remaining 25–30% in the scheme of assessment is obtained by assessment of the extension work in each examination board's syllabus.

STATEMENTS OF ATTAINMENT

Each Attainment Target is sub-divided into *Statements of Attainment* which are grouped at different levels. Only levels 4 to 10 are stated for the purposes of assessment at Key Stage 4, although levels 1 to 3 will have been assessed at Key Stage 3. The Statements of Attainment state what you should be able to do at a particular level (see later breakdown of Attainment Target 2 into Statements of Attainment and levels). For example, in Attainment Target 2, Life and Living Processes, one of the Statements of Attainment for level 5 reads "be able to name and outline the functions of the major organ systems in mammals and in flowering plants".

PROGRAMMES OF STUDY

Each Attainment Target is accompanied by a *Programme of Study* which sets out the essential information to be covered so that you meet the attainment targets. The Programme of Study is quite simply a description of how you, the learner, should approach the Statement of Attainment.

Each Examining Group may present its Biology syllabuses in slightly different ways but every syllabus will have a section on Aims, Assessment Objectives, Relationships between Assessment Objectives and Syllabus Content and the Scheme of Assessment, i.e. range of grades or levels it is possible to gain by sitting various combinations of terminal examination papers at different tiers of entry.

AIMS

The aims are a description of the educational purposes of following a course in biology at this level. Some of the aims are reflected in the assessment objectives; others are not, because they cannot readily be translated into objectives that can be assessed. It is hoped that you will attain these aims during your study of the biology course. An example of a long term aim would be:

> "become confident citizens in a technological world, able to take or develop an informed interest in matters related to biology".

ASSESSMENT OBJECTIVES

The assessment objectives are a list of the abilities you should develop as a result of studying biology. You will be assessed on these during the course and in the terminal examinations set by the Examination Group. There are two main groups of Assessment Objectives: those associated with Attainment Target 1, which are the Investigative and associated objectives (Scientific Investigations). These will be assessed by your teachers through a range of practical investigations you will be required to carry out during your course. The other assessment objectives relate to knowledge, skills and understanding which will be assessed through the terminal written examinations.

ATTAINMENT TARGET 1 — SCIENTIFIC INVESTIGATION

25% of the marks for the final examination are awarded for practical skills you develop and which are assessed by your teacher throughout the course. You are expected to develop skills which enable you to explore and investigate the world of biology and to develop a fuller understanding of biological phenomena, theories to explain these and the procedures of biological investigation. This should take place through assessed activities which require a more systematic and quantified approach which develops an increasing knowledge and understanding of the subject. You will be expected to use the knowledge, skills and understanding specified in the syllabus to carry out investigations in which you:

a) ask questions, predict and hypothesise
b) observe, measure and manipulate variables
c) interpret results and evaluate scientific evidence.

ATTAINMENT TARGET 2 — KNOWLEDGE, SKILLS AND UNDERSTANDING OBJECTIVES

Through the knowledge, skills and understanding in relation to the specified syllabus content you will be expected to demonstrate the ability to:

a) communicate biological observations, ideas and arguments effectively
b) select and use reference materials and translate data from one form to another
c) interpret, evaluate and make informed judgements from relevant facts, observations and phenomena
d) use biology to solve qualitative and quantitative problems.

3 TARGET RANGES OF LEVELS

Most of the Examination Groups offer terminal examination papers at three "tiers" of entry. You will only be entered for one tier, i.e. for one examination which covers a specified range of target levels. The decision on the most realistic tier of entry for you should be taken after discussion with your teacher and your parents. If, however, you enter for a paper which proves to have been targeted at the wrong level, there is provision made by the Examination Group for the award of one grade higher or lower than the target grades. If, for example, you entered for the highest level of paper, target levels 8–10, and failed to achieve level 8, you could still be awarded level 7 if your work merited such an award.

An example of one SCHEME OF ASSESSMENT offered by one of the Examination Groups for biology is shown below. Note that the Group offers three options targeted at overlapping ranges of levels.

Scheme of Assessment, e.g. ULEAC Biology Syllabus A			
FOUNDATION TIER (levels 4 to 6)			
(a)	Paper 1	Teacher assessment of Sc1	25%
(b)	Paper 2F	1.5hrs, assessment of core (Sc2)	45%
(c)	Paper 3F	1.5hrs, assessment of extension strands (v) & (vi)	30%
INTERMEDIATE TIER (levels 6 to 8)			
(a)	Paper 1	Teacher assessment of Sc1	25%
(b)	Paper 2I	1.5hrs, assessment of core (Sc2)	45%
(c)	Paper 3I	1.5hrs, assessment of extension strands (v) + (vi)	30%
HIGHER TIER (levels 8 to 10)			
(a)	Paper 1	Teacher assessment of Sc1	25%
(b)	Paper 2H	1.5hrs, assessment of core (Sc2)	45%
(c)	Paper 3H	1.5hrs, assessment of extension strands (v) + (vi)	30%

For this particular Examination Group provision is made for the award of one level above or below the targeted range of the foundation and intermediate tiers. Thus, at the foundation tier levels 3 or 7 could be awarded at the discretion of the Group, and levels 5 or 9 at intermediate tier.

4 ACHIEVING THE TARGETS

ATTAINMENT TARGET 1: SCIENTIFIC INVESTIGATION

We have already outlined the broad objectives of this Attainment Target. Here are the specific requirements for reaching each level.

Statements of Attainment

Pupils should carry out investigations in which they:

Level 4

a) ask questions, suggest ideas and make predictions, based on some relevant prior knowledge, in a form which can be investigated.
b) carry out a fair test in which they select and use appropriate instruments to measure quantities such as volume and temperature.
c) draw conclusions which link patterns in observations or results to the original question, prediction or idea.

“Achieving the levels for each Target”

Level 5

a) formulate hypotheses where the causal link is based on scientific knowledge, understanding or theory.
b) choose the range of each of the variables involved to produce meaningful results.
c) evaluate the validity of their conclusions by considering different interpretations of their experimental evidence.

Level 6

a) use scientific knowledge, understanding or theory to predict relationships between continuous variables.
b) consider the range of factors involved, identify the key variables and those to be controlled and/or taken account of, and make qualitative or quantitative observations involving fine discrimination.
c) use their results to draw conclusions, explain the relationship between variables and refer to a model to explain the results.

Level 7

a) use scientific knowledge, understanding or theory to predict the relative effect of a number of variables.
b) manipulate or take account of the relative effect of two or more independent variables.
c) use observations or results to draw conclusions which state the relative effects of the independent variables, and explain the limitations of the evidence obtained.

Level 8

a) use scientific knowledge, understanding or theory to generate quantitative predictions and a strategy for the investigation.
b) select and use measuring instruments which provide the degree of accuracy commensurate with the outcome they have predicted.
c) justify each aspect of the investigation in terms of the contribution to the overall conclusion.

Level 9

a) use a scientific theory to make quantitative predictions and organise the collection of valid and reliable data.
b) systematically use a range of investigatory techniques to judge the relative effect of the factors involved.
c) analyse and interpret the data obtained, in terms of complex functions where appropriate, in a way which demonstrates an appreciation of the uncertainty of evidence and the tentative nature of conclusions.

Level 10

a) use scientific knowledge and an understanding of laws, theories and models to develop hypotheses which seek to explain the behaviour of objects and events they have studied.
b) collect data which is sufficiently valid and reliable to enable them to make a critical evaluation of the law, theory or model.
c) use and analyse the data obtained to evaluate the law, theory or model in terms of the extent to which it can explain the observed behaviour.

ATTAINMENT TARGET 2: LIFE AND LIVING PROCESSES

Pupils should develop knowledge and understanding of:

i) life processes and the organisation of living things;
ii) variation and the mechanisms of inheritance and evolution;
iii) populations and human influences within ecosystems;
iv) energy flows and cycles of matter within ecosystems.

Statements of Attainment

Pupils should:

Level 4

a) be able to name and locate the major organs of the human body and of the flowering plant.
b) be able to assign plants and animals to their major groups using keys and observable features.
c) understand that the survival of plants and animals in an environment depends on successful competition for scarce resources.
d) understand food chains as a way of representing feeding relationships in an ecosystem.

Level 5

a) be able to name and outline the functions of the major organ systems in mammals and in flowering plants.
b) know that information in the form of genes is passed on from one generation to the next.

c) know how pollution can affect the survival of organisms.
d) know about the key factors in the process of decay.

Level 6
a) be able to relate structure to function in plant and animal cells.
b) know the ways in which living organisms are adapted to survive in their natural environment.
c) know that variation in living organisms has both genetic and environmental causes.
d) understand population changes in predator–prey relationships.
e) know that the balance of materials in a biological community can be maintained by the cycling of these materials.

Level 7
a) understand the life process of movement, respiration, growth, reproduction, excretion, nutrition and sensitivity in animals.
b) understand the life processes of photosynthesis, respiration and reproduction in green plants.
c) understand how selective breeding can produce economic benefits and contribute to improved yields.
d) know how population growth and decline is related to environmental resources.
e) understand pyramids of numbers and biomass.

Level 8
a) be able to describe how the internal environment in plants, animals and the human embryo is maintained.
b) know how genetic information is passed from cell to cell and from generation to generation by cell division.
c) understand the principles of a monohybrid cross involving dominant and recessive alleles.
d) understand that the impact of human activity on the Earth is related to the size of the population, economic factors and industrial requirements.
e) understand the role of microbes and other living organisms in the process of decay and in the cycling of nutrients.

Level 9
a) be able to explain the coordination in mammals of the body's activities through nervous and hormonal control.
b) understand the different sources of genetic variation.
c) understand the relationships between variation, natural selection and reproductive success in organisms and the significance of these relationships for evolution.
d) understand the basic scientific principles associated with a major change in the biosphere.
e) understand how materials for growth and energy are transferred through an ecosystem.

Level 10
a) understand how homeostatic and metabolic processes contribute to maintaining the internal environment of organisms.
b) understand how DNA replicates and controls protein synthesis by means of a base code.
c) understand the basic principles of genetic engineering, selective breeding and cloning, and how these give rise to social and ethical issues.
d) understand how food production involves the management of ecosystems to improve the efficiency of energy transfer, and that such management imposes a duty of care.

ADDITIONAL STRANDS

This will vary in content for the syllabuses of different exam groups.

Eg. *ULEAC Syllabus A*
— Use by humans of other organisms
— Human health & disease

ULEAC Syllabus B
— Microbiology and biotechnology
— Social Biology

NEAB
— Biotechnology
— Movement and support.

All the biology written terminal examination papers set by the various Examination Groups will be similar in style and format. They will contain compulsory, structured questions to be answered on the examination script. Spaces are left after the questions so that you may write your answer. The mark allocated for each question is indicated inside brackets at the end of the question. The papers in the tiers aimed at the highest level will require more extended writing of continuous prose, but those for the lowest tiers will contain more questions which require shorter answers often requiring no more than a word or a sentence.

The terminal examination papers, in total, will only account for 75% of the total mark available for the examination, the other 25% will be awarded for scientific investigations of a practical nature during the course. These marks will be awarded by your teachers and the marks moderated by the Examination Group so that standards are applied consistently across the country.

Marks for spelling, punctuation and grammar will be awarded in accordance with published regulations by the Examination Group.

For each written component including Scientific Investigation (SC.1) 5% of the total marks available are allocated to spelling, punctuation and grammar according to the three performance criteria below. The marks indicated assume a total mark of 100% for the examination.

Threshold Performance

(Level 1) Candidates spell, punctuate and use the rules of grammar with reasonable accuracy; they use a limited range of specialist terms appropriately.

1 mark allocated.

Intermediate performance

(Level 2) Candidates spell, punctuate and use the rules of grammar with considerable accuracy; they use a good range of specialist terms.

2–3 marks allocated

High performance

(Level 3) Candidates spell, punctuate and use the rules of grammar with almost faultless accuracy, deploying a range of grammatical constructions; they use a wide range of specialist terms adeptly and with precision.

4–5 marks allocated.

7 YOUR FINAL CERTIFICATE

In May 1993, the Secretary of State for Education announced that the final record of your performance in National Curriculum examinations at KS4 will be indicated by grades currently used to report GCSE examination results. Grades A*–G will be used for the final GCSE award. The A* (super-A) will be used to grade those achieving the equivalent of a level 10.

The chart below shows the link between the National Curriculum levels and GCSE grades.

GCSE Grades	*Key Stage 4 Levels*
A*	10
A	9
B	8
C	
	7
D	
	6
E	
F	5
G	4
U	3

In the scheme of assessment indicated earlier the Higher tier of papers gives you access to grades A*–B; the Intermediate tier gives you access to grades B–E; and on the Foundation tier only grades D–G are awarded. In exceptional cases one grade higher or lower may be awarded by the Examination Group at each of the tiers of entry.

There is no stated "pass mark" for each level or grade. This will be determined by the Examining Group after an awarding panel has considered the performance of all candidates, in all papers for each syllabus offered by the Examination Group. The degree of difficulty of the examination papers will vary from one year to the next and, therefore, so will the performance of the candidates entered for the examinations. The Groups determine the marks at which each grade is awarded after attempting to ensure that the same standards are maintained year on year.

EXAMINATION TECHNIQUES AND COURSEWORK

STRUCTURED QUESTIONS

EXTENDED WRITING

MODULE TESTS

INSTRUCTION WORDS USED IN EXAM QUESTIONS

LANGUAGE AND LITERACY SKILLS AND MATHEMATICAL REQUIREMENTS

DIAGRAMS

COURSEWORK

REVISION TECHNIQUES

GETTING STARTED

Just as a tennis player learns new techniques and trains for a tournament, so there are a number of techniques which you can develop, which will help you to improve your performance in the examination, and gain more marks. For example, you can practise answering questions to a set time limit, and without using books to help you, before the exam. You will then be better prepared to cope with the actual conditions you will meet in the exam.

Examination questions are written by examiners to find out how much you know and understand. They also test your ability to handle information and to apply what you know to new situations. You are often given information in a question to help you answer the question, so a lot depends on reading the question fully, and making use of all the available facts. **You will only gain marks for answering the questions which are set by the examiner, not for answering your own version of the question!**

Coursework is the work that you will have completed at school or college prior to the written examination. This work will have been marked by your teacher and you may have had some feedback about how well you are doing. Remember that Coursework counts for 25% of your total assessment, and that the written Terminal Examination (together with Module Tests where applicable) counts for 75%.

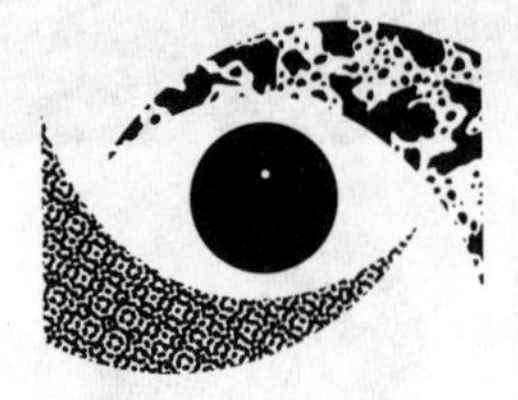

ESSENTIAL PRINCIPLES

The Terminal Written Examination is the examination you take at the end of your course of study, usually in May or June of Year 11 and is worth 75% of your marks, unless you are taking a modular course where the terminal exam is worth only 50%. As described in Chapter 1, all the Examining Groups set examination papers which are similar in style and format, consisting of compulsory, *structured questions* which are answered on the question paper in the spaces provided. The papers aimed at the higher levels provide more opportunity for *extended writing* and *calculations*.

1 STRUCTURED QUESTIONS

The structured questions are divided into sub-questions (a) (b) (c), etc. If you are unable to answer part (a) for example, you should still be able to go on to attempt (b) or (c) without doing (a). Sometimes the sub questions are further divided into say, (b) (i), (b) (ii), etc.; you will usually need to have answered (i) before doing (ii). The marks allocated for each part of the question are indicated in brackets. If a question asks you for "TWO changes" there are usually two lines for your answer with the numbers 1 and 2 on the lines. There may sometimes be a table for you to complete with your answers. The examiners are really trying to help you to write down what you know and understand!

Structured questions are usually based on a particular topic or theme from the syllabus, such as "osmosis", "digestion", or "reproduction". The questions are usually based on some "stimulus material" such as one of the following.

- A diagram of some apparatus which you may be familiar with; for example, a potometer.
- A diagram of something you have seen in a book; for example, the human body.
- A table (chart) showing information; for example about the heights of pupils in a class.
- A graph; for example, showing the rate of enzyme action at different temperatures.
- A photograph or sketch with some explanatory information.
- Some written information; for example, a newspaper article.

The question usually starts with an introductory sentence telling you what the question is about, followed by the "stimulus material" as described above. This is in turn followed by a series of short questions requiring a few words or a few sentences for an answer.

The diagram below shows a simple food web in a cabbage field.

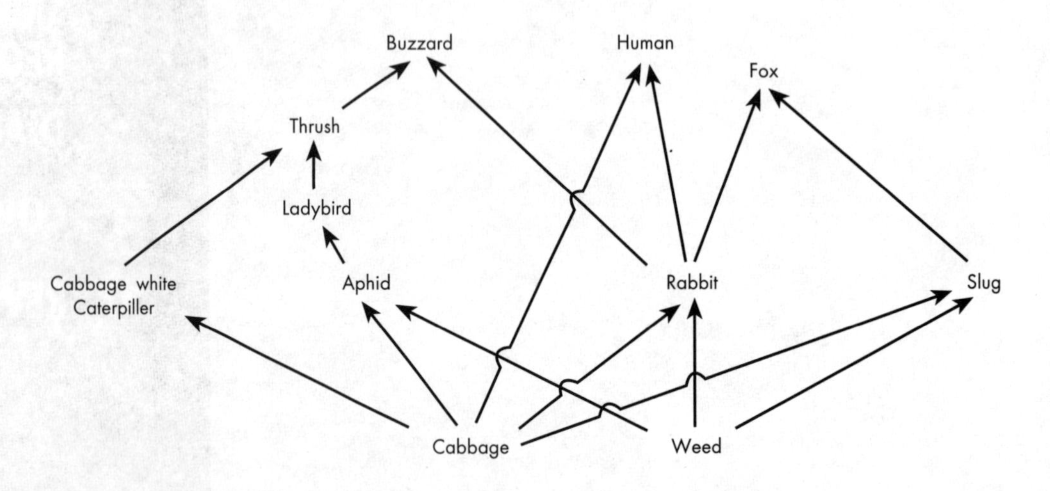

Fig. 2.1

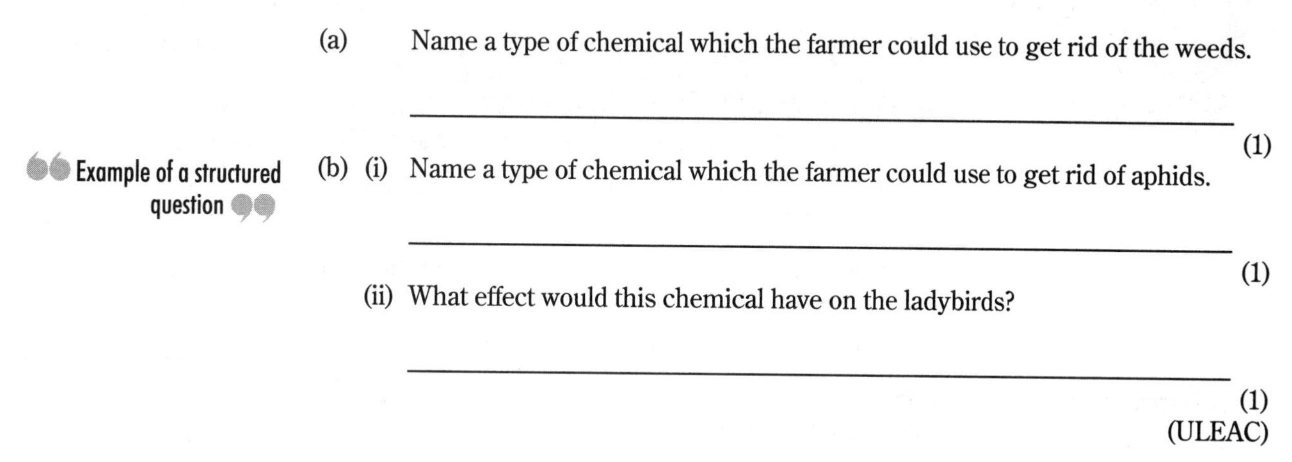

(a) Name a type of chemical which the farmer could use to get rid of the weeds.

______________________________ (1)

Example of a structured question

(b) (i) Name a type of chemical which the farmer could use to get rid of aphids.

______________________________ (1)

(ii) What effect would this chemical have on the ladybirds?

______________________________ (1)

(ULEAC)

At the end of the space for your answer you may see how many marks are being awarded for that part of the question. Your answer should aim to fill the space which is provided, but you can write a little more or a little less if you wish. Try and use as many scientific words as possible so that the examiner can see evidence of your knowledge and understanding in Science: Biology and can award you marks. Remember the general rule that one correct fact usually gains one mark.

If you are asked for the answer to a calculation, always include the units for any number which you write: for example, the length of a leaf is 10 cms, not just 10.

2 QUESTIONS INVOLVING EXTENDED WRITING

These questions appear more frequently on the intermediate and higher level papers. They are often aimed at assessing level 9 and 10 candidates. They usually require a longer answer in the form of a few sentences and may allocate up to 10 marks for your answer. These questions may ask you to "explain how something happens" or to "suggest reasons" for a particular observation. You may find it helpful to write the main points of your answer in pencil on the question paper before answering the question. Check that you have matched the number of points you have made to the marks allocated and remember to cross out any rough notes you have made.

If three marks are available, it is likely that you will be expected to make three distinct points. For example:

"Explain what is meant by *excretion.*" (3)

To be awarded three marks, you would need to include three distinct points in your answers. For instance, Excretion is a process by which the *toxic waste products* of *metabolism* are *removed from the organism's body.* Each part of the answer shown in italics would be worth one mark, though terms such as "toxic waste products", "metabolism" and "removed" may need to be explained more fully, to be absolutely sure of the marks.

Some questions give more marks to reflect the amount of work that you have to do, rather than the length of your answer. In answers to "longer" parts of structured questions you will usually be expected to *write complete sentences.* Keep them as "tight" as possible. Don't waffle; it will waste time and also reveal weaknesses in your answer. If parts of your answer contradict other parts, you will not be given credit for anything which is correct. This may seem unfair, but examiners want to avoid giving marks to candidates who simply "pepper" their answers with alternative responses, hoping that at least some of them will be correct! It is a good idea, therefore, to *develop a concise, accurate style.*

Examination papers remind candidates of the use of good, clear answers. This will often be stated in the instructions (the rubric) which appears before the questions.

The three main points to remember when writing *longer answers* to structured questions and (especially) essays are:

Points to remember when answering longer questions.

- **clarity**. Avoid ambiguous ("two meaning") answers. Use good English. If you cannot remember a biological word, describe it instead. You will not necessary lose marks for incorrect spelling, provided that the meaning is clear.
- **relevance**. Only include material which is *directly relevant* to the question. Do not be tempted to introduce additional material because you happen to have revised it

and want to use it! Irrelevant information may be confusing, will detract from your main points, lose marks and also may irritate the examiner!

- **organisation.** The sequence of separate points in your answer may be important, for instance to build up an argument in stages. This requires organisation, which means *planning* your answer in advance. This skill is particularly important in *essay questions*.

You may be given the opportunity of providing **diagrams** as part of your answer. The instructions for the examination paper will make this clear.

Diagrams can help.

Diagrams can be a very effective means of answering a question. In the "excretion" question above, a diagram would not be appropriate. However, in the following question a diagram might be a good idea:

"Explain briefly how plants respond to the stimulus of gravity."

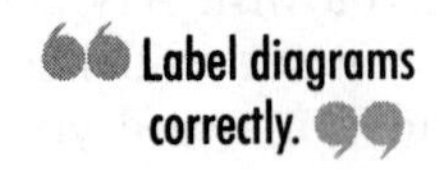

As with the written part of an answer, diagrams should be clear, relevant and well-organised. Avoid drawing sketchy, small, incompletely labelled diagrams. Simple diagrams are generally more effective and less time-consuming to draw than complex diagrams. If you decide to draw a diagram ensure that it *adds* to what you have written; it should not *duplicate* your written answer.

Essay Questions

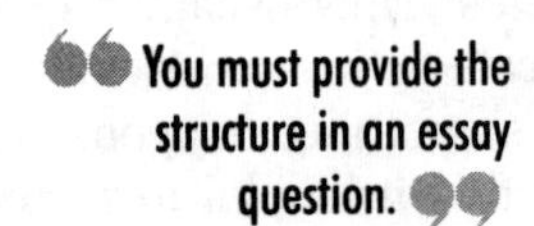

Essay questions can be very demanding, especially if you do not have a clear understanding of the topic which forms the basis of the question. Essay questions unlike multiple-choice and structured questions, are unstructured, and *the structure must be provided in the answer*. You will gain credit for both the *content* and the *organisation* of the essay. The comments already made for structured questions apply particularly for answers to essay questions. Plan how you're going to answer the question *before* you actually begin the answer. In other words, you need to *plan the answer*.

An essay plan can be written in note form before you start the essay. Single words will do. Establish the content of your answer, and also the *sequence* of separate points. You can even add to this plan while you are writing the essay, as additional points for inclusion occur to you later. You can also use the plan as a "checklist" to ensure that all your points are included. If time runs out, you may not be able to complete the essay; if so, leave the plan as part of your answer – label it "essay plan". You may be given credit for this.

In general, each main point in an essay can be expanded within its *own paragraph*, with each sentence being used to make subsidiary points, in support of the main theme for each paragraph. Ideally, each sentence should link with the next. If you achieve this successfully, the essay will "flow", and will be fairly easy to write!

The opening and closing paragraphs of an essay are often the most critical. The first paragraph should establish the general theme of your essay. It's often a good idea to begin with a *definition* of the subject you are writing on. The initial paragraph is often the most difficult to write. However, once you have managed this, the remainder of the essay should develop more easily.

The number and frequency of Module Tests vary from syllabus to syllabus. They seem to have a variety of question styles, involving the use of short answer and structured questions of the type already considered. However the module tests also make use of a variety of types of *multiple choice* (or *objective*) questions.

MULTIPLE CHOICE OR OBJECTIVE QUESTIONS

These questions test your knowledge and understanding of a wide range of topics from the syllabus. Each question has a "stem", which is the main part of the question. The stem may include diagrams or tables of figures, and you need to read the stem carefully to obtain as much information as possible before you read the list of suggested answers, called "options". These options are usually labelled A to E, and only one of them is the correct answer, which is the "key". The incorrect answers are the "distractors".

One way of approaching this type of question is to look for the correct answer and to ignore the distractors. Alternatively, you can work through the list of options and *reject each distractor* until you are only left with the "key". Look at the following example.

Fish is a good source of one of the following food groups

Example of multiple choice question

A Fat **B** Protein **C** Vitamins **D** Minerals **E** Carbohydrates

You may know the correct answer to this (key = **B**), or you may be able to reject each of the distractors.

In the exam, you may be able to use rough paper or the question paper to work out your answer. You mark the answer which you have chosen on a special answer grid using a soft HB pencil. You can then rub out an answer if it is wrong, and make a new mark. Remember, there is only *one* correct answer to each question, and every question carries one mark. If you get stuck on a question, leave it and go on to the next. You may have time to come back to that question later and at least have a guess at the answer before the exam ends. When you have completed as many questions as possible, you should *check* each of your answers carefully, making sure that you have put your answer against the number on the grid which corresponds to the question you have answered. Try to use all the time you have in the exam in this positive way instead of just doing nothing!

Check your answers

Sometimes the type of objective question may involve *sentence completion* or selecting *matching pairs* out of a range of possible answers.

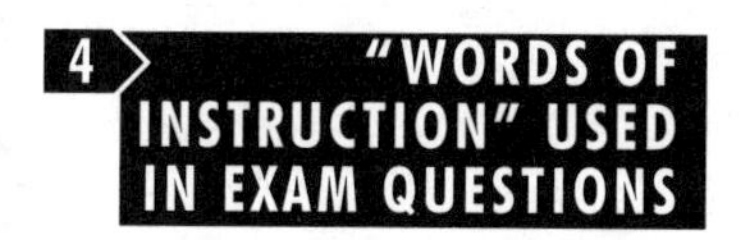

4 "WORDS OF INSTRUCTION" USED IN EXAM QUESTIONS

An examiner asks you to answer the question in a certain way by giving you help in the form of a "word of instruction" in each sub-question. For example, if you are asked to "Explain . . ." then writing a list will not qualify you for the marks. Some examples of the *words of instruction* which are often used are given below.

Define

means provide a *definition* or formal statement to show your understanding. Structured questions sometime begin with "define". Definitions for the main terms in biology should be learnt as part of your revision.
Example: "Define the term *digestion*".

"What is meant by" or "What do you understand by"

means provide a *definition* with some further *explanation* on the significance of the subject. This may, for instance, involve linking two definitions.
Example: "What is meant by the terms *recessive* and *dominant?*"

State

means that a *concise answer* is expected, with little or no supporting argument.
Example: "State *two* advantages to a plant in having a large number of root hairs."

List

means that you should provide a *number of points*, often consisting of single words. No elaboration is expected. You may number each item, if appropriate. If the question specifies a particular number of points, this should not be exceeded. In some structured questions, the list may be the basis for supplementary questions.
Example: "List three important functions of a skeleton."

Explain

means that you should include some *reasoning* in your answer, perhaps explaining a fact or piece of evidence provided in the question.
Example: "Explain how a food bolus passes along the oesophagus of a mammal."

Describe

means that you should state the *main points* of a topic, using words and (where appropriate) diagrams.
Examples: "Describe an experiment to show the effect of different intensities of light on photosynthesis."
"Describe, with the help of examples, how the components of a balanced diet are important in the functioning of the human body."

Outline

means that you should give only "essential" details in your answer, which should be *brief.*
Example: "Outline the sequence of events which results in a person rapidly removing their hand from a very hot object."

"Some useful words of instruction"

Suggest

means that you are required to provide a *suitable explanation* for a given situation, using your biological knowledge. There may not be a single, "unique" answer, but credit will be given for your answer if it is *reasonable, relevant* and *logical* to the question. You may be expected to apply your knowledge to a "novel" situation.
Example: "Suggest an explanation for the fact that, given the choice, maggots (housefly larvae) tend to prefer dark places compared with light places."

Give an account of

means that you should *describe a series of events*, referring to particularly important features as appropriate.
Example: "Give an account of the life cycle of a *named* amphibian from the time the egg is released by the female to the time the young amphibian assumes the adult form."

In revision and in the examination itself, you should identify the key term, if it exists, which tells you what sort of question you are about to answer.

LANGUAGE AND LITERACY SKILLS

As well as the ability to use correct spelling, grammar and punctuation described in Chapter 1, the following skills are required:

- recording and storage of information in appropriate forms
- using and understanding information gained from various sources
- communicating ideas to others
- summarising and organising information in order to communicate adequately
- using appropriate language to explain the results of observations in a variety of contexts
- using and interpreting scientific nomenclature, symbols and conventions.

MATHEMATICAL REQUIREMENTS

You are expected to have some skills in mathematics to answer some of the questions in a Biology: Science examination. They vary from one Examination Group to another but are basically as follows:

Some useful mathematical skills

Levels 4–6	National Curriculum Mathematical reference
add, subtract, multiply and divide whole numbers	2.2b, 2.3c
recognise and use expressions in decimal form	2.4c
make approximations and estimates to obtain reasonable answers	2.4e
use simple formulae expressed in words	3.4b
undertake mensuration of triangles, rectangles and cuboids	4.4d
understand and use averages	5.4e
read, interpret and draw simple inferences from tables and statistical diagrams	1.5b, 5.5c
find fractions or percentages of quantities	2.5b
construct and interpret pie-charts	5.5c
construct bar charts	5.5c
calculate with fractions, decimals, percentage or ratio	2.6a
solve simple equations	3.6b
substitute numbers for letters in simple equations	3.6b
interpret and use graphs	5.6
plot graphs from data provided, given the axes and scales	5.6

Levels 7–10	National Curriculum Mathematical reference
use appropriate limits of accuracy	2.7c
make approximation to a given number of significant figures or decimal places	2.7c
undertake mensuration of a cylinder	4.7d
choose by simple inspection and then draw the best smooth curve through a set of points on a graph	5.7a
recognise and use expressions in standard form	2.8a
manipulate simple equations	3.8a
select appropriate axes and scales for graph plotting	3.8c
determine the intercept of a linear graph	3.8c
understand and use inverse proportion	

A NOTE ABOUT GRAPHS

If you are asked to draw a graph in the exam there are some important points to remember:

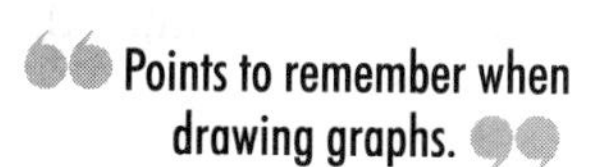

a) use the axes and scale if they are given in the question;
b) if the axes and scale are not given in the question then decide on a scale which will fit the figures given in the data;
c) the x axis goes along the bottom or on the horizontal, the y axis goes upwards or on the vertical;
d) the factor which changes regularly, such as time, goes on the x axis;
e) label your axes, and indicate the scale used;
f) write a heading on your graph;
g) draw either a smooth curve through the points on the graph or the best-fitting straight line.

Marks are usually given for:

- use of appropriate scale and correct axes;
- correctly labelled axes and title;
- accurate plot of points;
- best straight line or best smooth curve.

6 > DIAGRAMS

You may be asked to draw diagrams to show apparatus, or possibly diagrams to show biological structures; eg. the structure of an animal cell.

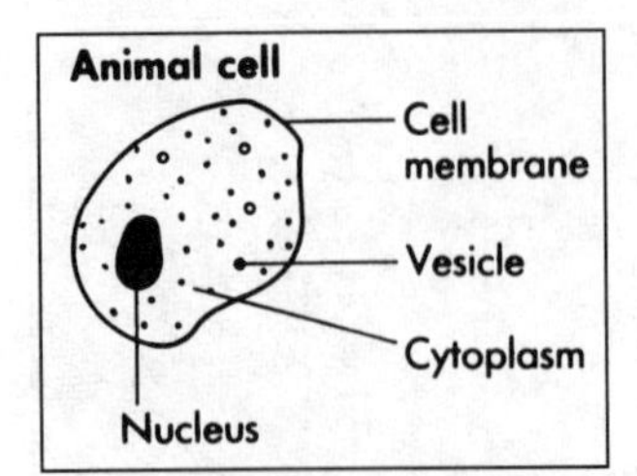

Fig. 2.2 How to use radial labelling lines around a diagram.

- Use a *sharp* HB pencil.
- Try to make the diagram fit the space allowed on the exam paper, or use about a third to a half page of A4 if answering on lined paper.
- State the *magnification/scale* (if relevant).
- Write a *heading* above the diagram to state what it is showing.
- Use *labelling lines* to label the different parts of the diagram clearly. For example, Figure 2.2 opposite shows how you can use radial *labelling lines* around a diagram. Alternatively you can use a *list of labels* at the side of a diagram.
- Include as much *accurate detail* on the diagram as possible.
- When you draw apparatus, only draw the *relevant parts* of the apparatus and omit standard equipment such as retort stands, Bunsen burners, etc. The diagrams in this book may be useful to study for guidance.

Look at the *number of marks* allocated for the diagram. Five marks may mean that it's better to only spend about five minutes on a diagram. You can always come back to it if you have extra time at the end.

7 > COURSEWORK

The Coursework components of all Biology: Science courses is worth 25% of your overall assessment for Key Stage 4. It should form part of your normal Science: Biology course and assess those skills which you have had a chance to develop during the course. The assessments will be based on whole investigations during the units of work you are studying for Attainment Target 2.

The emphasis in Coursework is that you, as a learner, take responsibility for making decisions in planning, carrying out and reporting investigations. Your teacher is there to assist you in developing these skills and to assess you, not to make the decisions for you.

Coursework is designed to assess the three *strands* in Attainment Target 1, as follows.

Candidates should use the knowledge, skills and understanding specified in the syllabus to plan and carry out investigations in which they do the following:

- ask questions, predict and hypothesise; (**strand 1**) for example this could mean:
  - suggesting your own questions
  - predicting the outcome of an experiment you have planned
  - planning a solution to a problem, deciding how to make a fair test

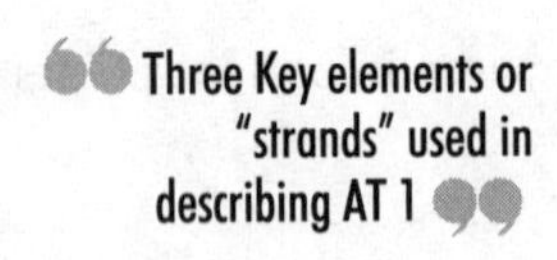

- observe, measure and manipulate variables: (**strand 2**) for example this could mean:
  - deciding what to measure
  - choosing the most appropriate apparatus
  - deciding how to organise your apparatus to achieve results
  - deciding how to record and display your results

- interpret their results and evaluate scientific evidence; (**strand 3**) for example this could mean:
  - deciding what your results mean
  - arriving at your own conclusions
  - evaluating what you have done

In similar terms these three "strands" are described by other Exam Groups as:

- Designing investigations/Predicting
- Carrying out investigations/Implementing
- Interpreting investigations/Concluding

For each level (1–10) there are *statements* describing what you should be able to do within each strand.

In *strand 1* for example, a level 8 candidate would be expected to:

“Some examples of the statements used to describe what you should be able to do to reach specified target levels for AT1”

"use scientific knowledge, understanding or theory to generate quantitative predictions and a strategy for the investigations".

In *strand 2* within the context of an investigation where one independent variable is given, a level 8 candidate would be expected to:

"take readings using the measuring instruments chosen to provide quantitative results to a high degree of accuracy"

In *strand 3* a level 10 candidate would be expected to:

"use and analyse the data obtained to evaluate the law, theory or model in terms of the extent to which it can explain the observed behaviour"

These statements are included here to give you an idea of how your teacher will be assessing your work in general terms. Your teacher will probably use more specific descriptive statements for a particular investigation. You should have an opportunity to discuss the quality of your coursework with your teacher and to improve on your level of achievement where appropriate.

LEVELS FOR AT 1

There are 10 levels described in *each* of the three "strands". These marks are described by one Examination Group as relating to the levels of achievement as follows:

Mark	*Level*
11–13	4
14–16	5
17–19	6
20–22	7
23–25	8
26–28	9
29–30	10

However, at this stage, this information can only be taken as an approximate guide to the relationship between marks and levels.

There are normally two different methods by which you can be assessed on practical skills. One way is by your teacher watching you carry out a particular practical, perhaps involving you in the handling of apparatus or in following instructions. The second method is by your teacher assessing what you have written during a practical investigation.

The work you hand in for assessment may include your observations and a presentation of your results, perhaps as a chart or graph. Your teacher can then use your written work to assess your ability to make and record observations. You are usually assessed on more than one occasion for a particular skill, so don't worry if you haven't done too well on any one particular piece of work. You may be assessed on the same skill at a later date, or you may be able to arrange this with your teacher. The best person with whom to discuss the standards you have reached on your practical assessments is your teacher at school. He or she may not be able to tell you the actual mark for any particular skill, but may be able to give you some guidance about how you can improve your level or performance in a particular skill area.

Points to remember when submitting coursework:

“Presentation is very important”

a) There should be a clear heading or title, and an introduction which describes the investigation, and shows that you understand what the investigation is about.
b) You should have your name, the date and your form or set, clearly written on the work.
c) Underline the headings and subheadings.
d) All diagrams, charts, graphs, photos, etc. should have a heading and labels.

e) List all relevant equipment and apparatus.
f) Describe any safety precautions which you have taken, for example wearing safety goggles.
g) Present your results as a chart or graph. Refer to, and make use of, your results when writing up your coursework.
h) Describe any problems you had during the investigation and suggest possible solutions.
i) Identify possible sources of error and suggest further investigations.
j) List any references which you may have used.

You will find it useful to keep all your coursework in a folder, as the examination boards usually look at the coursework from a random selection of about 10% of candidates from a centre. Your work may therefore go to an examiner, called a coursework moderator, who is responsible for ensuring that standards are similar between different schools.

8 REVISION TECHNIQUES

Learning biology is not just learning about biology! It also involves you in developing techniques for tackling revision and for taking the examination. The skills involved are, in most cases, useful for all your subjects, or at least all your sciences. So learning about good revision and examination techniques is not an additional burden to worry about; it should make things easier for you!

The fact that you are using this book means that you are motivated enough to prepare sensibly for the examination! Motivation – having a purpose – is a major factor in determining the quality of your revision. The purpose, of course, is to achieve the best possible grade in the examination.

In this section, brief guidance is given on three main areas: revision planning, revision sessions and examination preparation.

1 Revision Planning

In preparing for the examination, it is important to adopt an *active* and *positive* approach to the revision process. You can do much to improve your performance if you are clear about what will be expected of you. It is never too early to start revision!

- A *revision timetable* is an excellent way of organising your revision in a systematic way.
- Be *realistic* in planning revision. Set small, definite tasks which you can be sure of achieving. When you have completed a task successfully you will feel better motivated to continue a fairly demanding revision programme. This will obviously include all your examination subjects, not just biology!

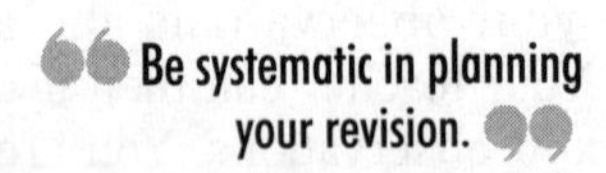

- Plan individual *revision sessions*. Each session should consist of fairly intensive work on a particular subject. The session should be short, say 30 minutes. Allow a 5 minute break between each session. If you plan 5 or 6 sessions for each subject every week, your revision programme will occupy about 16–20 hours per week in total.
- Set *definite goals* within each session. For instance, after reading through your notes on the human heart, say, write a *summary*. This could include drawing the main diagram(s), clearly labelled of course, and listing the main points. Balance your programme between subjects, and include plenty of time for *relaxation*, too!
- Spread your revision *evenly* between now and the actual exams. Avoid leaving too much work too near the exam. You may need to find out how much you can achieve each week, perhaps by having a "trial week" first. It's easy to over-estimate what can be done during each session.
- Keep a *work diary* to record what you have actually done. You could also have a revision planning session at the end of each week, to prepare for the next.
- Develop *regular working habits*. This will avoid time-wasting periods of indecision and worry.

2 Revision Sessions

There is no "right way" of revising. Use whichever technique seems appropriate for *you,* and for the particular subject you are revising. The important thing is that revision should be *efficient*, in other words, that it helps you to *learn*.

- *Vary* your revision programme as much as possible, for instance by balancing out different subjects and by using different revision techniques. This lively approach should keep you alert and interested in the revision programme.
- Be *active* rather than passive. You should be *doing* something during each revision session; this will increase your understanding of the subject. For instance, you could organise or re-organise your notes, so that they make sense to you. This will probably involve *condensing information* from classnotes and from textbooks (i.e. writing a *summary*).

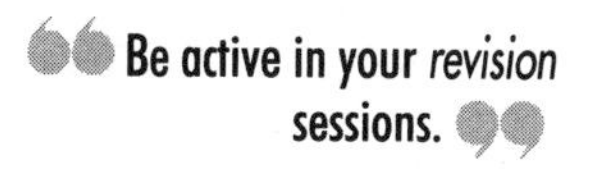

- Divide your syllabus into manageable "chunks". Each session should be devoted if possible to a *topic* or sub-topic; the sessions will therefore consist of a "logical unit" of learning.
- Use *headings and sub-headings*. Develop your own system for emphasising the relative importance for each heading, for instance with underlining, colours, a numbering system. Use brief summarised notes, from different sources of information.
- Write *revision notes* as lists, flow-diagrams, or so-called "spider diagrams" – whichever you prefer. The notes could be written on separate pieces of paper, or card. Focus on essential details only, but don't omit important ideas.
- Build up *links* between related ideas. You will actually need to do this when answering longer examination questions.
- *Memorise information*, for instance by repetition and association. Your revision notes should stimulate recall.
- Check your *understanding* by answering questions, especially as you approach the exam. Answer questions in outline, or in full. Work with others if you find this helpful. Practice with different types of questions; there are examples in this book and, of course, in sets of past questions.

3 Examination Preparation

You should aim to complete most of your revision before the exams actually start. Then you can spend any available time having a quick look at revision notes, and relaxing between exams.

- Avoid "overworking", as this can be counterproductive, making you confused, tired and increasingly anxious. Some anxiety is useful in giving you "positive anticipation" of the exam. However, too much anxiety can diminish your performance. You can reduce anxiety by developing an *exam strategy* in advance, and by various relaxation techniques.

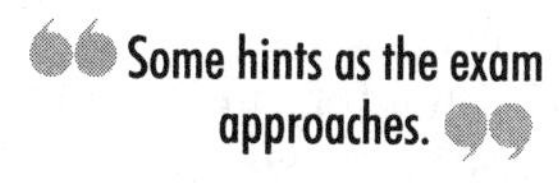

- Adopt a *positive attitude* to your work and to yourself. Revision helps you to prepare for the exams, and establishes a generally confident attitude.
- Remember that examiners are interested in what you *can do*, rather than what you cannot do. This should be your attitude, too. However, you should do what you can to come to terms with any topics which you do not fully understand.
- Check your *revision progress*, perhaps "ticking" topics as you revise them. This will give you a feeling of achievement, which will keep you motivated.
- Concentrate on *revision tasks*, rather than on yourself. However, you can design a revision programme which includes occasional "rewards" for all your hard work!
- Know in *advance* when and where the exams are to be held, and what equipment you will need. You should also know beforehand how the exam is organised, e.g. the style and number of questions for each paper.
- Plan an *exam budget* in advance. This means calculating how many minutes you should spend on each question, according to the number of marks available.

CHAPTER

ESSENTIALS OF LIFE

CHARACTERISTICS OF LIFE

ORGANISATION OF LIFE

BIOLOGICALLY IMPORTANT MOLECULES

ENZYMES

MOVEMENT OF SUBSTANCES ACROSS THE CELL MEMBRANE

STRAND (I) LIFE PROCESSES AND THE ORGANISATION OF LIVING THINGS

GETTING STARTED

This chapter and the next cover the materials involved in the core area of the Biology Key Stage 4 Syllabus, "Life and Living Processes". After considering background information on the "Essentials of Life" in Chapter 3 we cover all the key topic areas you should know in Chapter 4. By the end of Chapters 3 and 4 you should have met all the strand (i) statements of the National Curriculum outlined below.

- Pupils should extend their study of the major organs and organ systems and life processes.
- They should explore and investigate sensitivity; co-ordination and response, and should relate behaviour to survival and reproduction in plants and animals.
- They should investigate limiting factors in photosynthesis, and the use of photosynthetic products in plants.
- They should explore how the internal environments of plants and animals are maintained, including water relations, temperature control, defence mechanisms, solute balance, for example, *sugars, carbon dioxide, urea, and the human embryonic environment.*
- In the context of their study of the major human organs they should consider the factors associated with a healthy lifestyle and examples of technologies used to promote, improve and sustain the quality of life.
- They should consider how hormones can be used to control and promote fertility, growth and development in plants and animals, and be aware of the implications of their use.
- Pupils should have opportunities to consider the effects of solvents, alcohol, tobacco and other drugs on the way the human body functions.

In this chapter we are mainly concerned with "general" ideas and structures which underpin the key *topic areas* of Chapter 4. Many of these are needed to understand the higher levels (7–10) of strand (i). Don't worry if you don't understand them at first: when you have studied the *topic* they are linked to in Chapter 4, they will seem much clearer. Think of this as a **reference chapter** that you can use to find out more about these complex ideas. Review Questions and Examination Questions will appear in Chapter 4 near to the topics on which they are based.

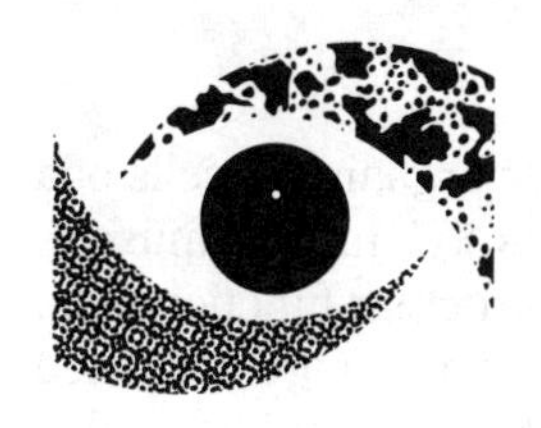

ESSENTIAL PRINCIPLES

Living things show considerable **diversity**. This is because they are *adapted* to different ways of life. However, there is also a remarkable degree of **uniformity** amongst organisms. This is apparent in two main respects:

1. All organisms demonstrate **characteristics of life**.
2. All organisms are part of an **organisation of life**.

1 CHARACTERISTICS OF LIFE

Try to memorise these characteristics of life.

It is very difficult to define life. However, it is possible to describe those features which are commonly associated with life:

1. **Movement.** Most organisms are capable of moving all or part of their bodies.
2. **Nutrition.** All organisms are capable of obtaining substances from their environment for growth, maintenance and energy.
3. **Respiration.** Living things need energy, released from the breaking down of certain food molecules inside living cells.
4. **Excretion.** Organisms need to remove the waste products of essential chemical processes which occur within them.
5. **Sensitivity.** All organisms can detect and respond to important changes in their internal or external environment.
6. **Growth.** Organisms increase in size, using materials which have been absorbed from their environment. Growth is often accompanied by an increase in complexity.
7. **Reproduction.** Every type of organism is capable of producing offspring which share many of their characteristics.

It is important to know these 7 characteristics of life. Some people use the mnemonic "MRS GREN" to help remind them of:

M ovement

R espiration

S ensitivity

G rowth

R eproduction

E xcretion

N utrition

You should note that living things show most if not all of these characteristics; non-living things demonstrate few if any of them. Other characteristics of life include the existence of **protoplasm** ("living material") and **cells, homoeostasis** and evolution. **Death** is also, in a sense, a characteristic of life.

2 ORGANISATION OF LIFE

Life is organised in different ways. For instance there is a **hierarchy** of scale:

Molecules

Biologically-important molecules include small, relatively simple **inorganic molecules** (e.g., water, oxygen, carbon dioxide) and also large, complex **organic molecules** (e.g. carbohydrates, proteins, fats and nucleic acids).

Cells

Cells are the structural and functional units of life. Most organisms consist of one or more distinct cells. Individual cells are subdivided by membranes into smaller units called **organelles**, each performing a particular function. **Multicellular** organisms consist of many cells, often of different types.

Organisms

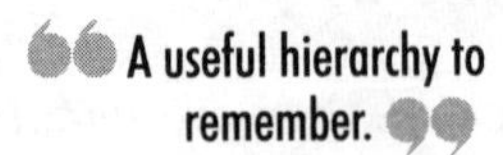

This is another word for "living things". Within multicellular organisms, cells of a particular type are grouped together in **tissues**, such as xylem and muscle. **Organs** are composed of different tissues and perform a particular function, e.g., leaf, stomach. Organs operate together as **systems** to perform a range of related and coordinated functions e.g., vascular system, the digestive system. Complex organisms, e.g. humans, have several systems.

Populations

Populations are groups of organisms of the same type; i.e., they belong to the same species. A species is a group of organisms which are capable of reproducing to form offspring with similar characteristics.

Communities

A community is a group of populations existing together in a common **habitat**. Organisms perform particular functions; they occupy a certain **niche**. Individuals interact with their living and non-living environment, to produce a stable **ecosystem**.

Molecules consist of combined **elements**. Only about 40 per cent of known elements occur in living things. Elements contained in organisms are of low atomic mass and are readily available in the environment.

The most common elements within living cells are oxygen, carbon and hydrogen (Fig. 3.1). These elements tend to be in a combined form with other elements as molecules.

ELEMENT	PERCENTAGE OF CELL MASS
Oxygen (O)	60%
Carbon (C)	21%
Hydrogen (H)	11%
Nitrogen (N)	3.5%
Calcium (Ca)	2.5%
Phosphorus (P)	1.2%
Chlorine (Cl)	0.2%
Fluorine (F)	0.15%
Sulphur (S)	0.15%
Potassium (K)	0.1%
Sodium (Na)	0.1%
Magnesium (Mg)	0.07%
Iron (Fe)	0.01%
Trace elements	0.02%

Fig. 3.1 Approximate composition of living cells

The actual amounts and distribution of biologically important molecules varies between, and even within, species. However, certain molecules are common to all forms of life. There are two main types of molecule: inorganic and organic.

(a) INORGANIC MOLECULES

Inorganic molecules are relatively small and simple. They contain a wide variety of elements. However, inorganic molecules do not contain the element carbon (carbon dioxide is an exception). Examples of inorganic molecules include:

(i) Water

Water is important as a habitat for aquatic organisms. It also has important functions within organisms:

- As a **component** of living material (protoplasm). Protoplasm is about 65 per cent water in animals and 80 per cent in plants.
- As a **transport medium** within cells and also between cells.
- To allow **chemical changes** to take place in solution. Water may also take part directly in some reactions, e.g., photosynthesis.
- To **dissolve respiratory gases** in land-living organisms.
- To **provide support**, especially in non-woody plants.

Functions of water *within* organisms.

Water is important in living systems because of its chemical and physical properties. The availability of water varies, for instance because of climate, and this can affect the distribution of organisms.

(ii) Oxygen

Oxygen is readily available in most environments as a gas or in dissolved form. Oxygen is available as a waste product of **photosynthesis** in plants. Oxygen is important in **aerobic respiration** in many living things.

(iii) Carbon dioxide

Some other inorganic molecules besides water.

Carbon dioxide is a waste product of **respiration**. Carbon dioxide levels in the environment tend to be low, however, because it is used by plants for **photosynthesis.**

(iv) Minerals

Minerals are important as components of organic molecules, for example iron in haemoglobin, magnesium in chlorophyll and iodine in thyroxin. Some are combined with each other, e.g., calcium and phosphate in bone. Minerals are obtained by plants from the soil and by animals from their diet. Inorganic molecules can form charged ions in solution.

(b) ORGANIC MOLECULES

Some important groups of organic molecules.

Organic molecules tend to be large and complex. Organic molecules contain a narrow range of elements, notably carbon as well as oxygen, hydrogen, nitrogen and sulphur.

Organic molecules fall into four main groups: carbohydrates, proteins, fats and nucleic acids.

CARBOHYDRATE GROUP	CHEMICAL STRUCTURE	FORMULA	EXAMPLES	FUNCTION
Monosaccharides (Simple sugars)	Consist of a single chemical group	$C_6H_{12}O_6$	Glucose Fructose Galactose	Soluble; structural units for making larger carbohydrates.
Disaccharides (more complex sugars)	Consist of two joined monosaccharides.	$C_{12}H_{22}O_{11}$	Maltose Sucrose	Soluble; similar to monosaccharides; sucrose is "a transport molecule" (plants).
Polysaccharides (large, complex sugars)	Consist of many joined monosaccharides (i.e., they are polymers).	$(C_{12}H_{22}O_{11})n$	Starch Glycogen Cellulose	Insoluble; used as food store in plants (starch) or animals (glycogen); used as structural material in plants (cellulose).

Fig. 3.2 Summary of common carbohydrates

(i) Carbohydrates

Carbohydrates contain carbon (C), hydrogen (H) and oxygen (O) in the ratio CH_2O. There are three main groups of carbohydrates (saccharides): **monosaccharides, disaccharides, polysaccharides**. These are summarised in Fig. 3.2

The structural unit of carbohydrates is the monosaccharide. Carbohydrates can be converted from one form to another, according to the organism's needs. For example, polysaccharides can be broken down during digestion to disaccharides and then monosaccharides; these can be further broken down during respiration to release energy. Polysaccharides can be made from simpler carbohydrates for growth (e.g., of cellulose cell walls) and for storage (e.g., of starch, glycogen). Polysaccharides are insoluble, so do not affect osmosis directly.

Carbohydrates can, like proteins and fats, be identified using standard tests. There are separate tests for so-called **reducing sugars** (monosaccharides and maltose) and **non-reducing sugars** (disaccharides other than maltose).

(ii) Proteins

Proteins contain carbon, hydrogen, oxygen, nitrogen and, in some cases, sulphur and phosphorus. The structural unit of proteins is the **amino acid**; these are joined together by **peptide bonds** to form short **peptides**, longer **polypeptides** or even longer protein molecules.

There are about twenty different amino acids in proteins. The precise way in which they are arranged during **protein synthesis** is determined by genetic information contained within **chromosomes**. The chain of amino acids is often folded into a three-dimensional shape; this may be even further folded. The precise shape of proteins is often important for their function.

Proteins have a wide range of functions:

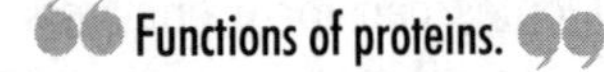
Functions of proteins.

- **Structure** of cells and tissues. This is more important in animals (50 per cent of dry mass) than in plants, whose main structural component is carbohydrate.
- **Movement**, by muscle fibres.
- **Control** of chemical processes by enzymes (see below) and hormones.
- **Prevention of disease**, by antibodies.
- As **energy providers**, especially in carnivores.

The relative number and type of proteins within organisms is characteristic of a given species, and can be used to show how different species are related.

(iii) Lipids

Lipids are *fats* (solids) and *oils* (liquids). They contain carbon, hydrogen and oxygen; the relative amount of oxygen is low, however. A typical fat consists of a *glycerol* unit with three *fatty acids* attached. The fatty acids may be *saturated* (having no double bonds) or *unsaturated* (having double bonds).

Lipids are important for:

- forming part of **cell membranes**.
- for **energy release**, during respiration.
- ***storing fat*** until needed, e.g., in seeds and in mammals, where it also protects organs and insulates.

(iv) Nucleic acids

Nucleic acids contain carbon, hydrogen, oxygen, nitrogen and phosphorus. The structural units of nucleic acids are **nucleotides**, which consist of a *sugar* (pentose), a *phosphate* and a *base*.

There are two types of nucleic acid:

- *DNA* is a large, complex molecule, twisted into a double helix shape.
- *RNA* is a shorter, single stranded molecule.

Both DNA and RNA carry genetic information.

4 ENZYMES

> Students are not always aware that enzymes are a type of protein.

Enzymes are biological **catalysts** which alter the rate of a reaction without themselves being changed. They are globular proteins.

Enzymes are made (by protein synthesis) within cells. **Extracellular enzymes** are secreted outside the cell where they have their affect; e.g., digestion in the gut cavity. **Intracellular enzymes** are retained within the cell and catalyse reactions there; e.g., respiration within mitochondria.

Enzymes are believed to work by forming a temporary "complex" with the chemical they are acting upon, the **substrate**, before forming the **product** or products. The way in which this could occur is called the "lock and key" model, shown in Fig. 3.3

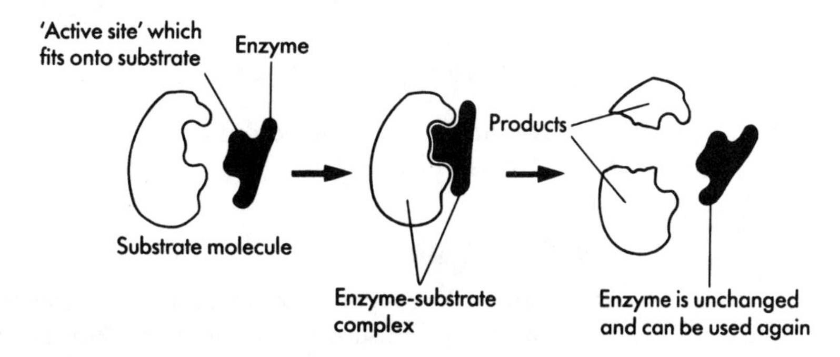

Fig 3.3 The role of the active site in enzyme activity

Fig. 3.3 shows a breaking down (*catabolic*) reaction; the "lock and key" model can also explain building-up (*anabolic*) reactions, in which two substrate molecules are combined into a single product.

Enzymes have certain **characteristics** which can be explained in terms of the "lock and key" model:

> Characteristics of enzymes.

(i) **Enzymes are specific for a particular substrate:**
The active site on the enzyme has to "fit" the substrate.

(ii) **Enzymes work within a narrow range of temperature** (Fig. 3.4)**:**
At high temperatures, enzymes are **denatured**; the shape of the molecule, including the active site, is distorted. At low temperatures, enzyme and substrate molecules are less liable to react together.

(iii) **Enzymes work within a narrow range of pH** (Fig. 3.5):
Extremes of pH may denature the enzyme; some enzymes are very sensitive to this. Acid or alkaline conditions may affect the other substances involved in the reaction.

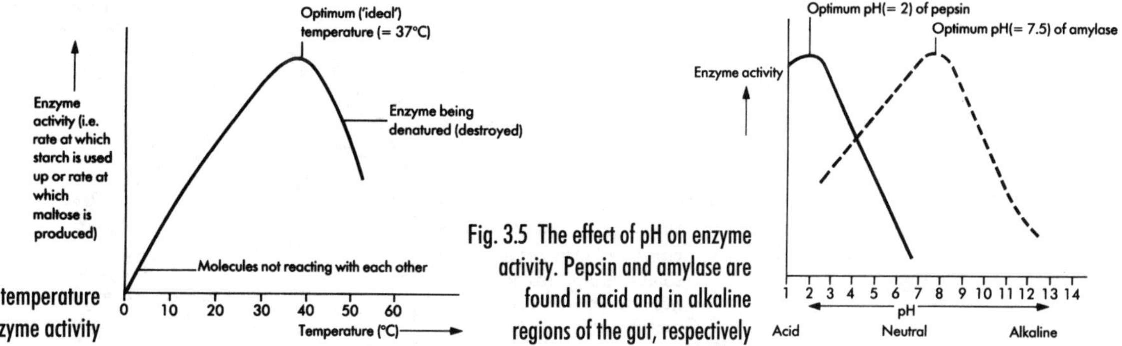

Fig. 3.4 The effect of temperature on enzyme activity

Fig. 3.5 The effect of pH on enzyme activity. Pepsin and amylase are found in acid and in alkaline regions of the gut, respectively

Organisms depend on enzymes for many important processes so it is important that they work efficiently. Organisms therefore maintain favourable reactions by regulating their internal environment. This is the purpose of *homeostasis*.

5 MOVEMENT OF SUBSTANCES ACROSS THE CELL MEMBRANE

Cells are not isolated from their external environment; substances enter or leave cells through their outer surface. The cell membrane has an important function in controlling this movement, and so determining the chemical composition of the cell. The relative concentration of substances within a living cell is often quite different from that of its non-living external environment. The cell's internal environment is maintained by both *passive* and *active* processes:

(i) **Passive processes.** These result from differences in the concentration of substances inside and outside the cell. Dissolved substances move by **diffusion** from high to low concentrations. Diffusion of water molecules is called **osmosis.** These passive processes do not require energy from the cell.

Example: oxygen from the environment will tend to diffuse into a cell where the relative concentration is low.

(ii) **Active processes.** Substances can be moved from low to high concentrations; this is the opposite situation to that of diffusion. This process is called **active transport** and requires energy from the cell.

Example: the uptake of some mineral ions into plant root cells.

Large or insoluble substances can enter or leave the cell by processes involving small, membrane-bound **vesicles**.

(a) PASSIVE PROCESSES

Passive processes involve diffusion, including osmosis which is the diffusion of water across a membrane.

(i) Diffusion

Diffusion is the random movement of substances from a region of high to a region of low concentration, down a **concentration gradient**. The rate of movement is determined by how "steep" the gradient is; diffusion is more rapid where the relative difference in concentrations is large. Diffusion is very important in the movement of substances over short distances, between and within cells. However, diffusion is not effective over longer distances (i.e., more than about 1 mm), so larger organisms may require **transport systems** as well.

(ii) Osmosis

"Many students find osmosis difficult to understand, partly because they may not realise that it is simply a form of diffusion, involving water molecules."

Osmosis is the movement of water molecules from a region of high concentration of water (a "dilute" solution) to a region of low concentration of water (a "concentrated" solution), through a **selectively-permeable membrane** (see Fig. 3.6).

Osmosis is a type of diffusion involving water molecules. These are small enough to pass through a cell membrane by simple diffusion. However, larger molecules such as

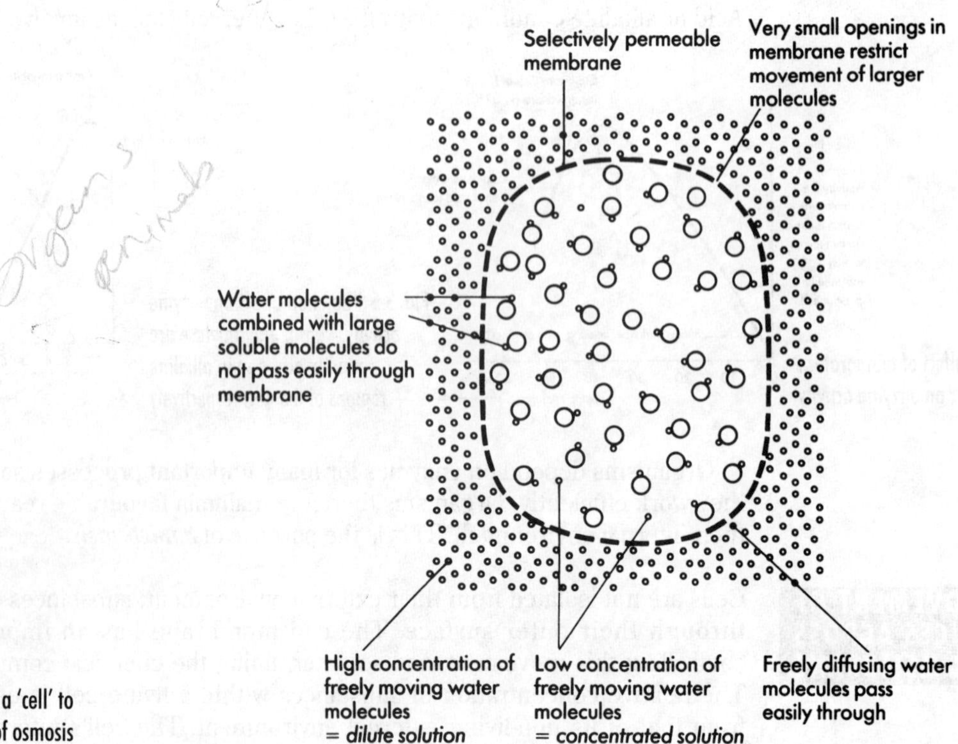

Fig 3.6 Diagram of a 'cell' to illustrate the principle of osmosis

glucose cannot easily cross a membrane (they require active transport). Large soluble molecules like glucose do not take part in osmosis directly, but they do determine the relative number (= concentration) of freely moving water molecules on either side of the membrane by combining with them.

Relatively more water molecules will tend to move from a dilute solution than from a concentrated solution across a membrane. A living membrane is important in osmosis because it is selectively permeable (semi-permeable); osmosis can therefore be regarded as *"restricted diffusion"*. Osmosis will not occur in cells whose membranes are not intact, for instance if the cell is denatured by heat, or plasmolysed.

Osmosis in cells. The concentration of dissolved substances is often higher inside than outside a cell. For this reason, water tends to enter cells by osmosis. Although this is a passive process, it can be controlled in various ways:

(i) the concentration of substances within the cell can be regulated by converting them to an insoluble, osmotically inactive form; e.g., glucose can be stored as starch in plant cells. The amounts of substances inside the cell can be altered by actively transporting them from the cell.

(ii) the concentration of substances immediately outside the cell in a multicellular organism can be controlled by various processes in homeostasis; this is particularly evident in animals.

(iii) the pressures resulting from the entry of water by osmosis can be resisted in plant cells by the cell wall. This maintains the cell shape and is important in support.

Osmotic changes in cells. The gain or loss of water from cells can be controlled by individual cells or by the organism as a whole. However this control may not always be possible. One consequence of a variation in water content of cells is a change in their shape. In some cells, for example the guard cells of plant leaves, this may be an important part of their function.

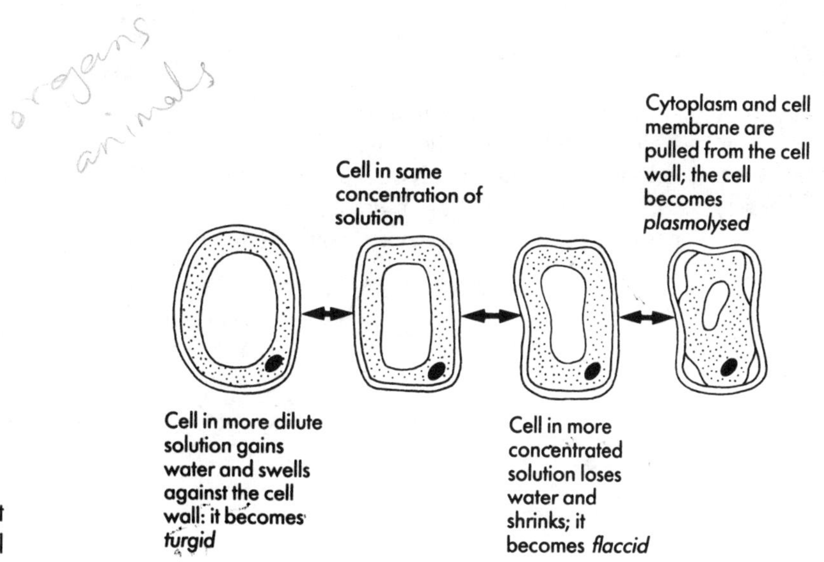

Fig. 3.7 Variation of water content in a plant cell

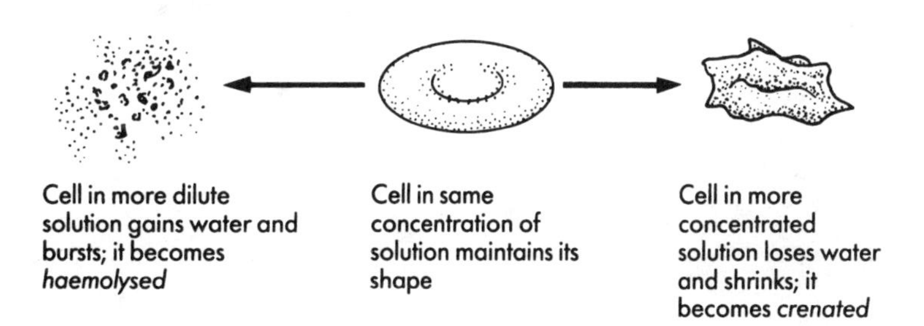

Fig. 3.8 Variation of water content in an animal cell

"Plant cells can avoid damage by swelling because, unlike animal cells, they have a cell wall. However, both types of cells can be damaged if too much water is lost."

The effect of variations in water content by osmosis on plant and animal cells is shown in Fig. 3.7 and Fig. 3.8. These variations may have a destructive effect on cells, so they are often avoided in plants and, in particular, animals. Major disruptions to the cell structure such as **plasmolysis** (plant cells), **haemolysis** and **crenation** (red blood cells) are often permanent and therefore serious for the organism as a whole.

(b) ACTIVE PROCESSES

Active transport

Active transport is the movement of substances across a cell membrane against a concentration gradient, using energy from respiration. Substances which are in a relatively low concentration on one side of a cell membrane can be moved across the membrane. This process is thought to involve **carrier molecules** within the membrane, and requires energy from **ATP molecules**, produced from respiration. Active transport occurs in cells lining the gut which take up glucose molecules into the blood, which already contains a relatively high concentration of glucose. Glucose is also taken up by cells lining the kidney nephron during reabsorption. Active transport occurs in root hair cells to move minerals from the soil into the root.

Note: This chapter underpins much of the specific topic-based work on life and living processes in Chapter 4. In that chapter you will find many past examination questions to practice, and to check your understanding of what you have read.

LIFE AND LIVING PROCESSES

ORGANISATION OF THE HUMAN BODY

ORGANISATION OF THE FLOWERING PLANT

CELLS

SURVIVAL IN A NATURAL HABITAT

LIFE PROCESSES IN ANIMALS

LIFE PROCESSES IN PLANTS

MAINTENANCE OF INTERNAL ENVIRONMENTS

CO-ORDINATION IN MAMMALS

GETTING STARTED

The material in this chapter is arranged in **topic groups**. Each topic group has a particular *theme* which links the various topics involving 'life and living processes'. You can often understand topics better if you also revise related topics elsewhere in the book. We have provided *Links with Other Topics* tables after each topic group to help you find your way around the book.

Each Topic Group concludes with *Review Questions* to test your understanding of the preceding sections. In many cases, the review exercises are based on actual examination questions, so that you get used to dealing with them. Answers to the review questions can be found at the end of the chapter. We have also included *Practice Examination Questions* at the end of the chapter. The answers to these, with examiners comments, are in a separate chapter at the end of the book.

The eight topic groups presented in this chapter cover the folllowing:

1. Organisation of the human body
2. Organisation of the flowering plant
3. Cells
4. Survival in a natural habitat
5. Life processes in animals
   - Movement
   - Respiration
   - Growth and development
   - Reproduction
   - Excretion
   - Nutrition
   - Sensitivity
   - Blood circulation
6. Life processes in plants
   - Photosynthesis
   - Respiration
   - Reproduction
7. Maintenance of internal environment in plants, animals and human foetus
   - Skin
   - Hormones
   - Foetus
   - Absorption in plants
   - Transport in plants
   - Homeostasis
8. Co-ordination in mammals
   - Nervous System
   - Hormonal System

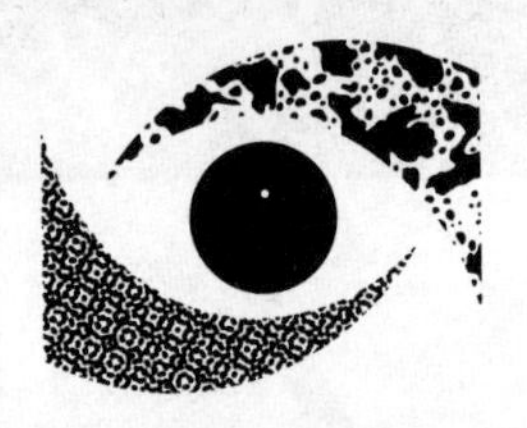

ESSENTIAL PRINCIPLES

1 ORGANISATION OF THE HUMAN BODY

"Level 4"

Within the human body are millions of cells organised into groups called **organs**. Each organ has specific functions within the body e.g. the **heart** pumps blood through the blood vessels. The organs work together so that the person is healthy and their body processes are carried out efficiently.

Key Points

- humans have many different organs, each with their own specific functions
- organs are grouped together to form systems
- there are 7 systems in the human body
- if all the organs are working properly, the person will be healthy

NAMES AND LOCATIONS OF THE MAJOR ORGANS

Inside the human body there are lots of different **organs** (Fig. 4.1)

e.g.	heart	stomach	kidneys	ovaries
	lungs	intestines	bladder	testes
	brain	liver	bones	uterus.

"You should be able to label a diagram like this."

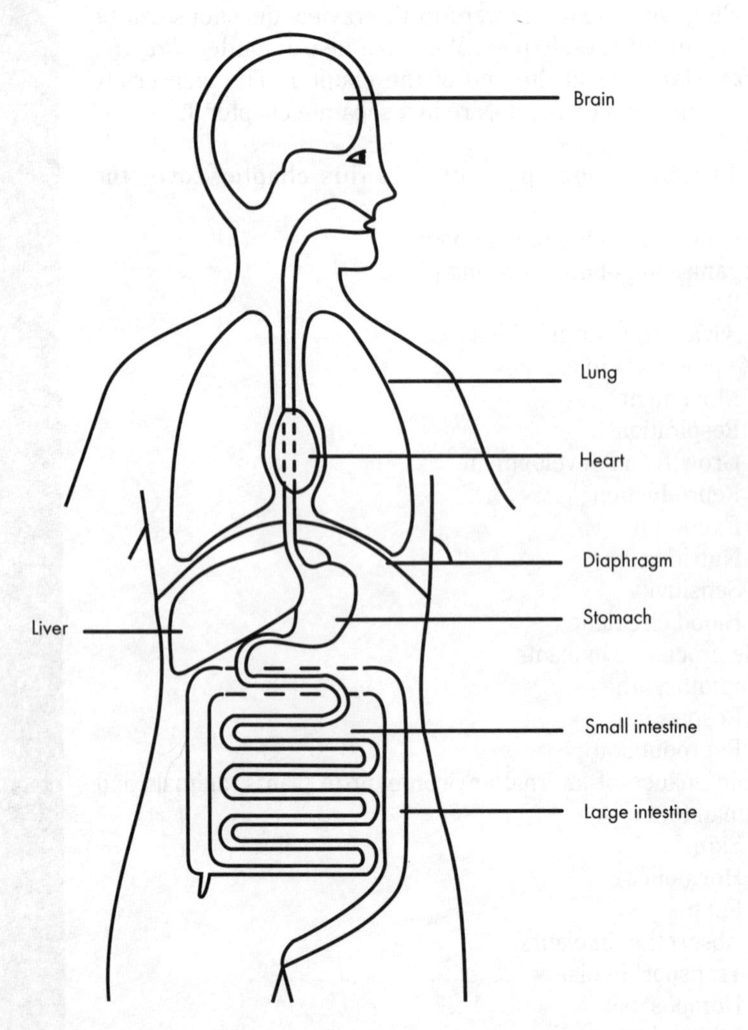

Fig 4.1 Diagram to show positions of organs in human body

Each organ has a particular **function** (job to do) within the body (Fig. 4.2), e.g. heart pumps blood around the body.

FUNCTIONS OF THE MAJOR ORGANS

ORGAN	FUNCTIONS
Brain	1. to control all the processes of the body, e.g. heart rate, body temperature 2. to make decisions 3. to think and remember
Lungs	1. to transfer oxygen from the air into blood 2. to transfer carbon dioxide from blood into the air } this is gas exchange
Heart	to pump blood around the body
Diaphragm	to move up and down so that air moves in and out of the lungs
Stomach	to mix food with digestive juices (enzymes)
Liver	1. to destroy poisons in the body, e.g. alcohol, drugs 2. to make bile (helps to digest fats) 3. to break down excess amino acids 4. to change glucose into glycogen or fat
Small intestine	1. to mix food with digestive juices (enzymes + bile) 2. to absorb digested food into the blood
Large intestine	1. to absorb water from faeces 2. to store faeces before they are egested
Kidneys	1. to remove urea from the blood 2. to remove excess salt from the blood 3. to remove excess water from the blood
Bladder	to store urine before it is excreted
Testes	1. to make sperm 2. to make the male sex hormone (testosterone)
Penis	to pass sperm into a woman's body during sex
Ovaries	1. to make ova (egg cells) 2. to make the female sex hormones (oestrogens)
Uterus	to provide a safe place for the foetus to develop
Bones, e.g. skull vertebrae limb	1. to protect soft organs e.g. brain, lungs 2. to help us to move

Fig. 4.2 The major organs and their functions

In humans, organs work together to form **systems**. There are 7 different systems (Fig. 4.3)

SYSTEM	FUNCTIONS	ORGANS/PARTS
circulatory	■ to transport substances around the body, in blood.	heart blood vessels
respiratory	■ to transfer oxygen into the body and remove carbon dioxide	lungs trachea diaphragm
urinary	■ to remove urea, salts and water from the body, in urine	kidneys bladder
nervous	■ to control all the activities of the body ■ to link all parts of the body to the brain, and pass impulses (messages) between them ■ to find out about our surroundings	brain sense organs, e.g. eyes nerves
digestive	■ to digest the food we eat ■ to absorb digested food into the blood	stomach intestines liver
reproductive	■ to make sperm and egg cells ■ to provide a safe place for the foetus to develop	testes } male penis } ovaries } female uterus }
skeletal	■ to move parts of the body ■ to protect soft organs, e.g. skull protects brain	bones muscles

Fig. 4.3 Systems of the human body

Links with Other Topic Groups

Related topics elsewhere in this book are as follows

Topic	Chapter and Topic number	Page no.
Cells	4.3	36
Movement	4.5	49
Respiration	4.5	52
Reproduction	4.5	65
Excretion	4.5	73
Nutrition	4.5	78
Sensitivity	4.5	90
The Nervous System	4.8	149
The Hormonal System	4.8	152

REVIEW QUESTIONS

Q1 (a) The diagram (Fig. 4.4) shows the human digestive system.

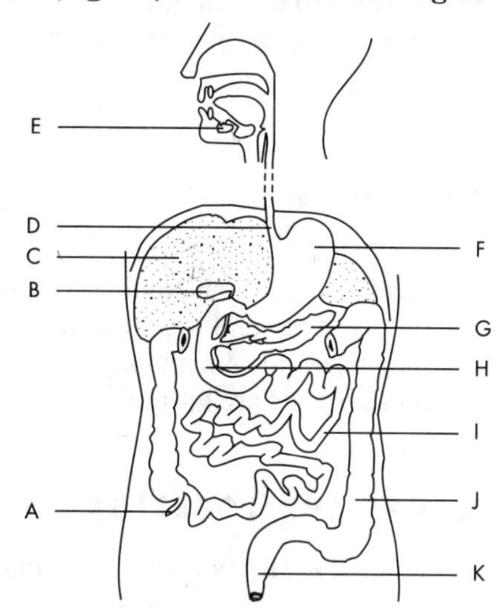

Fig. 4.4

Choose words from the box below to name the structures labelled **A, C, E, F** and **J.**

Salivary gland	Appendix	Liver
Stomach	Gall bladder	Pancreas
Small intestine	Large intestine	

A ______________________________

C ______________________________

E ______________________________

F ______________________________

J ______________________________

Now try to label/identify the other structures in the diagram.

B ________________ **H** ________________

D ________________ **I** ________________

G ________________ **K** ________________

(b) Describe the *functions of* each of the following systems

- circulatory ______________________________

____

- respiratory ______________________________

____

- urinary ______________________________

____

- nervous ______________________________

____

- digestive ______________________________

____

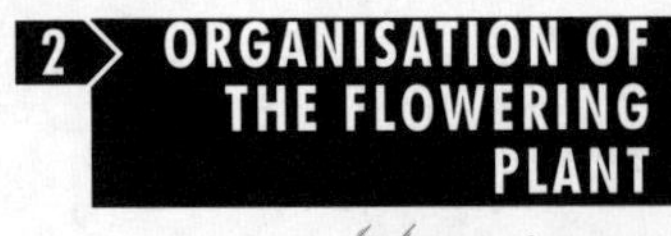

Level 4

Plants are a very important group of organisms to biologists. Animals rely on them for food, and for oxygen; in fact all life on Earth depends on plants. Biologists need to study plants to find out how their life processes work. Although they are very different to animals, they are just as successful.

Key Points

- plants have a totally different lifestyle to animals
- the structure and organisation of plants is suitable for their lifestyle
- they have only 4 main organs
- these organs are grouped into 2 systems

NAMES AND LOCATIONS OF THE MAJOR ORGANS

There are 2 **systems** in flowering plants

- Shoot system — all the parts of the plant which are above ground
- Root system — all the parts of the plant which are below ground

There are 4 main **organs** in plants:

- roots
- stems
- leaves
- flowers

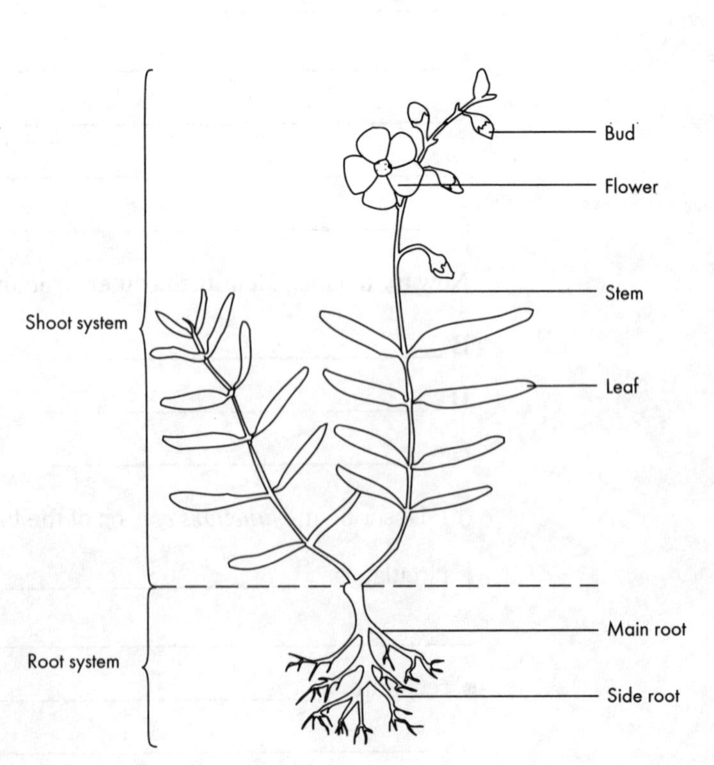

Fig. 4.5 Positions of organs and systems in a flowering plant, Rockrose *(Helianthemum sp.)*

FUNCTIONS OF THE MAJOR ORGANS

Each organ has its own specific **function**, in the same way that animal organs do. However, plants have a different mode of life to animals so they have a simpler pattern of organisation (Fig. 4.6)

ORGAN	FUNCTIONS
root	1. to collect water from the soil 2. to collect minerals, e.g. nitrate, from the soil 3. to anchor the plant in the soil
stem	1. to transport substances, e.g. water and sugars, around the plant 2. to support the shoot system
leaf	to photosynthesise (make sugar)
flower	to reproduce and make seeds, which will grow into new plants

Fig. 4.6 Functions of the major organs in a flowering plant

Links with Other Topics

Related topics elsewhere in this book are as follows

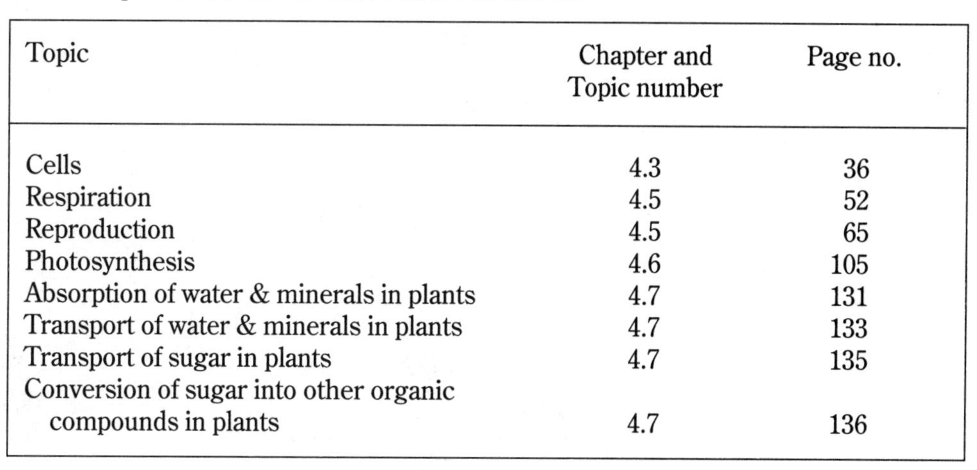

Topic	Chapter and Topic number	Page no.
Cells	4.3	36
Respiration	4.5	52
Reproduction	4.5	65
Photosynthesis	4.6	105
Absorption of water & minerals in plants	4.7	131
Transport of water & minerals in plants	4.7	133
Transport of sugar in plants	4.7	135
Conversion of sugar into other organic compounds in plants	4.7	136

REVIEW QUESTIONS

Q2 Here is a list of four parts of a plant and a diagram of a plant.

(a) Draw a line from each name to touch the right part of the plant diagram.

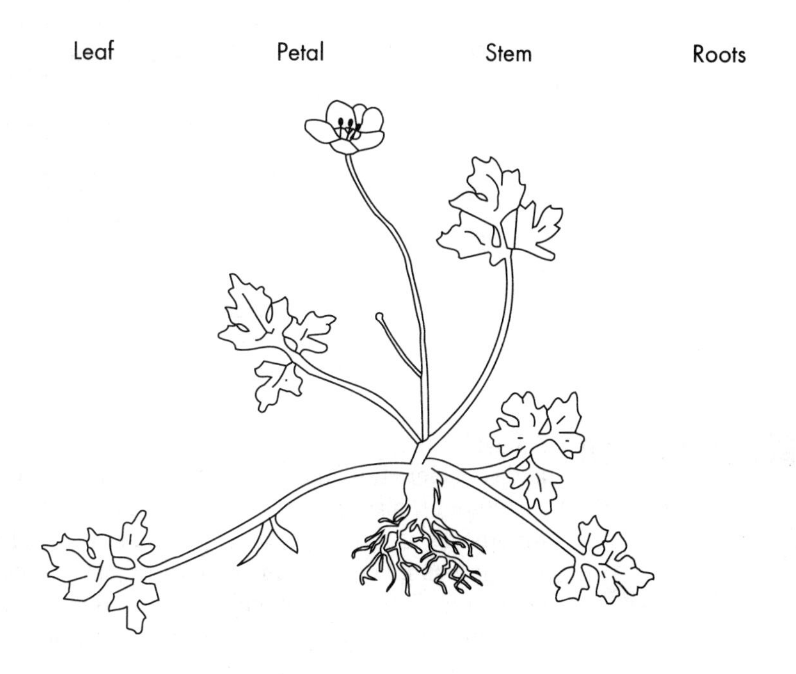

Fig. 4.7

(b) Complete the table to show **one** function of each part.

Part	Function (what it does)
Stem	
Leaf	
Roots	
Petal	

Q3 The diagrams below show certain organs of plants which are used as food.

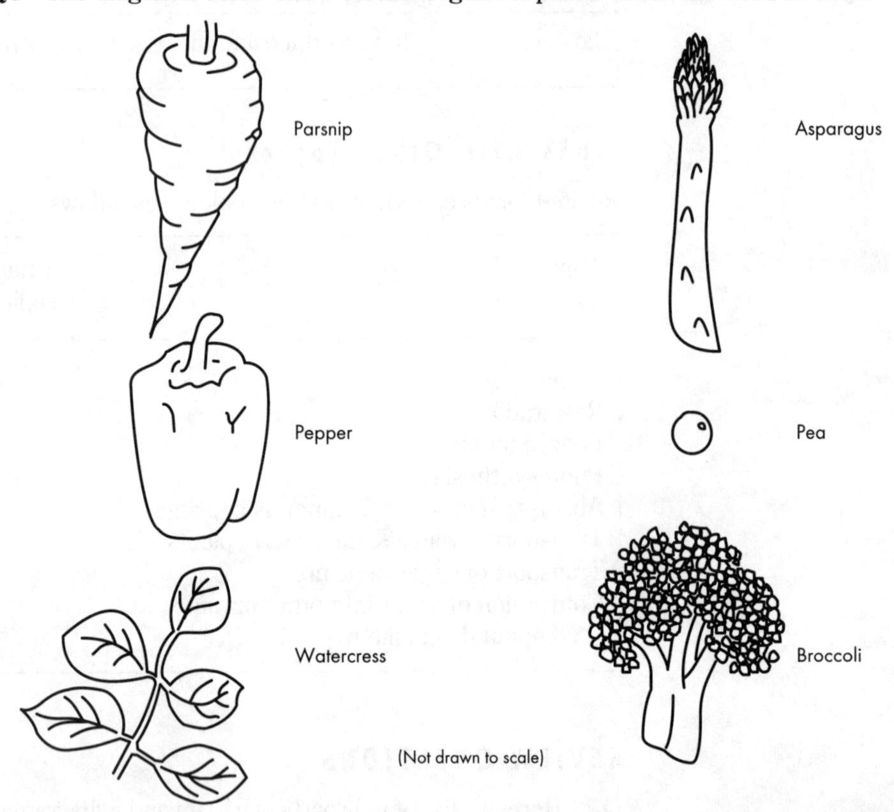

Fig. 4.8

Complete the table below by writing in the name of the plant next to the correct organ.

Name of organ	Name of plant
Stem	
Flower	
Seed	
Fruit	
Leaf	
Root	

3 CELLS

Level 6

All living things are made up of **cells**. This topic group looks at the structure of cells, and considers the functions of each of the cell components.

Although plant and animal cells have a similar structure, there are some important differences as a result of their different modes of life.

Some simple organisms consist of just one cell, but more complex organisms may contain millions of cells, and will show division of labour. This means that cells within an organism are different to each other, because they have become specialised to carry out a particular function.

Key Points

- cells are the basic *units of life* – all living things are made of them
- all cells contain a *nucleus*, cell *membrane* and *cytoplasm*
- plant cells are surrounded by a cellulose *cell wall*, and have a large sap-filled *vacuole*. Some will contain *chloroplasts*.
- cells may be *specialised* to perform a particular function e.g. palisade cells in leaves are adapted for photosynthesis; red blood cells in humans are adapted to carry oxygen.
- cells are organised to form *tissues*, tissues are grouped together to form *organs*, and organs work together as a *system*

STRUCTURE OF ANIMAL AND PLANT CELLS

All living things are made up of cells so they are sometimes called the units of life, or the building blocks of life.

You should know the structure of a typical animal cell and a typical plant cell, and be able to compare them.

Structure of an animal cell e.g. human cheek cell

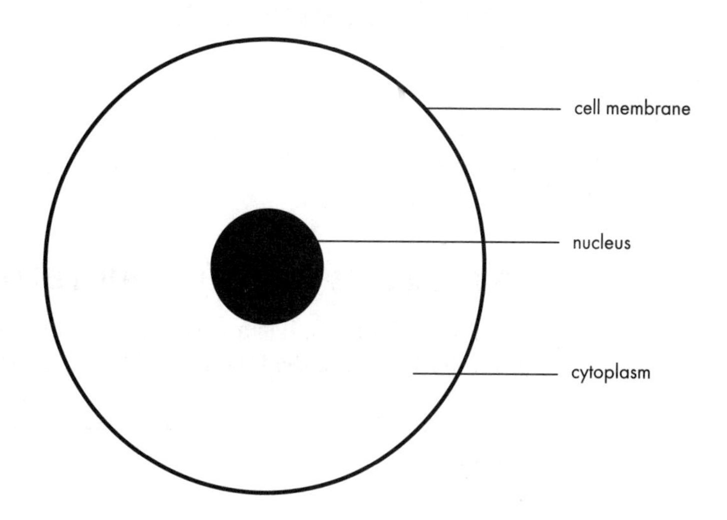

Fig. 4.9 Typical animal cell eg. cheek cell

“You should be able to draw and label both of these diagrams”

Structure of a plant cell e.g. leaf palisade cell

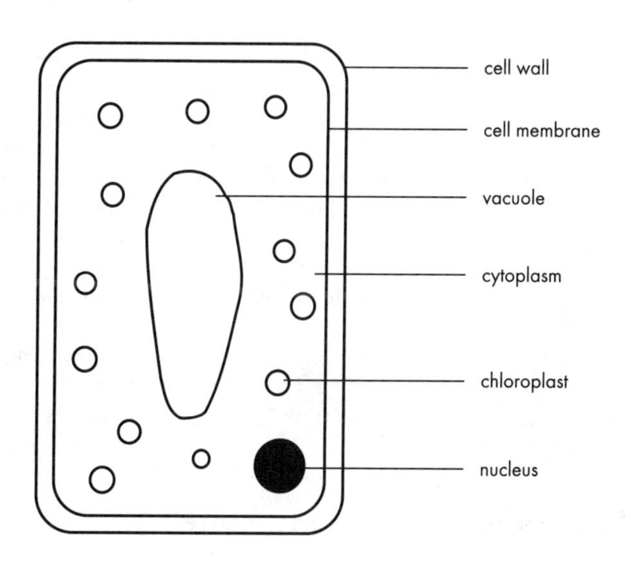

Fig. 4.10 Typical plant cell eg. palisade cell from leaf

These are the parts of the cell which can be seen using a light microscope which magnifies about 1000 ×.

If you observe a cell with an electron microscope which magnifies 500000 ×, it looks much more complicated, and you can see other cell components e.g. ribosomes, mitochondria.

Comparing animal and plant cells

- **Similarities**

1. They both have a nucleus.
2. They both have cytoplasm.
3. They both have a cell membrane.

- **Differences**

PLANT CELLS	ANIMAL CELLS
Always have a cell wall Usually have one large vacuole Often contain chloroplasts Often contain starch granules	Do not have a cell wall Do not have a large vacuole Never contain chloroplasts Never contain starch granules

COMPONENTS OF CELLS AND THEIR FUNCTIONS

Each part of the cell has a particular function to make sure that the whole cell works properly. These are the functions of the main parts of the cell (Fig. 4.11).

CELL PART	FUNCTIONS
nucleus	1. It controls all of the activities of the cell e.g. dividing, making new proteins 2. It contains chromosomes (made of DNA) which carry genetic information
cytoplasm	1. This is a jelly-like substance where most of the chemical reactions occur in the cell 2. It contains vacuoles and food stores e.g. starch grains, oil droplets
cell membrane	1. It controls what enters and leaves the cell. It is semi-permeable (lets some substances through, but not others). 2. It protects the cytoplasm and nucleus
cell wall *	This is a tough, outer layer made of cellulose. It is rigid and it keeps the cell the right shape
chloroplast *	1. It contains a green chemical, chlorophyll, which traps light energy 2. Photosynthesis occurs here
vacuole *	1. It stores water, sugar and minerals 2. It is important in supporting the plant (stops wilting)

Fig. 4.11 Functions of the cell parts

* found in plant cells only

SPECIALISED CELLS

Make sure you understand the link between structure and function

In a large, multi-cellular organism there are lots of different types of cells, each with its own function. This arrangement, with different types of cells carrying out different functions is called *division of labour*. The cells are specialised, or adapted, so that they can carry out their function efficiently.

Types of cells

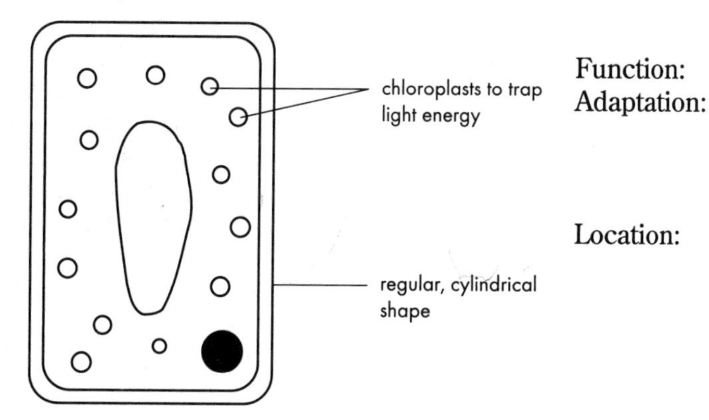

Fig. 4.12 Palisade cell

■ Palisade cell

Function: to make sugar by photosynthesis
Adaptation: 1. has lots of chloroplasts to trap light energy
2. cylindrical shape to pack together tightly
Location: close to the upper surface of leaves

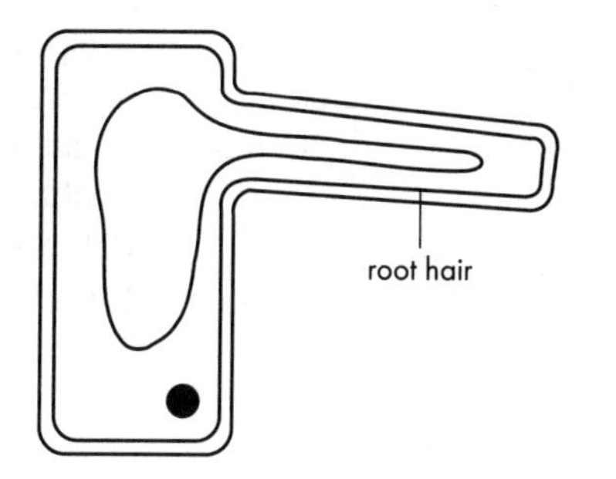

Fig. 4.13 Root hair cell

■ Root hair cell

Function: to absorb water and minerals from the soil
Adaptation: has a long root hair to increase its surface area so it can absorb water and minerals more easily
Location: close to the tip of the root

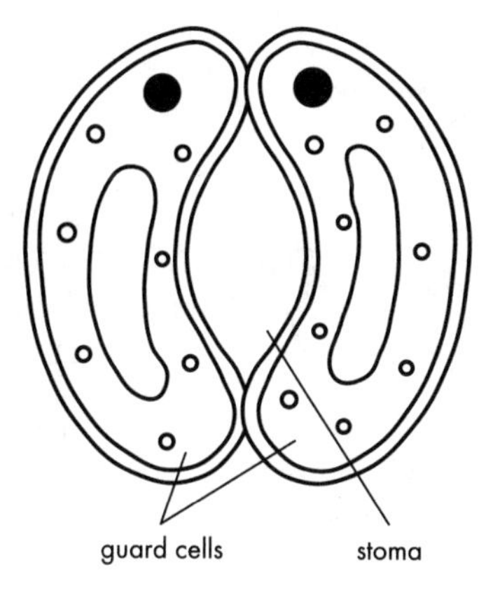

Fig. 4.14 Guard cells

■ Guard cells

Function: to open and close the stomata on the leaf
Adaptation: can change shape
Location: on the lower surface of leaves

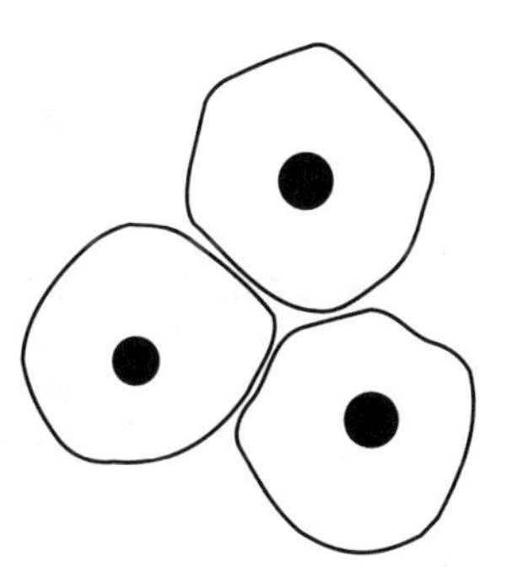

Fig. 4.15 Cheek cells

■ Cheek cells

Function: to line the inside of the mouth
Adaptation: thin, flat shape so they can fit together closely
Location: inside of mouth

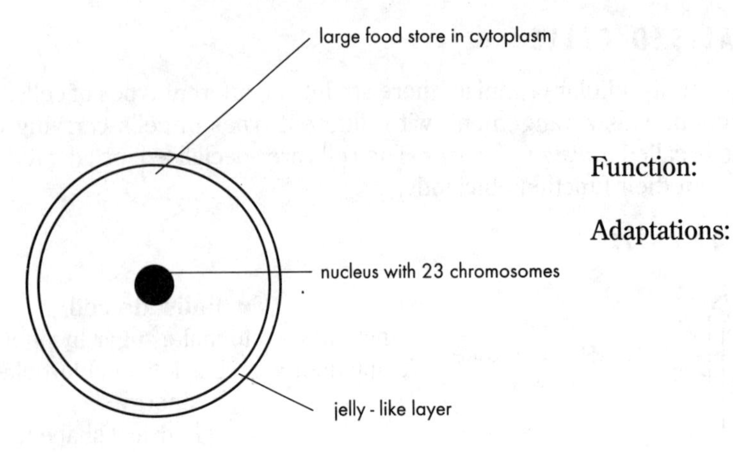

Fig. 4.16 Ovum

■ Ovum

Function: to develop into a baby when fertilized

Adaptations: 1. large food store in cytoplasm (gives energy for the growth of the zygote)
2. jelly-like layer surrounds cell membrane (prevents entry of more than one sperm cell)
3. nucleus contains 23 chromosomes

Location: produced in the ovary

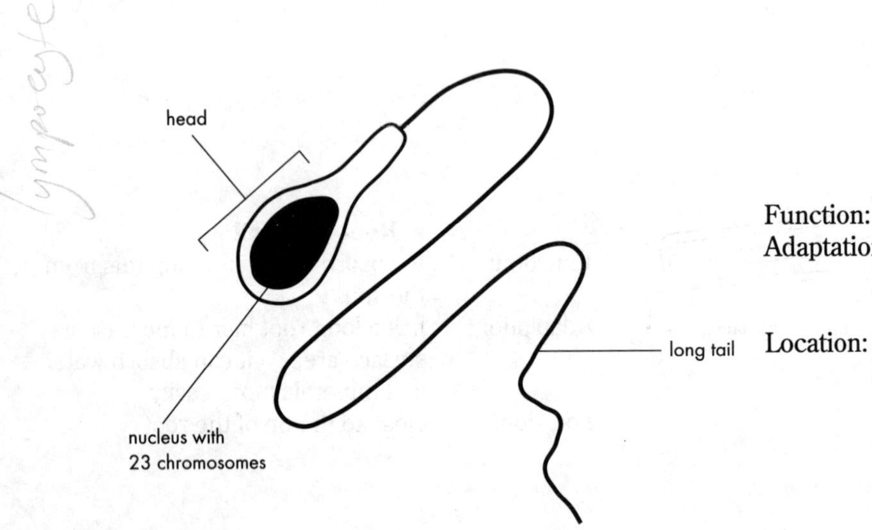

Fig. 4.17 Sperm cell

■ Sperm cell

Function: to fertilize an ovum

Adaptations: 1. has a head with a nucleus containing 23 chromosomes
2. has a tail which helps it to swim

Location: produced in the testis

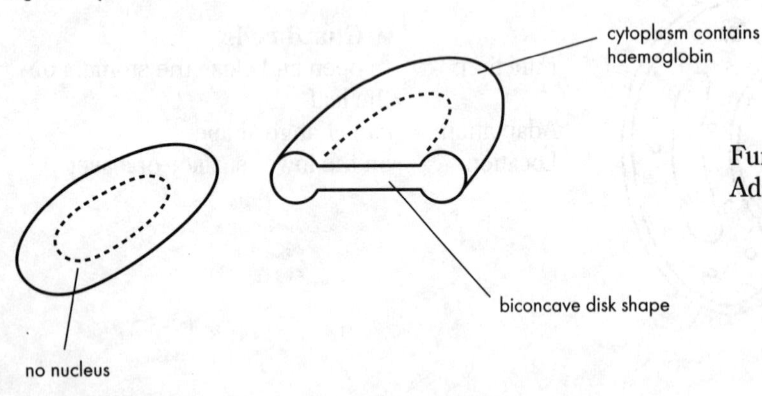

Fig. 4.18 Red blood cell

■ Red blood cell

Function: to carry oxygen

Adaptations: 1. biconcave disc shape gives it a large surface area so it can absorb more oxygen
2. it contains a red chemical, haemoglobin, which joins to oxygen
3. it has no nucleus so it can carry more oxygen

Location: in the blood

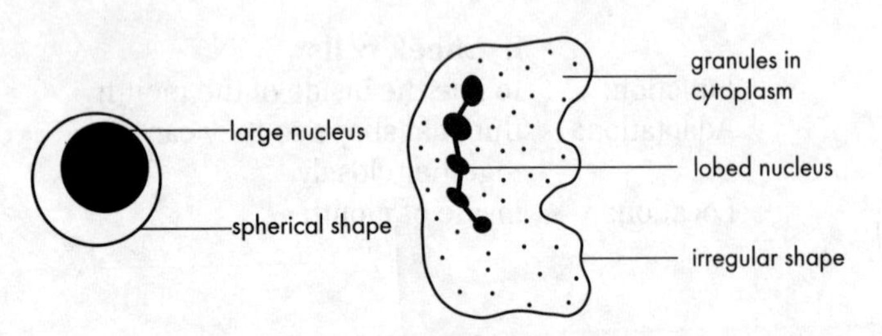

Fig. 4.19 a) Lymphocyte

Fig. 4.19 b) Phagocyte

■ White blood cells

Function: to fight disease by destroying pathogens

Adaptation: 1. lymphocytes (Fig. 4.19a) make chemicals called antibodies which destroy pathogens
2. phagocytes (Fig. 4.19b) can change shape to engulf and destroy pathogens

Location: in the blood

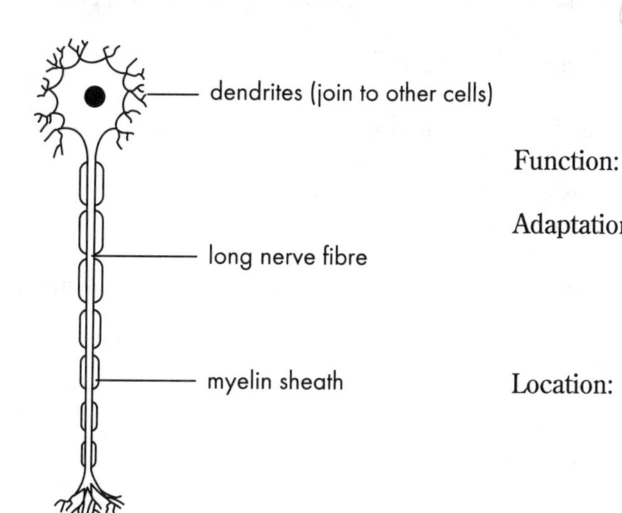

Fig. 4.20 Neurone

■ **Neurone (nerve cell)**

Function: to carry nerve impulses around the body

Adaptations: 1. has a long nerve fibre to reach other cells
2. nerve fibre has a covering made of myelin to speed up transmission of nerve impulses

Location: all parts of the body

HOW ARE CELLS ORGANISED WITHIN THE BODY?

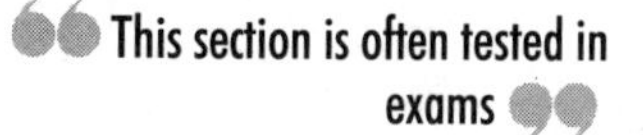

- A group of similar cells carrying out the same function is called a **TISSUE**; e.g. nerve cells make up nerve tissue.
- A group of tissues which work together to carry out a particular function is called an **ORGAN**; e.g. heart contains muscle tissue, nerve tissue, epithelial tissue.
- A group of organs which work together to carry out a particular function is called a **SYSTEM**; e.g. circulatory system is made up of heart and blood vessels.

Links with Other Topics

Related topics elsewhere in this book are as follows:

Topic	Chapter and Topic number	Page no.
Reproduction in animals	4.5	65
Blood circulation	4.5	96
Photosynthesis	4.6	105
Absorption of water and minerals by plants	4.7	131
Nervous system	4.8	149

REVIEW QUESTIONS

Q4 The diagrams below (Fig. 4.21) show four cells, some from animals and some from plants.

Note: The diagrams are not all drawn to the same scale.

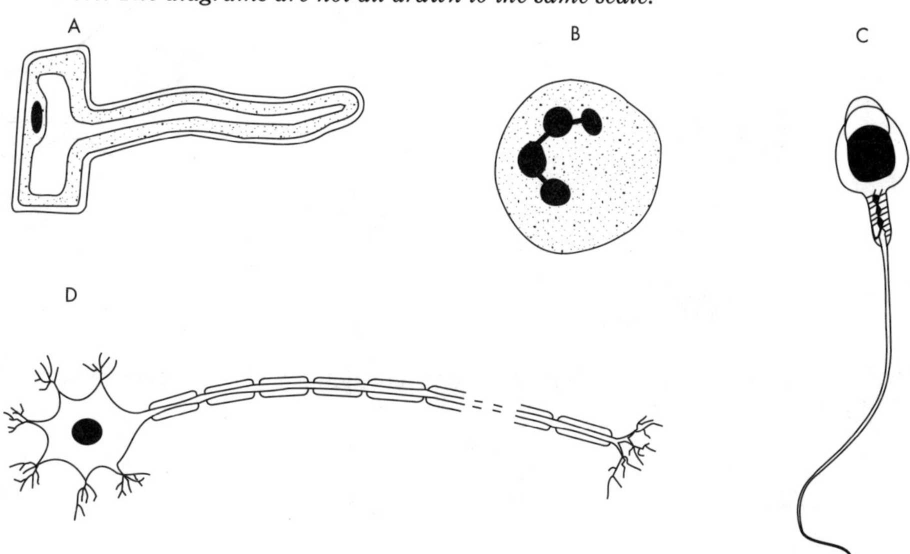

Fig. 4.21

(a) The list below gives the functions (jobs) of each cell.

Functions of each cell

1. Carries nerve impulses
2. Fertilises an ovum (egg)
3. Kills bacteria
4. Absorbs water

Name of cell	Letter A–D	Number of the function of the cell
White blood cell		
Neurone (nerve cell)		
Root hair cell		
Sperm cell		

Complete the table by matching the cells with letters from the diagrams and the number of their correct function.

(b) Give *one* feature that is shown by all the four cells.

____

(c) Give *one* feature, apart from shape, which is shown by cell A but not shown by the other three cells.

____

(d) Name the *two* cells which can move by themselves.

1. ______________________________________

2. ______________________________________

Q5 The diagram below (Fig. 4.22) shows *Pleurococcus* which is a single-celled green alga.

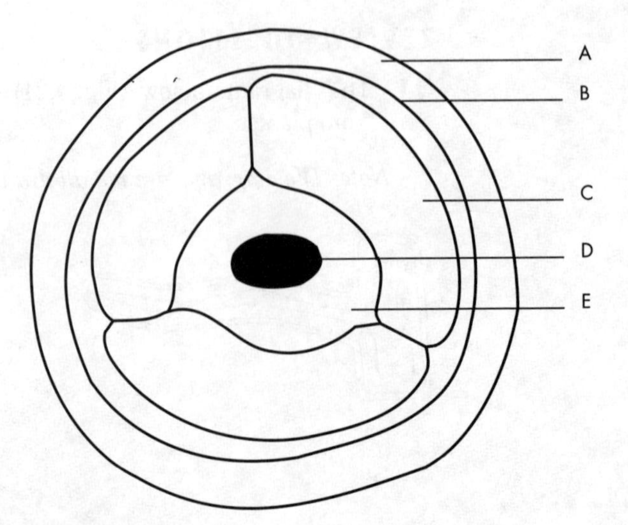

Fig 4.22

(*a*) Write down the letter which labels each of the following.

Cell wall __________

Cell membrane __________

Nucleus ____________

Cytoplasm ____________

Chloroplast ____________

(*b*) Write down *two* ways in which the structure of this cell is different from that of a human cheek cell.

1.____

____

2.____

____

4 SURVIVAL IN A NATURAL HABITAT

Level 6

In the natural world, organisms must compete with each other for the things they need to **survive** – these are called *resources*. The resources an organism needs will depend on its mode of life.

If an organism does not get the resources, it needs, it will die.

If it successfully competes for resources it can survive long enough to breed and produce offspring like itself. Organisms are adapted to their way of life, i.e. they have features which make them more likely to survive in their natural habitat.

Key Points

- resources are in short supply
- organisms must compete with each other to obtain resources
- unsuccessful organisms will die
- successful organisms may survive long enough to breed and produce offspring like themselves
- organisms have features which make them successful in their natural habitats

In order to be successful, organisms must survive long enough to breed and to produce offspring like themselves.

Competition and survival

Organisms must *compete* with others for the things they need to stay alive (we call these **resources**). Often these resources are in short supply, so only the most successful competitors will survive; less successful organisms will die.

Plants compete for

- light, to photosynthesise
- water
- mineral salts e.g. nitrates, phosphates
- space

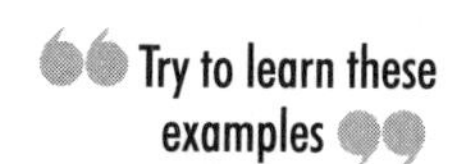

Animals compete for

- food
- space (sometimes called territory)
- a mate
- also, they must avoid being eaten by predators.

Adaptations for survival

Organisms are successful because they are *adapted* (suitable) for the habitat where they normally live. This means that they have certain features which help them to survive.

We will consider 4 different types of habitat and the ways that some plants and animals are adapted to survive there.

DESERT HABITAT

Think about why they are successful

Conditions: very little water
very hot during the day and very cold at night
sand may be blown by wind.

Successful plant: cactus

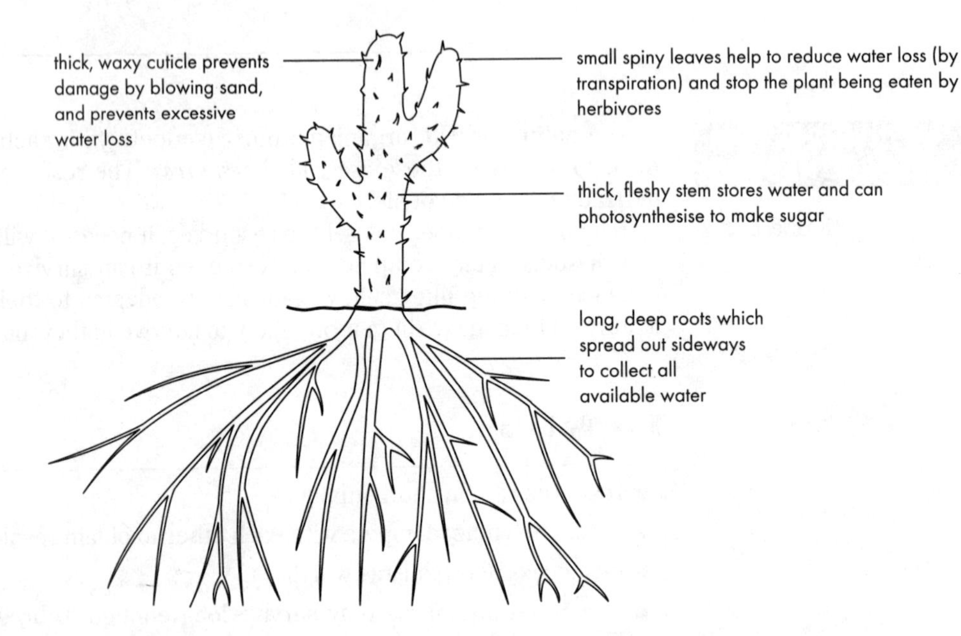

Fig 4.23 Cactus

Successful animal: camel

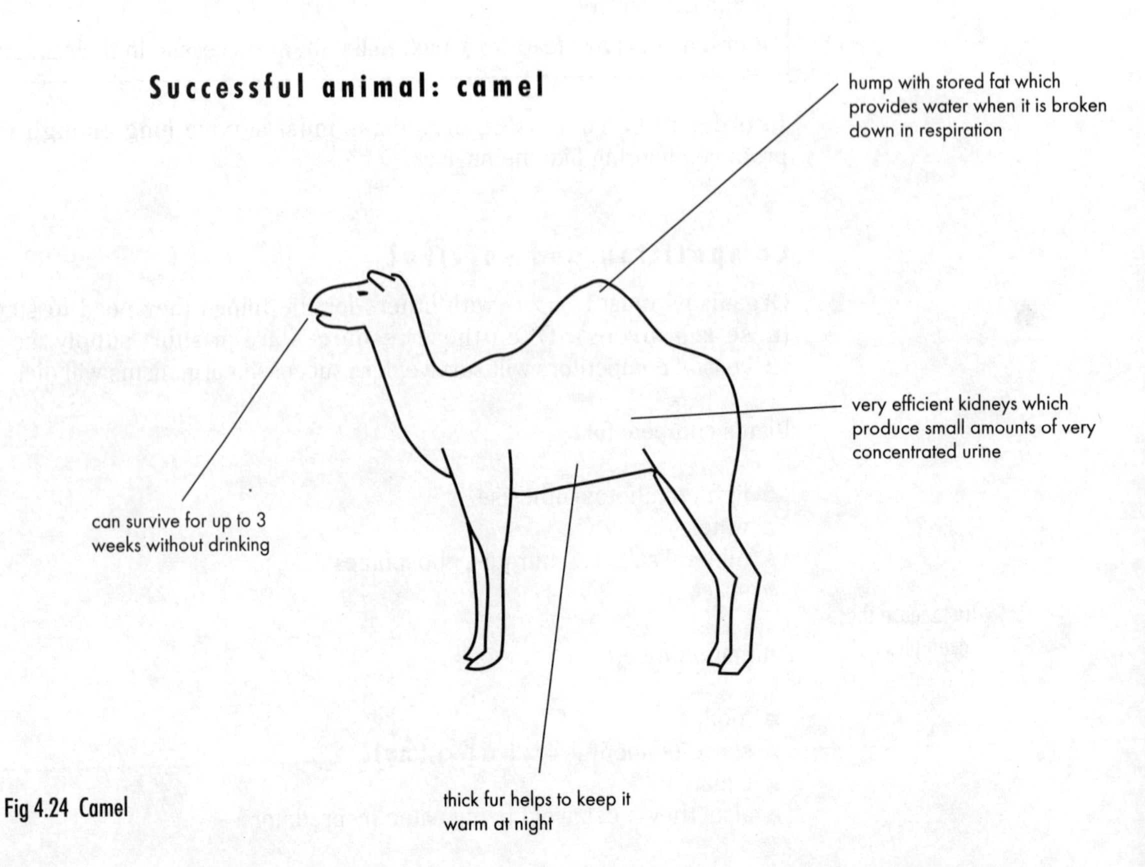

Fig 4.24 Camel

STREAM HABITAT (BRITAIN)

Conditions: freshwater, i.e. water with some minerals in it
temperature varies throughout the year
flow rate may be rapid or slow

Successful plant: water crowfoot

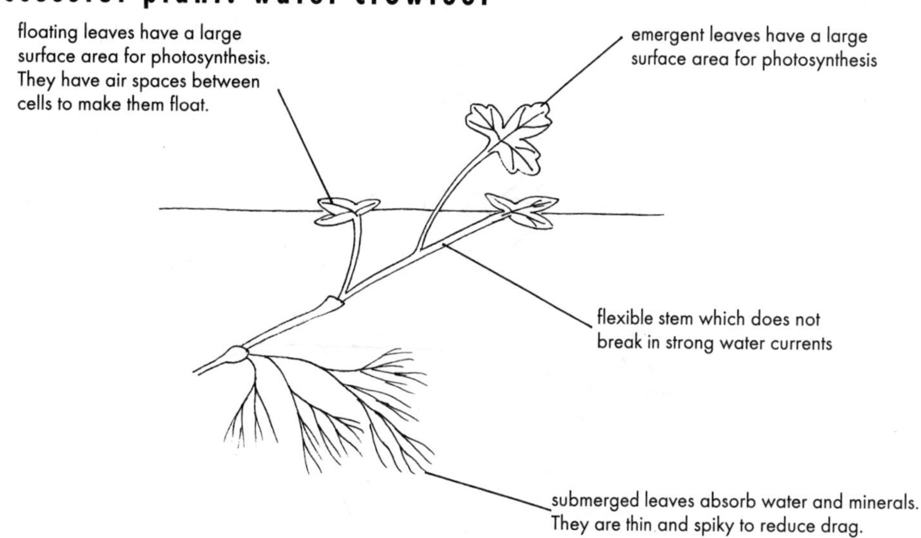

Fig 4.25 Water Crowfoot

Successful animal: mayfly nymph

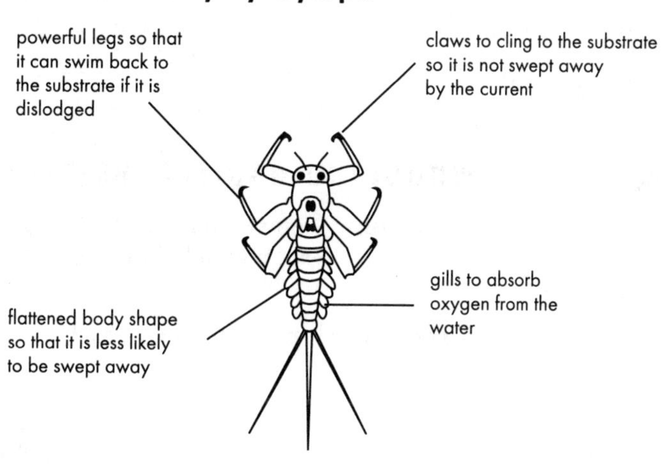

Fig 4.26 Mayfly nymph

POLAR HABITAT

"This is a very difficult place to live"

Conditions: very cold
very little light for parts of the year
very few minerals in soil

Successful plant: Lichen

This is a tiny plant which grows very slowly over a long period of time. It has no proper stem, root or leaves, but it has leaf-like structures called *thalli*. Each thallus is made of algal and fungal cells; the algal cells can photosynthesise to make sugars and the fungal cells can absorb minerals from the soil.

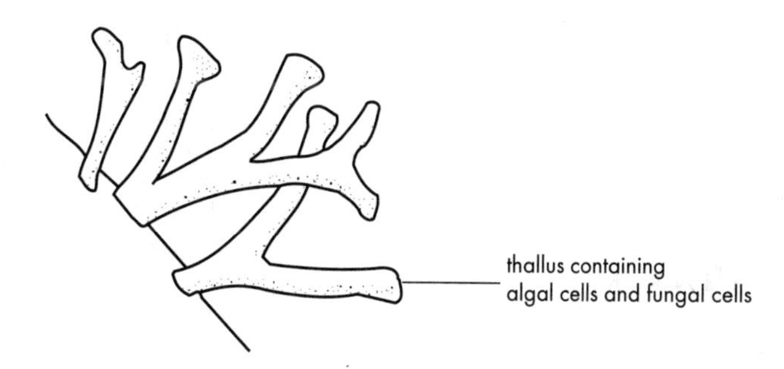

Fig 4.27 Lichen

Successful animal: Leopard seal

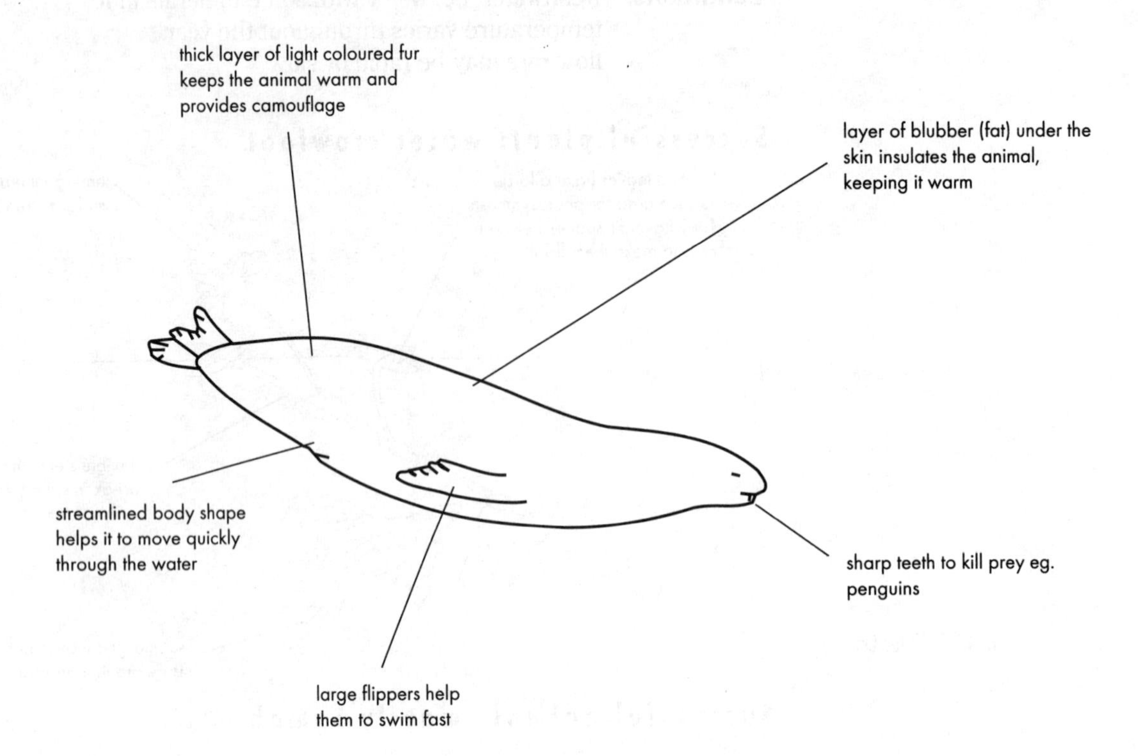

Fig 4.28 Leopard Seal

WOODLAND HABITAT (BRITAIN)

Conditions: temperature varies throughout the year
amount of light varies as leaf canopy grows in spring and summer

Successful plant: Bluebell

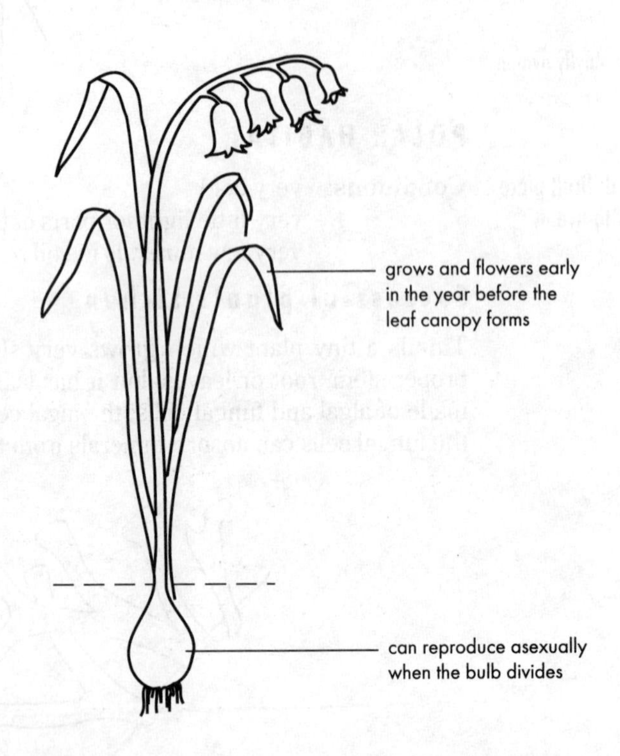

Fig 4.29 Bluebell

Successful animal: Dormouse

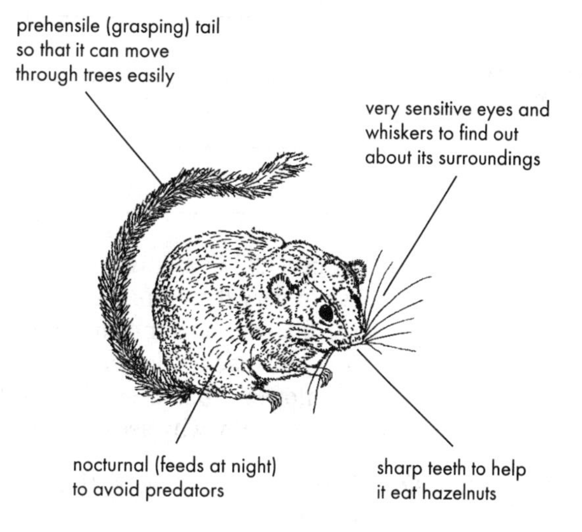

Fig 4.30 Dormouse

Links with Other Topics

Related topics elsewhere in this book are as follows

Topic	Chapter and Topic number	Page no.
Investigating ecosystems	7.1	275
Predator prey relationships	7.3	296

REVIEW QUESTIONS

Q6 Marram grass lives on sand dunes. The diagram below (Fig. 4.31) shows a section through a marram grass leaf, as seen through a microscope.

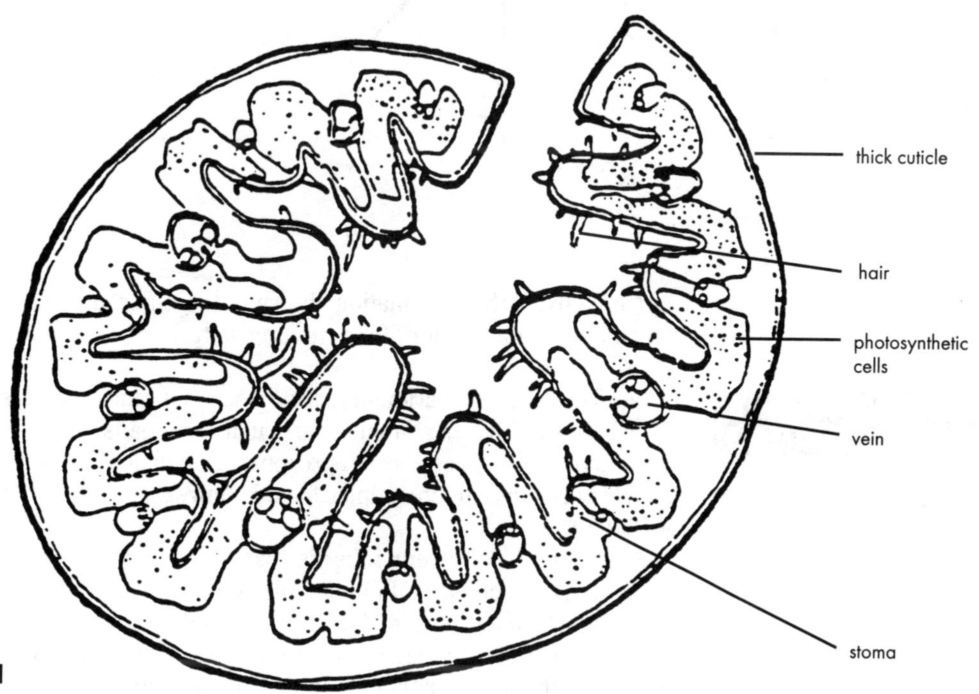

Fig 4.31

Explain THREE ways in which the leaf is adapted to reduce the amount of water vapour to the atmosphere.

1. ______________________________

____

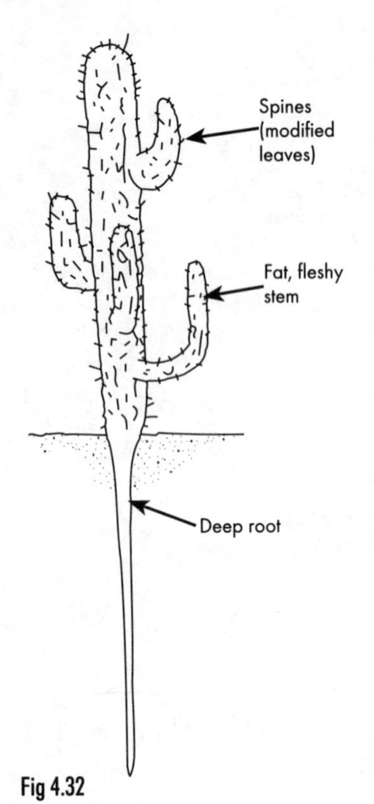

Fig 4.32

2. ____

____

3. ____

____

Q7 Figure 4.32 is a drawing of a cactus plant. These plants are highly specialised to live in very dry climates. The adaptations are labelled.
(*a*) Explain how the adaptations shown will help this desert plant survive.
(*b*) This cactus flowers once every 50 years, when it produces huge bright red flowers that are pollinated by birds. Usually it reproduces asexually. Explain carefully why asexual reproduction might be an advantage to this cactus.

Q8 The diagrams below (Fig. 4.33) show an animal which lives in fast-flowing streams.

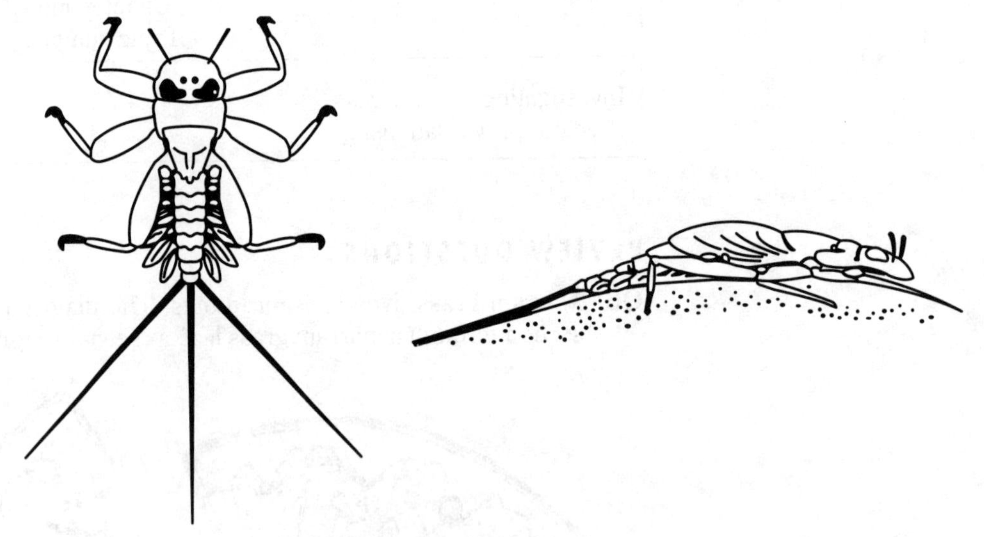

Fig 4.33

Using ONLY the information shown in the diagrams, explain THREE ways in which the animal is adapted for life in running water.

5 LIFE PROCESSES IN ANIMALS

In this Topic Group, various life processes in animals are described. The main emphasis will be on 'higher animals', particularly humans. However, you should remember that each of the processes represents a so-called *characteristic of life*; though will usually occur in a much simpler form in 'lower animals'. They also occur in plants (see Section 6).

Something else to remember is that the life processes are linked, and all are interdependent on each other. For example, food from nutrition, and energy from respiration, are needed to allow an animal to move, but both nutrition and breathing themselves need energy! The release of energy creates waste by-products, which are removed by excretion. Growth, development and reproduction are all part of a continuous series of processes, called life cycles.

In more advanced animals, each of the life processes described in this Topic Group are carried out by specialized cells, tissues and organs, often located in particular parts of the body (see Section 1 earlier). However, none of these parts is in isolation from the rest of the body. They are linked by networks such as the nervous system and the hormonal system (see Section 8) and the blood system.

MOVEMENT

Key Points

- Movement occurs as *locomotion*, in which the whole animal moves, and as other types of movements, involving just part of the animal.
- Movement is usually caused by the action of *muscles* against a *skeleton*. Muscles can either contract or relax. They are often arranged in pairs acting in different directions.
- Parts of the skeleton which are free to move in relation to each other are called joints. There are various types of joint. Movable joints, e.g. the elbow, have various structures to reduce friction.
- Natural joints which do not function properly can be replaced by artificial joints, e.g. hip joints. These must have many of the features of a healthy natural joint if they are to work effectively.

The Purpose of Movement

Animals need to move for a wide variety of reasons, e.g. to obtain food, to escape danger, to find a mate. Movement can involve the whole animal going from one place to another = *locomotion*. The type of movement involved varies between different species of animals, and also depends on what medium (land, water, air) the animal is moving through.

Movement may also involve just part of the animal. In this sense, an animal is always moving, since processes such as digestion, breathing, feeding and sensitivity include movement, and occur almost continually.

Muscles and Joints

Muscles

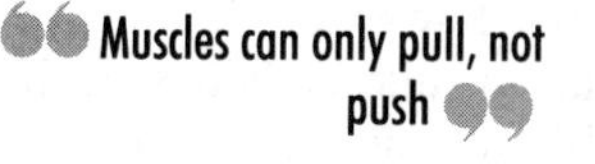

Most animal movement is achieved by the action of *muscles*. Muscles can either *contract* or *relax*. When a muscle contracts, it gets shorter. A muscle which is relaxed allows itself to get longer when pulled by another muscle. Muscles are often arranged in pairs – called *antagonistic pairs* – operating in different directions. When one muscle contracts, the other one relaxes (see below).

Muscles are attached to different parts of the animal's *skeleton*. Skeletons are support structures consisting of rigid components, such as bone. Although individual bones do not bend, many are attached to each other by movable *joints*, which allow the movement of one bone in relation to another.

Joints

There are three main types of joint:

(a) *Immovable joints*. Bones are fused, or held together by a protein called collagen. Example: the bones of the skull.
(b) *Partially movable joints*. Bones slide or glide over each other. The articulating ('rubbing') surfaces are covered by a layer of cartilage. Example: wrist, ankle.
(c) *Movable (synovial) joints*. There are two types, each allowing a different amount of movement:

- *ball-and-socket joint*, which allows movement in most directions. Several pairs of muscles are attached to each of the bones of the joint. Examples: hip, shoulder.
- *hinge joint*, allows movement in one plane only. Examples: elbow, knee, finger joints.

A good example of a movable (or *synovial*) joint is the human elbow, which allows the arm to bend and straighten (Fig. 4.34). The elbow joint consists of structures which occur in all movable joints:

- *Ligaments*, which attach bones to bones. They stretch, to allow movement.
- *Tendons*, which attach muscle to bone. They do not stretch, so the pull of the muscle is transmitted directly to the bone.
- *Cartilage*, which reduces friction between bones when they move against each other. Cartilage also acts as a 'shock absorber', to cushion the joint during vigorous movement, such as running and jumping.
- *Synovial fluid*, produced by *synovial membranes*, which lubricates the joint.

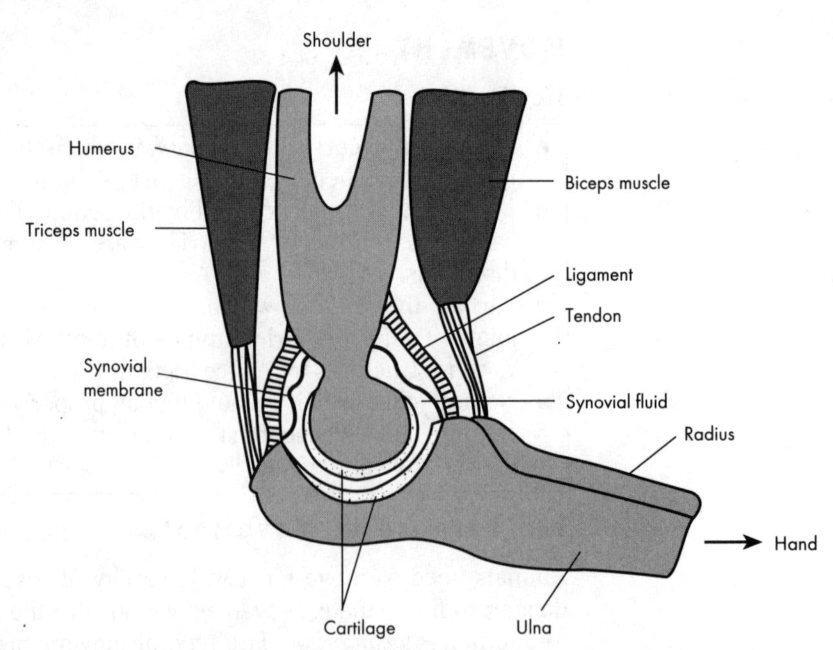

Fig 4.34 Human elbow

Movement at the elbow joint, like all movable joints, is produced by the operation of a muscle pair. They act 'against' each other. In the elbow, the muscle pair consists of the *biceps* and the *triceps*. Contraction of the biceps bends (*flexes*) the arm, contraction of the triceps straightens (*extends*) the arm. When either muscle contracts, the other normally relaxes. However, partial contraction of both muscles together can be used to 'lock' the joint.

Replacements joints

Natural joints are capable of withstanding massive strains. For example, a knee is subjected to approximately five times the body weight during mild activities such as jogging or climbing stairs. Certain sports impose even higher forces on the body's joints. Recovery from injuries to joints in young people can often be achieved by natural processes, although in some cases surgery may be needed.

Damaged joints in older people often do not heal naturally or by surgery, and may need to be replaced by *artificial joints*. These are commonly fitted in elderly people suffering from *arthritis*, or 'joint inflammation'. Examples of artificial joints are knees and hips. Hip replacements tend to be more successful than knee joints. This is because hips are ball-and-socket joints, and are more stable and better protected by surrounding tissues than the more exposed hinge joint of the knee.

Replacement joints consist of materials such as stainless steel, alloys and plastics. These materials are not usually 'rejected' by the body as part of the immune reaction. The joint is designed so that surfaces are smooth, to decrease friction, and do not produce fragments of 'debris'. The artificial joint is attached directly to the limb bones, using special glues and steel pins.

Practical Work

Practical 1: bones of the arm

Examine the bones of a human skeleton, or a model of a human skeleton, or the bones of a mammal's fore-limb.

(a) *Identify* each of the following bones: ***humerus, radius, ulna***. Observe how they form an elbow joint.
(b) *Explore* your own elbow, by placing your hand on your elbow and feeling the movement of the bones and muscles when you bend and straighten your arm. Feel the movement of the bones in your forearm when you make your arm twist from the elbow.
(c) *Draw* a simple diagram using what you have learned from (a) and (b), of the arrangement of the three bones at the elbow. Use Fig. 4.34 to help. Label the following:
– where the biceps is attached
– where the triceps is attached
– where the 'funny bone' is situated.

Practical 2: model arm

(a) Using the materials shown in Fig. 4.35, construct a model arm. The two pieces of cardboard represent bones of the arm. The elastic bands represent muscles. Observe what happens when you bend and straighten the model arm. Experiment with using the slots in different positions.

(b) List the biologically-important differences between your model arm and a real arm.

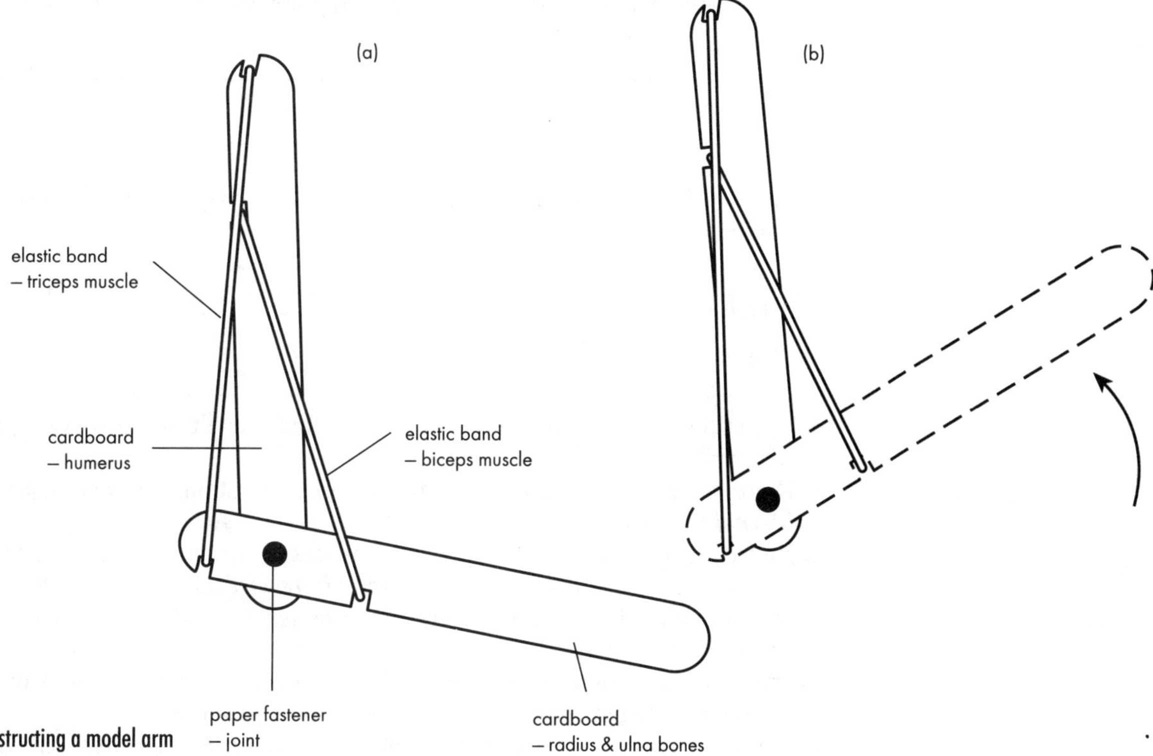

Fig 4.35 Constructing a model arm

Links with Other Topics

Related topics elsewhere in this book are as follows:

Topic	Chapter and Topic number	Page no.
Names and locations of the major organs	4.1	30
Functions of the major organs	4.1	31
The nervous system	4.8	149

REVIEW QUESTIONS

Q9 Fig. 4.36 is a diagram of the human arm. Fig. 4.37 is a transverse (cross) section of this arm taken through the line A–B.

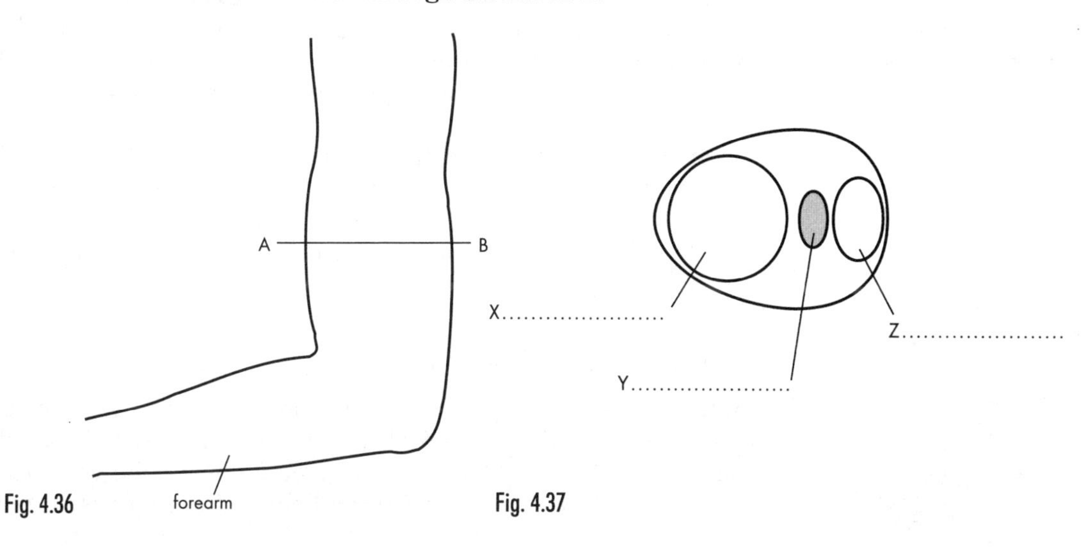

Fig. 4.36

Fig. 4.37

(a) (i) Label the structures **X, Y** and **Z** on Fig. 4.37
(ii) Which of these structures would increase in diameter when the forearm is raised?

____ *(4)*

(b) Name **two** other structures which might have been included in Fig. 4.37.

____

____ *(2)*

(c) By drawing on Fig. 4.36, show the positions occupied by the structures **X, Y** and **Z.**

RESPIRATION

Key Points

- Respiration consists of a series of events which result in the *release of energy in living cells.*
- There are two main processes involved in respiration: *internal respiration* and *external respiration.*
- Internal respiration involves chemical processes which release energy from the breakdown of glucose molecules. *Aerobic respiration* occurs when oxygen is present, and releases more energy than *anaerobic respiration*, which occurs in the absence of oxygen.
- External respiration involves the exchange of the *respiratory gases* oxygen and carbon dioxide between the organism and its environment.
- Respiratory gases are exchanged across a *respiratory surface*, e.g. the lungs. Gases are moved to and from the lungs by breathing movements.
- Smoking can cause serious damage to the lungs, and various chest diseases are considerably more common in smokers.

Respiration is an essential series of processes resulting in the release of energy within all living cells of all organisms. The release of energy occurs as the outcome of a sequence of chemical reactions often involving gases which are exchanged between the organism and its environment.

For convenience, the overall process of respiration can, as we have seen, be divided into two distinct components:

- **Internal respiration**, also known as cellular or 'tissue' respiration. Internal respiration is a *chemical* process involving the breakdown of organic molecules and the liberation of energy.
- **External respiration**, also known as gaseous exchange. External respiration, is a *physical* process involving an exchange of gases between the organism and its external environment. In larger, more complex and more active organisms there may be a specialised *respiratory surface* across which the exchange can take place. *Breathing* known also as *ventilation* or *respiratory movement* may promote gaseous exchange in certain organisms.

Make sure you understand the difference between breathing and respiration

Breathing is therefore *part* of respiration and should not be regarded as an alternative word for the same thing. Some organisms do not make any particular respiratory movements (for example, plants and very small animals). Also, the release of energy can occur without oxygen or carbon dioxide being involved, in 'anaerobic respiration'.

Internal Respiration

Internal respiration consists of the release of energy from the chemical breakdown of certain organic molecules such as glucose. Glucose is in fact the main respiratory substrate in respiration; other organic foods can be converted into glucose for respiration. The amount of energy liberated during respiration depends on how completely the glucose molecules are broken down. This in turn is determined by the availability of oxygen. Consequently, there are two types of respiration: *aerobic* (*with* oxygen) and *anaerobic* (*without* oxygen).

Aerobic respiration occurs in most living things, plant and animals. Aerobic respiration takes place in the *presence of oxygen* and results in the *complete breakdown of glucose* to water and carbon dioxide with the release of a relatively *large amount of energy*. This complete breakdown of glucose in the presence of oxygen is known as the *oxidation* of glucose. The process can be summarised as follows:

“Try to remember this word equation”

GLUCOSE	+	OXYGEN	→	WATER	+	CARBON DIOXIDE + ENERGY
$C_6H_{12}O_6$	+	$6O_2$	→	$6H_2O$	+	$6CO_2$

Note: the chemical formula which accompanies the word formula is not required by all examination boards.

The experimental evidence for respiration is based on the way in which substances are either used or produced during the processes of aerobic or anaerobic respiration. The uptake of oxygen and the production of energy (as heat) and carbon dioxide are relatively easy to observe. The use of glucose is difficult to demonstrate, however. The production of water from respiration (sometimes called 'metabolic water') cannot be shown in a simple experiment, although this water is important for many organisms.

External Respiration

External respiration involves the exchange of certain gases between the organism and its external environment. The *respiratory gases* involved are oxygen and carbon dioxide. Oxygen is needed for aerobic respiration, a process which occurs in most organisms. Also, carbon dioxide is produced as a waste product which can be poisonous (toxic) if allowed to accumulate, and which therefore has to be removed. The efficient exchange of gases during external respiration is an important part of the overall process of obtaining energy from aerobic respiration. Aerobic respiration yields about *20 times* more energy than anaerobic respiration.

(a) The Respiratory Surface

The surface through which respiratory gases are exchanged between the organism's internal and external environment is called the *respiratory surface*. The actual design of the surface depends on many factors, including habitat (aquatic or land) and size and activity of the organism. Examples of respiratory surface include lungs (mammals, birds, reptiles, amphibia), gills (fish, amphibia), tracheoles (arthropods), body covering (smaller animals).

(i) Characteristics of respiratory surfaces

All respiratory surfaces share certain *characteristics*, which in general promote the movement of gases by diffusion

- **Permeability.** Respiratory surfaces need to be thin, because diffusion does not readily occur over distances of more than about 1 mm.
- **Moisture.** Respiratory gases need to be dissolved within the tissues of organisms; a moist respiratory surface is particularly important in land-dwelling (terrestrial) organisms (they secrete mucus to maintain the moisture).
- **Large surface area.** The surface area available for the exchange of gases must be large in comparison with the volume (or mass) of the organism. The outer surface of a large, active animal would not be sufficient in area for gaseous exchange, even if it was permeable and moist. The reason for this is that large animals have a relatively small *surface area / volume ratio*; i.e. the proportion of body surface for a certain body volume.

“Your work on the blood circulation system will help you here”

■ **Transport system.** The rate of diffusion of gases across the respiratory surface is increased if a steep *concentration gradient* is maintained. For oxygen, this is achieved by carrying it away to the rest of the organism. In larger organisms, respiratory pigments such as *haemoglobin* are important in carrying oxygen. There is a diffusion gradient for carbon dioxide, too, though this operates in the opposite direction.

(ii) Experimental evidence for gaseous exchange

There are several experiments which can be used to confirm that oxygen is used and that carbon dioxide is produced during respiration. The relative rate at which respiratory gases enter or leave an organism can be used to determine the respiratory rate.

Practical 1: Experiment to demonstrate the relative amount of oxygen in inhaled and exhaled air in humans

This experiment involves the burning of a candle in inhaled and exhaled air (Fig. 4.38). The *combustion time* in the two situations depends on the amount of oxygen present and is used as an indication of the relative amounts of oxygen in each case.

It is normally found that a candle will burn for longer in ‘inhaled’ air than in exhaled air. Burning times of ‘inhaled’ air are typically 1–5 seconds, depending on the size of the jar; for exhaled air the times may be about 1–2 seconds. The results indicate that relatively more oxygen is contained in inhaled air than in exhaled air (the actual amounts are given in Fig. 4.46 below).

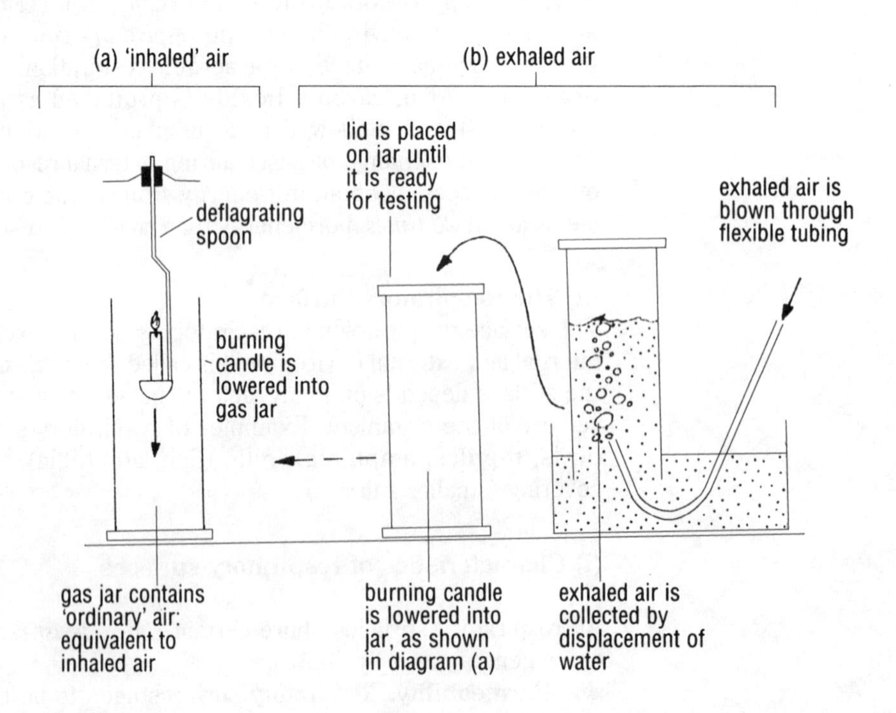

Fig 4.38 Experiment to compare oxygen in inhaled and exhaled air

Practical 2: Experiment to demonstrate the relative amount of carbon dioxide in inhaled and exhaled air in humans

The experiment is set up as shown in Fig. 4.39. Each of the two test tubes contain a *carbon dioxide indicator* solution such as bicarbonate indicator (or limewater). The arrangement of delivery tubes causes inhaled air to bubble through tube A whilst exhaled air bubbles through tube B.

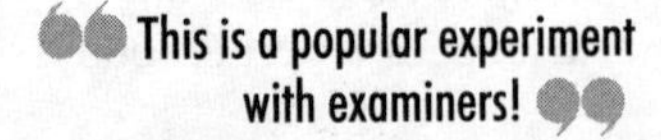
“This is a popular experiment with examiners!”

After a few seconds, tube B will show the presence of carbon dioxide; bicarbonate indicator will change colour from red to yellow; limewater will go 'milky'. Tube A will take much longer to give a positive result. This demonstrates that relatively more carbon dioxide is contained in exhaled air (the actual amounts are given in Fig. 4.46 p. 58).

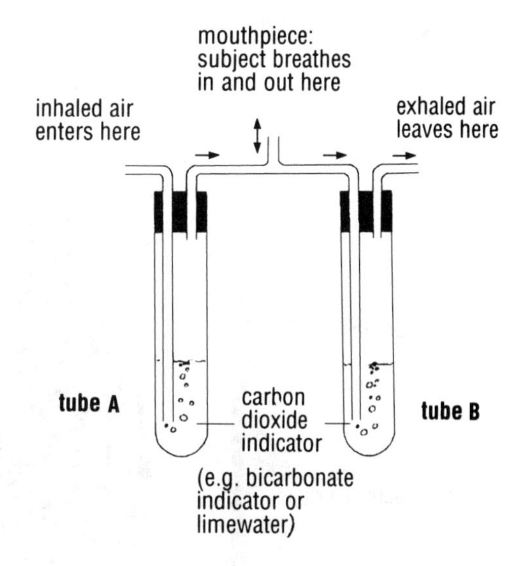

Fig 4.39 Experiment to compare carbon dioxide in inhaled and exhaled air

Practical 3: Experiment to demonstrate that carbon dioxide is given out by small organisms

Small organisms can be enclosed within the chamber into which carbon dioxide-free air is pumped (Fig. 4.40). Air can be pumped into the system by an aquarium pump, or pumped out using a filter pump. Any carbon dioxide which emerges from the chamber can be tested with a carbon dioxide indicator such as bicarbonate indicator or limewater.

The indicator in Tube 3 will show the presence of carbon dioxide, assumed to be from the organism, after several minutes.

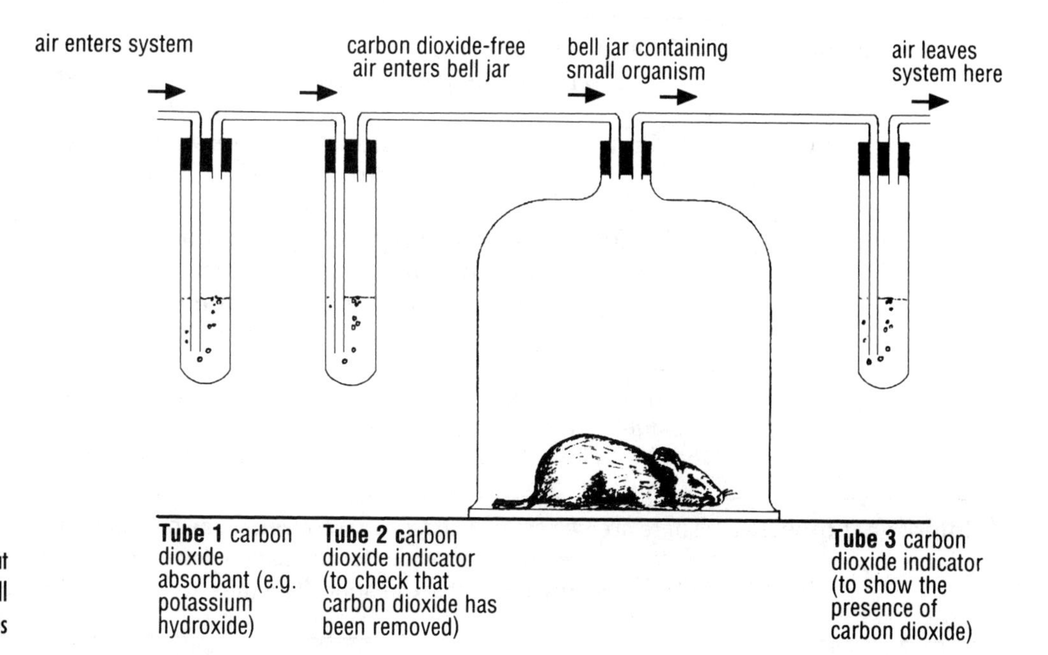

Fig 4.40 Experiment to show that carbon dioxide is given out by small organisms

(b) The Human Gas Exchange System

The *respiratory system* (shown in Fig. 4.41) is located in the thorax (chest cavity) and includes the *lungs*. The respiratory surface in humans is the lining of the lungs. Lungs are used by all large, air-breathing animals, including those mammals such as dolphins, seals and whales which live in water. The lungs have all the *characteristic features of respiratory surfaces*. They are *ventilated* by breathing movements. A brief description of the functioning of the main components of the respiratory system is given in Fig. 4.42.

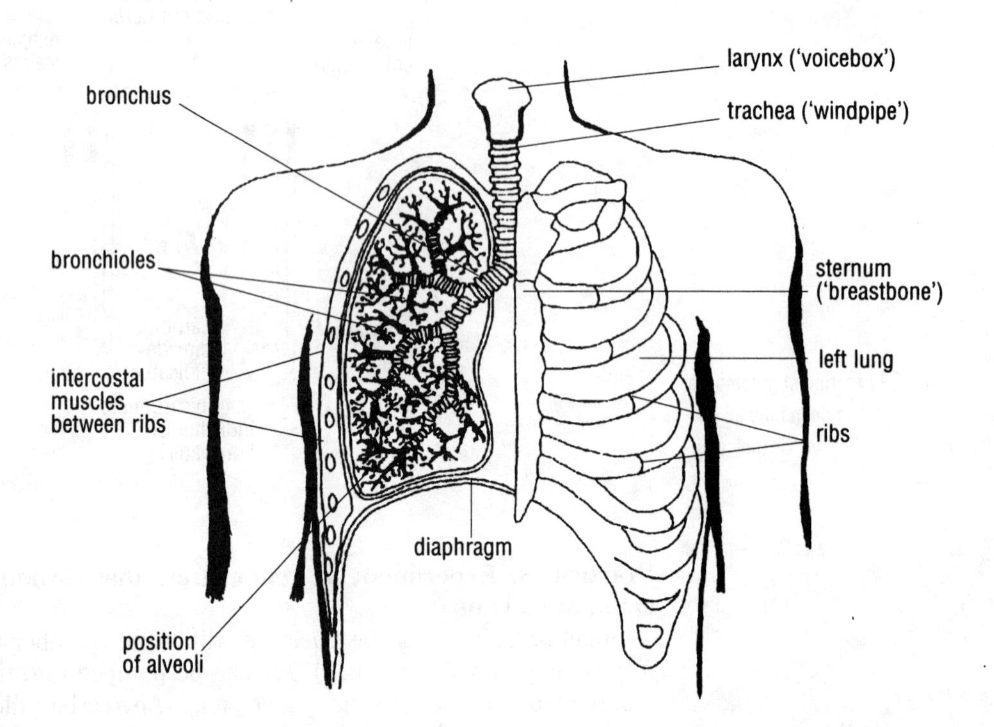

Fig 4.41 Human respiratory system (vertical section of right lung; front view of left lung)

COMPONENT	DESCRIPTION
Lungs	Consist of ***alveoli*** and ***bronchioles.*** Each lung is covered by a double ***pleural membrane***, enclosing ***pleural fluid***; these membranes protect the lungs.
Alveoli	Alveoli are the sites of gaseous exchange. The total number of alveoli in the lungs is about 700 million, giving a combined surface area of 80 m^2.
Bronchioles	Bronchioles consist of a branching network of tubes, carrying air to and from the alveoli. The wider bronchioles are strengthened with ***cartilage***.
Bronchi	Bronchi connect the bronchiole network with the trachea. Each ***bronchus*** is strengthened by ***cartilage***.
Trachea	The trachea (windpipe) connects the lungs to the mouth and nose cavities. ***Cilia*** (very fine hairs) cover the lining of the trachea; these beat rhythmically and move particles away from the lungs. ***Mucus*** secreted by the trachea, traps particles including microbes.
Ribs	The ribs protect the lungs (and heart) and are important in breathing movements. Ribs are free to move at their points of attachment to the ***vertebral column*** (backbone) and to the ***sternum*** (breastbone).
Intercostal muscles	There are two sets of ***antagonistic muscles*** between the ribs. ***External*** intercostal muscles are used for inhalation; ***internal*** intercostal muscles contract during exhalation. The muscles raise and lower the rib 'cage'.

COMPONENT	DESCRIPTION
Diaphragm	The diaphragm consists of a muscle sheet which, when relaxed, becomes ***domed*** in shape. Contraction of the diaphragm muscles causes it to flatten, increasing the chest volume.
Nasal cavity	Air that is breathed through the nose is ***filtered*** by hairs, moistened by ***mucus*** and warmed by blood capillaries lining the nasal cavity. Breathing can continue whilst ***chewing*** occurs in the mouth cavity, but is temporarily interrupted by ***swallowing***.
Epiglottis	The epiglottis is a muscular flap of tissue which automatically closes off the trachea during swallowing. This is an example of a ***reflex action***.

Fig 4.42 Summary of the main components of the respiratory system

The respiratory cycle

The respiratory (breathing) cycle is a rhythmic, alternating process involving inhalation (*inspiration*, 'breathing in') and exhalation (*expiration*, 'breathing out') (Fig. 4.43). Breathing rate in a resting condition is about 16 cycles per minute. The rate and depth of breathing vary according to the individual's need for energy (see below). The cycle is mostly coordinated by *involuntary control*.

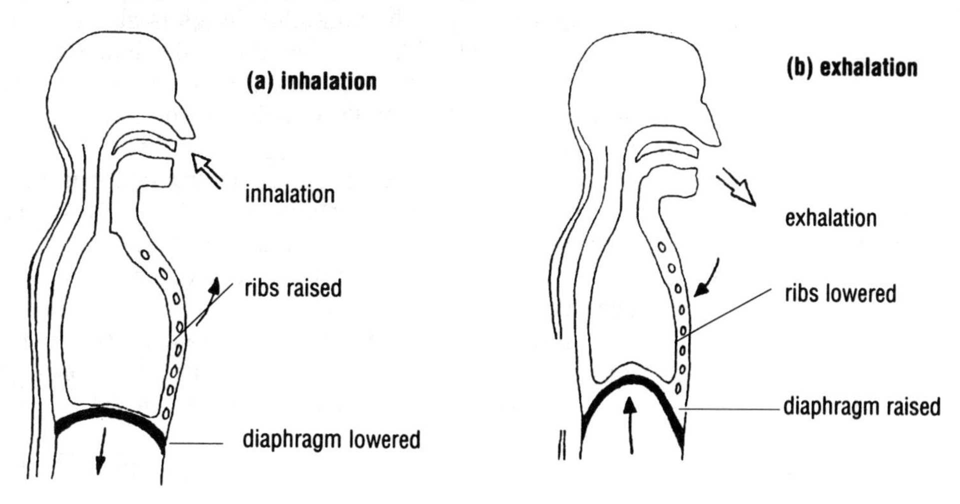

Fig 4.43 Breathing movements in humans (vertical section; side view)

Air is moved into and out of the lungs by changes in the *volume* of the thorax; this results in corresponding changes in *pressure* within the thorax. Breathing movements involve the contraction and relaxation of muscles associated with the ribs and diaphragm.

The *processes* of inhalation and exhalation are summarised in Fig. 4.44. Two models to illustrate the action of the ribs and the diaphragm are shown in Fig. 4.45.

Changes in air resulting from inhalation and exhalation are summarised in Fig. 4.46.

CHANGE	INHALATION	EXHALATION
Intercostal muscles	***External*** muscles contract, causing the ribs to move upwards and outwards.	***Internal*** muscles contract, causing the ribs to move downwards and inwards; gravity may assist this
Diaphragm	Contracts and flattens, pushing down on the contents of the abdomen below.	Relaxes and becomes domed; displaced contents of abdomen push from below.
Lungs	Become inflated against their elastic tendency.	Elasticity of lungs causes them to become deflated.
Volume of thorax	Increases	Decreases
Pressure in thorax	Decreases	Increases

Fig 4.44 Summary of inhalation and exhalation

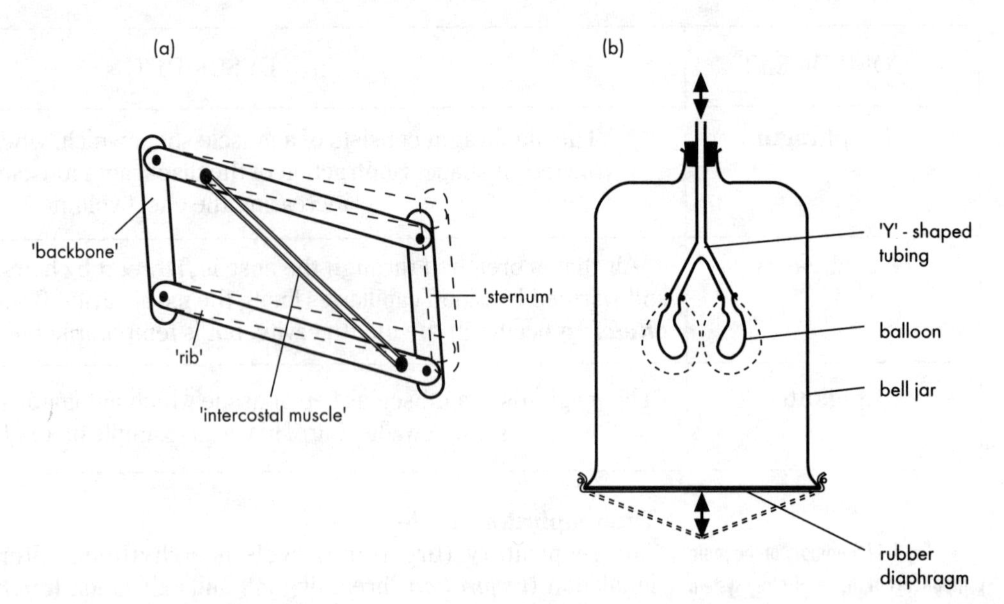

Fig 4.45 Models to illustrate the action of ribs (a) and the diaphragm (b) in breathing

When the demands for energy change, there is often a corresponding change in breathing *depth* and *rate*:

(i) **Depth of breathing**. Although individual measurements vary widely the average capacity of human adult lungs is about 5 litres). During normal (*'quiet'*) breathing, about 0.5 litres of air (the tidal volume) is breathed in. This is increased to 4.5 litres (the *vital capacity*) when more energy is needed. Some air (about 0.5 litres during exercise) normally remains in the lungs at all times; this is known as the *residual capacity.*

(ii) **Rate of breathing**. The breathing rate is normally about sixteen breaths per minute during 'quiet' breathing. Most (about 60 per cent) of the normal breathing movements in this situation are due to the diaphragm, rather than to the intercostal muscles. The rate can increase to about 30 breaths per minute when more oxygen is needed.

Despite increases in the depth and rate of breathing during exercise, the demands for oxygen may not be fully met. In this situation, respiration may become partially anaerobic for a while and an *oxygen debt* is caused.

Try to understand how these differences in gas composition occur

COMPONENT	EXHALED	INHALED AIR
Oxygen	16%	21%
Carbon dioxide	4%	0.04%
Water vapour	Saturated = 6.2%	Variable, depends on humidity; average = 1.3%
Temperature	37°C	Ambient

Fig 4.46 Relative composition of inhaled aand exhaled air

(c) Gaseous Exchange

An exchange of gases occurs at the *alveoli*; this results in a change in the air that enters the lungs (see Fig. 4.46). Approximately 70 per cent of inhaled air reaches the alveoli and can take part in gaseous exchange; the other 30 per cent of inhaled air occupies the 'dead space' of the impermeable tubes leading to the alveoli.

There are about 700 million alveoli in an average adult's two lungs; these provide a total surface area of about 80 m^2. This surface area is about forty-four times that of skin (1.8 m^2) and is necessary to meet the needs of humans which are comparatively large, active animals.

The structure of a single alveolus is shown in Fig. 4.47. Each alveolus has the characteristics of respiratory surfaces in general; they are permeable and thin (0.0001 mm), moist (with mucus), have a large surface area and are closely associated with a transport system. Steep concentration gradients of oxygen and carbon dioxide are maintained across the alveolus lining. The respiratory gases are carried to and from the lung by the blood in slightly different ways:

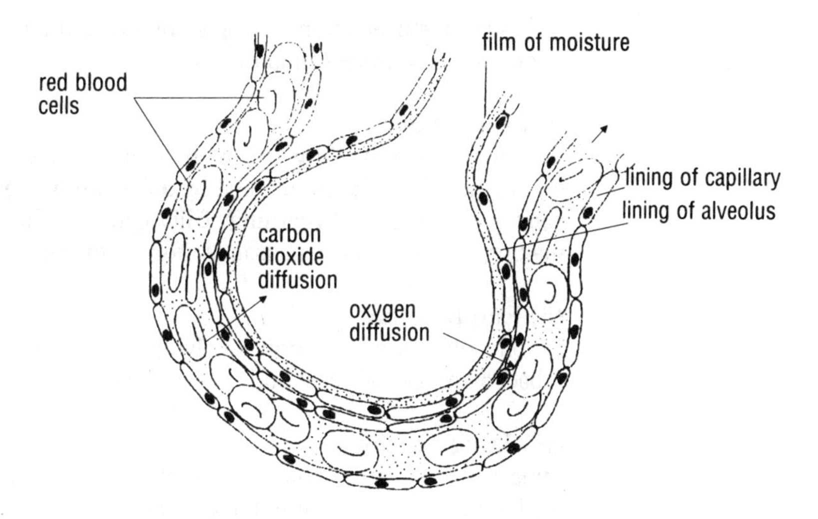

Fig 4.47 Structure of an alveolus, with capillary

(i) Oxygen

Oxygen is not very *soluble* in the liquid part of blood and is carried instead in *red blood cells*, where it becomes attached to ***haemoglobin***. Haemoglobin is converted to the unstable molecule *oxyhaemoglobin* when it combines with oxygen. Haemoglobin is re-formed again when the blood reaches tissues in the body where the oxygen concentration is low:

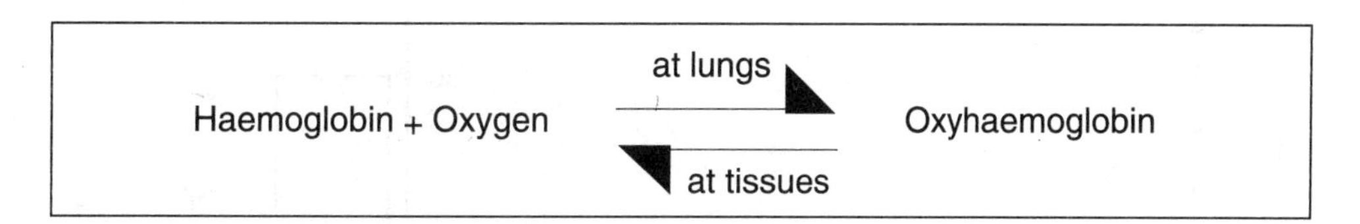

Red blood cells have a comparatively large surface area, allowing a more rapid uptake and loss of oxygen. The cell are pressed against the walls within the narrow capillaries; this increases the efficiency of oxygen uptake.

(ii) Carbon dioxide

About 90 per cent of the carbon dioxide is carried in the blood *plasma* as *bicarbonate ions* which may combine with hydrogen ions to form *carbonic acid*. The remainder of the carbon dioxide is carried by haemoglobin molecules.

Refer again to the comparison of inhaled and exhaled air in Fig. 4.46 above. The fact that a significant amount (16 per cent) of oxygen is exhaled is one reason why *expired air resuscitation* (E.A.R., the 'kiss of life') can be so effective in reviving an unconscious subject. Nitrogen is the most abundant gas (79 per cent) in air; it is not included in Fig. 4.46 because it does not take part in chemical reactions (it is inert) within the body, so it can be ignored for normal situations.

(d) Environmental Effects on Respiration

The quality of the air inhaled can have significant effects on an individual's health. Air that is inhaled through the nose is filtered, moistened and warmed before it reaches the delicate lung tissues. Air entering the body can cause damage, however, especially if it contains impurities. *Polluted air* can cause disease and, directly or indirectly, even death. *Environmental pollution* includes the presence in the atmosphere of gases from industry and internal combustion engines.

Self-pollution by individuals who smoke is a major problem in many countries in the world. Smoking increases the risk of certain respiratory and circulatory diseases for smokers. Smoking also affects others who breathe smoke-laden air; this is known as 'passive smoking'. Pregnant women who smoke may affect the growth of their unborn child because smoking reduces the oxygen available for the foetus. This is because *carbon monoxide* absorbed from the smoke displaces oxygen from haemoglobin molecules.

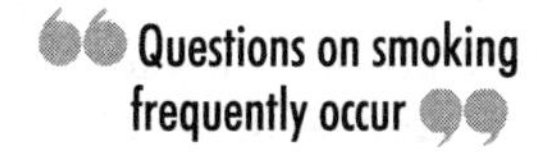
Questions on smoking frequently occur

There are three main diseases, involving the respiratory system, which occur more frequently in smokers than non-smokers:

(i) Bronchitis

Inflammation (swelling up) of bronchi, caused by irritants and infectious micro-organisms; this is accompanied by an accumulation of mucus. The narrowing of tubes causes difficulty in breathing. Cilia can be killed by substances in tobacco smoke, making the respiratory passages more vulnerable to disease.

(ii) Emphysema

Emphysema is a condition involving the breakdown of the alveoli walls. Irritants in tobacco smoke induce coughing, which damages already weakened lung tissue.

(iii) Lung cancer

Lung cancer is much more frequent in smokers than non-smokers; about 90 per cent of all lung cancers occur amongst smokers.

There is much evidence to indicate a strong relationship between smoking and diseases; life expectancy is, on average, decreased by increased smoking (Fig. 4.48), especially if smoking was started in earlier life. Smokers who give up increase their life expectancy, for instance by reducing the chances of contracting lung cancer (Fig. 4.49).

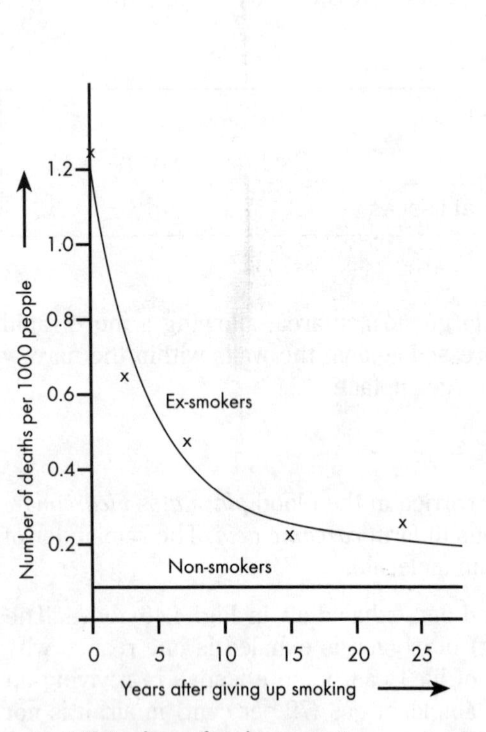

Fig 4.48 Death rates from lung cancer in ex-smokers and non-smokers

100
80
60
Proportion of 35 year old men expected to live beyond 65 years (%)
40
20
0
Non-smokers
1-14
15-24
25+
cigarettes per day
Smokers

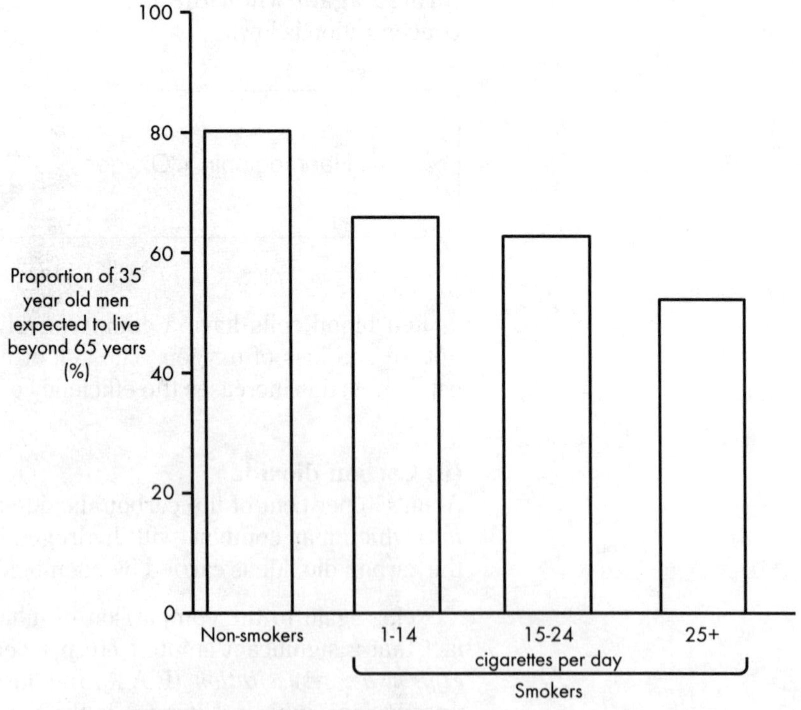

Fig 4.49 Proportion of men smokers and non-smokers (aged 35) who are expected to die or live beyond 65 years

(e) Mechanical Respirators

Mechanical respirators are used for the treatment of medical conditions which result in an inability of the patient to breathe properly. Certain diseases, such as poliomyelitis, can result in loss of muscle function in the chest region, and the patient requires assistance in ventilating. One type of mechanical respirator consists of an airtight metal or plastic container. This encloses the chest and, in some cases, the rest of the body except the head, which protrudes through a neck seal. An air pump applies a partial vacuum to the respirator at regular intervals, causing the chest to expand. The process is reversed, to allow breathing out. The principle is similar to that demonstrated in the model shown in Fig. 4.45.

Patients are also unable to breathe properly when they are unconscious, e.g. resulting from anaesthetics during an operation. In these cases, mechanical respirators periodically pump air into and out of the patient's lungs through a mask. Additional oxygen or anaesthetics may be added to the airstream.

Links with Other Topics

Related topics elsewhere in this book are as follows:

Topic	Topic number	Page no.
Names and locations of the major organs	4.1	30
Functions of the major organs	4.1	31
Blood circulation	4.5	96
Respiration in plants	4.6	112
Air pollution	7.4	303

REVIEW QUESTIONS

Q10 Fig. 4.50 shows the apparatus used for collecting some of the substances in cigarette smoke.

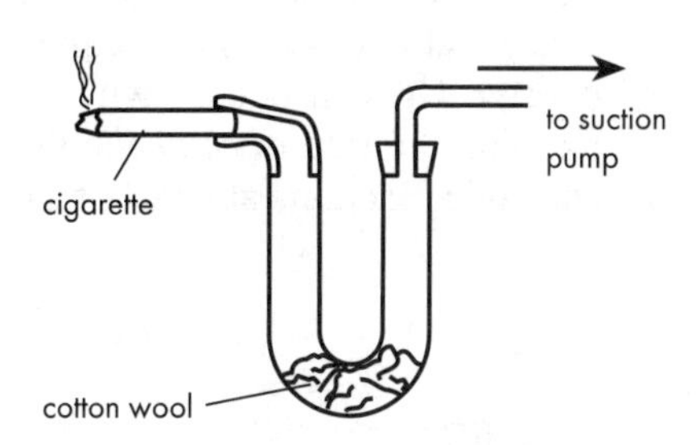

Fig 4.50.

As the cigarette burns, the cotton wool turns brown.

(a) (i) Name the substance which causes the cotton wool to change colour.

____

(ii) The cotton wool provides a large surface area on which this substance collects. What structures in the lungs does the cotton wool represent?

____

(iii) Explain how smoking affects the amount of oxygen taken up by the blood.

____

____

____

(iv) State two ways in which smoking can damage the smoker's health.

____

____

(b) What are the effects on an embryo if the mother smokes during pregnancy?

____

____

____

GROWTH AND DEVELOPMENT

Key Points

- *Growth* is a permanent increase in size of an organism.
- Patterns of growth vary between species. *Growth rates* are typically highest during early or intermediate ages.
- *Development* consists of a re-organization of tissues during particular stages of an organism's life. In higher animals, development in some parts of the body is more rapid than in others. The result is a change in *shape* as well as *size* during growth and development.

(a) Definition of Growth

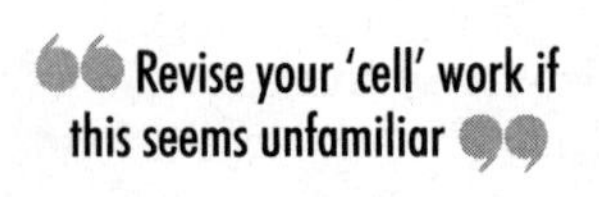

Growth is a permanent increase in the size of an organism, caused by the formation of new protoplasm. A 'permanent increase in size' is achieved by the synthesis of the protoplasm (cytoplasm and nucleus) of cells, which enlarge and may divide. The synthesis of new protoplasm involves chemical reactions which use the raw materials and energy obtained from nutrition. These raw materials are either assembled into structural and storage components, or broken down during respiration to provide energy. Growth occurs when the overall rate of *anabolic* ('building up') reactions exceeds the rate of *catabolic* ('breaking down') reactions (Fig. 4.51(a)). Anabolic and catabolic reactions are part of an organism's *metabolism*.

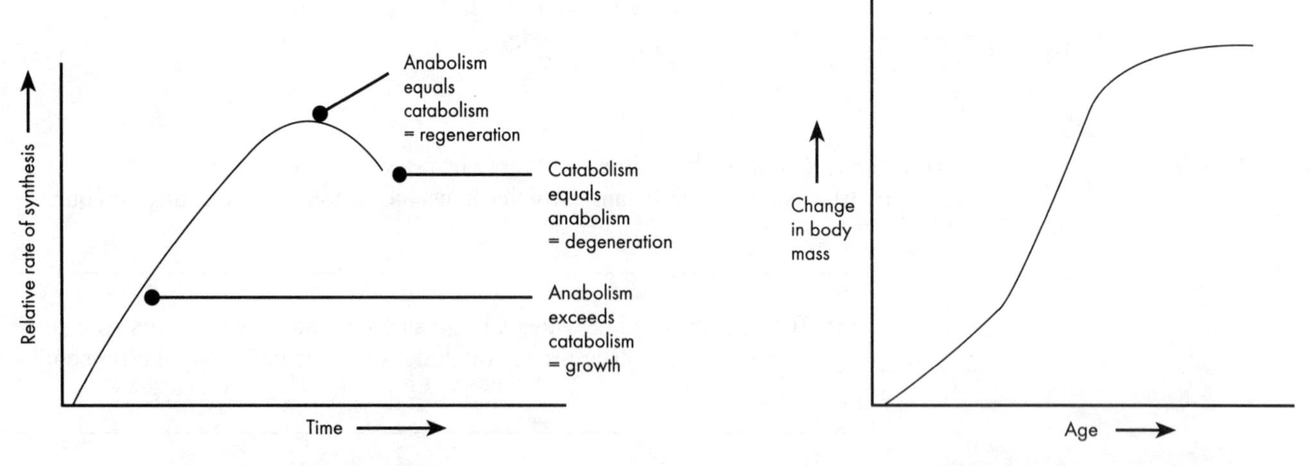

Fig 4.51 (a) Growth as a building process

Fig4.51 (b) Typical growth pattern

(b) Patterns of Growth

Measurements of growth tend to produce S-shaped, or *sigmoid*, curves when plotted as a line graph (Fig. 4.51(b)). Such curves of absolute growth are typical for most organisms, and shows that the period of most rapid growth occurs during an intermediate stage of the individual's life span. The sigmoid curve is also characteristic of the growth of populations (see Chapter 7).

Development

Development is an increase in complexity of an organism, and often occurs at the same time as growth. Development involves coordinated processes of *specialization* and *redistribution* of tissues. This produces a progressive change in shape and form of the individual.

In animals, including humans, most development actually occurs *before* much growth occurs.

Growth and development in some animals (particularly in 'warm-blooded' animals) is often completed in the earlier part of their life whilst for others it may be a continuous and potentially unlimited process. Although, unlike in plants, growth takes place *throughout* an animal's body this does not necessarily occur at a constant rate. Growth and development can occur unevenly, producing changing bodily proportions; this is called *allometric growth* and is characteristic of mammals. A more uniform pattern, called *isometric growth*, occurs in most other animals (and in plants); there is a progressive increase in size and mass which is not accompanied by much change in overall shape, e.g. in fish and reptiles.

Human development

A human is first formed as a single cell at conception and then rapidly grows by cell division. Most of the cell division occurs before birth, the fertilised egg dividing 44 times, to produce 2×10^{12} cells. A further four cell divisions produces the 6×10^{13} cells which make up an adult! The average number of cell divisions for any cell is therefore about 45.

Further growth occurs in size, rather than in number of cells. This can be part of a compensatory growth response when tissues such as muscle, bone and blood have additional demands imposed upon them.

Development mostly takes place during the early stages of growth, and there is an increase in diversity of cell type so that at birth an individual will have most of the required 1000 different types of cell present, arranged in a characteristically human form.

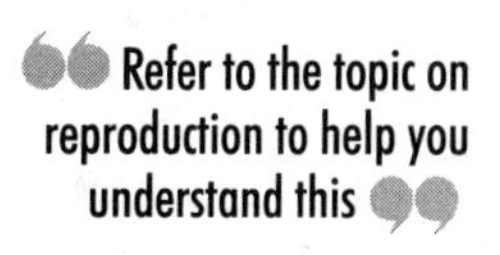

Growth rate is not constant in humans; the maximum rate occurs between conception and two years after birth. There is another, smaller increase during adolescence which begins earlier in girls, although the rate is generally more rapid in boys, who tend to attain a greater maximum average height (Fig. 4.52). Growth and development cause a steadily increasing difference between males and females, as *secondary sexual characteristics* appear. This represents a preparation for adulthood, including reproduction.

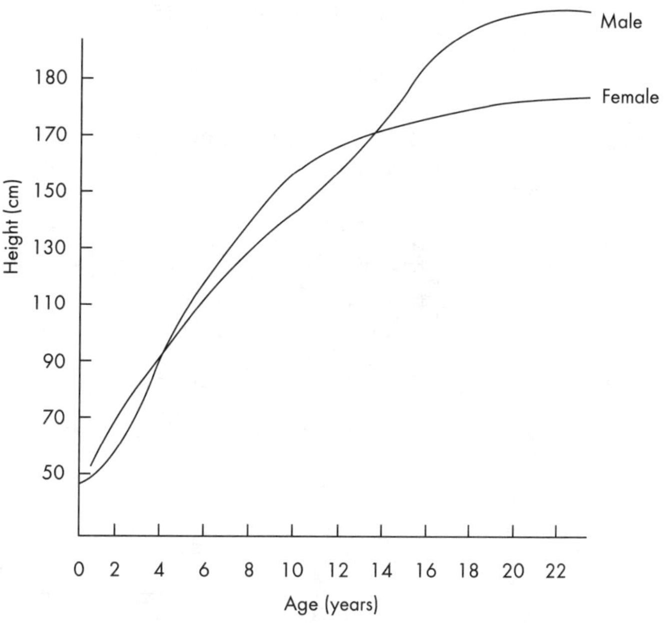

Fig 4.52

The uneven (allometric) growth in humans produces changes in the relative proportions of the body (see Fig. 4.53); this reflects the changing emphasis on different functions during an individual's life.

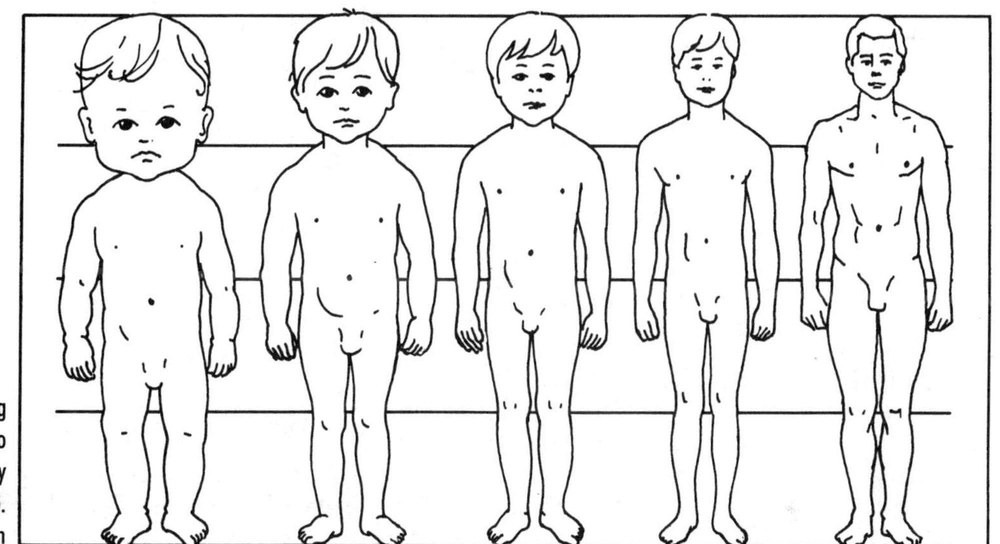

Fig 4.53 Human growth, showing allometric effect. Figures drawn to same scale, to show changing body proportions). Illustration by D.G. Mackean, after C.M. Jackson

Links with Other Topics

Revise your 'cells' work if this seems unfamiliar

Related topics elsewhere in this book are as follows:

Topic	Chapter and Topic number	Page no.
Specialised cells	4.3	39
Reproduction	4.5	65
The hormonal system	4.8	152
Sources of variation	6.1	203
Types of variation	6.7	223

REVIEW QUESTIONS

Q11 A boy's height was measured every 2 years from birth to 18 years. The information is given below:

Age (years)	birth	2	4	6	8	10	12	14	16	18
Height (cm)	50	90	105	118	130	143	150	164	180	183

(a) By how much did the boy grow between

(i) birth and 2 years ________ (ii) 4 and 6 years ________

(iii) 14 and 16 years ________

(b) Plot a graph below to show the change in the boy's height from birth to 18 years.

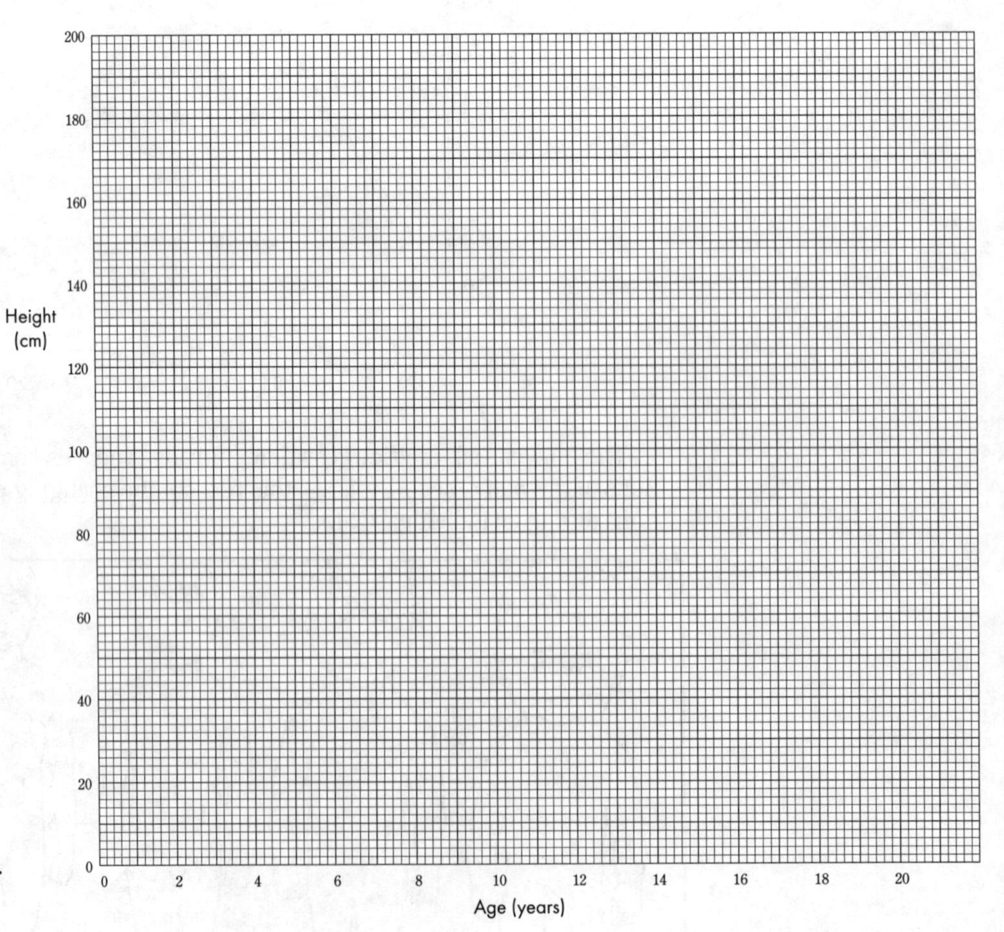

Fig 4.54.

(i) Use the graph to find the height of the boy at age 13 years. ______

(ii) Between which ages was the increase in height the most rapid? ______

Q12 The graph below shows the relative growth rates of different organ systems, expressed as a proportion of their weight in an adult human.

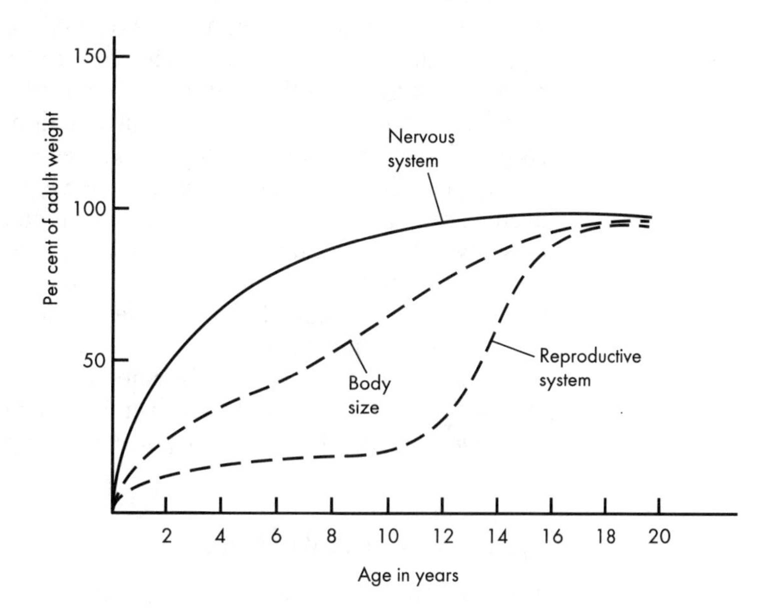

Fig 4.55.

(a) Between which ages was the growth rate most rapid

(i) for the nervous system ______________________

(ii) for the reproductive system ______________________

(b) Briefly explain the biological importance of the pattern of nervous system development shown in the graph.

____

____

REPRODUCTION

Key Points

- There are two main types of reproduction, *asexual* and *sexual*. Sexual reproduction involves the production of sex cells (*gametes*), which fuse during *fertilization* to form a new individual. Human reproduction is sexual.
- The reproductive systems of human males and females become active following sexual development. Males produce sex cells called *sperms*, females produce sex cells called *eggs*. Sperms are produced continuously between the ages of about 12 to 70 years. A single egg is usually released every month during the ages of about 12 to 45 years, during a *menstrual cycle*.
- Fertilization results from *copulation* (*sexual intercourse*), when a sperm fuses with an egg to form a *zygote*, which develops into a *foetus* during *pregnancy*.
- The foetus develops within the mother's *uterus* (womb), where it receives essential substances through the *placenta*.
- There are different methods of *contraception* available to prevent unwanted fertilization.
- Various *sexually transmitted diseases* can occur, and treatment or prevention exist for all of them.

The purpose of reproduction

Reproduction is the production of new, independent organisms with similar, though not necessarily identical, characteristics to their parents. Reproduction is one of the 'characteristics of life'. However, it is needed for the survival of a species rather than of individual organisms. A species is a group of similar organisms which may reproduce with each other but not with individuals from other species. An increase in *population* results if the rate of reproduction exceeds the rate of death within a species. In natural conditions populations tend to remain fairly constant.

Complete life cycles of organisms involve growth, development, reproduction and death. These processes are linked and continuous so that a species may continue to exist, even though individuals die. A lifespan is the time required to complete a life cycle and varies between species. Lifespans range from minutes (in bacteria) to several years (in mammals) to centuries (in some trees).

The more complex organisms in general have relatively long periods of growth and development. Life cycles may be interrupted by disease, predation or unfavourable changes in the environment. However, the extinction of a species may be avoided if at least *some* individuals have the opportunity to reproduce.

Types of reproduction

There are two main types of reproduction:

(a) **Asexual reproduction.** Part of a single 'parent' separates and becomes an independent organism, which is genetically identical to the parent. Asexual reproduction occurs in many plants and in simpler animals.
(b) **Sexual reproduction.** Two parents each produce special sex cells, called gametes, which become fused together during fertilization. This process forms an individual which combines characteristics of both parents but is different from each parent. This is important, because it increases variation amongst individuals.

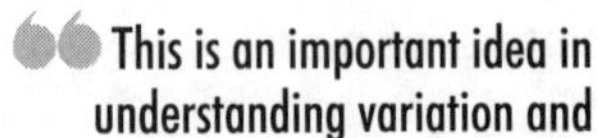
This is an important idea in understanding variation and natural selection.

Many organisms reproduce asexually but most organisms reproduce sexually. Some organisms are capable of both sexual and asexual reproduction depending, for instance, on environmental conditions. In all cases, reproduction involves the transfer of genetic information to the next generation.

Fig. 4.56 summarises the main features of asexual and sexual reproduction.

FEATURE	ASEXUAL	SEXUAL
Occurrence	Simpler, smaller plants and animals.	Complex, larger plants and animals.
Parent(s)	Only one parent involved.	Usually two parents involved.
Specialised structures needed.	Reproductive tissue may include various outgrowths, spores, etc.	Reproductive tissue is specialised for the production (by meiosis) and transfer of gametes.
Inheritance	Offspring are genetically identical to the parent, i.e. they are clones.	Offspring are genetically different from each parent. This increases variation.
Reproductive rate	Relatively fast	Relatively slow.

Fig 4.56 Comparison of asexual and sexual reproduction

Human reproduction

Humans, like all other vertebrate animals, reproduce only by sexual means.

(a) Sexual Development

Individual humans are either male or female, and this is determined by the inheritance of *sexual chromosomes*. Individuals of each sex are, however, similar in many respects until the *primary sexual development* of the *gonads* which are the sites of gamete production. Gonads also produce hormones which stimulate the development of *secondary sexual characteristics*.

The bodies of children undergo extensive changes which prepare them for the possibilities of reproduction as adults. The period of transition of children into adults is called adolescence and the onset of primary sexual development is called puberty. This is accompanied by the development of secondary sexual characteristics. Some of the physical changes which take place include:

Male: development of muscles, widening of shoulders, deepening of the voice, development of coarser hair on body surface, especially on face, under arms and in the pubic region.

Female: development of subcutaneous fat, development of breasts, widening of hips, slight deepening of the voice, development of hair under arms and in the pubic region.

In both males and females there is a general growth of body tissues, although this does not occur at the same rate throughout the body (this is called allometric growth). The outcome of secondary sexual development is, in effect, to advertise maturity to possible sexual partners. In females, development is also a preparation for pregnancy and birth if fertilization occurs. In general, the process of puberty begins earlier in females than in males, though there is much variation between individuals.

The function of the male and female reproductive systems includes the production of gametes and hormones. In males, there is a means of transferring gametes into the female. In females, provision is made for the possible development of one or more fertilized eggs (zygotes).

(i) Male reproductive system

“Try to learn & memorise the *function* of each of the structures shown in these diagrams”

The male reproductive system is shown in Fig. 4.57. The male gamete is the *sperm* (Fig. 4.58) and the gamete-producing structures (gonads) are the *testes*. Each testis consists of about 500 metres of sperm tubules which produce and store sperm. Between the tubules are cells which produce *testosterone*, the male sex hormone.

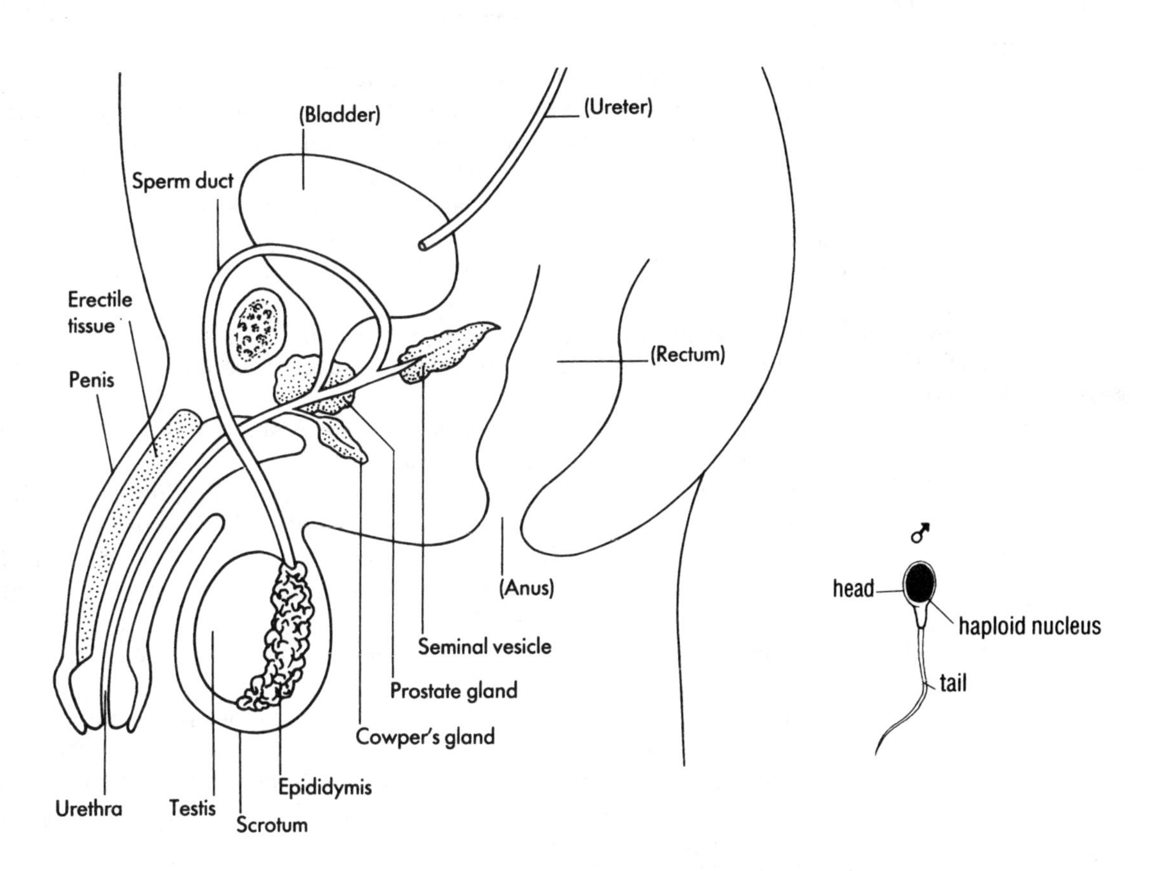

Fig 4.57 Human male reproductive system (section side view)

Fig 4.58 Male gamete sperm 0.05mm long

(ii) Female reproductive system.

The female reproductive system is shown in Fig. 4.59. The system is more complex than that of the male because, apart from the production of gametes and sex hormones, it may be the site of development if fertilization occurs.

The female gamete is the *egg (ovum)* (Fig. 4.60). Eggs (*ova*) are produced in the two *ovaries*, which are the female gonads. Each ovary is also responsible for producing various hormones including *oestrogen*, the female sex hormone.

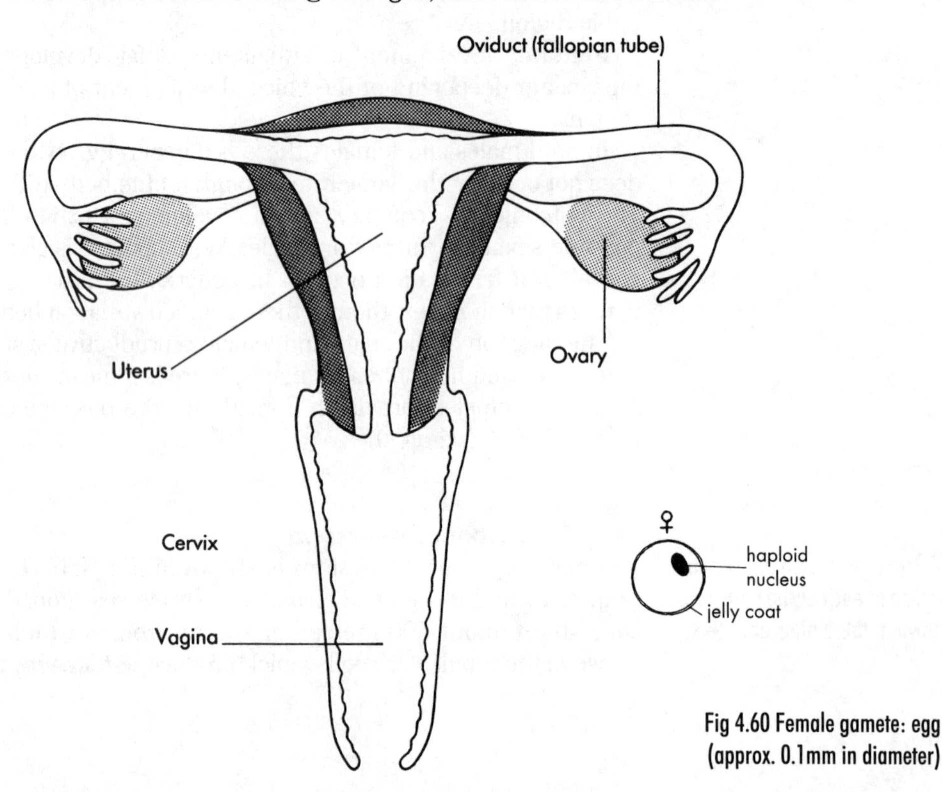

Fig 4.59 Human female reproductive system (section front view)

Fig 4.60 Female gamete: egg (approx. 0.1mm in diameter)

(iii) Female menstrual (oestrous) cycle

Females produce eggs as part of a regular *menstrual* or *oestrous cycle*, often of about 28 days. About 400 eggs are released between puberty (around twelve years) and *menopause* (around forty-five). This periodic and limited release of gametes is unlike the situation in males, who normally produce gametes continuously (millions per day) and over a longer period (from about age 12 to 70 years).

The menstrual cycle involves two important events:

> Remember, menstruation and ovulation are quite distinct processes, though one follows the other in a cycle.

(a) **Menstruation** ('period') at days 1–5. This is the shedding of the lining of the uterus (endometrium), which had previously been thickened to receive a possible fertilised egg.

(b) **Ovulation** (egg release) at days 13–15. A single egg is usually released from alternate ovaries during each cycle.

Both these processes can be interrupted by fertilization and *implantation*, and ovulation can be interrupted by the contraceptive pill (see below). If no fertilization takes place, the cycle will normally continue. However, the frequency and duration of the menstrual cycle can be affected by such factors as emotions (e.g. extreme stress or trauma) and an inadequate diet.

Changes within an ovary at various intervals during the menstrual cycle are shown in Fig. 4.61.

The relative amounts of the hormones including oestrogen and progesterone vary during the cycle. Oestrogen from the ovary tissue has various effects, including the thickening of the uterus lining following menstruation. Progesterone from the 'yellow body' (*corpus luteum*) maintains the thickened lining; a drop in progesterone levels causes menstruation to take place.

(b) Copulation and Fertilization

Copulation is a process which brings male and female gametes into contact. *Fertilization* (conception) is when they fuse and their chromosomes are combined, forming a diploid zygote.

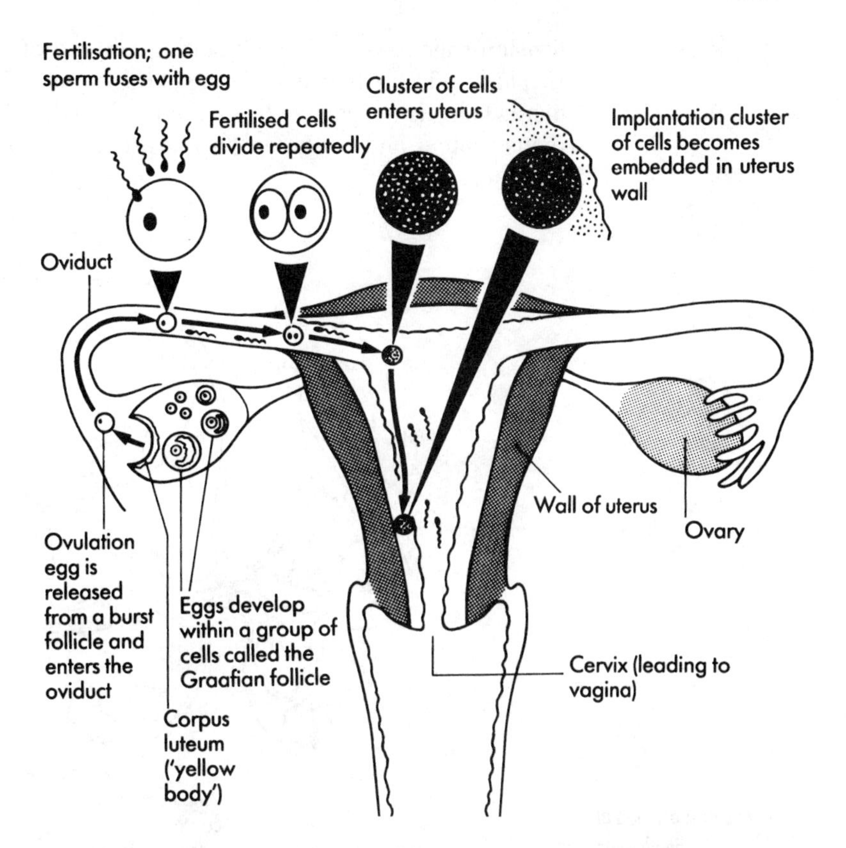

Fig 4.61 Summary of ovulation ('egg release' see menstrual cycle), fertilisation and implantation

The *penis* of a sexually excited male becomes erect as blood under pressure fills the spongy 'erectile tissue'. The *vagina* of a sexually excited female widens and becomes lubricated by the secretion of mucus. During copulation (also called mating, sexual intercourse, coition) the erect penis is inserted into the vagina. Movements of the penis may result in increasing physical excitement in the male and the female and can lead to a peak of excitement, called *orgasm*. Orgasm in the male is a reflex process which results in the ejaculation of about 5 ml *semen* by rhythmic contractions of the sperm duct. Orgasm in the female results in contractions of the vagina and uterus, which draws sperm towards the uterus.

The transfer of sperm (*insemination*) can also be achieved artifically, using a syringe.

Only a small proportion (i.e. several hundreds) of the 400 million sperm deposited at the cervix actually reach the site of fertilization in the oviduct. One sperm only is allowed successfully to penetrate any egg present in the oviduct. This must be within thirty-six hours of ovulation. The sperm will lose its tail, and the haploid nuclei of the sperm and egg will form the diploid zygote of a new life.

Twins

There are two ways in which twins (and other multiple births such as triplets etc.) are formed:

(a) **Identical twins.** An egg is fertilized as normal with a single sperm and the resulting zygote divides into two cells. However, the cells do not remain together; instead, they separate and continue independent development. Identical twins have the same genotype, so will for example, be of the same sex. Identical twins have been useful in studying inheritance.

(b) **Non-identical (fraternal) twins.** Two eggs are released at the same time, by one or both ovaries. Both eggs are fertilized separately by different sperm. Non-identical twins have the same genetic relationship as brothers or sisters; they need not be of the same sex.

(c) Pregnancy and Development

Pregnancy (*gestation*) is the interval between the conception and birth of an individual. This normally takes 40 weeks in humans.

Development involves both the growth and organisation of tissues; this prepares an *embryo* or *foetus* for a relatively independent existence as a baby.

It may take four days for the cilia on the inside of the oviduct to waft the zygote down to the uterus, during this time the zygote divides repeatedly by a type of cell division called *mitosis*. The resulting embryo of about 128 cells becomes embedded (*implanted*) in the thickened lining of the uterus, which supplies essential nutrients. The uterus lining is maintained by secretions from the uterus of progesterone, which prevents menstruation and ovulation during pregnancy. This early development is shown in Fig. 4.62.

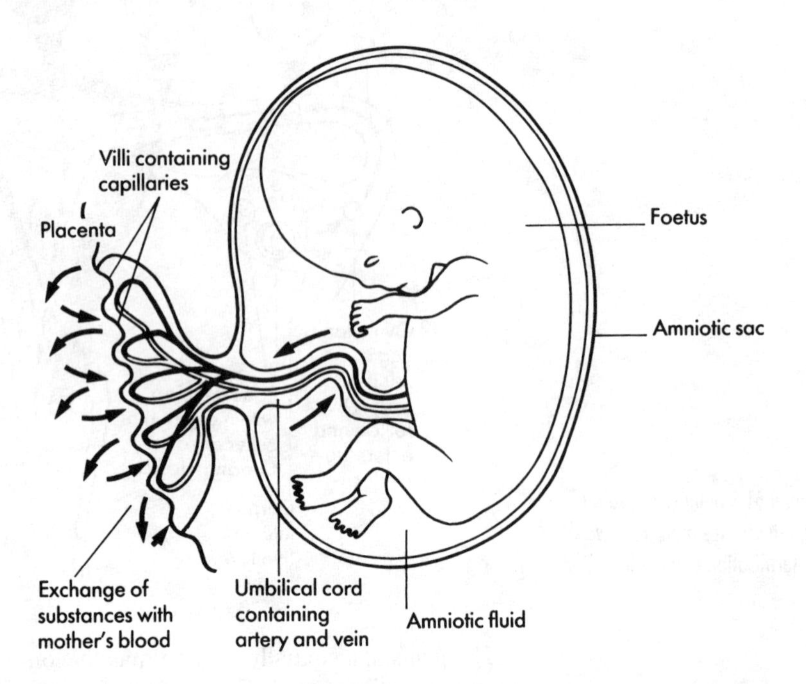

Fig 4.62 Exchange of substances at the placenta

The growing embryo gradually begins (at about 8 weeks) to show distinct human features and is now called a *foetus*. The foetus and the mother jointly form a *placenta* (from about 12 weeks). This allows materials to be exchanged between the blood of the foetus and mother; substances diffuse across a very small gap between foetal and maternal capillaries (Fig. 4.62). The surface area available for diffusion is increased by the presence of *villi*. However, their blood never actually mixes. This is important because the blood may be incompatible, and there are differences in pressure between the two systems. The mother's diet should be sufficient to support the needs of the growing foetus. In particular, the diet must contain enough protein, calcium, iron and vitamin D to support the development of the foetus. The baby will continue to make nutritional demands on its mother as a result of breast feeding.

Make sure you understand the function of the placenta

The foetus is connected to the placenta by an *umbilical cord* and is supported and protected in an *amniotic sac* containing *amniotic fluid*, which acts as a shock-absorber.

(d) Birth

Birth (*parturition*) is the transition of a dependent, 'parasitic' foetus into a relatively independent baby. Processes such as gaseous exchange, nutrition and excretion which have been undertaken by the placenta need to be taken over by specialised organs.

Birth normally commences after about 40 weeks with the baby in a head-down position; less usually the baby may be born feet first (breech birth). The baby is expelled by rhythmic involuntary contractions of the uterus, which increase in intensity and frequency during labour.

The cervix, vagina and even the hips expand to allow the baby through. The separate bones of the baby's skull, which are not yet fused, allow the head to be compressed. Amniotic fluid is released as the baby is born, and the placenta (afterbirth) is delivered shortly afterwards. The umbilical cord is then tied and cut.

The sudden temperature drop at birth causes a reflex response in the baby, which begins to breathe. This involves an inflation of the lungs as the pulmonary circulation comes into use.

(e) Parental Care

Mammals in general, and humans in particular, show a considerable degree of parental care. This corresponds with the long period of development of these complex animals.

Babies are normally breast-fed (*suckled*) from about 24 hours after birth. Initially babies receive colostrum, which is easier to digest than milk and also contains antibodies. The mammary glands develop during pregnancy to produce milk; this is a characteristic feature of mammals. The production of milk is stimulated by the suckling action; this is another example of a reflex response. Human milk contains the correct balance of nutrients, delivered at a suitable temperature. Babies are gradually *weaned* into solid food.

(f) Contraception

“Questions on this are very common!”

Sexual partners may want to remove the possibility of conception (fertilization) from sexual intercourse. This is called *contraception* (birth control, family planning). It is achieved by preventing the egg and sperm from meeting. There are various ways in which this can be done; these are summarised in Fig. 4.63. Contraception is one means of limiting an increase in human populations.

METHOD	HOW IT WORKS	ADVANTAGES	DISADVANTAGES
Contraceptive pill	Contains hormones which prevent ovulation.	Very effective.	Possible side effects; possible to forget.
Cap (diaphragm)	Blocks path of sperm at cervix.	Simple, effective if used with spermicidal cream.	May be incorrectly fitted.
Intra-uterine device (IUD) (coil)	Fitted in uterus; prevents implantation.	Once fitted, does not require frequent attention.	May cause pain or heavy bleeding.
Condom (sheath)	Rubber sleeve, fits over erect penis; retains semen.	Simple, effective, may prevent sexually transmitted diseases e.g. AIDS.	May be damaged; may not be carefully removed.
Withdrawal	Penis withdrawn from vagina before ejaculation.	Does not require any preparation.	Semen may be released before ejaculation. (∴ unreliable method)
Rhythm method (safe period)	Intercourse avoided during 'high risk' time around ovulation.	Relatively 'natural'; acceptable to Roman . Catholic church	Menstrual cycles may be irregular. (∴ unreliable method)
Sterilization	Male: sperm ducts surgically cut (vasectomy). Female: oviducts surgically cut.	Totally effective.	Permanent, irreversible.

Fig 4.63 Summary of some common methods of contraception

(g) Sexually Transmitted Diseases

Sexually transmitted diseases are generally passed from one person to another during sexual intercourse. The diseases are spread by direct physical contact (contagion). The chances of spreading such diseases is increased if one or both partners have sexual relationships with many others, i.e. they are promiscuous. The general name for sexually transmitted diseases is *venereal disease* (V.D.), and includes *gonorrhoea* and *syphilis*, caused by bacteria. The virus HIV (Human Immune Deficiency Virus) can also be spread during sexual intercourse involving an infected person. The virus causes a weakening of the body's normal defence systems, and may result in AIDS

(Acquired Immune Deficiency Syndrome) developing. A summary of these sexually transmitted diseases is given in Fig. 4.64.

In practice, the spread of sexually transmitted diseases can be avoided to some extent by the use of condoms.

DISEASE	PATHOGEN INVOLVED	TRANSMISSION	SYMPTOMS	TREATMENT AND PREVENTION
Gonorrhoea	*Neisseria gonorrhoeae* (bacterium)	Direct sexual contact	Burning sensation during urination. Discharge of pus (Symptoms may not be obvious in infected women).	Treatment: Antibiotics, e.g. penicillin, streptomycin. Prevention: avoidance of sexual contact with infected persons
Syphilis	*Treponema pallidium* (bacteria)	Direct sexual contact	Temporary painless ulcer on penis (males); on labia, or within vagina, uterus (females). Secondary rash develops over body later; also disappears. Disease may eventually affect heart, liver, bones, brain.	Treatment: Antibiotics, e.g. penicillin. Prevention: avoidance of sexual contact with infected persons.
AIDS	H.I.V. (virus)	Enters body via the blood-stream e.g. during sexual intercourse in some circumstances, using contaminated syringes or infected blood. Can be passed from mother to baby via placenta or milk.	Gradual breakdown in body's normal defence system against disease. May lead to pneumonia, various cancers.	Treatment: None at present. Prevention: avoidance of sexual contact with infected persons. Use of sterile syringes, etc.

Fig 4.64 Examples of the transmission, symptoms and treatment of certain sexually transmitted diseases

Links with Other Topics

Related topics elsewhere in this book are as follows:

Topic	Chapter and Topic number	Page no.
Names and locations of the major organs	4.1	30
Functions of the major organs	4.1	31
Growth and development	4.5	62
The internal environment around the human embryo	4.7	129
The hormonal system	4.8	152
Genes in inheritance	6.1	199
Sources of variation	6.1	203
Meiosis	6.3	219
Trends in human populations	6.4	222

REVIEW QUESTIONS

Q13 Most babies born in Britain 'weigh' about 3 kg at birth. A few babies 'weigh' more than 5 kg at birth. Some babies can 'weigh' as little as 1 kg. One reason for small babies is that they are born early or prematurely.

(a) Suggest **two** more reasons why the birth 'weights' of babies can be so variable.

(i) ______________________________

(ii) ______________________________

(b) Rakesh and Rinku are twin brother and sister. At birth, Rakesh 'weighed' 2.7 kg and his sister 'weighed' 2.6 kg.

(i) Suggest **one** reason why the twins were nearly the same 'weight'.

____

(ii) Suggest **one** reason why the twins' 'weights' were slightly different.

____

(c) The following warning is taken from a cigarette packet:

WARNING By HM Government
SMOKING CIGARETTES DURING
PREGNANCY CAN SERIOUSLY HARM
YOUR BABY'S HEALTH AND MAY
LEAD TO PREMATURE BIRTH

A mother who smokes may affect the development of her baby in the womb.

Give **one** reason why the development of the baby might be affected.

____

EXCRETION

Key Points

- *Excretion* is the removal of the waste products of *metabolism* from the body.
- The main waste products in humans are *carbon dioxide* and *urea*. The main organs of excretion are the *skin, liver, lungs* and *kidneys*. These organs also have other functions.
- The kidney is composed of thousands of tiny 'filtration units' called *nephrons*. Each nephron 'cleans' the blood flowing through in two main processes; *filtration* of small molecules, and *reabsorption* of useful molecules. The remainder are waste molecules, mainly urea, which leave the body as a solution called *urine*.
- The kidney is also responsible for regulating the amount of water in the body, by varying the amount that is reabsorbed after filtration.
- Failure of the kidney can be treated by the use of an 'artificial kidney', or *dialysis machine*, or by kidney transplantation.

The Purpose of Excretion

The removal of the waste products of metabolism is called *excretion*. The chemical processes of metabolism may lead to the formation of waste products, which are not required by the organism and which might be harmful. Some substances, such as alcohol, are *toxic* when they enter the body and they may be removed directly or after being made less harmful (e.g. in the liver). In practice, excretion in animals is also taken to include the removal of excess water and minerals absorbed from the diet.

Excretion is necessary for two main reasons:

(i) to maintain the composition of an organism's fluids, including water content and pH.
(ii) to prevent the accumulation of poisonous (toxic) wastes which might otherwise interfere with metabolism.

"Candidates often confuse these three terms, each of which has a very different meaning"

It is important to be able to distinguish *excretion* from *secretion*, which is the production of useful substances from cells. *Egestion* (p.88) is not really excretion because faeces contain undigested food which has not been absorbed or taken part in metabolism. Faeces do contain bile pigments which are the products of metabolism.

The main waste products of metabolism in animals are *carbon dioxide*, mineral salts and nitrogen-containing compounds such as *urea*. The main organs of excretion are the *skin, liver, lungs* and the *kidneys*. The kidneys are also involved in the regulation of water content in the organism (*osmoregulation*).

The Kidney

The kidney (Fig. 4.65) has two main functions; excretion of waste products and osmoregulation (the regulation of the water content of body fluid). Waste products leave the body in solution with water. Excretion therefore provides an opportunity for the concentration of body fluids to be adjusted.

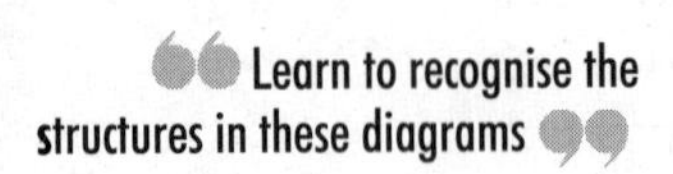

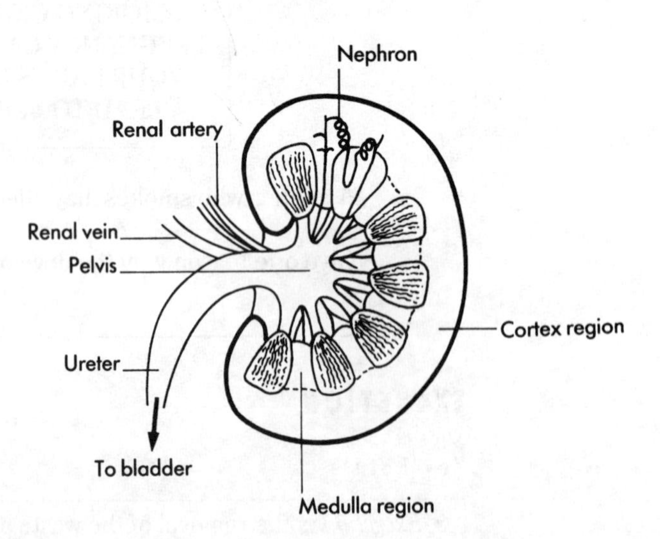

Fig 4.65 Vertical section through a mammalian kidneys

The structural and functional unit of the kidney is the *nephron* (Fig. 4.66), or *kidney tubule*. Each of the two kidneys in humans contains about one million nephrons. The nephron is a narrow tubule about 3 cm long; the combined length of the nephrons within each kidney totals about 30 km. This provides a very large surface area. There are also about 160 km of blood vessels in each kidney. The main function of the kidney is to filter blood. This is achieved during two processes:

(i) Filtration

Blood from a narrow branch of the renal artery enters the *glomerulus*; smaller molecules in the blood are forced under pressure into the surrounding *Bowman's capsule*. This process of filtration involves such molecules as water, urea, glucose, amino acids and also minerals.

(ii) Reabsorption

About 99 per cent of the filtered fluid (filtrate) in the nephron is reabsorbed. The useful components re-enter the blood by active transport, diffusion, and osmosis. The

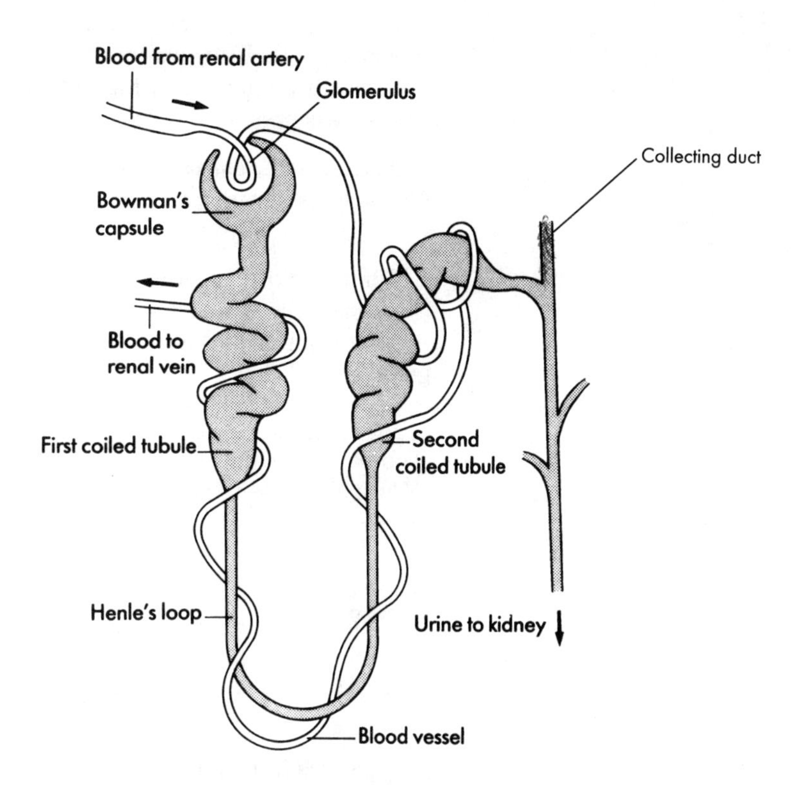

Fig 4.66 Structure of nephron

active transport uses energy released by respiration. Oxygen is therefore used by the kidney; carbon dioxide is produced.

Substances which are normally totally reabsorbed are glucose and amino acids. This occurs mostly in the first coiled tubule. Water is reabsorbed throughout the nephron; minerals are reabsorbed in the two coiled tubules. The relative amounts of water and minerals which are returned to the blood can be varied and the process is controlled by hormones, including *ADH* (antidiuretic hormone, or ***vasopressin***). This control allows the body to adjust the concentration of its fluid (See section 4.7).

“This may seem complicated! Read this section carefully”

Urea is not reabsorbed. Instead, it forms a solution in water called ***urine***; this is only about 1 per cent of the solution originally entering the nephron during filtration. This empties from nephrons into collecting ducts; it accumulates in the pelvis of the kidney, then the urine drains down the ***ureter*** to the ***bladder*** where it is temporarily stored (about 400 ml) before being released at intervals through the (the release of urine is called ***micturition***).

The exact composition of urine depends to a large extent on how much water has been gained or lost by the body (Fig. 4.67). This, in turn, is affected by such factors as external temperature; in hot weather less urine is formed because more water is lost by sweating, but the urine is more concentrated.

Gain (cm^3)		Loss (cm^3)	
		From kidney:	
Absorbed in gut:		**Urine**	1500
Drink	1500	From skin:	
Food	700	**Sweating**	500
From respiration:		From lungs:	
Metabolic water		**Exhalation**	400
	300	From egestion:	
		Faeces	100
Totals	2500		2500

Fig 4.67 Relative daily gain and loss of water in a human (an average adult male, resting, in moderate warmth

Practical: analysis of 'urine'

1. Doing the test

As part of your practical work on excretion, you may be provided with artificial 'urine' to test. These may contain either normal or abnormal amounts of either glucose or protein, or both. They can be tested by dipping in special test strips, which have ends coated with a chemical which changes colour, according to how much of the test substance is present:

'Clinistix' test for *glucose*.
'Albustix' test for *protein*.

2. What do your results mean?
Think carefully whether you would expect to find glucose and/or protein in 'normal' or 'abnormal' urine. Summarise your results in a table, something like this (Fig. 4.68):

Tube of 'urine'	Glucose present/absent?	Protein present/absent?	Is the 'urine' normal or 'abnormal'? Explain the results.
Tube 1 Tube 2 Tube 3			

Fig 4.68

disease

Kidney Malfunction

(a) Diabetes
Although not strictly speaking a malfunction of the kidney itself diabetes (***diabetes mellitus***) is a condition caused by a lack of sufficient levels of the hormone ***insulin***; this results in an increase in glucose concentration in the blood. Excess glucose filtered out of the blood is not necessarily all reabsorbed in the kidney; the presence of glucose in the urine is an indication of diabetes. The disorder can be treated by low-sugar diets and also by regular injections of insulin (See section 4.7).

(b) Kidney failure
Kidneys may fail because of disease or because blood pressure drops too low to maintain filtration. There are two main ways in which the failure of kidneys can be treated:

Artificial kidney. The ***artificial kidney*** or ***dialysis machine*** performs some of the functions of a normal kidney (Fig. 4.69). Blood from the patient is temporarily diverted from an artery in the arm through the machine and is then returned to a vein in the arm. Blood in the machine is passed over a ***dialysis membrane***, which functions as a synthetic living membrane, which is ***selectively permeable***. This membrane is surrounded by a solution containing the 'ideal' concentrations of substances. Certain small molecules such as urea and salts pass from the blood through the dialysis membrane by diffusion because they are above a certain concentration. Useful substances such as glucose, amino acids and some salts and water are retained in the blood. A patient requiring dialysis treatment needs to undergo this process every 2–3 days; between treatments toxic substances build up, and the person may feel ill. Artificial kidneys are relatively expensive and not necessarily available to all patients with kidney failure.

Kidney transplantation. A kidney from a suitable donor can be transplanted into the patient, normally in the lower abdomen. The tissues of the donor kidney must be similar to the patient's own tissues, otherwise an ***immune reaction*** may result and the transplanted kidney will be rejected. Kidneys may be donated by someone closely related to the patient; the donor can live healthily with just one kidney. Kidneys may also be obtained, for instance, from the victims of road accidents. Kidney transplants allow patients to conduct a relatively independent life; however, there may not be enough suitable donor kidneys available to meet the need for them.

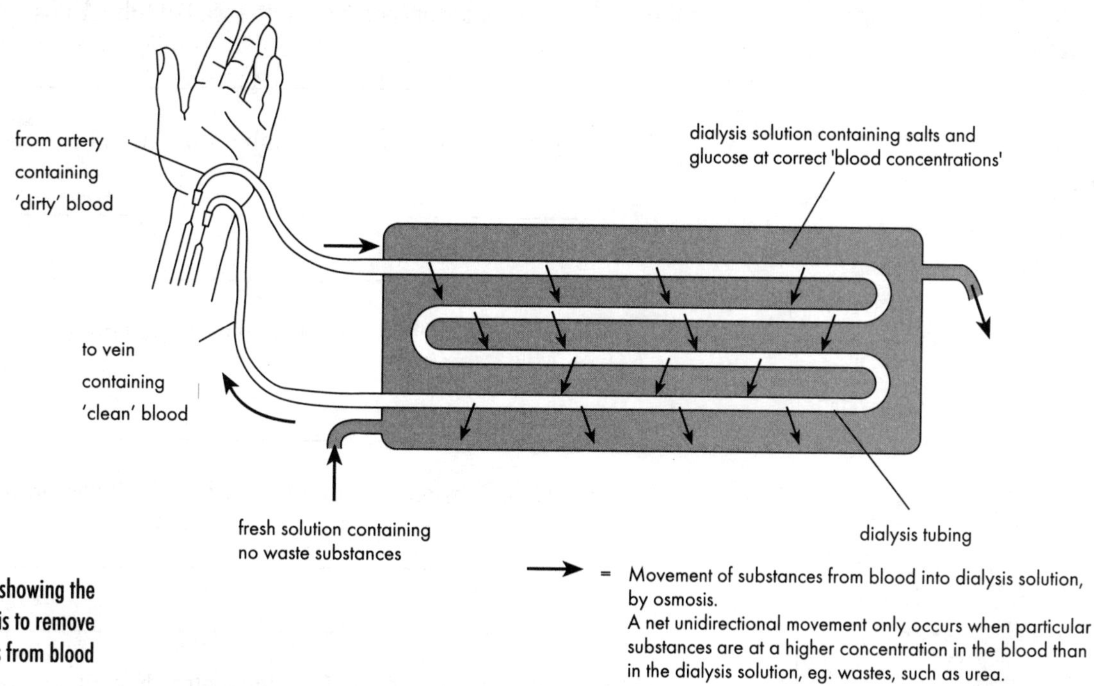

Fig 4.69 Diagram showing the principle of dialysis to remove wastes from blood

Links with Other Topics

Related topics elsewhere in this book are as follows:

Topic	Chapter and Topic number	Page no.
Names and locations of the major organs	1.1	30
Function of the major organs	1.1	31
Respiration	1.5	52
Blood circulation	1.5	96
Homeostasis	1.7	128/138

REVIEW QUESTIONS

Q14 (a) (i) What are the **two** main functions of the kidneys?

____

____

(ii) Name the blood vessel which carries blood from the aorta to a kidney.

____

Fig. 4.70 shows how a kidney machine works.

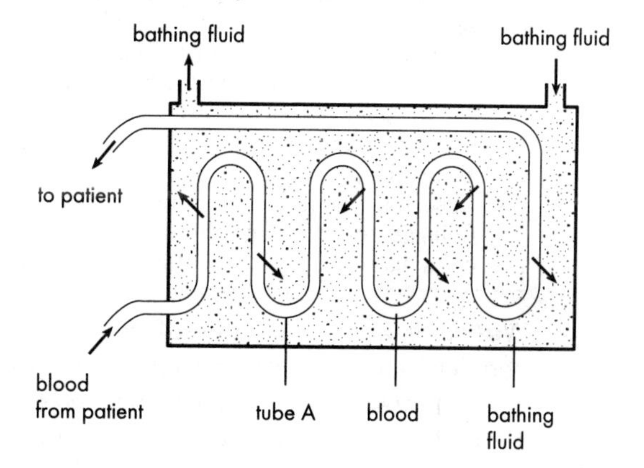

Fig 4.70

(b) (i) What particular property must the wall of tube **A** possess?

____

(ii) Suggest why tube **A** is coiled rather than straight.

____

____

(c) (i) Name a constituent of the blood plasma that does **not** pass into the bathing fluid.

____

(ii) Name **two** substances, other than salts, which pass into the bathing fluid.

____

____

(iii) Explain why the bathing fluid must already contain essential salts before it enters the machine.

____

____

____

NUTRITION

Key Points

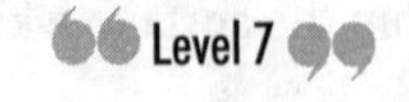

- *Nutrients* are types of chemicals present in food, and required by the body. The main nutrients are ***proteins, lipids, carbohydrates, vitamins*** and ***mineral ions. Fibre*** and ***water*** are also essential in a balanced diet. A ***balanced diet*** consists of food of an appropriate type and in suitable proportions to meet an individual's needs.
- Particular foods are good *sources* of particular nutrients. Insufficient amounts of particular nutrients can lead to *deficiency*. Nutrient deficiency may be accompanied by *deficiency symptoms*.
- *Digestion* is the process by which food is prepared for absorption into the body's cells. This is achieved by reducing the size of food particles and by making food more soluble. There are two main types of digestion: *physical digestion* and *chemical digestion*. Physical digestion, e.g. by chewing, reduces food to smaller fragments. Chemical digestion, mostly involving *enzymes*, chemically changes food to make it easier to absorb.
- Enzymes are *biological catalysts*, which increase the rate of chemical change, but without becoming changed themselves. Examples of enzymes in digestion include *lipases* (digest lipids), *proteases* (digest proteins), and *amylase* (digest a type of carbohydrate called starch).
- The *gut* (or *alimentary canal*) is a long muscular tube within which digestion takes place. The lining of the first part gut secretes enzymes and acids, for digestion. The lining of the second part of the gut is adapted to absorb digested food. Food is kept moving through the gut by muscular contractions, called *peristalsis*.
- Digested food passes through the lining of the gut into either the *lymphatic system* (lipids) or the *blood system* (all other nutrients). The lining of the small intestine (part of the gut) has many internal projections, called *villi*, which provide a large surface area for absorption.
- Absorbed nutrients are mainly carried by the blood system to each living cell in the body. Cells absorb nutrients from the blood according to their needs. Nutrients are used for the cell's particular activities; this is called *assimilation*.

Nutrients are the groups of chemicals which are present in food and which are required for the body's various activities. The main groups of nutrients are ***proteins, lipids, carbohydrates, vitamins*** and ***mineral ions***. Other substances which are needed, though often not classified as nutrients, are ***fibre*** and ***water***.

The amount of each type of nutrient depends on the type of nutrient and on the body's needs. Nutrients which are required in relatively large amounts, the *macro nutrients*, are lipids, proteins and carbohydrates. Nutrients needed in relatively small amounts, the *micro nutrients*, are vitamins. Mineral ions can be either macro nutrients or micro nutrients, depending on what type they are.

Proteins

Try to remember the main differences between proteins, lipids and carbohydrates

Proteins are absorbed as *amino acids*. There are two nutritional groups of proteins. *First class proteins* ('high biological value') contain all or most of the eight *essential amino acids*. Another twelve or so amino acids can be formed within human tissues. *Second class proteins* ('low biological value') contain fewer essential amino acids. Proteins have two main functions:

- **Growth and repair.** Proteins are needed to make new living material within cells.
- **Metabolism.** Enzymes and hormones are types of protein. Both are important in controlling chemical processes within the body.

disease

Proteins are not normally used much for respiration, since they are too important for other functions in the body. If they are used in respiration, the energy yield is 17 kJ/g. Excess proteins in the diet cannot be stored. Instead, they are broken down in the liver to carbohydrates and urea. If the diet does not contain enough protein, a deficiency disease can occur, e.g. in Kwashiorkor
Main food sources:

- First class proteins: cheese (dairy products), eggs, meat.
- Second class proteins: peas, beans (pulses).

Lipids

Lipids consist of *fats* and *oils* in food. During digestion, they are broken up into units called *glycerol* and *fatty acids*. Lipids have three important functions:

- **Respiration.** The energy yield is 39 kJ/g. This is more than twice that of carbohydrates (or proteins). However, lipids are more difficult to digest.
- **Component of cell membranes.** Lipids form part of the *phospholipid* molecule.
- **Insulation and protection.** Lipids are stored under the skin for insulation, and around organs for protection.

Main food sources: butter, vegetable oil.

Carbohydrates

Carbohydrates occur as three main types: *monosaccharides, disaccharides* and *polysaccharides* (Fig. 4.71). The 'structural units' or 'building blocks' of carbohydrates are the monosaccharides, which is the form in which digested carbohydrates are absorbed.

CARBOHYDRATE GROUP	STRUCTURE	CHEMICAL FORMULA	EXAMPLES	FUNCTION
Monosaccharides (Simple sugars)	Consist of a single chemical group.	$C_6H_{12}O_6$	Glucose Fructose Galactose	Soluble; structural units for making larger carbohydrates
Disaccharides (more complex sugars)	Consist of two joined monosaccharides.	$C_{12}H_{22}O_{11}$	Maltose Sucrose	Soluble; similar to monosaccharides; sucrose is 'a transport molecule' (plants).
Polysaccharides (large, complex sugars)	Consist of many joined monosaccharides (i.e., they are polymers).	$(C_{12}H_{22}O_{11})n$	Starch Glycogen Cellulose	Insoluble; used as *food store* in plants (starch) or animals (glycogen); used as *structural material* in plants (cellulose).

Fig 4.71 Summary of common carbohydrates

There is one main function of carbohydrates in the body:

- **Respiration.** The energy yield is 17 kJ/g. Carbohydrates are relatively easy to digest and are readily available in food, especially from plants and their food derivatives.

Main food sources:
- Monosaccharides: fruit, honey
- Disaccharides: cane sugar, milk (lactose)
- Polysaccharides: flour, potatoes. Cellulose is important as roughage (see below).

Practical: Investigation of the carbohydrate content of foods

There are various tests which you can perform to investigate the presence or absence of carbohydrates in food. The tests for starch and reducing sugars are summarised in Fig. 4.72.

Type of food	*Food test*	*Positive result*
Carbohydrate: starch	Add *iodine solution* to the solid or liquid food.	Colour change: orange/brown ➡ blue/black.
Carbohydrate: reducing sugar e.g. maltose, glucose	Add *Benedict's solution* to liquid food in a test tube; solid food should be ground up in a little water. Heat the tube in a beaker of boiling water.	Colour changes: blue ➡ green ➡ yellow ➡ red

Fig 4.72 Summary of tests for carbohydrates

Vitamins

Vitamins are required in small ('trace') amounts in the diet; they are micronutrients which are in many cases necessary for *enzymes* to work properly. The exact way in which many vitamins function is not fully understood, however. Vitamins have no energy value and are not digested. Most vitamins are not formed in the tissues of the organism that needs them; instead, they are obtained from other organisms that can make them. A lack of sufficient vitamins in the diet results in deficiency diseases. There are two main groups of vitamin:

Water-soluble (present in certain vegetables and fruits)
Example:

Vitamin C (ascorbic acid), needed for formation/repair of tissue, including skin, teeth and bones.
Food sources: green vegetables, potatoes, citrus fruits (e.g. lemons).
Deficiency: scurvy (tissue not formed/repaired).

Fat-soluble (present in animal and vegetable fats)
Example:

Vitamin D (calciferol), essential for the absorption of calcium and phosphorus in the gut; also needed for the formation of teeth and bone.
Food sources: fish, egg (yolk), milk, liver. Also formed by the action of sunlight on skin.
Deficiency: rickets (poor bone formation during development).

Minerals (salts)

Minerals are inorganic molecules which can be important in the formation of more complex organic molecules. Heterotrophic and autotrophic organisms share many requirements for minerals because minerals often perform similar functions in plants and animals. In animals, certain minerals may be available from protein, carbohydrate, and lipid molecules; these contain carbon, hydrogen, oxygen, nitrogen, phosphorus and sulphur. About ten additional minerals are needed in a balanced diet. Minerals can be divided into two main groups according to the amount in which they are required:

Macronutrients (needed in relatively large amounts)
Example:

> **Calcium**, important in the formation of teeth and bones.
> *Food sources*: milk, cheese, fruit.
> *Deficiency* brittle teeth and bones.

Micronutrients (trace elements) (needed in relatively small amounts)
Example:

> **Iron**, important as a component of the haemoglobin molecule.
> *Food sources*: Liver, egg (yolk), spinach.
> *Deficiency*: anaemia (insufficient haemoglobin in blood).

Water

Water is a major component (65–70 per cent) of the human body. It is essential in chemical reactions and as a means of carrying substances around the body. About half of all water absorbed is present in food, the rest is consumed as a liquid. The amount of water taken into the body corresponds to the amount lost from the body and is under involuntary control.

Roughage (fibre)

Roughage is an essential part of the diet, but it is not digested or absorbed. Roughage consists mainly of *cellulose*. Humans do not produce an enzyme to break this down, so it is *egested* without being significantly altered. However, roughage provides bulk which presses against the gut walls, especially in the large intestine. This stimulates the movement of food by *peristalsis* (see below). *Food sources*: vegetables (especially uncooked). *Deficiency*: constipation.

The balanced diet

Exam. questions on this topic occur frequently

The quality and quantity of food consumed by an individual is known as its *diet*. A *balanced diet* contains the correct type of food, in the correct proportions, to meet the needs of a particular body. The main components of a balanced diet are *proteins, lipids, carbohydrates, vitamins* and *mineral ions*. *Fibre (roughage)* and *water* are also required.

The relative quantity of each of these nutrients in a balanced diet will vary according to the activity, size, growth rate, age and sex of the individual. It will also be determined by 'external' factors such as climate, since an important function of food is to provide energy, often in the form of heat.

An individual who does not receive a balanced diet – expecially over a long period – may be in a state of *malnutrition*. There are two main types of malnutrition:

- **Undernutrition.** Approximately two-thirds of the world's human population are suffering from undernutrition. These people, mainly living in developing countries, are 'malnourished' in the sense that they do not receive enough food, or they do not receive a suitable range of foods.
- **Overnutrition.** People living in parts of the world where food is more readily available may consume more food than they need in a balanced diet. If this continues, an individual may become *obese* (overweight). Obesity may contribute directly or indirectly to other health problems, such as heart disease.

Practical: Analysing diet data tables
You may be asked to answer questions about the energy value and nutrient content of a particular diet, reading information from a data table. All the answers to the questions will be contained in the table; you will not be expected to memorise information like

this! However, you should read the questions carefully, so that you understand what is being asked for, and where to find the answer within the data table. Note that you may find a calculator useful for some questions, although the calculations will not be too complicated. An example of a diet data table is provided in the Review Sheet at the end of this Topic.

Practical: Food additives

Food additives include *preservatives, artificial colouring* and *artificial flavouring*. They are added to food to make it more acceptable and economical. Food additives are included in many processed foods, although the long-term effects of some of them are not fully understood. Some additives are known to cause allergic reactions.

Examine the labels of a range of processed foods, and prepare a table summarising their food additives. Using reference materials, find out the exact function of any four additives. Briefly describe the relative advantages and disadvantages of each of the four additives.

Digestion

Digestion is the process by which food is prepared for absorption into the cells of an organism. This is achieved by reducing the size of food particles and also by making food more soluble. Digestion consists of a combination of physical and chemical processes.

(a) Physical digestion

Physical digestion reduces food into smaller fragments which (a) are small enough to pass through the gut, and (b) have an increased surface area for the action of enzymes. The main processes involving physical digestion are *chewing* in the mouth, and *peristalsis* in the gut. Both are muscular processes which break up the food.

- **Ingestion.** Before food can be digested, it has to be taken into the mouth. Animals use senses including sight, smell and taste to decide whether food is acceptable or not. This is to prevent potentially harmful food from entering too far into the body. In many *carnivore* (meat eating) animals, teeth are used to kill and hold their prey.
- **Chewing.** Chewing (or *mastication*) is the first stage of digestion as a whole. The combined movements of the teeth, cheeks and tongue are involved in chewing. This action also mixes food with *saliva*, which contains the enzyme *amylase*. Food particles are often formed into a rounded lump (or *bolus*), which can be swallowed easily.

Teeth

Make sure you are familiar with the structure of teeth

Teeth have an important role in physical digestion. The relative number, arrangement and types of teeth vary according to the particular animal's diet. Teeth are mounted in jaws within the animal's skull, and the way in which the jaws move is

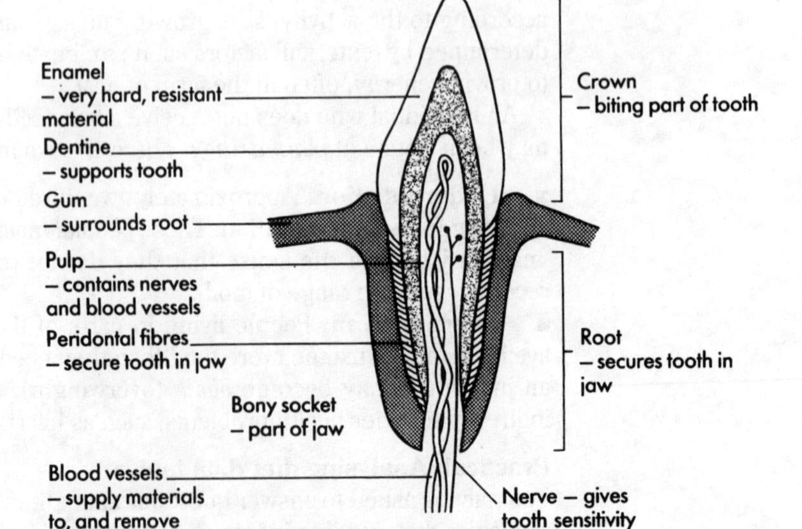

Fig4.73 Structure of a 'typical' tooth; canine (vertical section)

another adaptation to diet. Teeth in animals other than mammals tend to be fairly simple, as their main function is to seize and retain food. Teeth in mammals are capable of several different actions.

Types of teeth

The four main types of teeth in mammals, including humans, are ***incisors*** (for cutting), ***canines*** (for piercing and grasping), ***premolars*** and ***molars*** (for crushing and slicing). These teeth all have the same basic structure, though their shape and position in the jaw vary according to their function (Fig. 4.73).

Arrangement of teeth

In mammals, teeth develop in two stages: ***deciduous*** (temporary, 'milk') teeth and ***permanent*** (adult) teeth (Fig. 4.74). Deciduous teeth appear first, when the young mammal is ***weaned*** onto solid food. Having two sets of teeth allows the older animals gradually to acquire larger and more numerous teeth as the jaws increase in size during growth.

Type of tooth	*Deciduous teeth*	*Permanent teeth*
Incisor	8	8
Canine	4	4
Premolar	8	8
Molar	0	12
Maximum number of teeth	20	32

Fig 4.74 Summary of teeth development in humans

The arrangement of teeth in mammals can be expressed in a *dental formula*.

Only half of the teeth are shown for the upper and lower jaws because each *half-jaw* is symmetrical (a 'mirror image') with the other half. The total numbers of each type are represented with those for the upper half-jaw being written over numbers for the lower half-jaw. The dental formula for an adult human would be written:

$$\frac{\text{(upper half-jaw)}}{\text{(lower half-jaw)}} \quad - \; i\,\frac{2}{2} - c\,\frac{1}{1} - pm\,\frac{2}{2} - m\,\frac{3}{3} - \text{ total for complete jaws} = 32$$

i = incisor, c = canine, pm = premolar, m = molar

Action of teeth

Teeth are brought against or past each by jaw movements; this action forms part of the process of *chewing* (*mastication*). Chewing also includes movements of the cheeks and tongue, which position food between the teeth. Secretions contained in *saliva* help bind food into a bolus (lump of food) and lubricate it for swallowing. Teeth are essential in the first part of the digestion of solid food, before swallowing takes place. Teeth in humans can, however, be damaged by certain types of diet or by neglect.

There are two main types of dental disease; both are caused by ***plaque***, an accumulation of food (especially sugar) and bacteria on the exposed surfaces of teeth:

Dental caries (tooth decay)

Acid produced as a waste product by bacteria in plaque dissolves through the enamel layer of teeth. Decay can spread more rapidly through the softer dentine and then into the pulp cavity. Infection of the pulp may cause an inflammation, *abscess*, of the gum.

Periodontal disease

Spread of plaque between teeth and gums may cause an inflammation, *gingivitis*, around the roots. Periodontal fibres become destroyed as the gums recede, resulting in teeth becoming loose and possibly being lost.

The prevention of both diseases may be achieved by increased *oral hygiene.* For instance, by regular brushing, the avoidance of sugary foods and periodic dental checks. Enamel can be made more resistant by the presence of *fluoride*; this is naturally present in the drinking water in some areas. Many water authorities in the UK add fluoride to drinking water and this *fluoridation* seems to be associated with a significant reduction in dental caries. However, some people object to being given fluoride without choice. Fluoride is present in many toothpastes.

Muscular movements of the gut

The gut is, in many respects, basically a muscular tube. The muscles are mostly under involuntary control. Muscular movements of the gut cause food to be physically broken down and also mixed with digestive enzymes, for chemical digestion (see below). Muscular movements also move food through the gut; this is achieved by *peristalsis.* Peristalsis is caused by the alternate contraction of the circular and longitudinal muscles contained in the gut wall (Fig. 4.75).

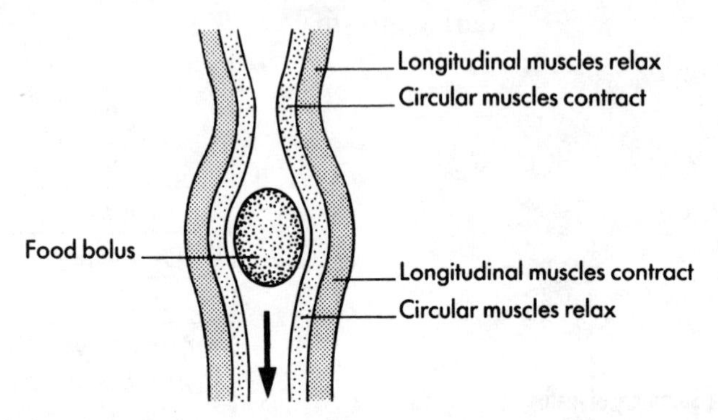

Fig 4.75 Peristalsis in the gut

Muscles are also used to control the movement of food through the gut. For instance the *epiglottis* is a muscular flap which closes during *swallowing* to prevent solid and liquid food from entering the trachea; *sphincters* are circular muscles which, when contracted, prevent food moving in the wrong direction or at the wrong time, for example, at the entrance and exit of the stomach and at the anus.

Comparison of Herbivorous and Carnivorous Digestion

Herbivores e.g. sheep, rabbits, are animals whose diet consists mainly of plant material. *Carnivores* e.g. cats, foxes, are animals whose diet consists mainly of animal material. *Omnivores,* including humans, are animals whose diet is mixed. Animals belonging to one of these groups often exhibit characteristic adaptation of the gut; these are summarised for herbivores and carnivores in Fig. 4.76. In particular, the skulls of herbivores and of carnivores are quite distinctive. Some other important adaptations, such as sight are described elsewhere. Omnivores tend to have intermediate characteristics.

Adaptation	**Herbivore:** e.g. sheep *(Ovis aries)*	**Carnivore:** e.g. cat (*Felis* spp.)
Feeding frequency	**Continuous:** also includes re-chewing partially digested food (*cud*).	*Occasional.* Food is more 'concentrated'.
Teeth	Incisors and canines absent in upper jaw. Gap *(diastema)* between front and back (cheek) teeth allows grass to be manipulated by the tongue. Premolar and molar teeth are ridged, for grinding. Teeth continue growing in adult.	Large canines used for seizing and killing prey. Some *(carnassial)* premolar and molar teeth are shaped for slicing; others, nearer the jaw attachment, are used for grinding.
Jaw attachment	Loose jaw attachment, allowing sideways jaw movement.	Tight jaw attachment, preventing dislocation.
Stomach	'Stomach' consists of several chambers, allowing digestion of cellulose by symbiotic microbes.	Simpler stomach arrangement.
Length of gut	Relatively long	Relatively short

Fig 4.76 Some adaptations of herbivores and carnivores to diet

(b) Chemical Digestion

Chemical digestion involves the breakdown of food mainly by the action of enzymes. (see chapter 3.)

This breakdown results in large, insoluble molecules being made small and soluble so that the food can be absorbed. Most foods consumed by holozoic animals e.g. humans, consist of large molecules which are insoluble in water. The *gut* (*alimentary canal*) (see Fig. 4.77) is a specialised tube which has various glands associated with it. These glands secrete substances which are involved in digestion.

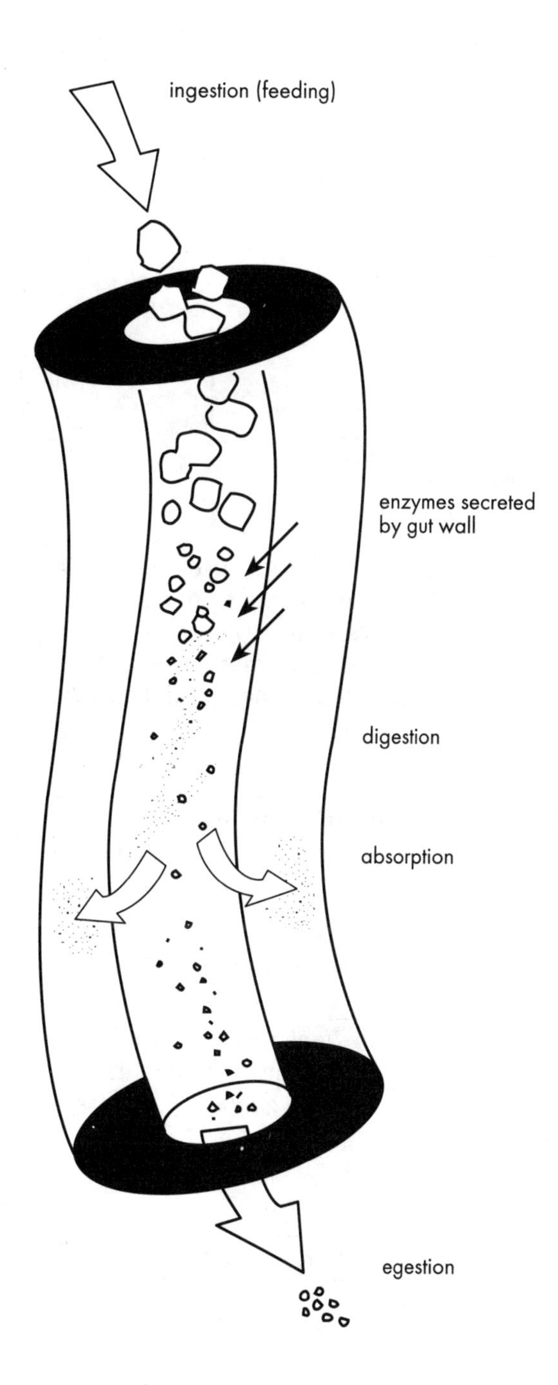

Fig 4.77 Summary of main processes in obtaining nutrients

Enzymes involved in digestion are made in cells in the gut wall, or in specialised glands which empty their contents directly into the gut through tubes, or *ducts*. Enzymes digest food in the gut cavity, or *lumen*. Digestion is *extra-cellular*, enzymes are secreted onto food outside living cells. The gut is essentially a hollow tube which is continuous with the environment at each end (see Fig. 4.78) and food does not pass through a living membrane until it is ready for absorption in the ileum.

❝ It is worth becoming familiar with the layout of the gut so that you can recognise the relative position of different parts of the gut. ❞

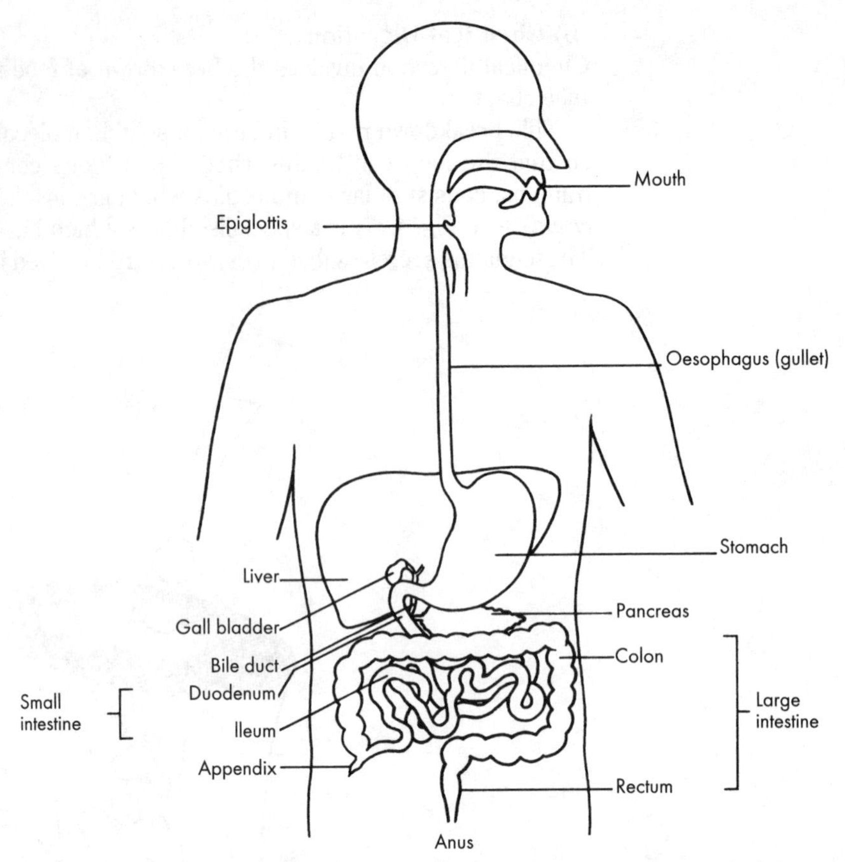

Fig 4.78 Structure of the human gut

The gut provides optimum conditions for enzyme activity by maintaining a favourable temperature and pH. The pH (acidity or alkalinity) in each region of the gut is kept fairly constant by additions of acid or alkali. Chemical digestion occurs as a progressive process as food is moved through the gut; the main stages of this process are summarised in Fig. 4.79.

Region of gut	Secretions	Examples of chemical digestion
Mouth	*Saliva* (from three pairs of *salivary glands*) contains *mucus* and the enzyme *salivary amylase.*	amylase Starch ➡ maltose (pH 7)
Stomach	*Gastric juice* (from stomach wall) contains *hydrochloric acid* and enzymes including *pepsin.*	pepsin Protein ➡ peptides (pH 2)
Duodenum	*Intestinal juice* (from duodenum wall) contains enzymes, including *maltase* and *amylase.*	maltase Maltose ➡ glucose (pH 8.5) amylase Starch ➡ maltose (pH 7)
	Pancreatic juice (from pancreas) contains alkaline secretions and also enzymes, including *amylase, trypsin, lipase.*	lipase Fats ➡ fatty acids + glycerol (pH 7)
	Bile (from liver) contains *bile salts*	Bile salts *emulsify* lipids; they reduce lipids into small droplets, increasing the surface area for digestion.

Fig 4.79 Summary of the main stages of digestion in humans

Each region of the first part of the gut tends to be specialised for a particular chemical process. Enzymes can be grouped according to the type of food that they break down:

Carbohydrases, e.g. amylase, maltase, break down carbohydrates.
Proteases, e.g. pepsin, trypsin, break down proteins.
Lipases, e.g. lipase, digest fats and oils.

Food is mixed with enzymes by muscular movements of the gut, which also cause physical digestion (see above) to take place. Food remains briefly in the mouth before being swallowed; conditions in the stomach are very acid (pH 2) and any digestion of starch by amylase is temporarily prevented. Food is held (by sphincter muscles) in the stomach for about four hours. This allows protein to be digested. A mucus lining prevents the lining of the stomach (which contains protein) from being digested, or damaged by the acid conditions. The acidic mixture in the stomach is called *chyme*. This is later neutralised in the duodenum by alkaline secretions from the pancreas; the mixture then becomes known as *chyle*. Digestion is mostly completed in the duodenum.

Absorption

Most absorption of nutrients occurs in the *ileum*, which is adapted by having the blood and lymphatic *transport systems* to carry absorbed materials to the rest of the body. Another important adaptation is the *large internal surface area* of the ileum. This is achieved in several ways:

(i) *Length*. The ileum is a comparatively long part of the gut. In humans the ileum is about 5 m; this represents about 45 per cent of the total length of the gut.
(ii) *Villi*. Numerous (about 30 million/mm^2). Each villus contains a *capillary network* and also a *lacteal*, a branch of the lymphatic system.

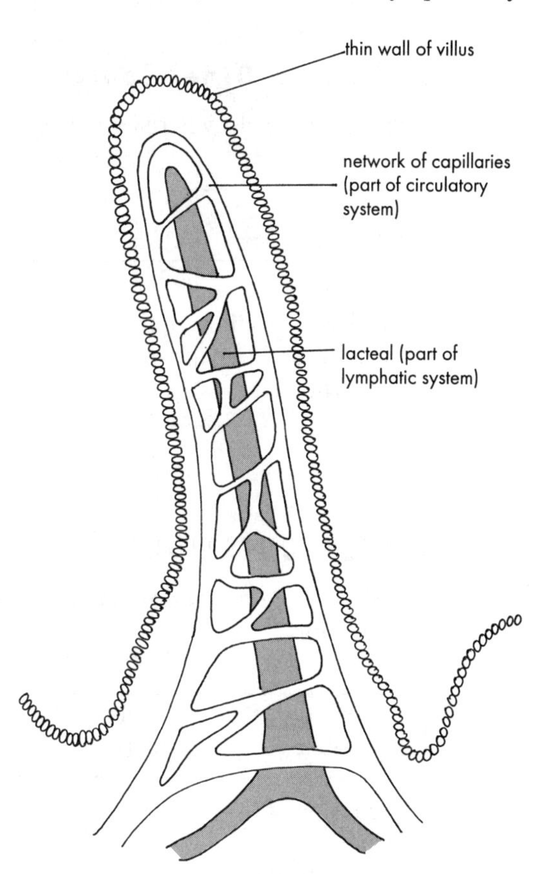

Fig 4.80 Structure of one villus

Nutrients are absorbed through the lining of the ileum by *diffusion* or *active transport* depending on the size of molecules and concentration gradients. Digested lipids are absorbed into the lymphatic system; all other nutrients, which are now water soluble, enter the blood system. Water enters the blood mostly in the *colon* (large intestine). Nutrients are *assimilated* (*utilised*) within the body in various ways, according to needs.

Assimilation

Assimilation occurs when nutrients are used, or modified by body cells. Absorbed food substances travel to the liver in the *hepatic portal vein*; many substances are modified while they are in the liver.

- **Sugars** (from digestion of carbohydrates)
  - are used to provide energy for all body cells, in the process of respiration
  - excess sugar is changed to glycogen, and stored in the liver and muscles, or converted to fat and stored under the skin and around body organs.
- **Amino acids** (from digestion of proteins)
  - are used inside body cells for growth and repair
  - excess amino acids are broken fown in the liver. this process is called *deamination*,
  - and it produces the highly toxic product ***urea*** which must be excreted from the body.
- **Fatty acids and glycerol** (from digestion of lipids)
  - are used to provide energy for all body cells in the process of respiration
  - excess fats are stored under the siin and around body organs.

Egestion

After most water has been absorbed, the contents of the gut are known as *faeces*. These consist of any undigested material, mainly *cellulose*, combined with microbes, mucus and dead cells from the lining of the gut. This is passed out of the body through the anus during the process of *egestion*, or *defecation*. This should not be confused with *excretion* which involves getting rid of the waste products of metabolism. Faeces are temporarily stored in the *rectum* and then egested at intervals.

Links with Other Topics

Related Topics elsewhere in this book are as follows:

Topic	Chapter and Topic number	Page no.
Enxymes	3.4	25
Names and locations of the major organs	4.1	30
Functions of the major organs	4.1	31
Photosynthesis	4.6	105

REVIEW QUESTIONS

Q15 The table below 4.80 shows the food value of a school lunch eaten by a 16-year-old girl.

Food eaten	Protein in g	Carbohydrate in g	Fat in g	Iron in mg	Vitamin C in mg
Sausages	9	5	24	1	0
Chips	8	70	20	2	20
Baked beans	10	20	1	3	4
Apple pie	5	60	25	1	1
Ice cream	2	20	12	0	0
Fizzy drink	0	30	0	0	0

Fig 4.80

(a) (i) In this meal, which food gave the girl most protein?

____

(ii) Name *one* other food not eaten in this meal which is rich in protein.

____

(iii) Why does the girl need protein?

____

(b) (i) The total energy value of this meal is 6600 kJ.
In one day the girl needs 9600 kJ.
If she ate this meal, how many *more* kJ would she need in that day?

____

(ii) What would happen if she eats much more than 9600 kJ of food every day?

____

____

(iii) A lot of energy comes from fat.
Name the *two* foods in this meal which gave her most energy.

____

____

(c) The girl needs 14 g of iron and 25 mg of Vitamin C each day to keep healthy.

(i) How much of her daily iron did this meal give her?

____

(ii) How much of her daily Vitamin C needs did this meal give her?

____

(iii) What will happen if she does not have enough iron and Vitamin C?

Not enough iron may cause ____________________________

Not enough Vitamin C may cause ____________________________

(d) Why should she eat fibre (roughage) every day?

____

SENSITIVITY

Key Points

- Sensitivity allows an organism to be aware of its environment. Changes in the environment are detected as *stimuli*. Sense organs are specialized to respond to particular stimuli. Information is relayed to the brain, which uses the information to organise an appropriate *response*.
- The *brain* is the central information processing centre in higher animals. Different parts of the brain are responsible for different types of information. Some brain processes are *voluntary* (under conscious control, usually optional). Other processes are *involuntary* (not under conscious control, usually essential).
- In humans, the use of 'recreational' drugs is widespread in society. All such drugs have their effect on the brain. Some drugs are habit-forming and addictive, sometimes resulting in unpleasant clinical, personal and social side-effects.
- The *eye* is the *sense organ* responsible for detecting light stimuli. The receptive part of the eye is the retina. The *retina* is composed of two types of sensory cells; rods and cones. Rods are sensitive to low-intensity, monochromatic light. Cones are sensitive to high-intensity light, and can distinguish colours. Information concerning light direction, brightness and colour is relayed to the brain, where a sense of vision is created.
- Common *defects* of the human eye include short-sight and long-sight. These conditions can be corrected by the use of spectacles.

The purpose of sensitivity

Sensitivity (*irritability*) in organisms means detecting and responding appropriately to changes in the internal and external environment. The response of an organism is *adaptive* because the organism becomes adapted to a new situation in such a way as to improve its chances of survival.

A change in an organism's internal or external environment which may result in a response is called a *stimulus*. A stimulus is detected by a *receptor* which relays the information to a coordinator. An appropriate impulse may then be sent to an *effector muscle* which then produces a response. This arrangement occurs, in various forms, in all organisms.

Sensitivity is a characteristic of life and occurs in all organisms. Sensitivity in more advanced plants tends to involve growth responses under the control of hormones. Animals generally have a more complex arrangement for achieving sensitivity. Many animals are coordinated by a nervous system as well as by hormones (see Section 4.8, page 145).

The central nervous system

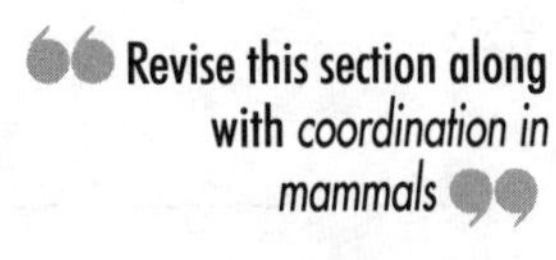

The Brain

The *brain* is a much enlarged part of the central nervous system in vertebrates which coordinates many of the activities of the body. In particular, the brain organises information from the sense organs (see below), many of which may be closely situated in the head region. Some of the main regions of the brain are shown in Fig. 4.81.

The cerebral cortex, cerebellum and medulla have quite distinct functions:

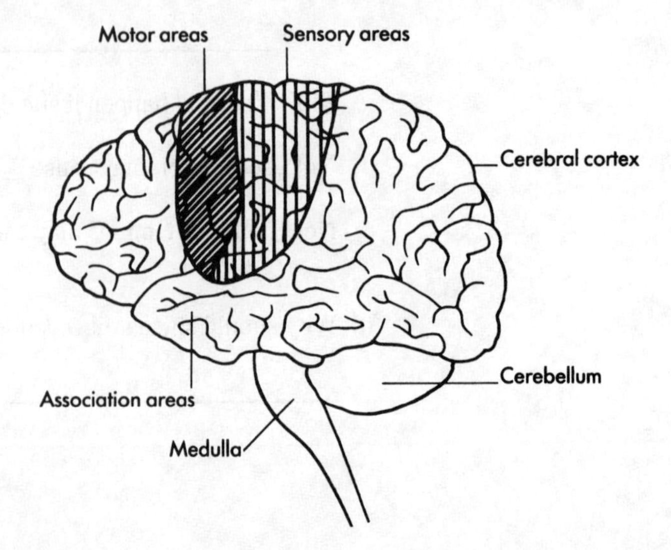

Fig. 4.81 The main regions of the human brain (surface view)

- **The cerebral cortex**
  The *cerebral cortex* coordinates many complex, voluntary processes; this region is particularly well developed in higher vertebrates, especially humans. The cerebral cortex is the folded outer 3 mm of the cerebral hemispheres (cerebrum), and consists of grey matter within which many connections are made between adjacent neurones.
  The cerebral cortex contains three main regions:
  sensory areas, which receive input from the sense organs (see below);
  motor areas, which send out instructions to effectors such as muscles and glands;
  association areas, which interpret and analyse information. The association areas contain brain tissue which determines memory and learning.
- **The cerebellum**
  The *cerebellum* coordinates learned processes which may become involuntary, such as speech, balance and movement. The cerebellum is important in organising the activity of voluntary (skeletal) muscle, for instance in limb movements such as walking.
- **The medulla**
  The *medulla* coordinates involuntary, automatic processes such as breathing, heart rate and swallowing.

Drugs and the Nervous System

A *drug* is a chemical which is introduced into the body and which affects the activity of the body. The three main groups of drugs which affect the nervous system are sedatives, analgesics (painkillers or pain relievers) and stimulants.

(i) **Sedatives (depressants)** induce a state of relaxation and diminished anxiety. Examples include barbiturates, alcohol and heroin.
Alcohol is a commonly-available and socially-acceptable drug in many countries. However, it interferes with coordination and reaction time; this could be a serious problem in someone who is driving a car or operating dangerous machinery. Alcohol causes behaviour to be less restrained, and can precipitate violent impulses in some people. Alcohol dilates skin capillaries, and the resulting loss of heat could be dangerous in cold conditions. Long term and excess drinking can result in dependence (alcoholism). Physiological effects include damage to the liver (cirrhosis).
Heroin is a powerful depressant. Additional short-term effects include vomiting and spread of infection amongst drug users sharing needles; hepatitis, septicemia and HIV (AIDS) can be spread in this way. Long term effects of heroin include dependence and withdrawal symptoms, damage to kidneys, liver and brain, and dependence of babies during pregnancy. Social problems include crime (e.g. theft) to sustain drug supplies.
(ii) **Analgesics (pain-killers)** affect the part of the nervous system which produce a sense of pain. Examples include aspirin and paracetamol.
(iii) **Stimulants** reduce feelings of fatigue and induce a sense of alertness. Examples are caffeine and amphetamines.

Many drugs are used in medicine and some, such as alcohol, are socially acceptable to many people for recreational purposes. Other such drugs are illegal in many societies. Drugs which affect the nervous system may result in increased tolerance, so that the drug needs to be taken in greater quantities to produce a certain effect. Such drugs can be habit-forming, or addictive, and may cause withdrawal symptoms if not readily available.

The sense organs

Sense organs consist of groups of sensory cells which detect a particular stimulus. Sense organs allow an organism to be aware of changes in the internal and external environment; there are both internal and external sense organs. Many sense organs are arranged on the outer surface of the body, particularly in the head region. Sense organs may be complex structures which have other, non-sensory, cells associated with them. Examples of sense organs and the stimuli to which they respond are given in Fig. 4.82.

Sense organ and sensory structure	Stimulus
EXTERNAL SENSE ORGANS	
Eye (retina)	Light
Ear (cochlea)	Sound vibration
Ear (vestibular apparatus)	Gravity and changes in body position
Tongue (taste buds)	Chemicals, taste
Nose (olfactory tissue)	Chemicals, smell
Skin (various receptors)	Heat, cold, touch, pressure, pain
INTERNAL SENSE ORGANS	
Muscle stretch receptors	Muscle movement
Hypothalamus (part of brain)	Blood glucose levels, blood concentration

Fig. 4.82

The eye

Questions on the structure and functioning of the eye are fairly common – make sure you understand this section

Each eye is a spherical structure composed of layers and open on one side to admit light. The structure of the eye is shown in Fig. 4.83. The function of various components of the eye are summarised in Fig. 4.84.

The eye can respond to variations in light conditions, for instance in the distance, direction and brightness of an object. The eye is able to respond to different light conditions by the activities of muscles inside and outside the eye. The contraction and relaxation of these muscles also provides information which the brain can use to

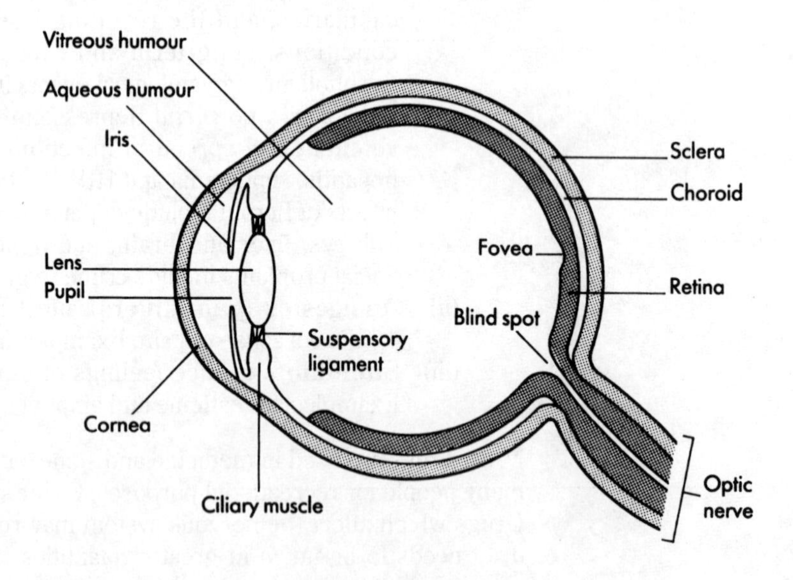

Fig. 4.83 Human eye (vertical section) showing the main components

Structure	Function
Eyelid	protects eye by *blinking* (a reflex action). Blinking washes eye surface with **tears** (they contain a mild antiseptic), prevents the contact of harmful substances with the eye surface, and protects the retina against bright light.
Conjunctiva	Protects part of the cornea.
Cornea	*Refracts* (bends) light entering the eye.
Aqueous humour	Supplies nutrients to the lens and cornea.
Iris	Controls the amount of light entering the eye by adjusting the size of the **pupil**.
Pupil	Aperture which allows light into the eye.
Lens	Allows fine focusing by changing *shape*.
Suspensory ligament	Attaches lens to **ciliary muscle**.
Ciliary muscle	Changes the shape of the lens by altering the tension on the **suspensory ligaments**.
Vitreous humour	Maintains the shape of the eye.
Retina	Contains light-sensitive **rod** and **cone cells** which convert light energy into a nerve impulse.
Fovea	Very sensitive region where most light is focused (also called **yellow spot**).
Optic nerve	Carries nerve impulses to the brain, where they are interpreted.
Choroid	Supplies retina with nutrients; also, contains pigments which absorb the light (about 80%) not absorbed by the retina.
Sclera (sclerotic coat)	Protects and maintains the shape of the eye.

Fig. 4.84 Summary of main functions of main eye components

estimate the distance of an object. This is important in animals, for instance in allowing them to judge distances from prey or predators. The overlapping fields of vision from the two eyes is called *binocular vision* and also allows distances to be estimated. Predator animals, for example the owl and fox, have good binocular vision for hunting. Prey animals, for example the rabbit and wood mouse, have good *monocular vision* (side vision) to increase awareness of possible predators.

- **Focusing (accommodation)**

Most (about 70 per cent) of the bending (refraction) of light occurs as light passes through the cornea. The angle at which light enters the lens is different for distant and near objects. However, the eye can allow for this by movements of the lens; this is called **focusing** or **accommodation** and is a reflex action.

The lens is important in focusing and this is achieved by changes in its shape. This is brought about by a ring of ciliary muscles. The lens is pulled into a relatively thin shape (made *less convex*) by a tension of the suspensory ligaments; this tension is caused by the outward pressure of the vitreous humour and also by the contraction of radial ciliary muscles. The lens thickens (is made *more convex*) if this tension is reduced by a contraction of the circular ciliary muscles.

Focusing depends on the flexibility of the lens and ciliary muscle and this flexibility may be lost to some extent in humans as they become older. The shape of the eye is very important in focusing because it determines the distance beween the lens and the retina, where the image is formed. The eye may become slightly distorted in shape as it grows. There are two main eye defects which involve difficulties in focusing; they are long sight and short sight, shown in Fig. 4.85.

Eye defect	Cause	Correction
Long sight	**Short eyeball: near objects cannot be focused** 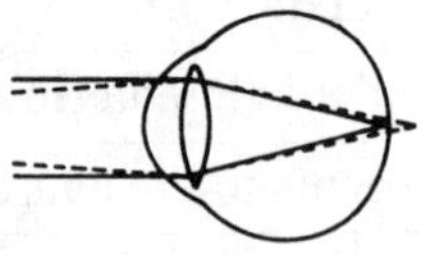	**Converging lens** 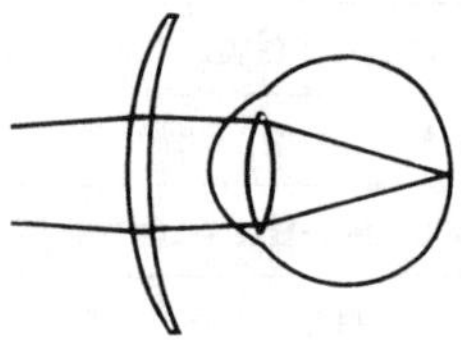
Short sight	**Long eyeball: distant objects cannot be focused** 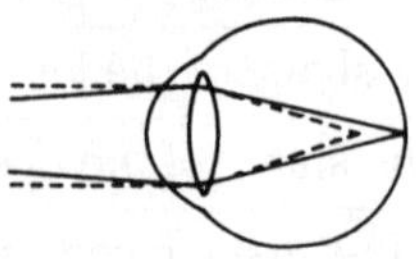	**Diverging lens** 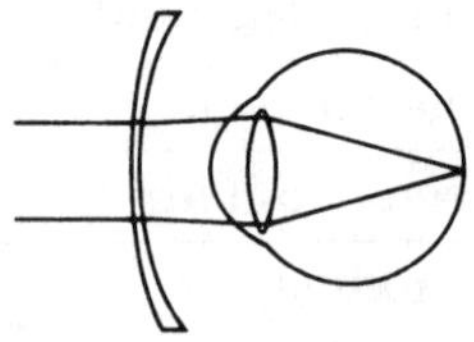
	> = focused	>· = not focused

Fig. 4.85 Human eye defects

Adjustments to changing light intensity

The amount of light entering the eye can be controlled by changes in the size of the pupil aperture, for instance a narrowing of the pupil in bright light: this protects the retina, which might otherwise be temporarily 'bleached'. Opening and closing of the pupil is a reflex action and is achieved by contractions of muscles within the iris. There are two types of *antagonistic muscles* in the iris (Fig 4.86); they are the radial muscles, which widen the pupil, and the circular muscles, which narrow the pupil.

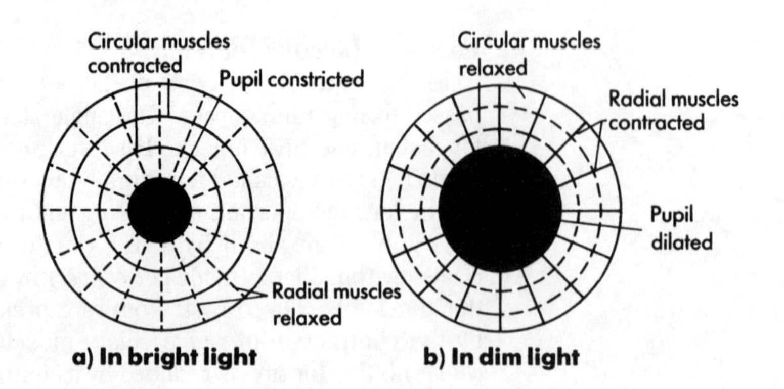

Fig. 4.86 Variations in pupil size in different light intensities

The retina

The retina is the part of the eye which is sensitive to light. An inverted image is formed on the retina because of the bending (refraction) of light by the front part of the eye. The function of the retina is to convert the information carried by light into nerve impulses, which are transmitted to the brain. The sensation of sight is formed in the brain, which interprets visual information in terms of experience. The retina consists of two main types of light-sensitive (*photo-sensitive*) cells; the *rods* and the *cones* (Fig. 4.87). These contain pigments which are 'bleached' in light, generating a nerve impulse. The pigment in rod cells is derived from vitamin A and is important for vision in dim light.

Comparison	Rods	Cones
Structure		
Distribution	120 million cells spread throughout retina; much less in fovea.	Six million cells concentrated in the **fovea**.
Sensitivity: light intensity	Sensitive to **low** light intensities.	Sensitive to **bright** light intensities only.
Sensitivity: colour	Not sensitive to different colours (sensitive to **monochromatic** light only); used mainly in dim light.	Sensitive to colour. Different types of cones respond to the **primary colours red, green** and **blue**; the relative amounts of each type of colour allows humans to detect about 200 different colours.
Nerve connections	Rods share **common nerve connections**. This increases sensitivity to dim light, and decreases **acuity** ('visual accuracy').	Cones have **individual nerve connections**. This decreases sensitivity to dim light and increases acuity.

Fig. 4.87

Links with other topics

Related topics elsewhere in this book are as follows:

Topic	Chapter and Topic number	Page number
Names and locations of the major organs	4.1	30
Function of the major organs	4.1	31
Homeostasis	4.7	137
Coordination	4.8	145
The nervous system	4.8	149
The hormonal system	4.8	152

REVIEW QUESTIONS

Q16 Fig. 4.88 shows a diagrammatic section of the eye.

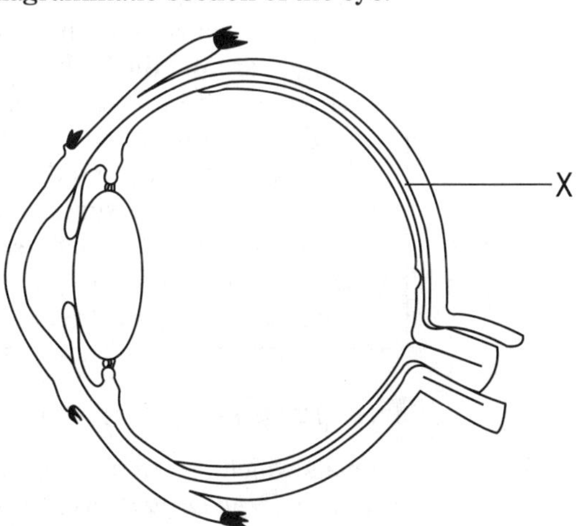

Fig. 4.88. Human eye (vertical section) showing the main components

(a) (i) Label clearly on Fig. 4.88 the following:

pupil, retina, suspensory ligament, cornea, conjunctiva

(ii) On the Fig. indicate clearly

(1) by means of the letter Z, the position of the muscles which alter the focal length of the lens;

(2) by means of the letter Y, the position of the muscles which control the size of the pupil.

(b) (i) State the name of the region of the eyeball which is labelled X.

____

(ii) State two functions of this part of the eye.

____

(c) (i) State the name of the process by which images of objects at different distances from the eye are focused on the retina.

____

(ii) Describe briefly how this is achieved when you look from a distant object to a nearby object.

____

____

____

BLOOD CIRCULATION

Key Points

- The blood system is one of the main transport systems within the human body. One function of the blood system is to carry substances to and from specialised tissues throughout the body. Another important function is in the prevention of disease.
- The main components of the blood system are blood, blood vessels, and the heart.

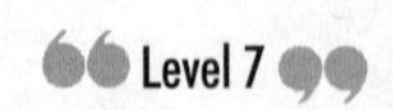

- Blood consists of a liquid plasma and cells. There are two main types of blood cells; red cells and white cells. Red cells are mainly responsible for carrying oxygen around the body. White cells provide resistance to infectious diseases.
- There are three main types of blood vessel. Arteries carry blood from the heart to tissues throughout the body. Capillaries allow an exchange of substances between the blood and the tissues. Veins carry blood to the heart.
- The heart is a muscular pump, which contracts rhythmically to circulate the blood. The heart is divided into two halves, each consisting of two chambers. One half of the heart pumps oxygen-rich (oxygenated) blood. The other half pumps oxygen-poor (deoxygenated) blood.

The purpose of transport systems

Animals which are *very small* (less than about 1 mm across) or which are thin have a *relatively large surface area to volume ratio.* Substances can be exchanged directly between cells within the organism and the environment surrounding the organism. Individual cells can obtain or remove substances over short distances by active transport and diffusion.

Larger animals have a *relatively small surface area to volume ratio*. Larger animals tend to be fairly complex and active, performing a wide range of activities. Such animals have many cells which do not have immediate access to the other parts of the body which supply or receive materials. Specialised tissues are used to allow mass flow of substances over large distances. The specialised system consists of three main components:

(i) **transport fluid**, e.g. blood, lymph
(ii) **vascular system**, e.g. blood or lymph vessels
(iii) **pumping mechanism**, e.g. the heart.

The **vascular system** can be 'open' or 'closed'. In an *open* vascular system, the transport fluid drains into body spaces before re-entering vessels; this form of circulation occurs in arthropod animals for example the insects. In a *closed* vascular system, the transport fluid remains within vessels during much of the circulation. Certain vessels, e.g. capillaries, are permeable, and substances enter or leave the circulation through the walls of the vessels. The closed system occurs in many animals including vertebrates. The advantage of closed circulatory systems is that they allow a much higher pressure to develop in the transport fluid, such as blood; this additional pressure is used to circulate the fluid more efficiently. A closed system also allows the distribution of the fluid within the animal to be controlled more easily.

THE CIRCULATORY SYSTEM

The three main components of the circulatory system are *blood, blood vessels* and the *heart*. Each of these is described below for a mammal such as the human.

Blood

Blood is a transport fluid which also has an important function in protecting the body against infection. Blood consists of two main components (these can be separated when blood is spun in a centrifuge):

plasma (55 per cent of the blood volume)
cells and cell fragments (45 per cent of the blood volume)

The total volume of blood in an average adult human is about 5.5 litres; this represents about 10 per cent of the body mass. The exact composition of the blood varies throughout the body as substances are added or removed in different regions.

Plasma

Plasma is the liquid part of blood; it consists of water (91 per cent) which contains dissolved and suspended particles:

- **water**. Water is a solvent; it also carries heat around the body. The amount of water in the blood is important in determining blood pressure (see below) and in osmoregulation, and is controlled by various processes including the activities of the kidney.
- **blood proteins**. Blood proteins are made in the liver. They are important in various processes and include:
  fibrinogen and *prothrombin*; important in blood clotting; plasma with these proteins removed is known as serum
  globulin; some globulins function as antibodies, others are used to carry hormones and vitamins in the blood-albumen; used in osmoregulation
- **control proteins**. Hormones ('Chemical messengers') and enzymes are important in regulating various metabolic processes.
- **dissolved food and wastes**. Food includes glucose, amino acids, fatty acids, glycerol, vitamins and minerals; wastes include urea and carbon dioxide.
- **dissolved gases**. As well as waste carbon dioxide, plasma contains small amounts of nitrogen and oxygen; most of the oxygen is not very soluble and most is carried within red blood cells (see below).

Cells

There are two main types of cells contained in blood. Blood also contains *cell fragments*, called *thrombocytes*. Each type of cell is quite different in structure and function (Fig. 4.89).

Cell type	Site of manufacture	Function	Relative no./mm³ of blood*	Structure
Red cell	Bone marrow	Carry oxygen and some carbon dioxide	5 000 000	
White cells:			7 000	
i) Phagocyte	Bone marrow	Engulf bacteria	4 900	
ii) Lymphocyte	Lymphatic system	Antibody production	1 680	
Thrombocyte (platelet)	Bone marrow	Part of blood clotting mechanism	250 000	
*1 mm³ of blood is about one small drop.				

Fig. 4.89 Comparison of different types of blood cells

(i) **Red cells.** Red cells, (or **erythrocytes**, red blood corpuscles) are formed (at the rate of about 1 million per second) in the bone marrow of the ribs, sternum and vertebrae. The outer membrane of the red cells may include certain proteins which determine the individual's blood group. In the foetus, red cells are mostly made in the liver and spleen. Mammalian red blood cells have a relatively short life time (about three months) because they lack a nucleus; the cells are broken down in the spleen and liver, and the iron and some of the protein is re-cycled. The nucleus is lost from each cell during its formation, creating a depression in the central part of the cell; this produces the characteristic *biconcave disc* shape of red blood cells. (Fig. 4.89). The shape provides a large surface area for the exchange of gases. The total surface area of all red blood cells in the human body is estimated to be about 3500m²! Exchange is further promoted by the fact that red cells need to squeeze through *capillaries* one after the other; the relatively slow movement and the close contact between the cells and the vessel wall allows diffusion to occur more readily.

Each red cell contains about 300 million haemoglobin molecules, which combine with oxygen in a reversible way to form oxyhaemoglobin. Haemoglobin increases (by about ten times) the capacity of the blood to carry oxygen. Red cells also carry carbon dioxide as bicarbonate ions. The uptake or loss of oxygen or carbon dioxide is determined by the relative concentration inside and outside the red cell.

The rate of production of red cells can be increased during growth, when new tissues require the supply of oxygen or the removal of carbon dioxide. Additional red cells will be needed as a result of blood loss. Individuals living at high altitude also make extra red cells, because the amount of oxygen available is relatively low. In any situation in which insufficient iron is available in the diet, fewer red cells than normal will be made. This condition is called *anaemia* and occurs because iron is needed in the manufacture of haemoglobin.

❝ Candidates often confuse the functions of different blood cells. Re-read this section if necessary ❞

(ii) **White cells (leucocytes).** White cells are generally less numerous (by about 700 times) and also larger (by about twice) than red cells. White cells are, however, very thin and can, unlike red cells, enter or leave the circulation through small gaps in capillaries (see below). Each white cell has a nucleus. In general, white cells form part of the body's internal defences against disease, or immunity. There are several types of white cell, but two types occur in relatively large numbers: these represent about 94 per cent of all white cells.

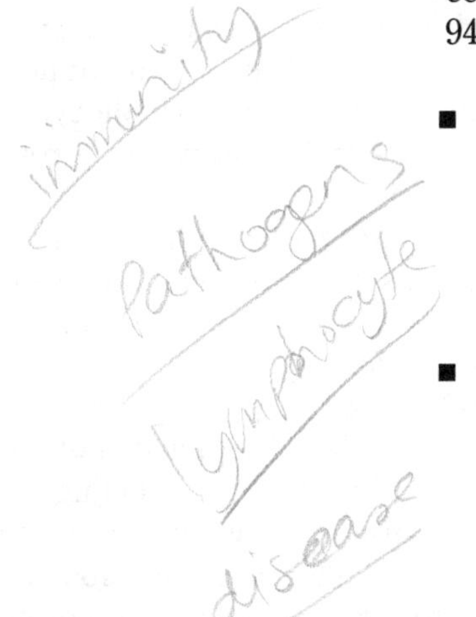

- **Phagocytes** (70 per cent). Phagocytes (or polymorphs, granulocytes) have a lobed nucleus (see Fig. 4.89) and the cell shape is variable. Phagocytes surround, engulf and destroy bacteria; they are not effective against viruses. The engulfing process of phagocytes is very similar to that in *Amoeba*. The bacteria may have previously been clumped together by the action of lymphocytes.
- **Lymphocytes** (24 per cent). Lymphocytes have a large, rounded nucleus and a fairly regular cell shape (Fig. 4.89). Lymphocytes produce *antibodies* (or *antitoxins*) in the presence of 'foreign' or '*non-self*' proteins (or *antigens*) in the blood, such as *pathogens* (disease-causing microbes). Antibodies are effective against many viruses, as well as bacteria. Antibodies have different effects, for example they may clump pathogens together (*agglutination*) or split pathogens open. Antibody production is a complex process (Fig. 4.90) involving the recognition of an antigen and the synthesis of an appropriate antibody by lymphocytes. This is called the *immune response* and may involve chemical 'memory'; the lymphocytes 'remember' which antibodies have proved effective against which antigens. The 'memory' is genetic and can be inherited by other lymphocytes. This is one of several types of immunity, some of which are summarised in Fig. 4.91.

"Some students find this difficult to understand. You may need to re-read this section"

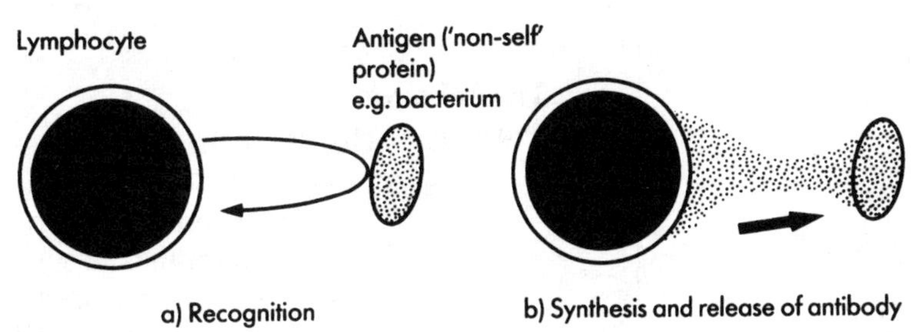

Fig. 4.90 Recognition and response in the immune system

Type of immunity	*Description and examples*
Non-specific immunity	Involves mechanisms which respond generally to non-self-proteins. Examples: the action of phagocytes also the action of chemicals such as **interferon** and **lysozyme** produced by the body.
Specific immunity	Involves mechanisms which respond specifically to non-self proteins.
Active immunity	Is based on antibody production and action within one organism; the **immune response.**
Active natural immunity	Involves the immune response resulting from a naturally occurring **infection** by a pathogen (disease-causing) organism. Examples: immunity to German measles (rubella), whooping cough.
Active induced (acquired) immunity	Involves the immune response resulting from an artificial introduction, e.g. by **vaccination (inoculation)**, of a modified type of antigen. Examples: cowpox, polio vaccine; much 'weaker' than the actual pathogen.
Passive immunity	Is based on antibodies produced by *another* organism which then have their action within a particular individual. This provides immunity for a fairly limited period.
Passive natural immunity	Involves the transfer of antibodies from a mother to a foetus across the **placenta**. Examples: measles and polio.
Passive induced (acquired) immunity	Involves the transfer of antibodies by **injection**, e.g. of **serum**. Example: diphtheria.

Fig. 4.91 Summary of immunity

Most types of specific immunity only provide protection against a particular type of disease. Immunity may not be effective for the remaining lifetime of an individual, especially in the case of passive immunity or active induced immunity. Lymphocytes may lose the 'memory' of how to make certain antibodies. Another problem is when the immune system itself gets attacked and is prevented from producing a range of antibodies. This has occurred in the case of AIDS (Acquired Immune Deficiency Syndrome). AIDS is caused by a virus which invades certain lymphocyte cells which are then unable to produce antibodies for what would otherwise be relatively mild infections.

The immune system unfortunately operates against transplanted tissue including blood and various organs from a donor organism (the kidney is one example, see Section 5, page 78). Tissue rejection is avoided by the use of immuno-suppressant drugs, though the patient is then susceptible to infections.

Thrombocytes. Thrombocytes **(platelets)** are cell fragments rather than complete cells; however they may contain nuclei, (see Fig. 4.89). The function of thrombocytes is to restrict blood loss at wounds by taking part in the blood clotting (coagulation) mechanism, by producing substances which cause blood vessels to constrict and by releasing thromboplastin (thrombokinase). The release of thromboplastin sets in motion a complex series of reactions which result in the conversion of fibrinogen to fibrin at the site of the wound (Fig. 4.92). Fibrin binds together the edges of the wound and causes the formation of a clot which prevents further blood loss and pathogen entry.

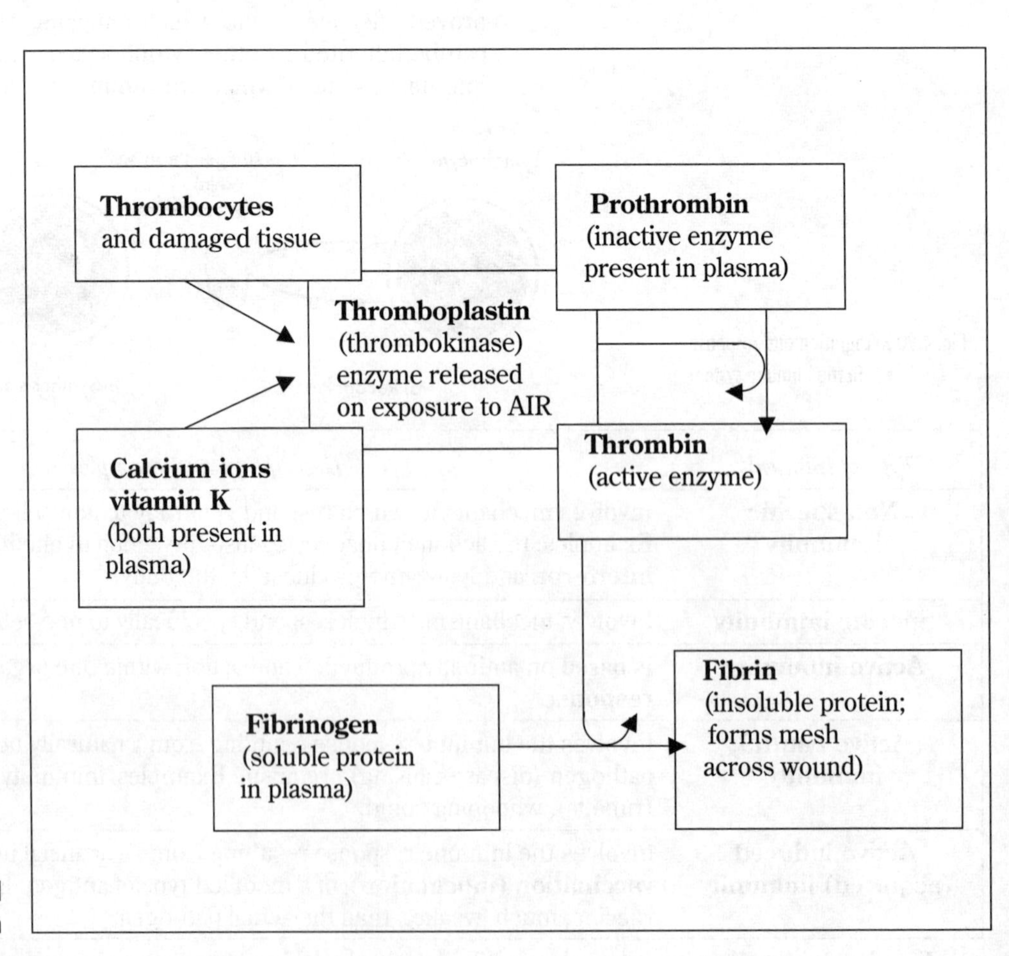

Fig. 4.92 Summary of the blood clotting mechanism

Blood Vessels

Blood vessels are tubes which form the circulatory system and which carry blood around the body in a particular direction. There are three main types of blood vessel; the *artery, capillary* and *vein*. These are compared in Fig. 4.93.

Comparison	Artery	Vein	Capillary
Cross-section (not to scale)	Muscle; Fibrous coat; Lumen		Single cell
Internal (lumen) diameter	Fairly narrow; can expand (= pulse)	Fairly wide.	Very narrow; red blood cells squeeze through.
Wall structure	The wall is relatively thick and also elastic, to withstand pressure.	The wall is relatively thin; there are **valves** to keep blood moving in one direction	Wall is composed of a single cell layer; gaps between cells allow exchange of materials with surrounding tissues.
Blood direction	Blood flows away from the heart.	Blood flows towards the heart.	Blood flows from arteries to veins.
Blood pressure	High.	Low.	Falling.
Blood flow rate	Rapid, irregular.	Slow, regular.	Very slow.

Fig. 4.93 Comparison of the main types of blood vessel

The structure of each of type of blood vessel is an adaptation to its function. The function of artery and vein is to carry blood to or from the capillaries respectively. The function of the capillary is to allow materials in the blood to be exchanged with surrounding tissues; no living cell is more than about 0.5 mm from the nearest capillary. The relatively high pressure of blood in capillaries forces some of the plasma through the permeable capillary walls (Fig. 4.94), which sometimes have special pores. The plasma which has leaked out is known as *tissue fluid*; this contains nutrients and oxygen, as well as antibodies and hormones.

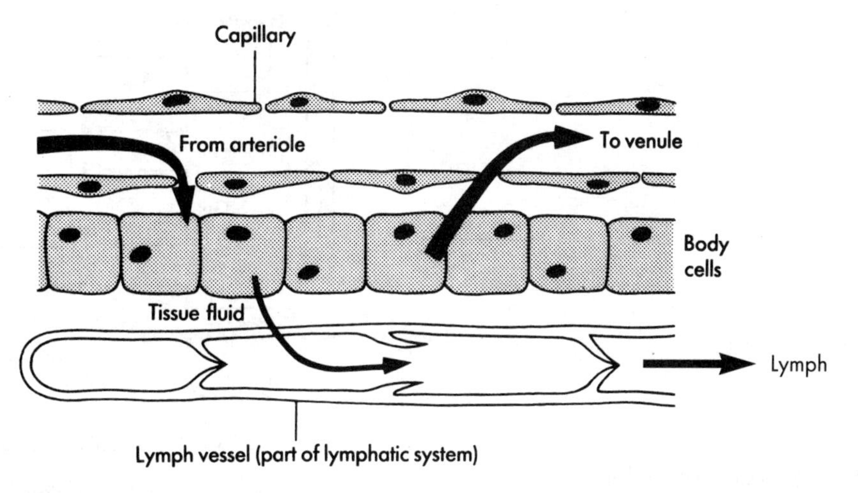

Fig. 4.94 Exchange of materials betwen capillaries and surrounding tissues

White cells may also move from the capillary into the spaces between cells. Larger protein molecules and red blood cells remain within the capillary. Tissue fluid allows an exchange of substances between the blood and surrounding tissues; wastes are carried from tissues back into the circulation. Most of the tissue fluid re-enters the capillary (by osmosis), but excess tissue fluid drains into the lymphatic system where it becomes known as *lymph*. Blood pressure drops in the capillaries during tissue fluid formation. The drop in pressure also occurs because the total space inside a group of capillaries is much greater than the space of the artery that supplies them.

Arteries divide repeatedly into smaller *arterioles* which eventually lead to the capillaries. These re-join to form *venules* and then *veins*. This sequence is shown in Fig. 4.95.

The lymphatic system collects excess tissue fluid and returns it to the blood.

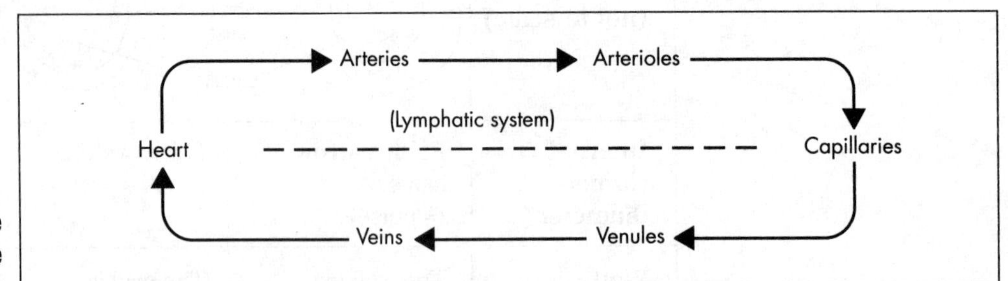

Fig. 4.95 Summary of the sequence of blood vessels in the circulation

The double (*dual*) circulation, typical of mammals, is shown in a simplified form in Fig. 4.96, and is composed of the ***pulmonary circulation*** serving the lungs, and the ***systemic circulation*** serving the rest of the body. Most arteries, except pulmonary arteries, contain oxygenated blood and most veins, except pulmonary veins, contain deoxygenated blood. The fundamental distinction between arteries and veins is that arteries carry blood *away from the heart*, and veins carry blood *towards the heart.*

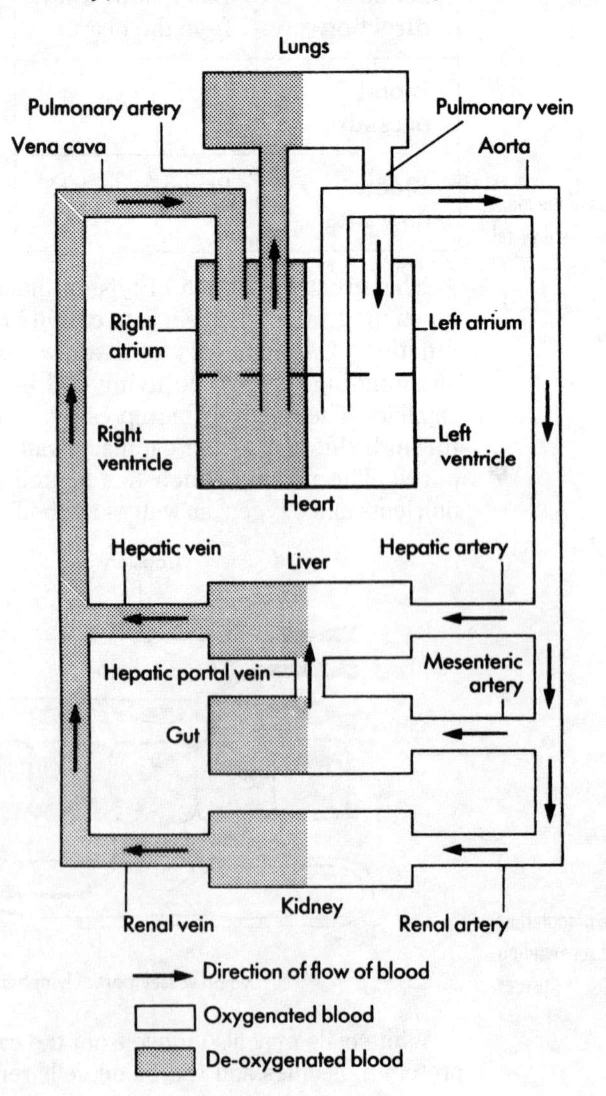

Fig. 4.96 Simplified diagram of the blood system in mammals, including humans

The heart

The heart is a muscular pump which keeps blood moving through the circulatory system. The heart in mammals such as humans consists of two fused pumps, which pump either *oxygenated blood* (*left side*) or *deoxygenated blood* (*right side*). Each side of the heart is further divided into two compartments (Fig. 4.97) the *atria* (*auricles*) and *ventricles*, which allows blood to be pumped in two stages.

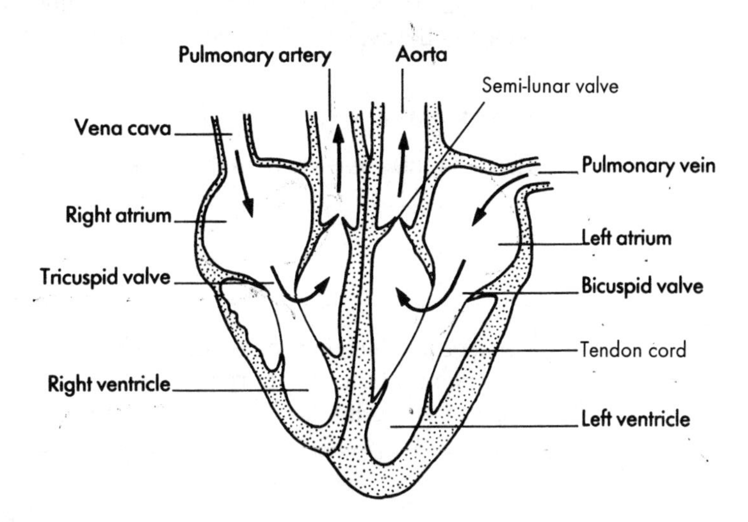

Fig. 4.97 Vertical section through a human heart

Questions on heart structure and function are very popular with examiners!

The heart beats in a continuous series of rhythmic muscular contractions; each repeating sequence is called a *cardiac cycle*. The frequency of cardiac cycles and the power of each contraction depends on the body's need for blood, and is regulated by the brain, by nerves and hormones (Section 4.8). The heart can also regulate its own activity by a small patch of tissue, called the *pacemaker*, embedded in the wall of the right atrium. The average adult rate of the cardiac cycle is about 72 beats per minute. This can be increased dramatically, for instance to supply and remove more materials to exercising muscle. Blood from all the veins in the body drains into the two atria of the heart at the beginning of each heart beat. Both atria contract together (Fig. 4.98), forcing blood into the ventricles. These then contract, pumping blood into arteries leading to the lungs or the rest of the body. The ventricles have thicker muscular walls than the atria because they need to push blood over greater distances; the left ventricle has particularly thick walls because it pumps blood around most of the body.

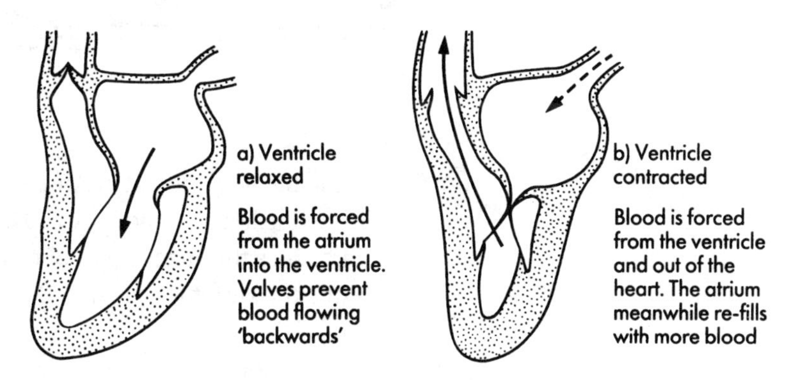

Fig. 4.98 The cardiac cycle (vertical section; left side)

Blood entering arteries under pressure causes them to expand against the elastic walls which then recoil, forcing blood around the circulation. Blood pressure is lost when blood has passed through the capillaries. The low-pressure blood is prevented from flowing backwards by the presence of valves in the walls of veins. This is very similar to the arrangement in the *lymph vessels*. The movement of blood towards the heart is assisted by pressure from surrounding *skeletal muscle* when it contracts; this may be particularly important when gravity is acting on the blood returning to the heart.

The heart muscle itself requires a constant supply of blood. This is provided by the *coronary artery*. If the flow of blood through the coronary artery is reduced the heart muscle cannot function properly; this may result in a coronary thrombosis (heart attack), caused by a blood clot which blocks the flow of blood to the heart muscle. The problem begins with *atherosclerosis,* the build up of fatty deposits on the lining of the arteries. The chances of heart disease can be significantly reduced by avoiding too much fat (especially animal fat) in the diet. Other possible causes of heart disease include obesity, smoking, excess alcohol, severe stress and insufficient exercise.

Links with other topics

Related topics elsewhere in the book are as follows:

Topic	Chapter and Topic number	Page number
Names and locations of the major organs	4.1	30
Functions of the major organs	4.1	31
Specialized cells	4.3	36

disease

REVIEW QUESTIONS

Q17 Examine Fig. 4.99 which shows the relationship between part of the blood system and some cells in the body.

→ circulation -

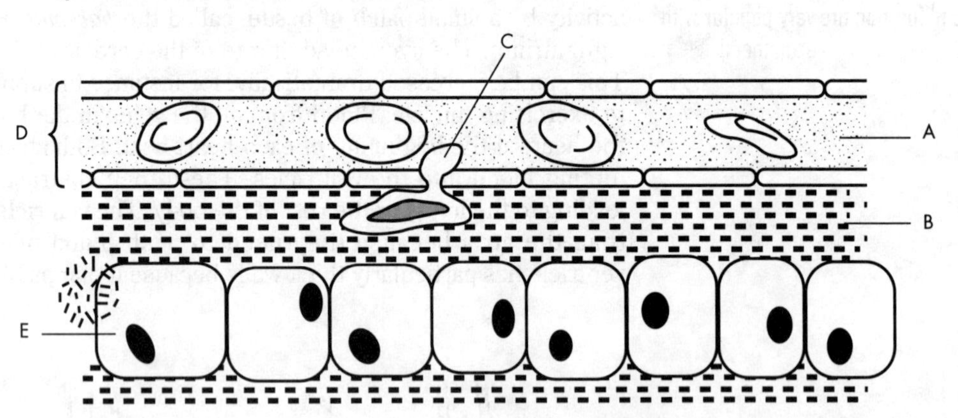

Fig. 4.99

(a) (i) Name the liquid found in Area **A**. __________

(ii) Name the liquid found in Area **B**. __________

(iii) What does the shaded part of cell **C** represent? __________

(iv) Name the structure labelled **D**. __________

(b) State two observations from Fig. 4.99 which helped you to name the structure labelled **D**.

____

____

(c) Cell **E** is being invaded by bacteria. Describe how the blood might help to control the infection.

____

____

____

6 > LIFE PROCESSES IN PLANTS

This Topic Group focuses particularly on processes which are very distinctive in plants, and which are important in relation to other topics in your biology course. However, note that many other life processes in plants are not covered here (and are not required by your syllabus). Plants are capable of performing all the characteristics of life, i.e. movement (of parts of the plant body), excretion (mostly of oxygen), sensitivity (e.g. to light and gravity, using hormones).

Photosynthesis is a very important topic in biology; it is also one that causes problems for some examination candidates! One reason that photosynthesis is stressed is because it has a central role in ecology. Green plants are able to make their food using simple molecules from the environment, and the sun's energy. This is the starting point for all *food chains* (see Chapter 8).

Respiration in plants is similar in principle to respiration in animals. The basic chemical reaction for respiration is the opposite to that of photosynthesis. In plants, the two processes occur together in light, though the rate of photosynthesis usually exceeds that for respiration.

Reproduction can be asexual (without sex cells) or sexual (with sex cells), and many types of plant are capable of both. Sexual reproduction in higher plants occurs within flowers. Pollen is transferred from the male to the female part of the flower, during *pollination*. Fertilisation then occurs, resulting in seed and fruit production. Pollination and *seed dispersal* in some plants can be dependent on animals.

PHOTOSYNTHESIS

Key Points

"Level 7"

- Photosynthesis is the process by which plants obtain carbohydrates for *nutrition*. *Carbon dioxide* from the atmosphere is combined with *hydrogen* to make *glucose* and other carbohydrates. The hydrogen is provided by *water*, which has been split into hydrogen and oxygen. *Oxygen* is released as a waste product. The energy needed to split water molecules is provided by light, which plants 'capture' using the green pigment *chlorophyll*.
- The main site of photosynthesis in plants is the *leaf*. The structure of the leaf is well-adapted to its many functions. Leaf cells contain chlorophyll in small 'packets' called *chloroplasts*. These can be seen under the microscope.
- Carbohydrates in plants are stored mostly as *starch*. The presence of starch in plant tissues can be tested with iodine, which produces a blue-black colour. This test can be used to confirm that photosynthesis has occurred in a leaf. Another test for photosynthesis is the production of oxygen bubbles from water plants.

Photosynthesis involves the incorporation of simple inorganic molecules into complex organic molecules, normally using light energy. This process consists of a series of enzyme-controlled reactions. The light energy for photosynthesis is absorbed and converted to chemical energy by *chlorophyll* molecules, contained in *chloroplasts*. The possession of chlorophyll molecules is a characteristic of all organisms capable of photosynthesis. The overall process of photosynthesis can be summarised in the following word equation:

"You should make a point of learning this equation. Do not, however confuse it with the one for aerobic respiration"

$$\text{Carbon dioxide} + \text{water} + \text{light energy} \xrightarrow{\text{chlorophyll}} \text{glucose} + \text{oxygen}$$

The overall reaction is, in many respects, the reverse of that for *aerobic* respiration (see Section 4.5); aerobic respiration and photosynthesis are complementary processes.

Light energy absorbed by chlorophyll is used to 'split' water molecules into the separate elements of hydrogen and oxygen. This process is called *photolysis* ('light-splitting'). The *hydrogen* is combined with carbon dioxide to form the carbohydrate *glucose*. This can then be converted into other molecules, including *sucrose* which can be transported around the plant, and *starch*, which can be stored for future use. Starch is therefore regarded as evidence of photosynthesis and can be tested for (see below) to confirm that photosynthesis has taken place. *Minerals* are used to make other organic molecules from carbohydrates; for example, nitrogen can be used to make proteins. The *oxygen* produced by photolysis is a waste product of photosynthesis. Oxygen can be used in aerobic respiration or released into the atmosphere, depending on the rate of photosynthesis (see below).

The site of photosynthesis

Photosynthesis takes place within chloroplasts; these organelles are particularly numerous in leaf cells. Leaves can be regarded as the main organs of photosynthesis. Leaf structure (Fig. 4.100) is well adapted for this function. The role of the main components of the leaf in relation to photosynthesis is summarised in Fig. 4.101.

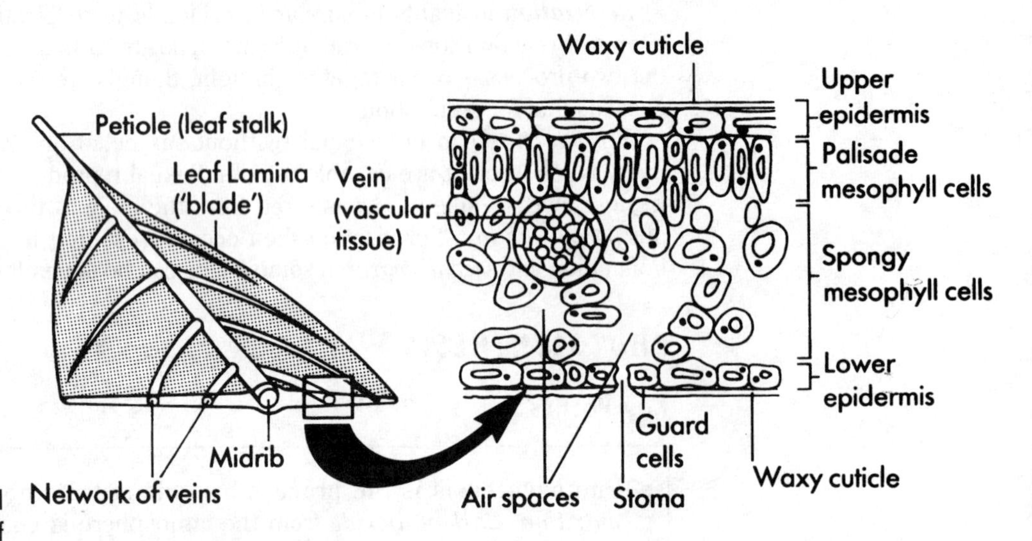

Fig. 4.100 Structure of a typical leaf

Component	Description	Adaptation
Chloroplasts	Comparatively large membrane-bound organelles which contain pigments such as *chlorophyll*. These pigments absorb light energy and convert it to chemical energy.	Chloroplasts contain complex molecules, such as pigments and enzymes, necessary for photosynthesis. The relative number and distribution of chloroplasts corresponds with the amount of light available to the leaf.
Palisade mesophyll cells	Vertically arranged 'column' (palisade) shaped cells, containing many (up to about 100) chloroplasts; main site of photosynthesis in 'broad-leaved' *dicot* plants. Absent in 'narrow-leaved' *monocot* plants.	The vertical arrangement of cells reduces the number of cross-walls which would interfere with the passage of light.
Spongy mesophyll cells	Loosely packed rounded cells, surrounded by air spaces. Main site of gas exchange surrounded by air spaces. Main site of photosynthesis in monocot plants.	Air spaces surrounding the cells are continuous, via the *stomata* with the atmosphere around the leaf. Cells are surrounded by a film of water in which gases entering the cells dissolve.
Epidermis and waxy cuticle	Epidermis cells maintain shape of leaf and produce a waxy cuticle layer. This protects the leaf from excess water loss and the entry of disease-causing microbes.	The epidermis and waxy cuticle are well-adapted to preventing excess water leaving the plant and also to preventing the entry of microbes. Light is allowed to enter the leaf, however, as most epidermis cells do not contain chloroplasts.

Component	Description	Adaptation
Guard cells and stomata	A pair of guard cells surround each *stoma* (pore); leaves have many stomata. *Stomata* are the main route for the movement of gases into and out of the leaf. Each stoma can be opened or closed by changes in the shape of the guard cells.	Unlike other epidermis cells, guard cells contain chloroplasts and can photosynthesise. This is thought to be important in their opening and closing. These cells have *uneven thickening* of their walls which causes them to change shape when water is gained or lost by osmosis.
Veins (vascular bundles)	Veins contain *vascular tissue* which conducts water and minerals to the leaf and removes products of photosynthesis to the rest of the plant. A network of veins supports the leaf and keeps it flat.	Leaf cells need a supply of water for photosynthesis. Minerals are necessary for the formation of certain molecules, such as proteins. Many molecules made during photosynthesis will be needed by other parts of the plant so are transported away through veins. Leaves of most plants are kept flat to maintain a sufficient surface area for the absorption of light and for the exchange of gases with the environment.

Fig. 4.101 Summary of the main leaf components involved in photosynthesis

The total *surface area* of leaves of a particular plant is an important factor in deciding how much photosynthesis the plant can undertake. There are two main reasons for this. Firstly, leaves absorb light for photosynthesis. Secondly, leaves absorb carbon dioxide for photosynthesis and allow waste oxygen to escape. Leaves are thin enough for diffusion distances to be relatively short.

Plants often turn their leaves towards the most intense source of light. The leaves are normally positioned so that they do not shade each other. The total surface area available will normally allow the plant to make enough materials for growth and development. However, leaves are also the main region of water loss from the plant. The pores (stomata) through which water is lost from the plant need to be opened for at least part of each day to allow oxygen and carbon dioxide to pass through for respiration and photosynthesis.

Factors affecting photosynthesis

The *rate* of photosynthesis is controlled by various *internal* and *external* factors. If any one of these is in short supply it is known as a *limiting factor* since it will determine the overall rate of photosynthesis. Note that, although water is needed in photosynthesis, this is not likely to be a limiting factor since a wilted plant will not in any case be functioning normally.

The main factors affecting photosynthesis are chlorophyll, light, carbon dioxide and temperature. Relatively simple experimental methods are available for studying the effect of various factors on photosynthesis. The presence of starch is often used to show that photosynthesis has taken place since starch is formed quite rapidly from the glucose which results from this process. Before the experiment is conducted the plant is destarched. *Destarching* is the removal of starch contained in a plant by placing the plant in darkness for about 48 hours, for instance by covering it with black polythene. The plant uses up any starch that is present in its leaves during this time. The way in which leaves, for example, may be tested for starch is shown in Fig. 4.102.

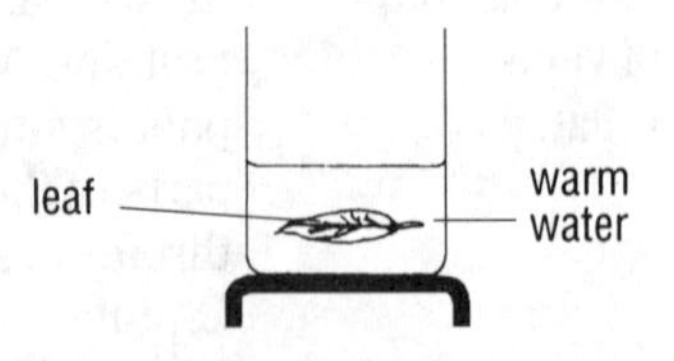

(a) Leaf is boiled in water (about 2 mins). (Purpose: to break down cell walls and to stop the action of enzymes within the leaf.)

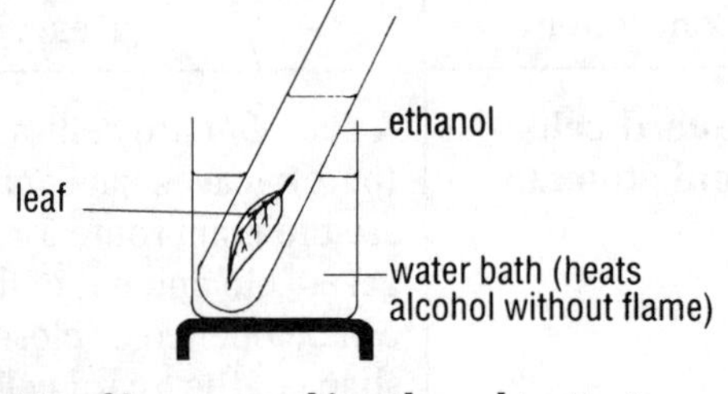

(b) Leaf is warmed in ethanol (until leaf is colourless) CAUTION: ETHANOL IS INFLAMMABLE; NO FLAMES SHOULD BE USED AT THIS STAGE. (Purpose: to extract the chlorophyll, which would obstruct observation later. Chlorophyll dissolves in ethanol but not in water).

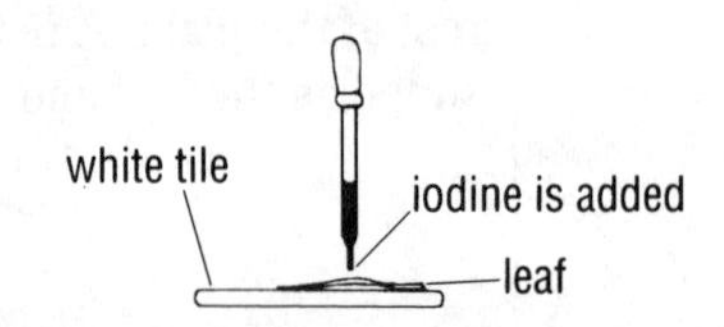

(c) Leaf is dipped into the warm water (briefly) (Purpose: to soften the now brittle leaf.)

(d) Leaf is placed on white tile and iodine added (Purpose: iodine shows the presence *blue-black* or absence *orange-brown* of starch; colours are shown against the white tile.)

Fig. 4.102 Testing a leaf for starch

“This common photosynthesis experiment causes confusion with some students. Make sure you understand the significance of the presence or absence of starch”

Chlorophyll

Chlorophyll is essential for photosynthesis; it is necessary for converting light energy into chemical energy for splitting water. Chlorophyll molecules are formed in the presence of light and are broken down in darkness. A plant kept in prolonged darkness (i.e. more than about ten days) will look yellow rather than green; this condition is known as *chlorosis*. An example of chlorosis occurs if insufficient light is available during germination, resulting in etiolated (tall and weak) seedlings. Chlorosis also occurs if certain minerals, especially *magnesium*, are lacking.

- **Experiment to demonstrate the need for chlorophyll in photosynthesis**

The experiment involves the use of *variegated* leaves. Such leaves have an uneven distribution of chlorophyll; some areas of the leaf may lack chlorophyll completely. Examples of plants which may have variegated leaves are privet, ivy, geranium and laurel. The presence of starch is used to confirm that photosynthesis has taken place.

A variegated leaf is exposed to light whilst still attached to the rest of the plant. The leaf is then removed from the plant and tested for starch, as described above. The distribution of starch after testing is compared with that of chlorophyll before testing. If possible, the leaf should be sketched before testing and again afterwards. Possible results are shown in Fig. 4.103. The distribution of starch is found to correspond very closely to that of chlorophyll. This suggests that chlorophyll is necessary for the formation of starch, during photosynthesis.

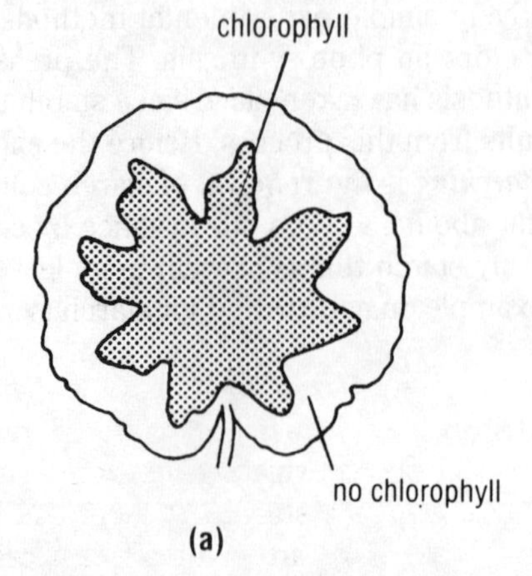

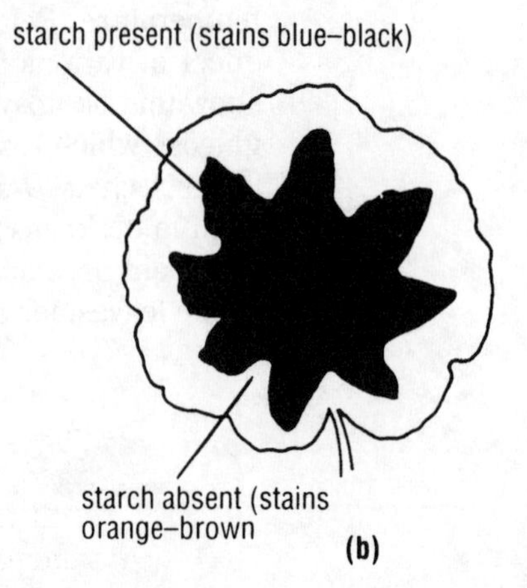

Fig. 4.103 A variegated leaf (a) before and (b) after testing for starch

Light

Light provides the energy for splitting water, which provides hydrogen for photosynthesis. Light may be a limiting factor (see above) in dim light, for instance at dawn and dusk. Light is a highly variable factor in the environment, for instance depending on the time of day or, in many areas of the world, the time of year. Light varies in three main ways:

(i) **Intensity**. The amount of light available will depend on the time of day and seasons, weather and shading. Small plants may be shaded by other, taller plants. Some plants, e.g. bluebell, growing in woodland complete most of their life cycle before the deciduous trees above them form leaves.

(ii) **Duration**. Photosynthesis can only occur when sufficient light is available. For instance in the UK the light available during 24 hours will be about 17 hours during the summer but only 8 hours during the winter. For this reason, deciduous plants lose their leaves at those times of the year when the duration of light is short, and when temperatures are relatively low. Evergreen plants do not lose their leaves in this way.

(iii) **Wavelength**. The rate of photosynthesis depends also on the type of light available; for instance, blue and red light is more effective than green light, which is reflected from leaves. It is possible to show the effect of varying the intensity and wavelength of light on the rate of photosynthesis (see below) and also the effect of the absence of light.

- **Experiment to demonstrate the need for light in photosynthesis**
  The leaf used in this experiment remains attached to the rest of the plant until it is tested for starch. The leaf is destarched (see above) and is then partially covered with a 'mask' from material through which light cannot pass; aluminium foil or card is suitable. A simple pattern (stencil) can be cut in the mask to make the experiment more interesting (Fig. 4.104). The leaf is placed in strong light for several hours, then tested for starch (see above). Those parts of the leaf which were exposed to light are found to contain starch. Those parts that were covered are found to contain no starch.

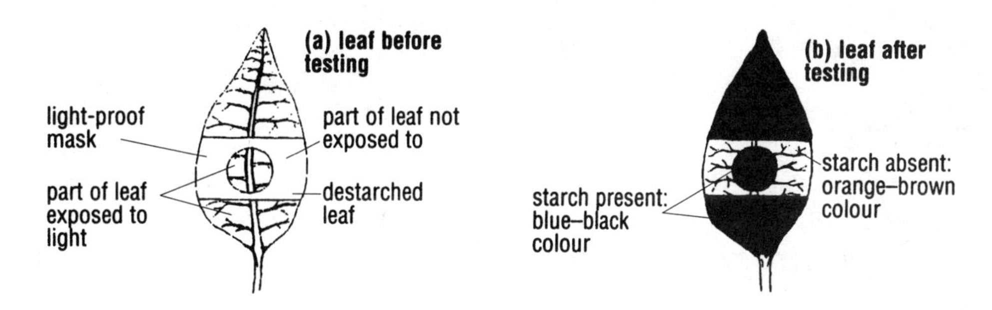

Fig. 4.104 A partially covered leaf before and after testing for starch

Carbon dioxide

Carbon dioxide is present in low concentrations (normally about 0.04 per cent) in air. For this reason, carbon dioxide is often a limiting factor (see above) in photosynthesis; this can limit the rate of photosynthesis. In greenhouse crops the concentration of carbon dioxide is raised artificially up to about 0.2 per cent; this increases the growth rate or *yield* of the plant.

“This idea is important in understanding global warming (Ch.7)”

- **Experiment to demonstrate the need for carbon dioxide in photosynthesis**
  Carbon dioxide can be removed from an enclosed atmosphere by an absorbant such as potassium hydroxide. In this experiment, a leaf of a destarched plant is enclosed in a carbon dioxide-free atmosphere whilst still attached to the rest of the plant (Fig. 4.105). Another leaf is enclosed in a similar way, but without a carbon dioxide absorbant being present; this is for comparison and is called a control.

  The plant is then exposed to light for several hours, to allow photosynthesis to take place. Both leaves enclosed in the flasks are then tested for starch, as described above. The leaf which was kept in an atmosphere free of carbon dioxide is found not to contain starch; the control leaf does contain starch. This experiment indicates that carbon dioxide is necessary for photosynthesis.

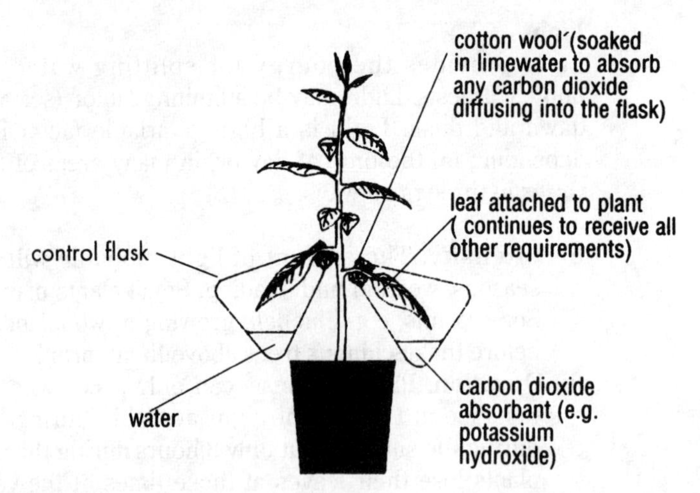

Fig. 4.105 Apparatus to show that carbon dioxide is needed for photosynthesis

Temperature

Temperature affects photosynthesis because the process is *enzyme*-controlled. There will therefore be an increase in the rate of photosynthesis as temperature increases. However, if the optimum temperature is exceeded the enzymes may be destroyed and photosynthesis will cease. Sources of light, especially the sun, are also sources of heat so variations in light and temperature tend to occur together. Both are kept constant and relatively high within glasshouses, where economically-important plants are grown independently of fluctuations in climate. The effect of this and of raising carbon dioxide concentrations (see above) is to raise the average yield of food plants.

The rate of photosynthesis

The rate of photosynthesis can be determined for various conditions. This is most easily done by using an aquatic plant such as Canadian pondweed (*Elodea canadensis*); the plant produces from its cut stem oxygen bubbles which are clearly visible (Fig. 4.106). The experiment can also be used to demonstrate that oxygen is produced during photosynthesis.

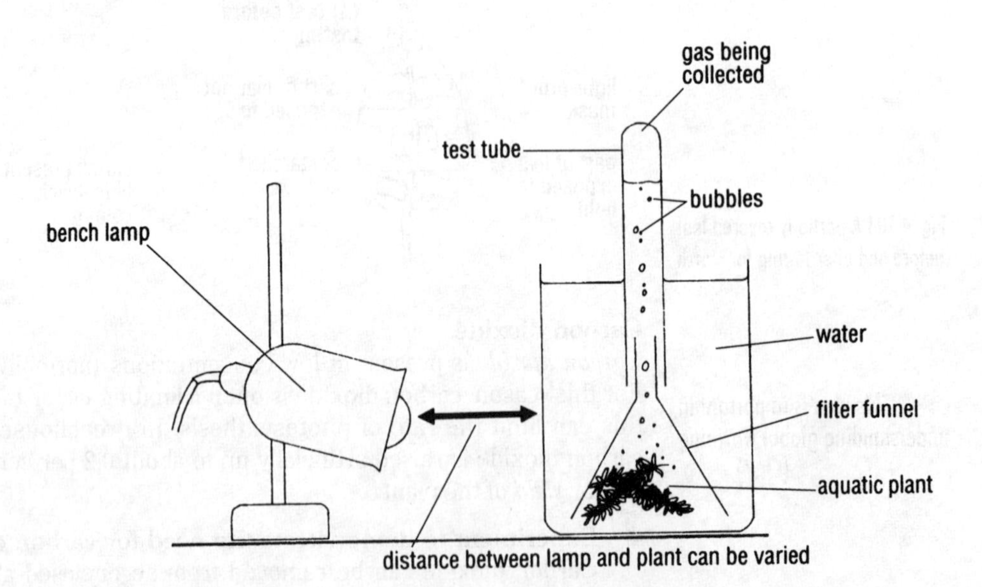

Fig. 4.106 Experiment ot demonstrate the effect of light intensity on the rate of photosynthesis

Questions on this are often set!

In this experiment, the intensity of light reaching the plant is varied by moving the lamp towards or away from the plant. The number of bubbles per minute is counted and noted. This is repeated twice more for each light intensity and an average (mean) value calculated. Only the larger, slower moving bubbles should be counted. If they are too numerous, especially at higher light intensities, dots can be made with a pencil on paper and the dots counted afterwards. The results of this experiment can be plotted as a line graph; the average number of bubbles per minute is taken as the *rate of photosynthesis* (vertical scale), *decreasing* distance is taken as increasing *light intensity*. An example of such a graph is shown in Fig. 4.107.

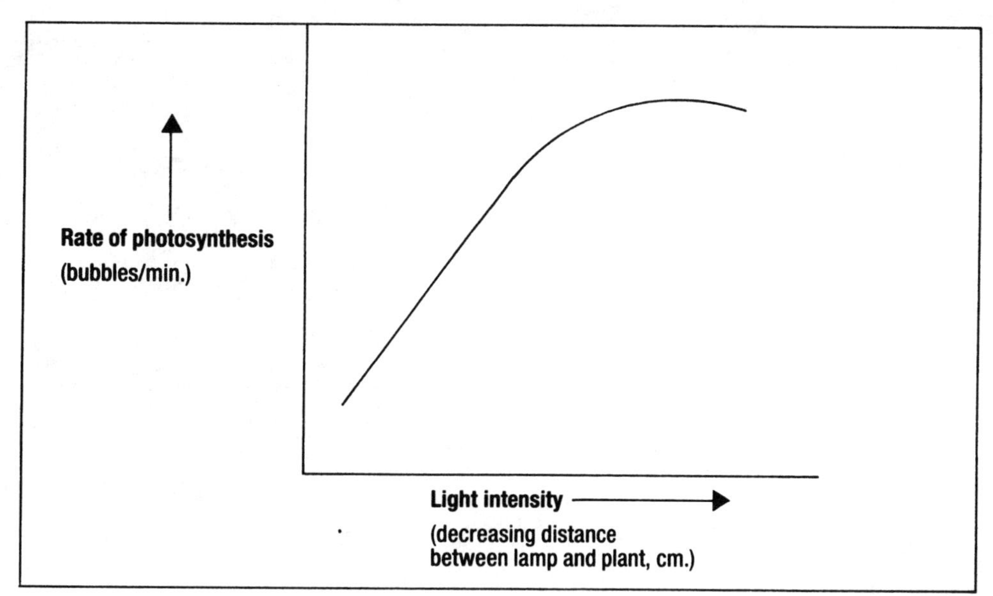

Fig. 4.107 Graph to show the effect of light intensity on the rate of photosynthesis

If sufficient bubbles are collected they can be tested with a glowing splint which should be re-lit; this confirms that the gas collected is rich in oxygen. The amount of oxygen produced by the plant during a particular time can be measured by using a 10 ml measuring cylinder instead of the test tube shown in Fig. 4.106. Other variations include:

- **Carbon dioxide concentration**
  Different amounts of sodium hydrogen carbonate (sodium bicarbonate) can be dissolved in the water shown in Fig. 4.106. This releases carbon dioxide in solution.
- **Colour of light**
  Different wavelengths of light can be used by inserting coloured filters between the lamp and the plant.
- **Temperature**
  The temperature of the water shown in Fig. 4.106 can be varied by adding hot or cold water. Note that varying light intensity in the experiment described above may in any case affect the temperature of the water; this is a *criticism* of the experiment described.

Links with other topics

Links with related topics in this book are as follows:

Topic	Chapter and Topic number	Page number
Names and locations of the major organs	4.2	34
Functions of the major organs	4.2	35
Components of plant and animal cells	4.3	36
Nutrition (animals)	4.5	81
Respiration (plants)	4.6	112
Resources in the environment	7.3	296
Global warming	7.4	305

REVIEW QUESTIONS

Q18 The diagram (Fig. 4.108) shows a tree growing in a field. The sun is shining, so photosynthesis is taking place rapidly in the leaves.

(a) Complete each of the boxes with the name of a suitable substance to show what happens during photosynthesis.

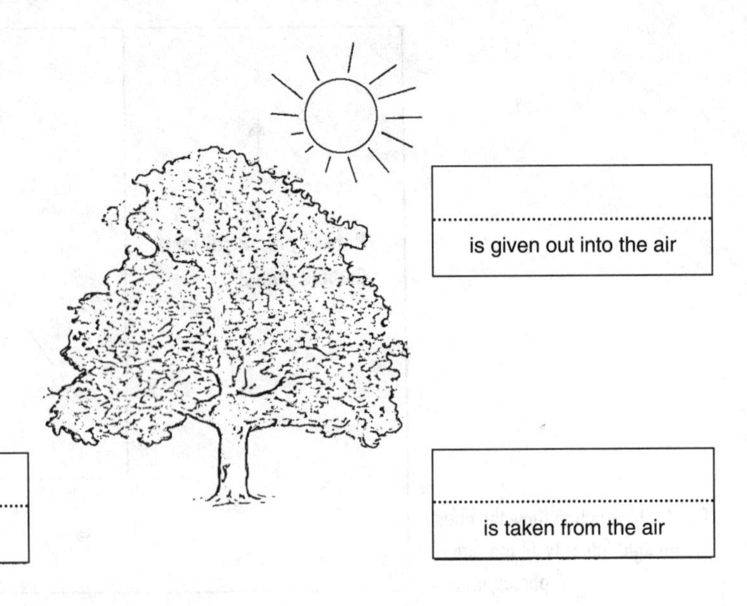

is taken from the soil

Fig. 4.108

(b) Complete the following sentence:

Starch is made in the leaves from __________ which is produced during photosynthesis.

The diagrams (Fig. 4.109) show both surfaces of a leaf taken from the tree.

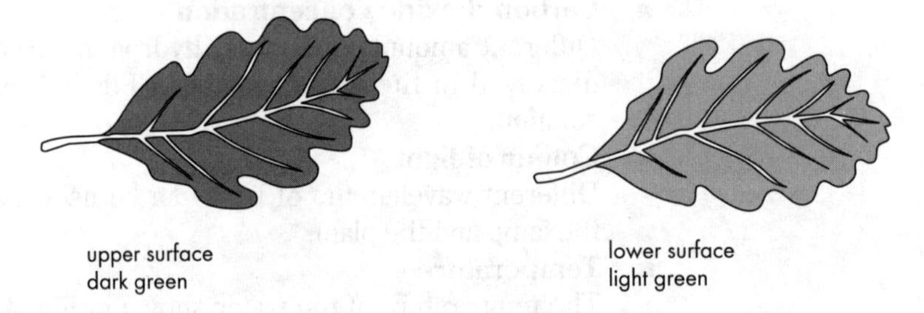

Fig. 4.109

(c) Name the green pigment present in the leaves.

____

(d) What is the function of this pigment?

____

(e) Explain why the upper surface of the leaf is a darker green than the lower surface.

____

RESPIRATION IN PLANTS

Key Points

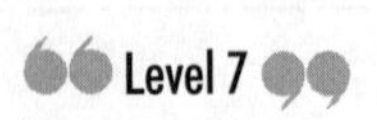

- Respiration occurs in all living plant cells. In many respects respiration is the opposite of photosynthesis, which occurs in plant leaves. Respiration results in a reduction in the plant's supplies of carbohydrates. Photosynthesis results in an increase in the plant's carbohydrate reserves. Growth can only occur if there is an overall gain in carbohydrates within the plant.
- Respiration involves the production of *carbon dioxide*. Photosynthesis involves the uptake of carbon dioxide. The relative rates of respiration and photosynthesis can be demonstrated in an experiment with a carbon dioxide indicator solution.
- Respiration also involves the production of *heat*. This can be demonstrated in germinating seeds.

Photosynthesis and respiration in plants

The chemistry of respiration in plants is similar in many respects to respiration in animals (described in section 5). In other words, *glucose* is broken down to release *energy*. If *oxygen* is present, i.e. in *aerobic respiration*, the glucose is broken down completely, with *carbon dioxide* and *water* as waste products.

Respiration is a continuous process in all plants; *photosynthesis* occurs in green plants when sufficient light is available. The overall processes of aerobic respiration and photosynthesis are, in many respects, opposite to each other. This can be shown by combining the equation for aerobic respiration with that for photosynthesis (see below): (Fig. 4.110):

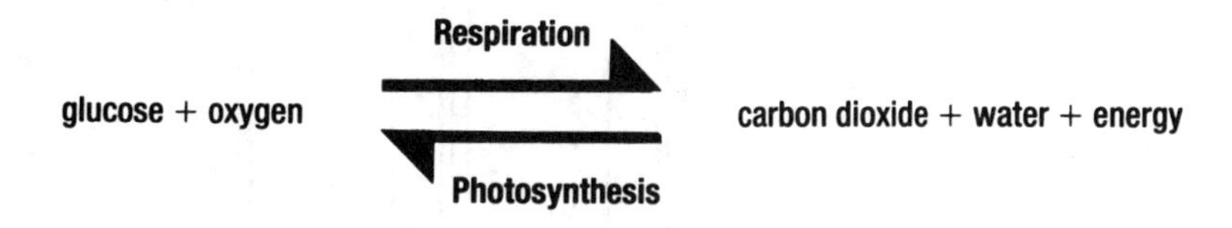

Fig. 4.110 Respiration and photosynthesis as related processes

"Remember that respiration and photosynthesis are not alternative processes in plants; both can occur together"

The energy is released from respiration in the form of heat and chemical energy. The energy used in photosynthesis is light energy, which is converted by chlorophyll molecules into chemical energy. The source of light energy most commonly used by plants for photosynthesis is the sun. Photosynthesis is therefore very important in allowing the sun's energy to be temporarily 'trapped' by living things.

If the average rates of photosynthesis and respiration within a plant were the same there would be *no overall gain* in materials for growth; *growth* can only occur if new materials are made available through nutrition, such as photosynthesis. In relatively *bright light*, the rate of photosynthesis will be greater than the rate of respiration, so there will be an overall gain in organic molecules. These molecules can be used for growth, or to provide energy in respiration.

In *dim light*, the rates of respiration and photosynthesis may be very similar, so that there is no overall gain or loss of glucose, oxygen or carbon dioxide in the plant tissues. This is known as the *compensation point*. The compensation point can be shown on a graph; the relative rates of respiration and photosynthesis can be expressed in terms of gain or loss of oxygen or carbon dioxide (Fig. 4.111).

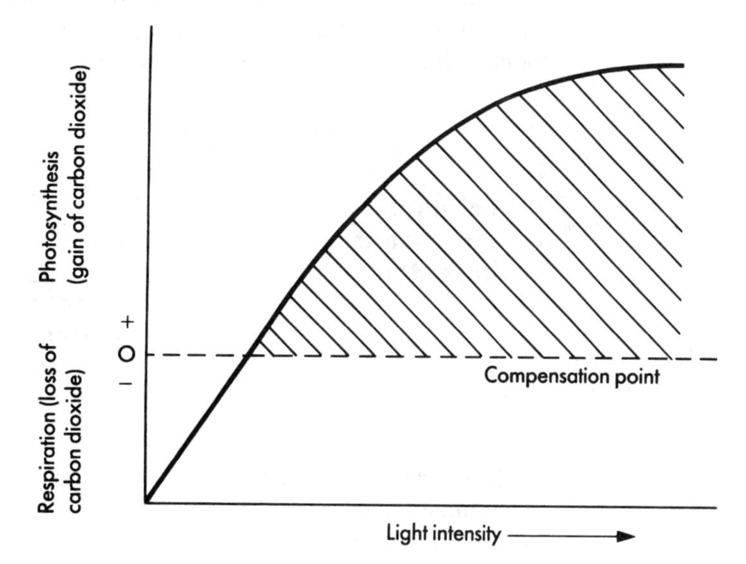

Fig. 4.111 Graph to show compensation point. The shaded part of the graph (beyond the compensation point) represents a net gain of glucose for the plant. This can be used for growth.

Experiment to demonstrate the production of carbon dioxide from plant respiration

This experiment (Fig. 4.112) uses the same set-up as the one described in Section 5 (Fig. 4.40). However, instead of an animal being put into the 'respiration chamber', a potted plant is used. The bell jar is completely enclosed by a light-proof covering, such as cooking foil, so that the plant is in darkness. This ensures that only respiration is occurring, as photosynthesis would absorb any carbon dioxide produced. A plastic bag encloses the pot of soil, to prevent any carbon dioxide from soil organisms from affecting the experiment.

Even if you don't do this experiment it is worth understanding; exam questions on this occur

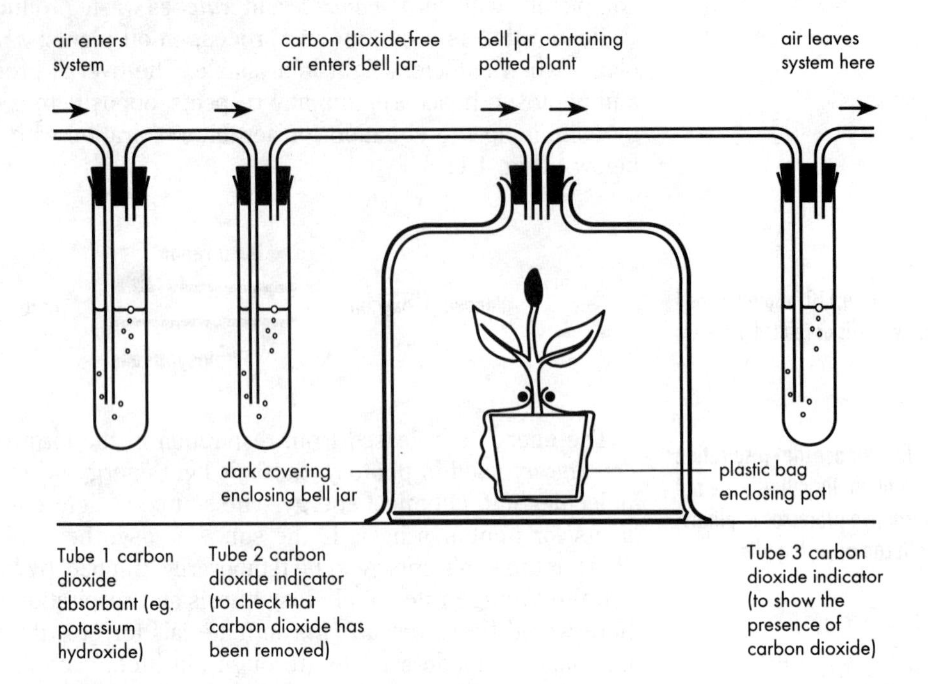

Fig. 4.112 Experiment to show that carbon dioxide is given out by potted plant

Experiment to investigate the relative rates of respiration and photosynthesis in leaves

This experiment is based on the relative loss or gain of carbon dioxide from leaves in different light conditions. Changes in carbon dioxide concentration can be monitored by using *bicarbonate indicator*. This shows colour changes in the presence of different amounts of carbon dioxide. The gas dissolves in the indicator to form a weak acid (carbonic acid) which causes an alteration of acidity or alkalinity (Fig. 4.113):

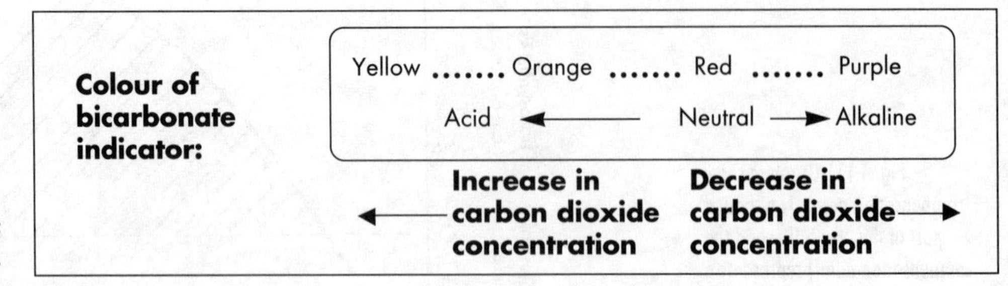

Fig. 4.113 Colour changes of bicarbonate indicator

The bicarbonate indicator is 'equilibrated' with normal atmospheric air before the experiment begins; air is bubbled through the solution, for instance by an aquarium pump. The solution should be *red* at the start of the experiment.

Leaves are placed in tubes containing bicarbonate indicator, as shown in Fig. 4.114. The lower surfaces of the leaves should face inwards, so that gas exchange with pores (stomata) on the underside of the leaves can occur more easily. One tube is set up in darkness, the other in light. Two *control* tubes are set up without leaves, for comparison. All the tubes are left for about two hours.

Respiration occurs in *both* Tubes A and B, resulting in the production of carbon dioxide. However, in Tube B photosynthesis is *also* occurring and this process is more rapid than respiration in bright light; the carbon dioxide concentration is therefore lowered in Tube B.

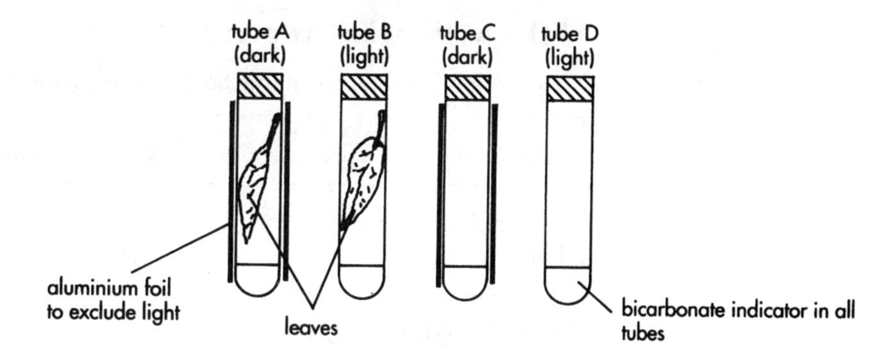

Fig. 4.114 Experiment to investigate respiration and photosynthesis

TUBE	COLOUR OF INDICATOR AT END OF EXPERIMENT	CONCLUSION
A	Yellow	Carbon dioxide concentration is high; produced by respiration.
B	Purple	Carbon dioxide concentration is low; used in photosynthesis.
C	Red	Carbon dioxide concentration remains constant; no respiration or photosynthesis is occurring because no living tissue is present.
D	Red	

Experiment to demonstrate the production of heat from plant respiration

The production of heat by living things is regarded as evidence of respiration. In this experiment, germinating seeds are insulated from temperature changes in the environment by being placed in a vacuum flask (Fig. 4.115). A *control* consisting of boiled (killed) seeds is set up for comparison. In both cases, the seeds are soaked briefly in a weak antiseptic solution (e.g. TCP) to kill any surface micro-organisms which might interfere with respiration.

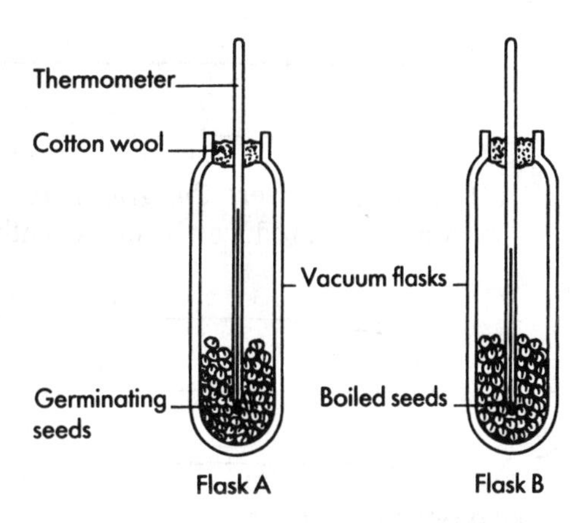

Fig. 4.115 The use of a control in an experiment to show that germinating seeds release heat from respiration

After a few days, the temperature in flask A is expected to be higher (by about 5–10°C) than that in flask B. This is because the seeds in flask A were living, and can respire. The temperature in each flask can be measured at regular intervals (e.g. 2 h), when convenient, over 48 hours; this allows estimates of the rate of respiration to be made.

Links with other topics

Links with related topics in this book are as follows:

Topic	Chapter and Topic number	Page number
Respiration (animals)	4.5	53
Photosynthesis	4.6	105

REVIEW QUESTIONS

Q19 Fig. 4.116 below shows data obtained relating to the rates of oxygen release and uptake in plants. Temperature was constant throughout the period of the experiment.

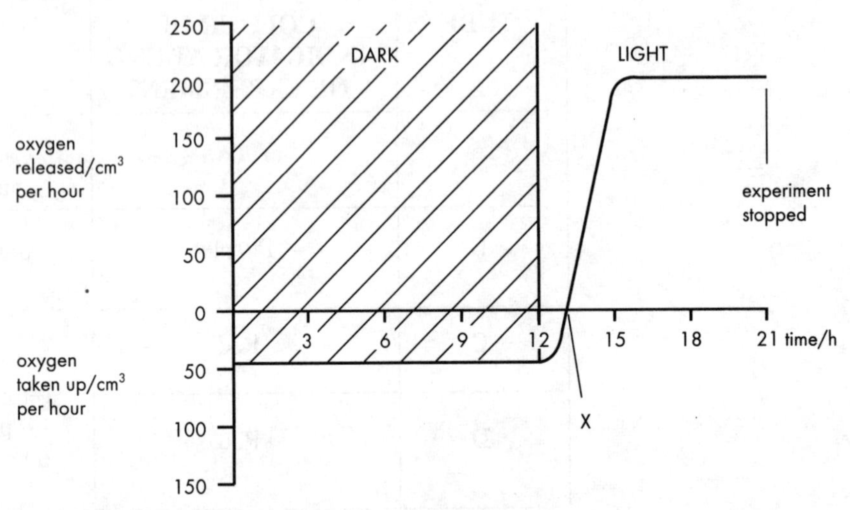

Fig. 4.116

(a) For how many hours was the plant photosynthesising?

(b) For how many hours was the plant respiring?

(c) Calculate the amount of oxygen used in respiration during this experiment (show your working).

(d) At point X where the graph crosses the zero line, what volume of oxygen was being produced by photosynthesis? Explain your answer.

REPRODUCTION

Key Points

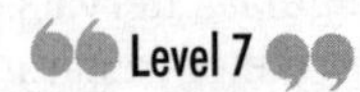

- Plants reproduce *asexually* (without using sex cells) and *sexually* (using sex cells). The organs of sexual reproduction in advanced plants are *flowers*. The different parts of a flower are adapted to their particular function.

- The male part of flowers produce sex cells which are contained in pollen grains. These are transferred to the female part of flowers, during *pollination*. Various methods are used for pollination, including wind pollination and insect pollination.
- Following successful pollination, *fertilization* occurs. This involves male and female sex cells becoming fused, forming a *seed*. The remainder of the flower develops into a *fruit*. Plants use a wide range of methods to disperse seeds.
- Following dispersal, if conditions are suitable, seeds *germinate* into a new plant.

The flower

The *flower* is the reproductive organ of sexual reproduction in the flowering plants. The flower is a modified shoot, with modified leaves mounted on a *receptacle* and attached to the rest of the plant by a flower stalk, or pedicel. The numbers and arrangements of flowers (*inflorescences*) on a single plant vary considerably. Composite flowers like dandelion are made of individual flowers called *florets*.

Components of the Flower

The arrangement of components within a flower is often characteristic for a particular species and can be used for identification and classification. Flower design is determined primarily by the method of *pollination* (see below). Although there are many variations, certain features frequently occur. Most flowers are *hermaphrodite* (or *monoecious*), that is they have both male and female parts present. Flowers commonly consist of four main components which are arranged in concentric layers or *whorls*. From the outside these whorls are:

(i) **Calyx.** This consists of *sepals* which protect the flower in bud, and in some flowers attract insects.
(ii) **Corolla**. This consists of *petals* which, often being conspicuously coloured and scented, attract insects for pollination; *nectaries* (sugar sacs) may be an added inducement. Petals are fused in some flowers. The collective name for the calyx and corolla is *perianth*.
(iii) **Androecium.** This is the male part of the flower consisting of *stamens,* each of which is composed of a *filament* supporting an *anther*. The anther produces *pollen*.
(iv) **Gynaecium** or **pistil**. This is the female part of the flower consisting of *carpels*. Each carpel is made up of a swollen *ovary* and an extended *style* which terminates in a *stigma* that receives pollen.

“You can help to remember the structure of the flower by drawing and labelling your own diagram”

These components are shown in their relative positions in a generalised flower (Fig. 4.117). This flower is not, however, to be seen as representative of flower structures in general because they are modified in so many different ways.

If you are familiar with the names and functions of the main parts of a flower, you will more easily gain a greater understanding of *particular* examples of flowers.

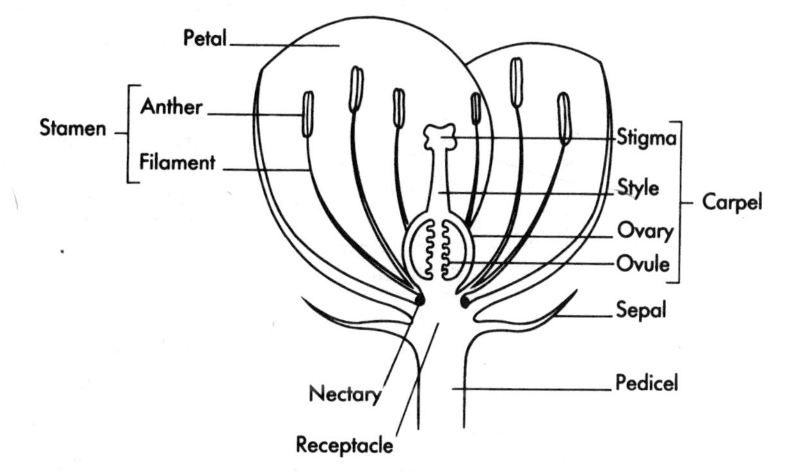

Fig. 4.117 Structure of a generalised flower (vertical section)

Pollination

Pollination is the transfer of pollen from an anther to a stigma. Pollen can be transferred *between* plants, by *cross-pollination*, or *within* plants, by *self-pollination*. Cross-pollination involves different plants (*outbreeding*), so it increases genetic variability (see Chapter 6). Self-pollination involves the same plant (*inbreeding*), and variability is therefore reduced. For this reason plants often avoid self-pollination and encourage cross-pollination, the mechanisms involved sometimes being very elaborate. Some of the main methods are presented in Fig. 4.118. You should note, however, that self pollination may take place in hermaphrodite plants if cross-pollination does not occur; e.g. in isolated plants. Some species (e.g. groundsel) are self-pollinated.

vector

MECHANISM	DESCRIPTION	EXAMPLES
Unisexual (dioecious) plants	Separate male and female plants.	holly willow
Unisexual (dioecious) flowers	Separated male and female flowers on same bisexual plant.	oak hazel
Protogyny	Female parts mature before the male parts in a bisexual flower.	bluebell plantain
Protandry	Male parts mature before the female parts in a bisexual flower.	white deadnettle dandelion
Self-incompatibility	Pollen grain from the plant does not successfully produce a pollen tube.	sweet pea clover
Heterostyly	Male and female structures are positioned differently in different insect-pollinated flowers.	primrose

Fig. 4.118 Mechanisms of avoiding self-pollination

Methods of pollination

The *actual transfer* of pollen during cross pollination is achieved in various ways. The two main methods are *insect pollination* and *wind pollination*, and flowers tend to be adapted to one or the other. This is reflected by structural differences summarised in Fig. 4.119.

FEATURE	INSECT-POLLINATED	WIND-POLLINATED
Petals	Large, conspicuous, brightly coloured	Small or absent.
Nectar	May be present	Absent
Scent	May be present	Absent
Anthers	Small, enclosed within flower	Large, hanging outside flower
Filament	Short, rigid	Long, flexible
Pollen grains	Sticky, rough and relatively large. Adhere to insect body	Smooth, light and relatively small. Produced in large quantities, to offset losses.
Stigma	Relatively small; enclosed within flower	Feathery, large surface area exposed on outside of flower to collect pollen.

Fig. 4.119 Comparison of insect and wind pollinated flowers

- **Insect Pollination** is common amongst herbaceous plants, e.g. buttercup, dandelion. Insects and insect-pollinated plants provide benefits for each other. The life cycles and the distribution of insects and the plants they pollinate correspond very closely. Each is adapted to the other. This is an example of *co-evolution*.

  Insects are a convenient *vector* (carrier) of pollen since they are small, numerous and highly mobile. Some plants are pollinated by various insects and the flowers have a fairly open plan, e.g. buttercup. Other plants are pollinated mainly by a particular type of insect, for instance with a certain length of proboscis (tongue). Such plants are fairly closed, with fused petals, e.g. white deadnettle (Fig. 4.120). The white deadnettle flower incorporates all of the characteristic features of insect-pollinated plants. These features are summarised in Fig. 4.120.

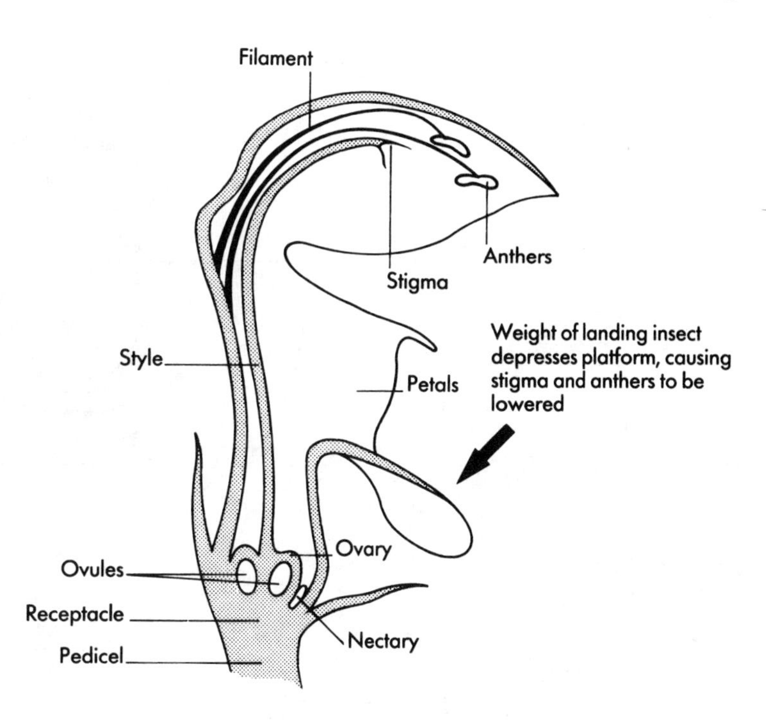

Fig. 4.120 Insect pollinated flower (white deadnettle)

- **Wind Pollination** is common amongst trees and grasses, e.g. willow, cocksfoot. Wind is a convenient vector (carrier) because it is relatively independent of seasons. The main characteristics of wind-pollinated flowers are given in Fig. 4.119. Clearly insects can pollinate wind-pollinated flowers, too.

  Some terrestrial (land) plants are pollinated by other animals such as mammals (bats, possums) or birds. Aquatic plants such as Canadian pondweed are pollinated by water. Flowers can also be pollinated artificially, in *artificial selection*.

Fertilisation

“Remember that pollination and fertilisation are not the same; pollination occurs before fertilisation, and fertilisation may not take place at all!”

When pollination is successfully completed, the pollen grains germinate to produce a *pollen tube*. This carries a *tube nucleus*, and also two *male nuclei* which fuse with *female nuclei* within the ovule (Fig. 4.121). There is a *double fertilisation*, so an *embryo* zygote and an *endosperm* (food) zygote are formed.

The ovule will form a *seed* which will, if conditions are favourable, germinate after disperal. Seeds are enclosed within *fruits*, which are important in dispersal.

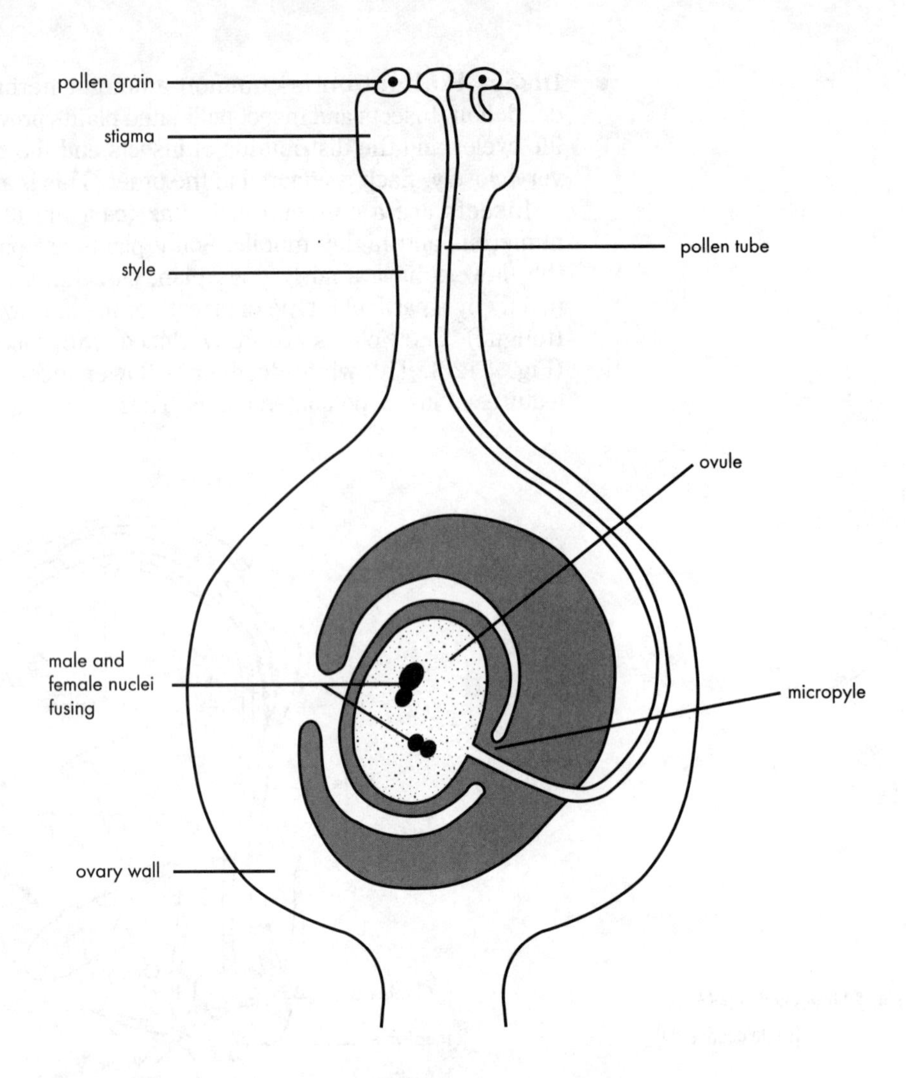

Fig. 4.121 Fertilisation in a generalised carpel

Fruit dispersal

A *fruit* is the fertilised ovary of a flower, containing one or more seeds. The main functions of the fruit are to provide food and protection for seeds, and to assist with seed dispersal.

> Students often confuse pollination and seed dispersal because both may involve wind or animal vectors

The development of a fruit is accompanied by certain changes in the flower. The ovary wall, now called the ***pericarp***, may swell considerably; to protect, nourish and, eventually disperse the seeds enclosed within it. In 'false fruits', e.g. the apple the swollen receptacle performs a similar function.

Other floral parts, for instance the sepals, petals and androecium, may wither away. The stigma and style of each carpel may also degenerate, though in some flowers they are involved in dispersal.

Dispersal is important because:

- Seeds which grow near the parent plant may *compete* for similar resources such as water, mineral ions and light.
- It helps *avoid overcrowding* which increases the chance of the spread of disease.
- It provides an opportunity to *colonise* new areas.

There are two main types of fruit, *fleshy* (succulent) or *dry*. Some examples of each of these, together with their methods of dispersal, are presented below:

Fleshy fruits. These are often brightly coloured, scented and nutritious. Once the fruit has been eaten, dispersal involves seeds being either spat out or egested, unharmed, with faeces. Examples of 'true' fruits include plum and blackberry. 'False' fruits include apple and strawberry.

Dry fruits. These have more varied means of dispersal:

Splitting (dehiscent) fruits spring open, sometimes quite violently, as they dry out, throwing seeds away from the parent plant. Examples include pea, violet, lupin.

Wind-borne fruits have adaptations to delay a fall to the ground. These include wings, e.g. in sycamore (Fig. 4.122) and ash, and 'parachutes', e.g. in dandelion and willow. In some species, small, light seeds are thrown from the plant as it sways in the wind. Examples of this 'censer' mechanism include poppy and foxglove.

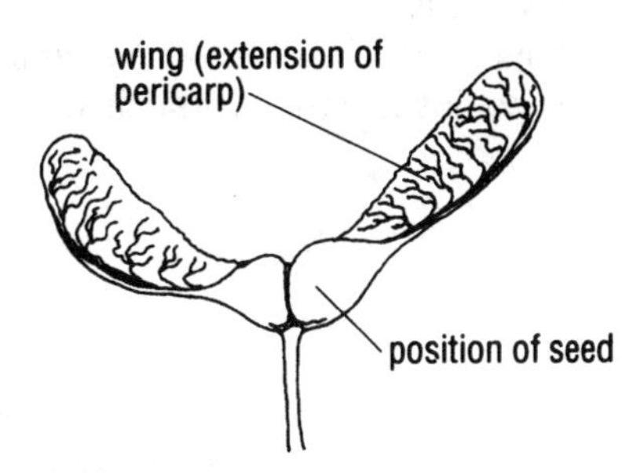

Fig. 4.122 Sycamore fruit

Nuts are carried by birds and mammals to hiding places and then, in some cases, forgotten. Examples are oak and hazel.

Buoyant fruits of plants living in or near water are dispersed by water currents. Such plants include water lily and coconut.

Seed germination and plant development

Plant growth and development begins when the seed, containing the embryo plant, is subjected to conditions which favour germination. Materials for growth are provided initially by food stores originally established by the parent plant. Food is stored in seeds in one of the two main ways, and this is even used to classify flowering plants into two groups as in Fig. 4.123.

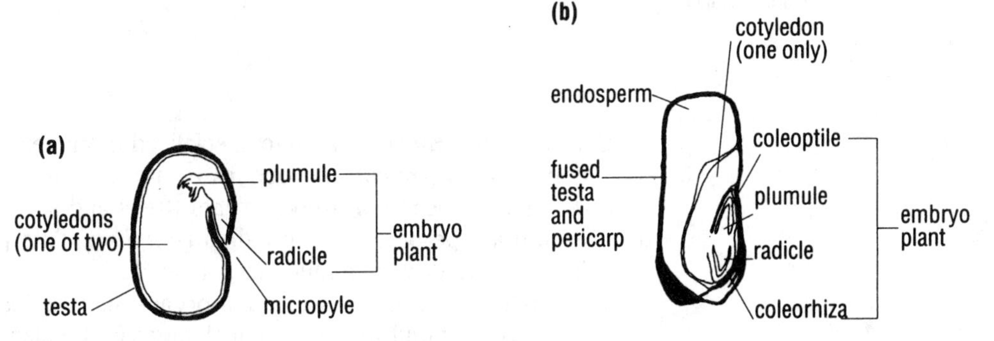

Fig. 4.123 Classification of seeds (a) Dicotyledonous seed; broad bean (Vicia faba) (b) Monocotyledonous seed: maize (Zea mays)

The end of seed dormancy and the beginning of germination is marked by a massive uptake of water (*hydration*), so that the proportion of water is raised from 10 per cent to about 90 per cent of fresh mass. Water triggers *hydrolytic* enzymes which break down the large molecules of stored food, such as starch, into smaller, mobile, molecules which can take part in growth and energy release. Soluble molecules also have a role in the further uptake of water by *osmosis*.

During germination there will be an increase in fresh mass due to water uptake, but an overall decrease in dry mass because some food is used up in respiration.

Early development in many plants occurs below ground, so is dependent on food stored within the seed. The cotyledons may remain below ground (= *hypogeal* germination) (Fig. 4.124) or may emerge and take part in photosynthesis (= *epigeal* germination).

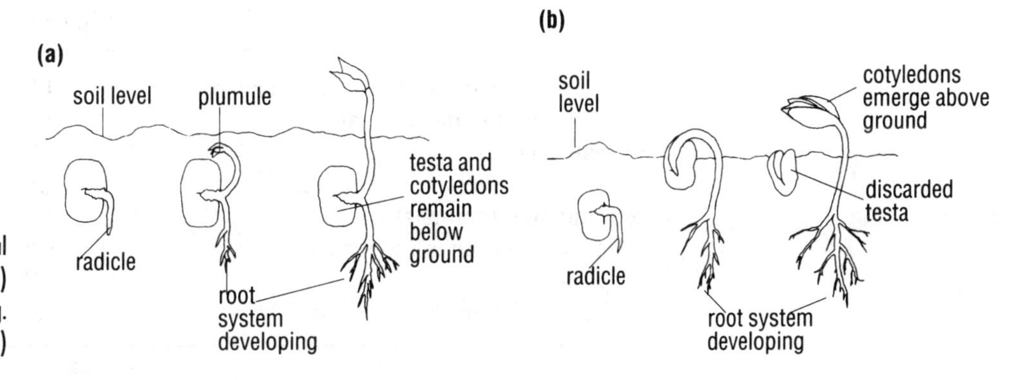

Fig. 4.124 (a) hypogeal germination (e.g. broad bean) (b) epigeal germination (e.g. French bean)

Growth and development in plants is confined mainly to regions near the stem and root tips, around the circumference of stems and roots, and in lateral buds. Cell numbers are increased by rapid cell division (*mitosis*) in the *meristematic* tissues of these regions. Once produced, a cell will then increase in size by forming a cell wall and vacuole.

This occurs in *zones of elongation* and is a very economic type of growth because it does not involve the synthesis of much new protoplasm (Fig. 4.125).

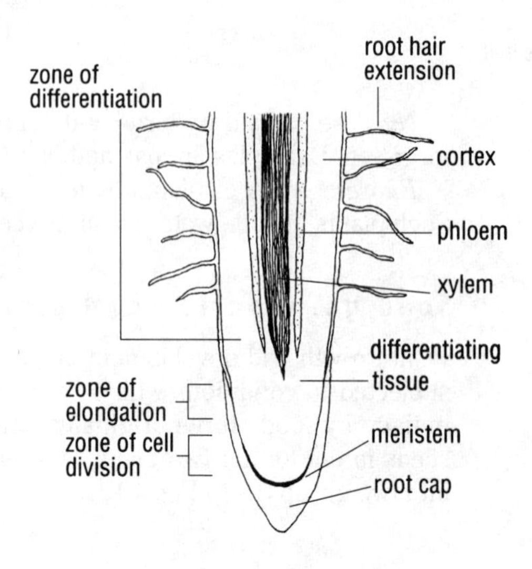

Fig. 4.125 Detail of root tip (vertical section)

Cells may then *differentiate* into specialised tissue, e.g. xylem, phloem, cortex. This is called *primary growth* and occurs in herbaceous (non-woody) plants. Further *secondary growth* occurs in woody plants (trees and shrubs), which form extra woody tissue from a type of meristem called *cambium*. This provides additional support, allowing plants to grow to a much larger size.

The production of new tissues and organs such as leaves, flowers, fruits and food stores may correspond with seasonal changes in the plant's environment, especially in temperate regions. This growth and development is accompanied by a general increase in dry mass. In general, growth and development in plants are continuous processes occurring throughout the plant's life, but confined to particular regions within the plant. The coordination of these processes is controlled by various plant hormones, which affect both cell division and growth. The result is a branching form, providing a large surface area for the exchange of substances with, and the absorption of light from, the environment.

Links with other topics

Topic	Chapter and Topic number	Page number
Names and locations of the major organs	4.2	34
Functions of the major organs	4.2	35
Specialised cells	4.3	36
Growth and development (animals)	4.5	63
Reproduction (animals)	4.5	67
Selective breeding-influences	6.2	209
Crop plants	6.2	210
Types of variation	6.7	238

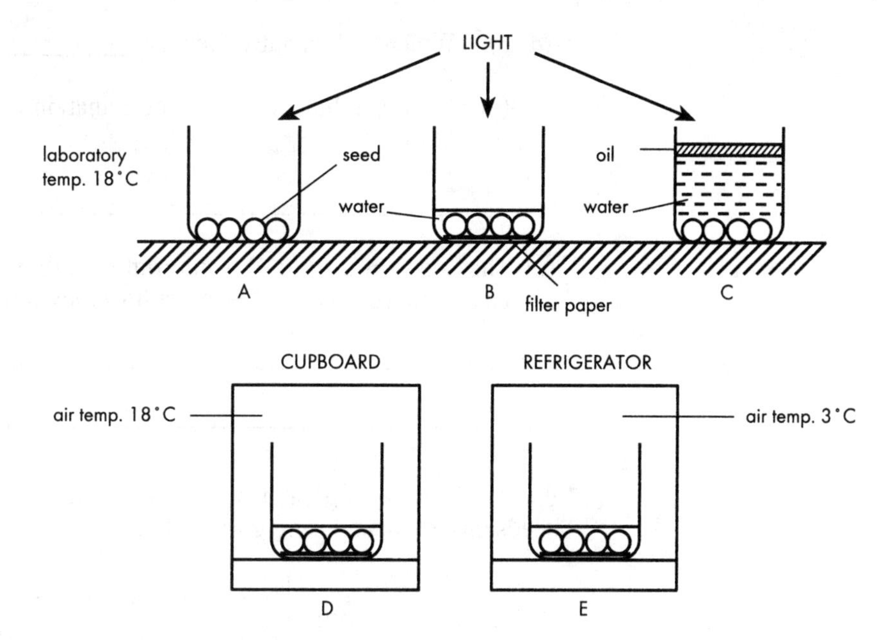

Fig. 4.126

REVIEW QUESTIONS

Q20 Fig. 4.126 above shows an experiment set up to determine the conditions necessary for seed germination.

Ten seeds (only four being shown in the figure) were placed in each beaker and the beakers labelled **A, B, C, D** and **E**. The following conditions were provided:

Beakers **B, D** and **E** – water added to just cover the seeds.

Beaker **C** – boiled and cooled water added and the water surface covered with a thin layer of oil.

Beaker **D** – placed in a dark cupboard in the laboratory.

Beaker **E** – placed in a refrigerator without light.

Beakers **A, B** and **C** – placed on a laboratory bench in the light at the laboratory temperature.

(a) State the letters of the beakers in which the seeds would have germinated after seven days.

___

(b) (i) Why was the water boiled and cooled before covering the seeds in beaker **C**?

___

(ii) What method, other than using boiled water, could be used to exclude the same factor?

___

(c) If the seeds did not germinate in the refrigerator, one could conclude that light might be an important factor in germination. Which beakers would demonstrate this conclusion to be incorrect? Give your reasons.

___

___

(d) (i) Which is the control beaker? ______________________

(ii) What conditions vital for germination are available to the seeds in this beaker?

____

(e) The seeds in beaker **E** would differ slightly from those in beaker **A** at the end of the seven days. Describe how they would differ.

____

____

Q21 Fig. 4.127 shows apparatus set up by a student in an attempt to see whether pea seeds give off heat during germination.

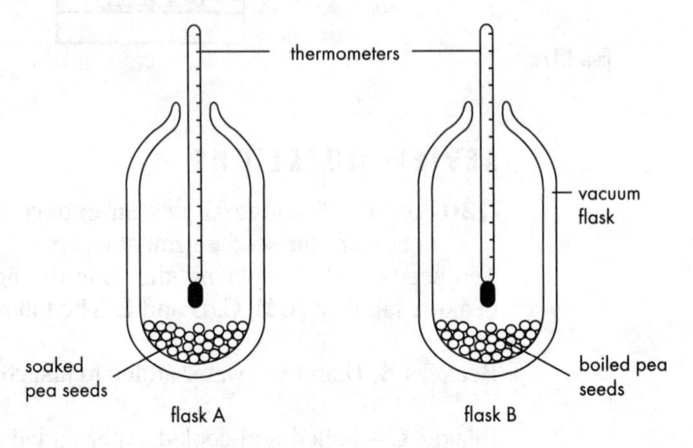

Fig. 4.127

(a) (i) Why were the seeds in flask **A** soaked?

____

(ii) What was the purpose of flask **B**?

____

(iii) Why had the peas in flask **B** been boiled?

____

Fig. 4.128 shows the changes in temperature in the flasks over a 5 day period.

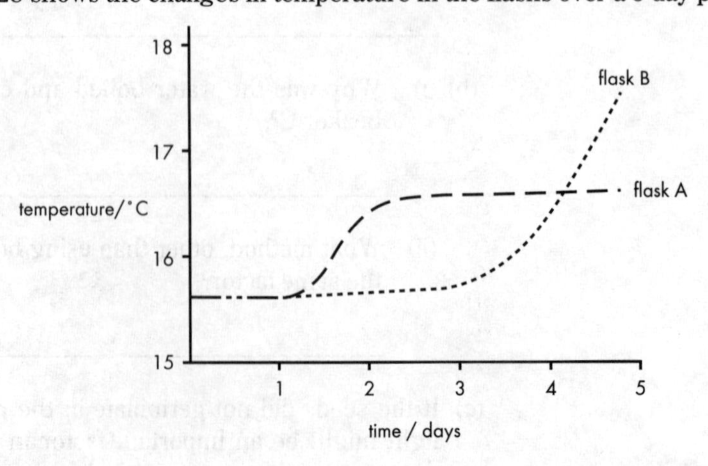

Fig. 4.128

(b) (i) Suggest a reason for this increase in temperature in flask **B**.

____

____

(ii) How might this increase in temperature have been prevented?

____

(c) The student had expected a far larger increase in temperature in flask A. Suggest ways in which he could have improved his experiment to obtain a larger rise in temperature.

____

____

Many organisms can control conditions within themselves (their 'internal environment') even when conditions in their external environment (their surroundings) are changing. This type of control is called *homeostasis*; it is an advantage to organisms because it allows them to be relatively independent of their surroundings and allows their body processes to operate more efficiently. We will consider the following examples of homeostasis:

- the role of the skin in maintaining human body temperature
- the role of hormones in maintaining human water balance and sugar balance
- the protection, respiration and nourishment of the foetus, and removal of its waste products
- absorption of water and minerals by plants
- transport of water and minerals in plants
- transport of sugar in plants
- conversion of sugar into other organic compounds in plants
- the regulation of homeostatic processes by negative feedback.

ROLE OF SKIN

Key Points

- humans are *endothermic* i.e. maintain a constant body temperature
- skin is responsible for maintaining body temperature at 37°C
- *blood flow* through the skin is regulated as the diameter of capillaries changes. This is *vasodilation* (when they become wider) or *vasoconstriction* (when they become narrower)
- *sweat* is produced in hot conditions
- skin *hairs* move to trap a layer of air close to the skin in cold conditions.

The skin is a large organ covering the entire body surface. It has 2 main functions:

(a) regulation of body temperature
(b) detection of stimuli (i.e. it is a sense organ and can detect touch, pressure, pain and temperature).

It consists of 2 layers, called the *epidermis* and *dermis* (Fig. 4.129).

Epidermis
The function of the epidermis is to protect the dermis. In some parts of the body, e.g. soles of feet, palms of hands, it is very thick. The top layer of cells (*cornified layer*) is tough and dead. These cells are continually worn off and replaced from beneath. The *Malpighian layer* is constantly dividing to make new cells, which move towards the top of the epidermis to replace the cells which have been worn off.

Cells in the Malpighian layer contain a chemical called *melanin*, which protects the skin from ultra-violet rays (in sunlight). If you are exposed to U-V rays you make more melanin (get a tan), but in extreme cases the dermis may still be damaged.

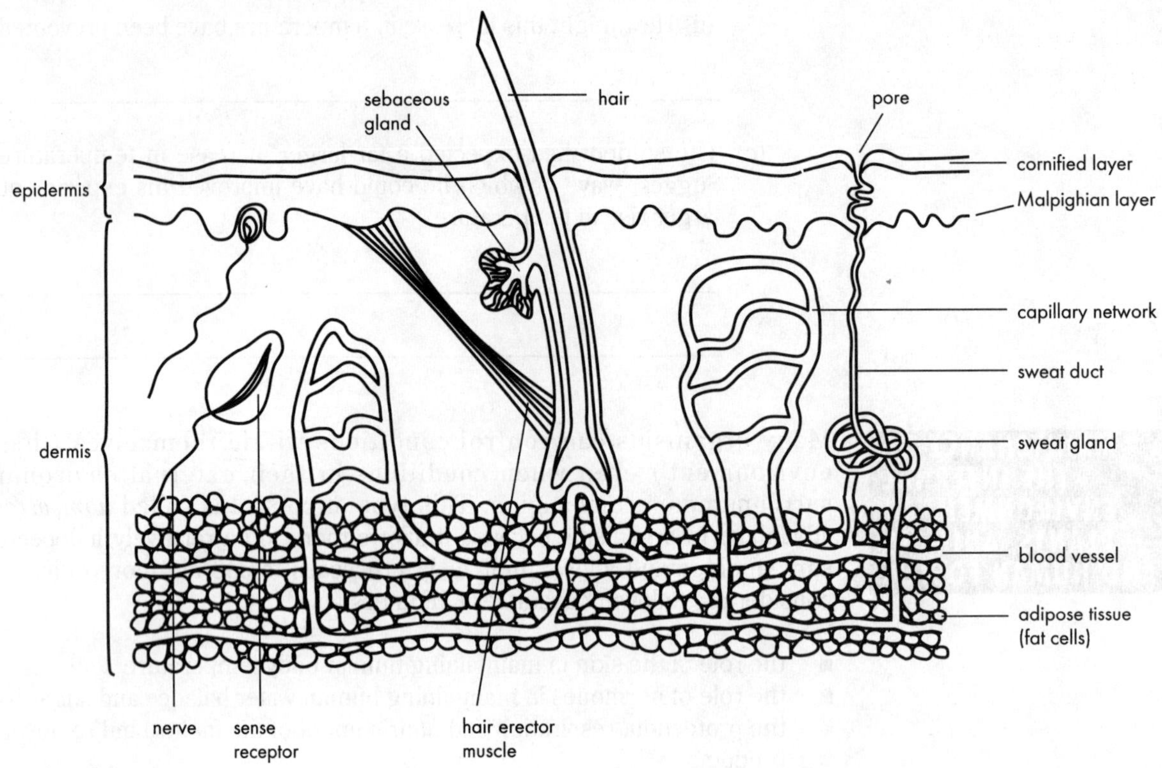

Fig. 4.129 Section through human skin

Dermis

This contains many different structures:

- hairs with sebaceous glands (make oil) and hair erector muscles
- sweat glands with pores on the skin surface
- sense receptors e.g. for touch, pain, pressure etc, and nerves
- blood vessels
- adipose tissue, made up of fat cells
- connective tissue, made up of packing cells (this is not shown in the diagram).

How does the skin control body temperature?

Do not get these terms mixed up

Mammals and birds are *endothermic* (warm blooded): this means that they have a relatively constant body temperature which does not depend on the temperature of their environment. This is an advantage because they can operate effectively in fairly low temperatures, so they can live in cold regions. Animals which cannot maintain their body temperature are described as *ectothermic* (cold blooded), e.g. fish, amphibians, reptiles. Humans have a constant body temperature of 37°C, even when the external temperature varies widely. Ectotherms have a body temperature which varies with the external temperature (Fig. 4.130).

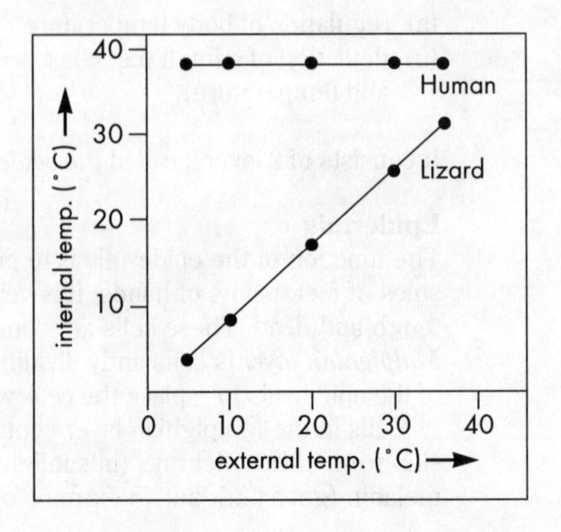

Fig. 4.130 The relationship between internal and external temperature in ectothermic and endothermic animals

Capillaries do not move their diameter changes to alter blood flow

The skin controls body temperature in 3 ways:

(a) Changes in skin capillaries
There are millions of capillaries in the dermis layer of the skin. These can change in size to alter the volume of blood flowing through them.

- **In *HOT* conditions** *VASODILATION* occurs
- capillaries close to skin surface widen
- more blood flows close to surface of skin
- heat is lost through skin surface
- skin feels warmer and looks redder

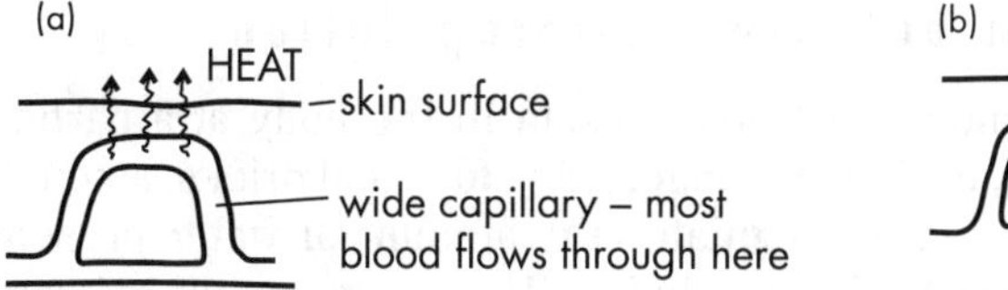

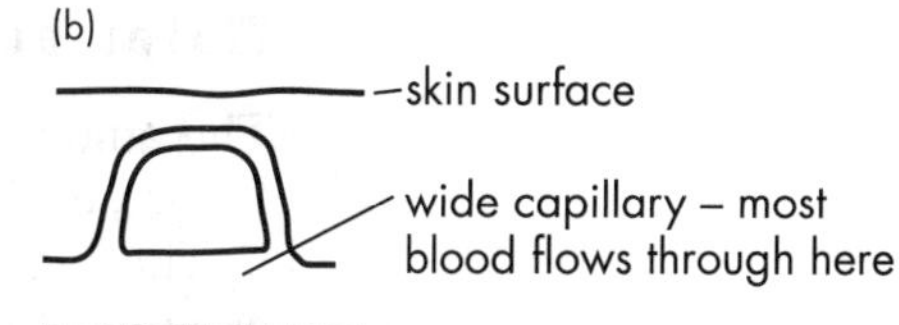

Fig. 4.131 Change in diameter of capillaries alters blood flow through the skin (a) Hot conditions. (b) Cool conditions.

- **In *COOL* conditions** *VASOCONSTRICTION* occurs
- capillaries close to skin surface narrow
- less blood flows close to surface of skin
- less heat is lost through skin surface
- skin feels cold and looks pale

(b) Sweating
Sweat glands in the dermis produce a liquid (mainly water and salt) called sweat. This moves up the sweat duct and onto the surface of the skin. It evaporates (changes into a vapour) and cools the skin, because it takes away some of the body heat when it evaporates. (Fig. 4.132). In very hot conditions a person may lose up to 30 litres of water per day as sweat.

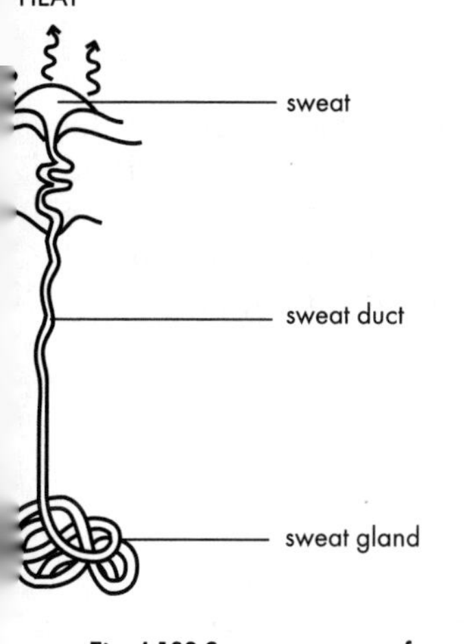

Fig. 4.132 Sweat evaporates from the skin surface

(c) Movement of Skin Hairs
This is not very important in humans as they have relatively few hairs on their skin, but it is important in other mammals e.g. dogs. In warm conditions the hair erector muscles are relaxed, and hairs lie flat against the skin (Fig. 4.133).

In cold conditions the hair erector muscles contract and hairs move upright, trapping a layer of air close to the skin. This makes an insulation layer which prevents heat loss (similar to the effects of wearing clothes).

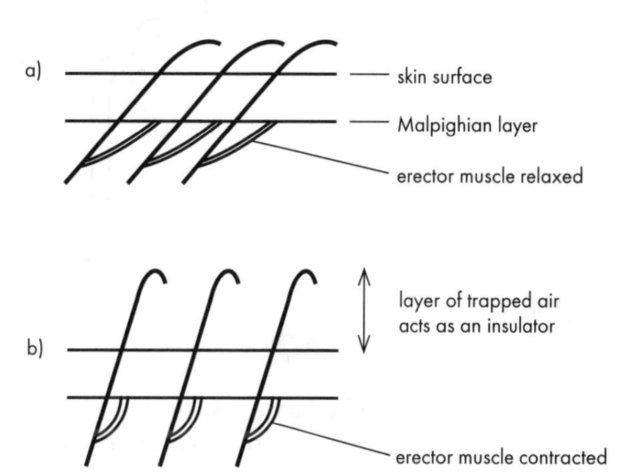

Fig. 4.133 Position of skin hairs (a) in warm conditions (b) in cool conditions

REMEMBER: heat loss is also affected by
(a) surface area : volume ratio (small animals lose heat faster)
(b) thickness of fat layer under skin
} see topic 4.4

ROLE OF HORMONES

Key Points

Level 8

- hormones are chemicals produced by endocrine glands
- water balance in the body is achieved by regulating the amount of water in *urine*
- water balance is controlled by *ADH* (hormone)
- sugar balance in the body is achieved by regulating the amount of sugar converted to glycogen and stored in the *liver* and *muscles*
- sugar balance is controlled by ***insulin*** and ***glucagon*** (hormones)

Water balance (osmoregulation)

The amount of water present in the body at a particular time depends on several factors, e.g. amount ingested in food and drinks, amount lost in sweat (Fig. 4.134).

The kidney regulates the amount of water present in the body by varying the amount of urine produced. This is known as ***osmoregulation*** and is vital for good health.

GAIN (cm^3)		LOSS (cm^3)	
ingested in drinks	1500	lost in urine	1500
ingested in food	700	lost in faeces	100
made inside cells	300	lost as sweat	500
(by respiration)		breathed out	400
TOTAL	*2500*	TOTAL	*2500*

Fig. 4.134 Typical daily water gain and loss (adult human male).

ADH (anti-diuretic hormone) is very important in osmoregulation. It is released by the ***pituitary gland*** (at the base of the brain). It travels all around the body in the blood and affects cells in the second coiled tubule and collecting duct of the ***nephron*** (these are the target organs). It makes these cells more permeable to water, so more water is absorbed from the nephron into the blood, and there is less water in the urine (Fig. 4.135).

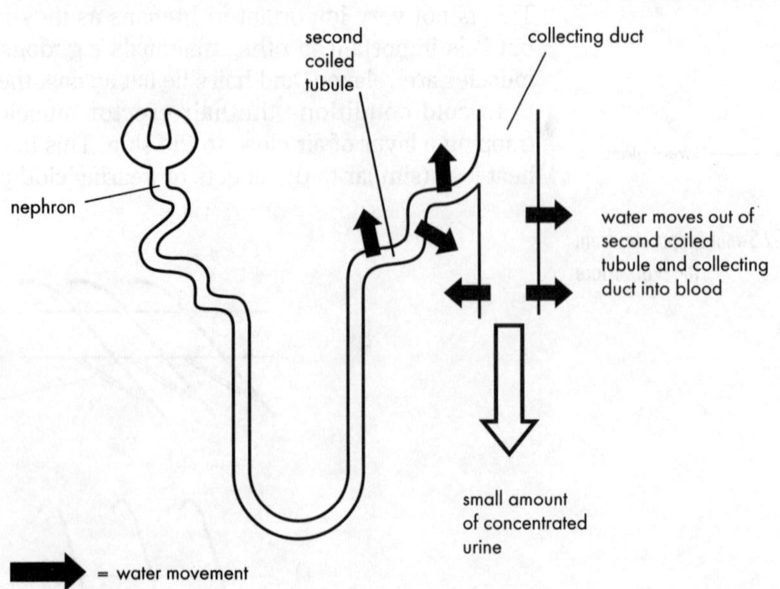

Fig. 4.135 Effect of ADH on the nephron

The pituitary gland produces ADH if a person has ingested very little water or if they have lost a lot in sweat or in faeces.

Sugar balance

The amount of sugar in the blood must be very carefully controlled, because most body cells need a supply of sugar to respire and make energy. When carbohydrate is eaten it is digested, and absorbed through the walls of the ileum into the blood. The amount of sugar in the blood is regulated by 2 hormones, insulin and glucagon.

Insulin

Insulin is made in the *pancreas* and travels all round the body in the blood. It affects cells in the *liver* and *muscle* (these are the target organs), causing them to absorb glucose from the blood and convert it to glycogen. The glycogen is then stored in liver and muscle cells.

Insulin is produced whenever blood sugar levels are too high, e.g. after a meal containing carbohydrate, and it *LOWERS* the blood sugar level.

Glucagon

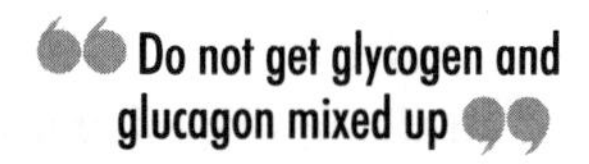

Glucagon is made in the *pancreas* and travels all round the body in the blood. It affects cells in the *liver* (this is the target organ) causing them to convert glycogen into glucose. This glucose then enters the blood and can be used by cells all over the body for respiration.

Glucagon is produced whenever blood sugar levels are too low, e.g. after a long period without food, and it *RAISES* the blood sugar level.

Some people cannot produce enough insulin and are *diabetic*. They are likely to have the following problems:

- they do not have glycogen stores in their liver so must eat small amounts regularly to keep blood sugar levels up
- they cannot store excess sugar (by converting it to glycogen) so this excess sugar is lost in urine
- blood sugar level fluctuates (changes) much more than in a non-diabetic person

Most diabetics can overcome these problems by regulating their diet or by having insulin injections, and can live normal, healthy lives.

HOMEOSTASIS IN THE FOETUS

Key Points

- materials passing between the foetus and the mother must pass through the *placenta*
- the foetus needs *oxygen* for respiration, and *nutrients* so that it can develop properly – it gets these from the mother
- the foetus makes waste products, e.g. *urea* and *carbon dioxide* which would harm it if they built up – it passes these substances to the mother
- the foetus is protected by *amniotic fluid*, and by *antibodies* (chemicals which fight diseases)

While the foetus is developing inside the uterus, it is dependent on its mother for all the things it needs to stay alive.

The foetus is connected to the mother through an organ called the *PLACENTA*; here the mothers blood and the foetal blood are very close together (although they do not mix) and substances can be exchanged between them. Fig. 4.136)

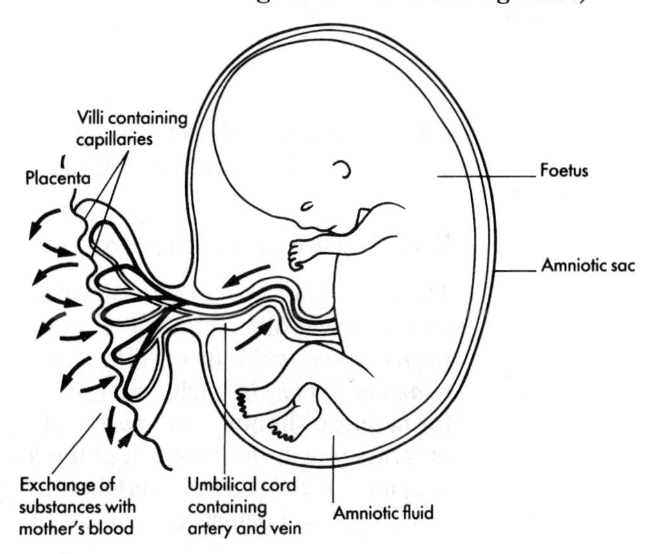

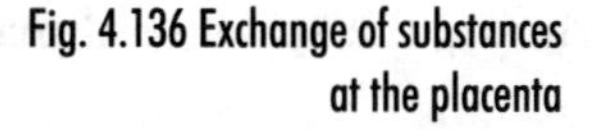
Fig. 4.136 Exchange of substances at the placenta

Protection of the foetus

The foetus is protected in 2 main ways:

(a) it floats inside a bag of fluid (*amniotic fluid*) and is protected from bumps as the mother moves around

(b) *antibodies* can pass across the placenta from the mother's blood into the foetal blood. These are chemicals made by the mother's lymphocytes which can protect the foetus from diseases while it is developing, and for a few months after it is born.

Respiration in the foetus

Respiration is a chemical reaction occuring in all cells of all living organisms to produce energy.

$$\text{Glucose} + \text{oxygen} \longrightarrow \text{energy} + \text{carbon dioxide} + \text{water}$$

The foetus has a high respiration rate as it needs a lot of energy for growth.

Oxygen passes from the mother's blood to the foetal blood for 2 reasons:

(a) there is a very low concentration of oxygen in blood in the umbilical artery, so oxygen diffuses into the foetal blood.

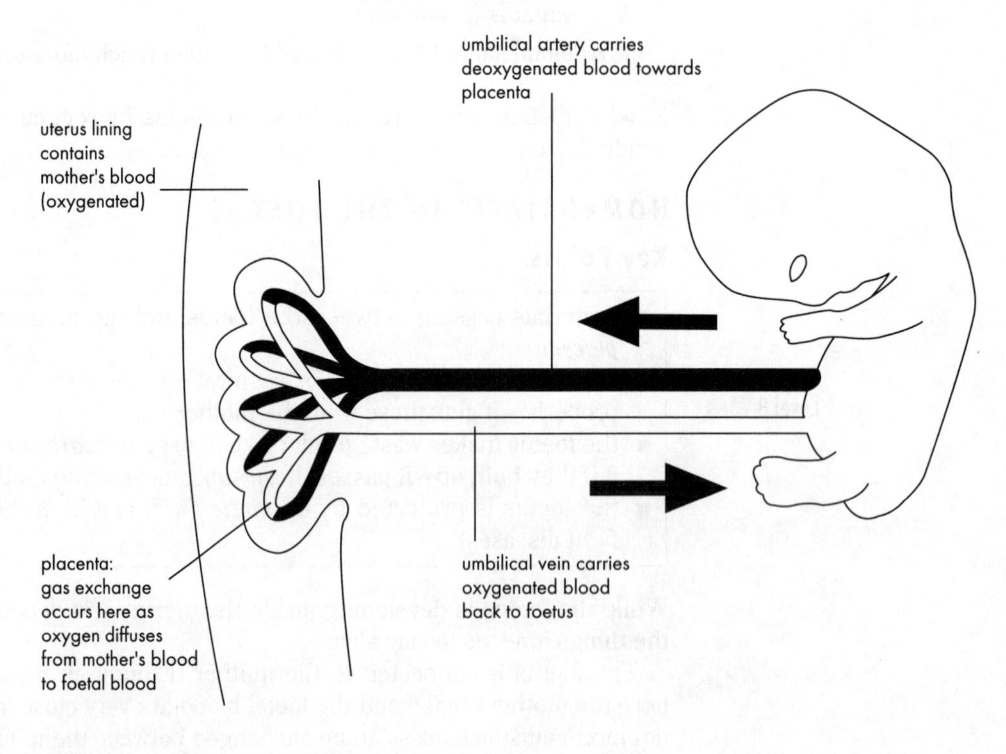

Fig. 4.137 Gas exchange in the foetus

(b) the foetal blood contains a different type of haemoglobin, which combines more readily with oxygen than the mother's haemoglobin does.

Nourishment of the foetus

The foetus needs nutrients in order to grow and to develop properly. All of the following will diffuse from the mother's blood to the foetal blood via the placenta: *sugars, amino acids* (needed to make foetal proteins for growth), *fatty acids, glycerol, vitamins, minerals* (including iron to make red blood cells and calcium needed for formation of bones) and *water*. There is evidence that a deficiency of nutrients prevents proper development of the foetus, e.g. a lack of folic acid (vitamin) is linked to spina bifida (when the vertebrae do not develop properly and the spinal cord may be exposed and damaged).

Removal of waste products

Chemical reactions inside the foetus produce toxic waste products which must be removed, otherwise the foetus will be poisoned. The main waste products are *carbon dioxide* (made in body cells by respiration) and *urea* (made in the liver by deamination of amino acids). These travel in the umbilical artery to the placenta, where they diffuse into the mother's blood, then are excreted from her body.

ABSORPTION OF WATER AND MINERALS BY PLANTS

Key Points

Level 8

- plants absorb water and minerals through their *roots*
- some root cells have root *hairs* which increase their *surface area* and therefore increase the *rate* of uptake
- water moves into roots by *osmosis*
- minerals move into roots by *diffusion* and *active transport*

How substances move in living things

Before we can study this topic, we must consider the 3 ways substances can move in living things.

(a) Diffusion

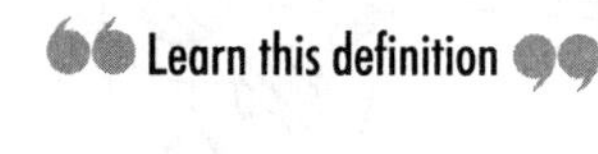
Learn this definition

This is the movement of a substance from a region where it is at a high concentration to a region of lower concentration down a concentration gradient.

The rate of diffusion depends on how steep the concentration gradient is, i.e. the difference in concentration between the 2 regions. It is only effective over short distances, e.g. 1 mm, but is very important in all types of living things
e.g. oxygen diffuses from the alveolus into the blood in humans.
carbon dioxide diffuses from body cells into the blood in humans
carbon dioxide diffuses into plant leaves for photosynthesis.

(b) Osmosis

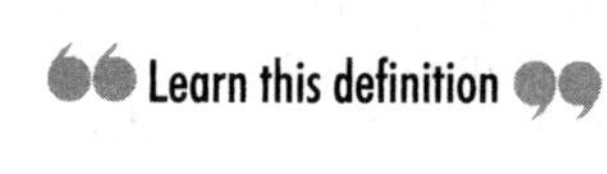
Learn this definition

This is the movement of water molecules from a region of high concentration of water (a dilute solution) to a region of low concentration of water (a concentrated solution) through a semi-permeable membrane. (Fig. 4.138)

Here we have a high concentration of water molecules on the outside and a lower concentration of water molecules on the inside, so water would move IN by osmosis.

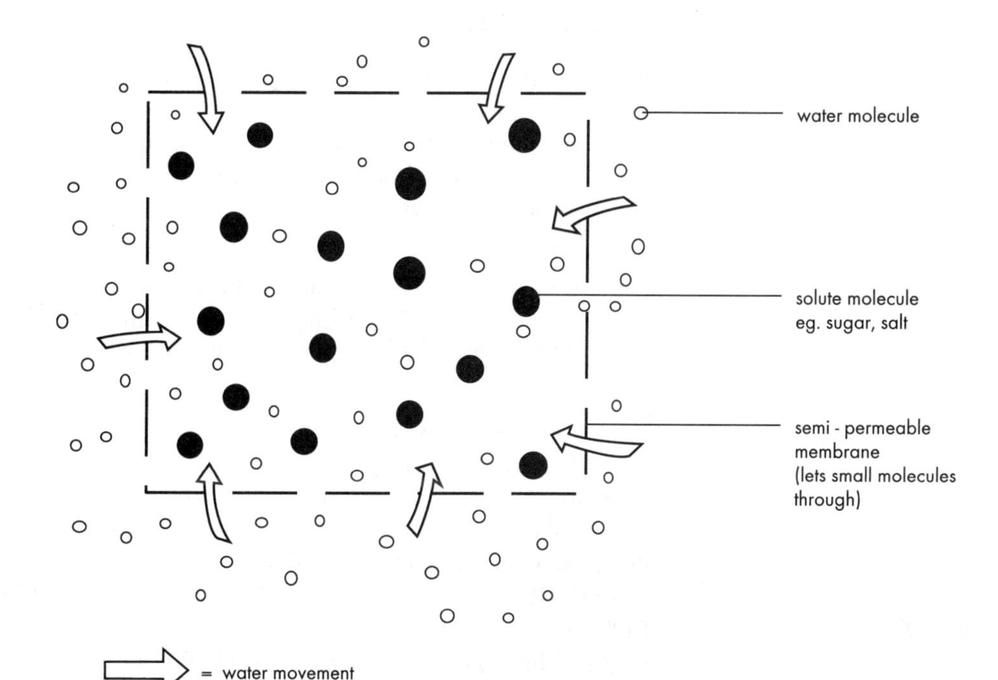

Fig. 4.138 Movement of water by osmosis

(c) Active Transport
This is the movement of a substance across a cell membrane against a concentration gradient, using *energy*. This process can only occur in living cells, and is thought to involve carrier proteins which can transport the substance across the membrane.

How is water absorbed?

Most water is absorbed by the region just behind the root tip, where there are *root hairs* to increase the surface area. There is a film of water around soil particles and the root hairs are in close contact with this.

The cytoplasm of the root hair cells contains a mixture of sugars and salts, so water molecules will move into the cell by *osmosis*. The solution inside these cells then becomes more dilute than that of cells closer to the centre of the root: the result is a *concentration gradient* across the root, causing water to move towards the centre. Eventually it passes into the *xylem vessels* and is transported upwards. (Fig. 4.139)

fertilizer

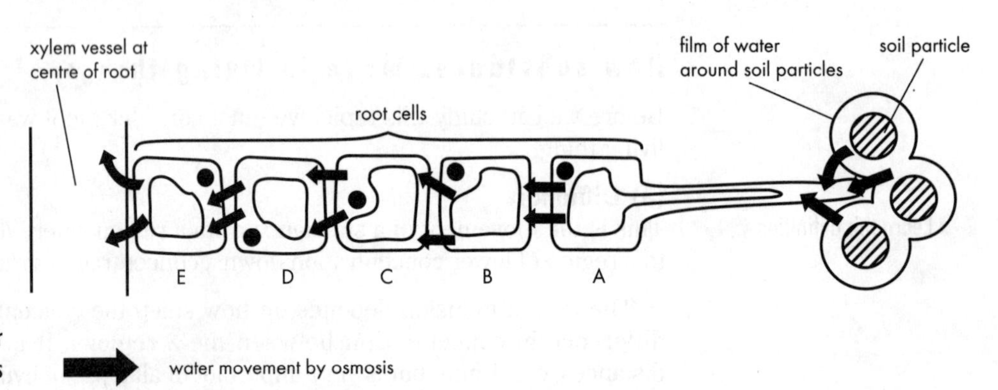

Fig. 4.139 Movement of water through roots by osmosis

Water moves into cell A by osmosis, because there is a higher concentration of water molecules in the soil than in the cytoplasm of cell A. This dilutes the cytoplasm of cell A, so water molecules move to cell B by osmosis. The cytoplasm of cell B is now more dilute, so water molecules move on to cell C. This continues until the water molecules reach the xylem vessels. These are long, narrow tubes, rather like drain-pipes. They are made of columns of dead cells which no longer have cytoplasm or end-walls.

How are minerals absorbed?

Plants need a variety of minerals in order to stay healthy, e.g. nitrate ions, phosphate ions, potassium ions, magnesium ions. These minerals are dissolved in the film of water surrounding the soil particles. Most minerals are absorbed by the region just behind the root tip where there are root hairs to increase the surface area. Minerals are absorbed in 2 ways:

(a) By Diffusion
If there is a higher concentration of a particular mineral ion in the soil than in the cytoplasm of the root cell, then ions will diffuse into the cell. This depends on a concentration gradient being present, and will allow cells to accumulate *small* amounts of minerals.

(b) By Active Transport
Ions are moved across the root cell membrane, into the cell, using carrier proteins. This process uses energy, so root hair cells have lots of mitochondria to generate energy. This allows root cells to accumulate *large* amounts of mineral ions, i.e. does not depend on a concentration gradient.

Once the mineral ions are inside the root cell, they move towards the centre of the root and into the *xylem vessels*.

Minerals are naturally present in soil, but the concentration falls as plants use them up. Extra minerals can be added to the soil by digging in *fertilisers*.

TRANSPORT OF WATER AND MINERALS AND PLANTS

Key Points

Level 8

- tubes called *xylem vessels* carry water to all parts of the plant
- water movement in plants depends on *transpiration* – this is the evaporation of water through the leaf stomata
- transpiration rate depends on temperature, air movement, humidity and time of day
- transpiration rate can be measured using a *potometer*
- minerals are transported in xylem vessels and phloem tubes

Water and minerals move upwards from the roots through xylem vessels to reach all parts of the plant. Xylem vessels are long, narrow tubes which run through the root system and shoot system.

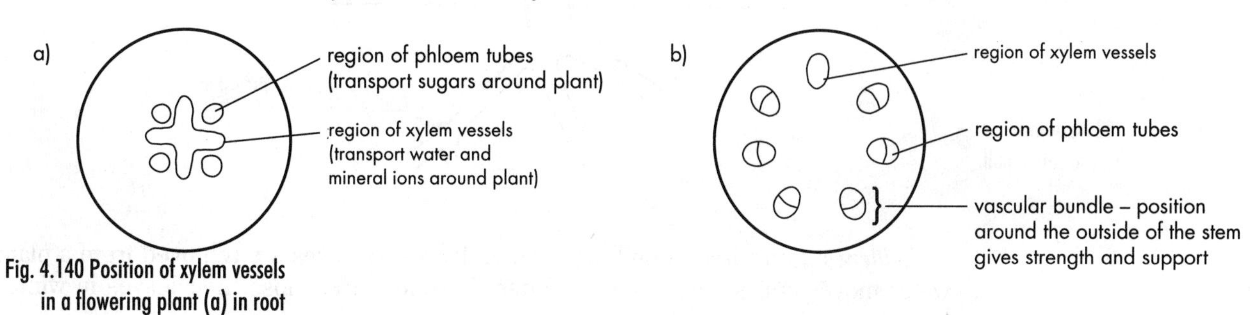

Fig. 4.140 Position of xylem vessels in a flowering plant (a) in root (b) in stem

Transport of water

Water moves up through the xylem vessels due to *transpiration*. Transpiration is the evaporation of water from the leaves of the plant, through stomata. (Fig. 4.141)

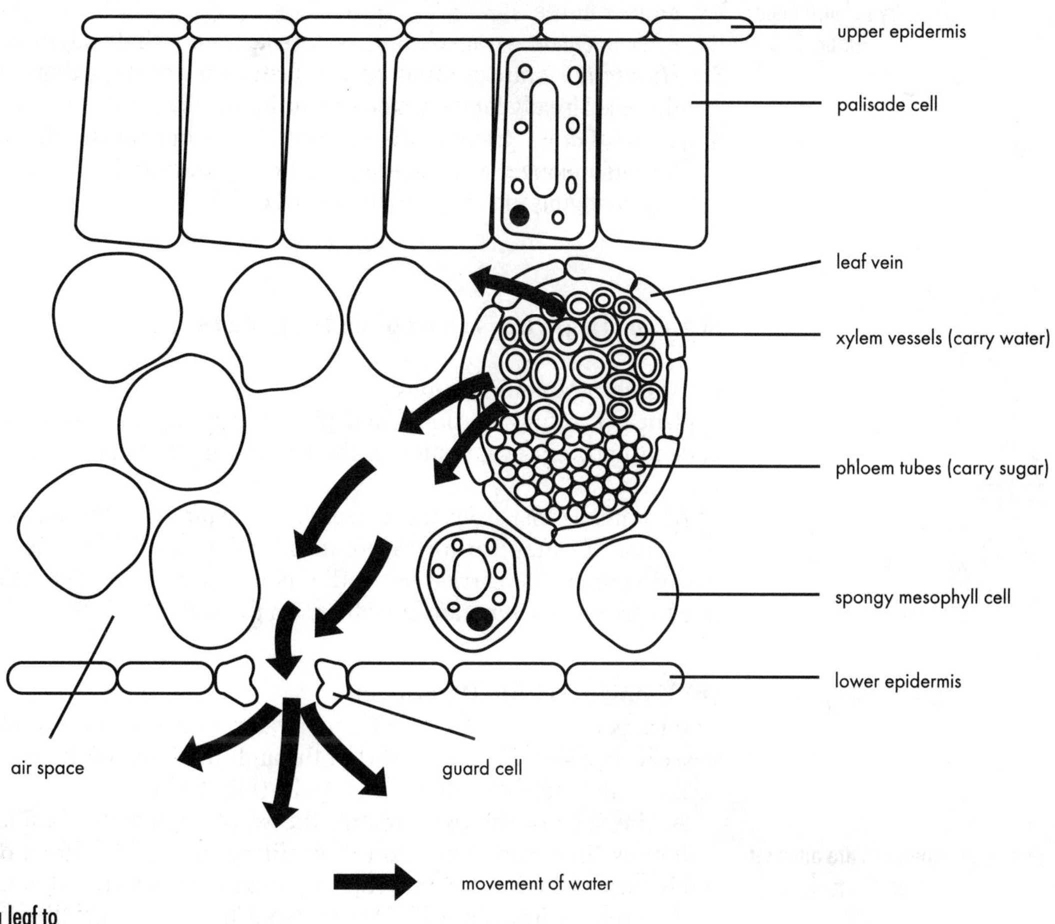

Fig. 4.141 Section through a leaf to show transpiration

In a healthy plant there is an unbroken column of water from the roots, through the stem, up to the leaves; all of this water is in the xylem vessels. As water evaporates from the leaves by transpiration, more water is pulled into the plant, and up through the xylem vessels to take its place. This process is called ***transpiration pull*** (Fig. 4.142).

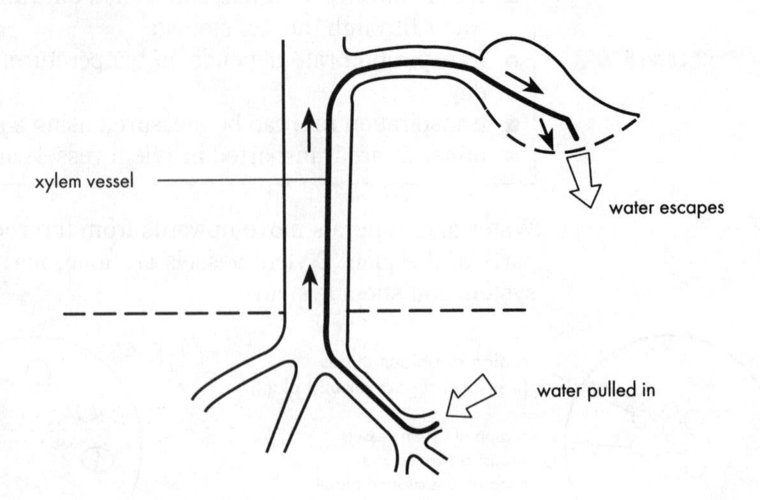

Fig. 4.142 Mechanism of transpiration pull

Evidence: transpiration pull depends on leaves. If leaves are removed from a plant, water movement is much slower. When deciduous trees lose their leaves in winter, there is very little water movement.

Factors affecting the transpiration rate

Try to learn these factors

1. *Temperature* -- transpiration occurs faster at warm temperatures than cool temperatures.
2. *Air movement* – transpiration occurs faster on windy days than on still days.
3. *Humidity* – transpiration occurs faster on dry days than on humid days (when there is already lots of water vapour in the air).
4. *Time of day* – transpiration occurs faster during the day than at night. This is because many plants open their stomata during the day and close them at night (by changing the shape of the guard cells).

Measuring the transpiration rate

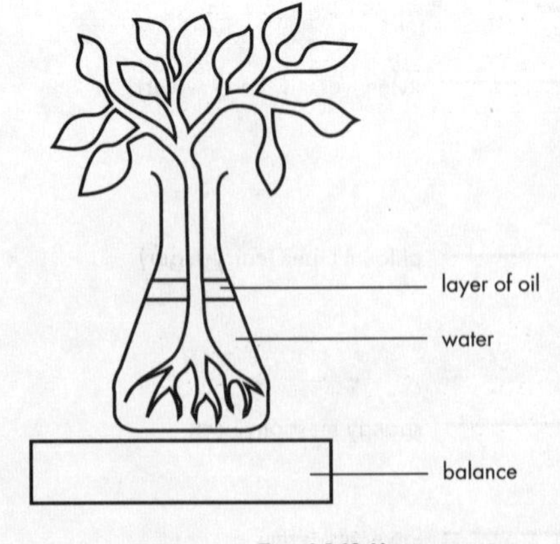

Fig. 4.143 Mass potometer

(a) Using a Mass Potometer

A plant is carefully uprooted and placed in a conical flask of water. A layer of oil is added to prevent evaporation of the water. The mass is measured at regular intervals (Fig. 4.143).

As water is lost from the leaves by transpiration, the mass of the apparatus will decrease. Transpiration rate can be calculated, i.e. mass loss per hour. This experiment can be carried out with a potted plant if the pot is covered in a polythene bag to prevent evaporation of water from the soil.

(b) Using a Bubble Potometer

The roots of a plant are cut off underwater so that no air bubbles get into the xylem vessels. The shoot is then inserted through the bung of the potometer, and the joints sealed with vaseline so that it is air-tight (Fig. 4.144).

Exam questions are often set on this topic

At the start of the experiment, the whole potometer is filled with water. An air bubble is then introduced to the capillary tubing. As the shoot transpires, the air bubble will move through the capillary tubing towards the shoot.

The *rate* of transpiration is calculated by measuring the distance moved by the bubble in a particular time, e.g. 10 minutes.

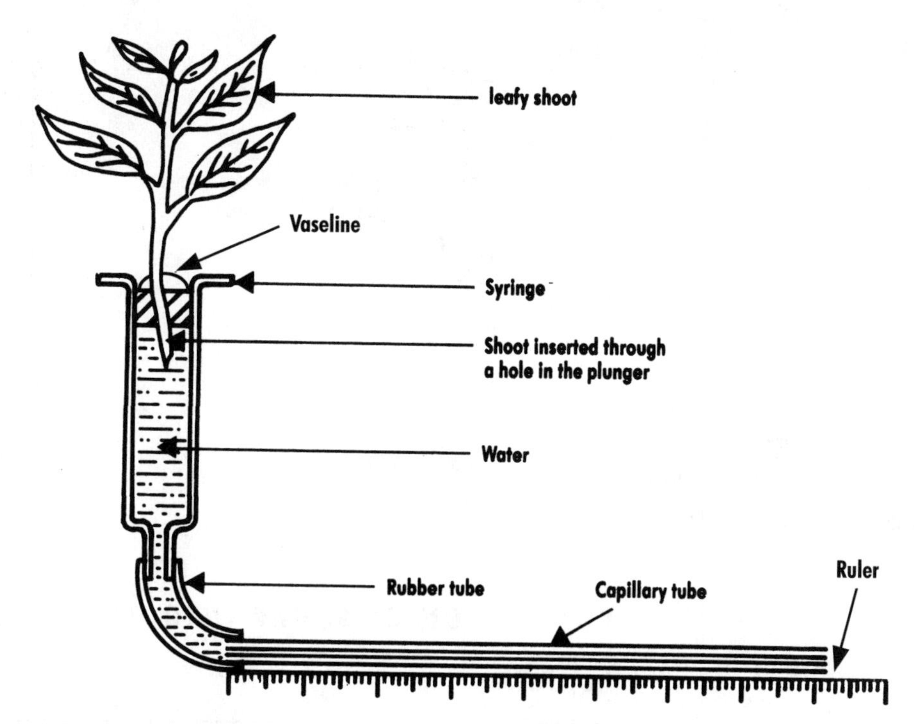

Fig. 4.144 Bubble potometer

TRANSPORT OF MINERALS

Mineral salts move with water because they are dissolved in it. However, they are also transported inside phloem tubes. Minerals are used by cells all over the plant, but especially at the growing points (tip of shoot and root).

TRANSPORT OF SUGAR IN PLANTS

Key Points

- tubes called *phloem tubes* carry sugar around the plant
- it is moved from the leaves (where it is made by photosynthesis) to other parts of the plant
- movement of sugar is called *translocation*

How sugar is moved around

Sugar is made in leaves by the process of photosynthesis. It is moved to other parts of the plant through phloem tubes, e.g. to the roots to be stored, to the tips of the roots and shoots to provide energy for growth. The movement of sugar solution through the phloem tubes is called *translocation*. Phloem tubes are found in the root and shoot (see Fig. 4.140) and are made up of living cells. Translocation is a complex process which seems to require energy.

Evidence for translocation

(a) Ringing Experiments

If a ring of tissue around the outside of the stem, including the phloem tubes, is removed, translocation cannot occur. Sugar solution which would normally move downwards in the phloem tubes accumulates above the ring (Fig. 4.145).

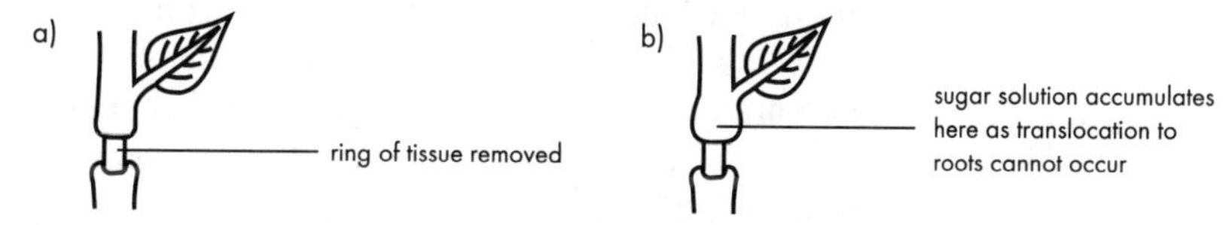

Fig. 4.145 Ringing experiment (a) immediately after removal of ring (b) several days after removal of ring

(b) Radioactive Isotopes

If part of a plant is provided with a radioactive form of carbon dioxide ($^{14}CO_2$), it will make radioactive sugar during photosynthesis. The route followed by this sugar in translocation can be observed by placing the plant on photographic film (the radioactivity causes 'fogging') (Fig. 4.146).

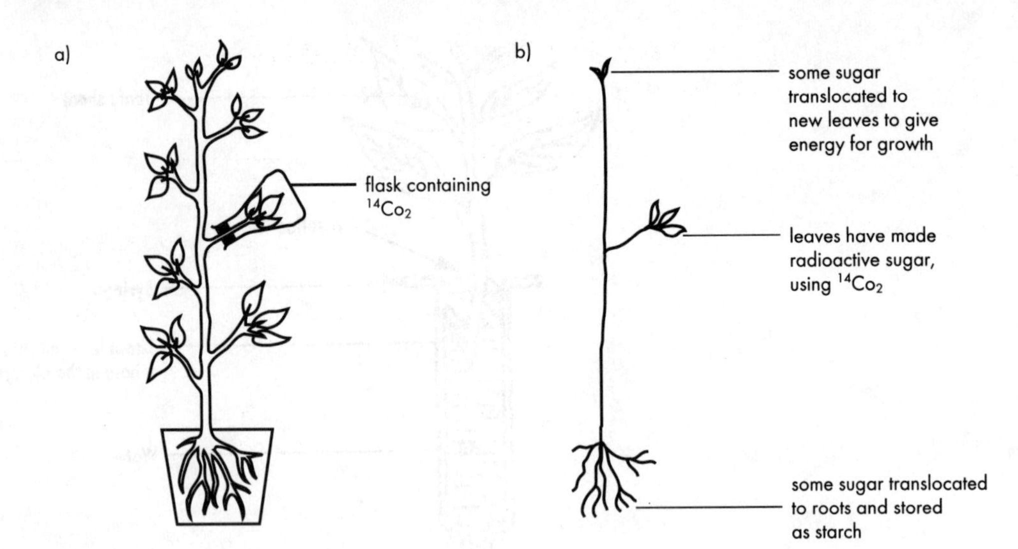

Fig. 4.146 Radioactive isotopes (a) some leaves are provided with $^{14}CO_2$ (b) appearance of photographic film

CONVERSION OF SUGAR INTO OTHER ORGANIC COMPOUNDS

Key Points

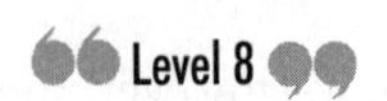

- sugar is used to make *energy*, through respiration
- sugar is changed into starch, and *stored*
- sugar is used to make *cellular components,* e.g. protein, cellulose, chlorophyll

Uses of sugar

The sugar made in photosynthesis can be transported to all parts of the plant by translocation. It can then be used in a variety of ways:

(a) To Make Energy by Respiration
All plant cells need energy to stay alive and to carry out essential processes.

$$\text{sugar} + \text{oxygen} \longrightarrow \text{energy} + \text{carbon dioxide} + \text{water}$$

(b) Changed into Starch for Storage
Plant cells cannot store large quantities of sugar because it affects their osmotic balance. The sugar is converted to starch, then stored in granules in the cytoplasm. Many plants store starch in their roots, leaves and seeds. It can then be converted back to sugar to provide energy for the plant.

(c) Changed to Cellulose
Cellulose is a complicated molecule made up of lots of molecules of sugar joined together in a precise way. All plant cells have cell walls made of cellulose. They are strong and rigid, but they are permeable (let substances pass through). New cellulose is needed wherever plant cells are dividing, i.e. at the growing points of the roots and shoot.

(d) Changed to Protein
Plants can make their own protein from sugar and minerals, especially nitrate. Proteins are needed for many processes within the plant, e.g. to make cell membranes, to make enzymes to control chemical reactions inside cells. If plants can't make protein, e.g. if they are short of nitrate, they cannot grow properly, and are very weak and unhealthy.

(e) Changed to Chlorophyll
Plants make chlorophyll from sugar and minerals, especially magnesium. Chlorophyll is needed to trap light energy for photosynthesis. If plants can't make chlorophyll, they have a yellow appearance (chlorosis) and will not grow properly, because they cannot photosynthesise efficiently.

MECHANISM OF HOMEOSTASIS

Key Points

Level 10

- *homeostasis* means 'controlling the internal environment'
- *negative feedback* is an important process in homeostasis, i.e. if there is a change away from the optimum (ideal) value, a process which will reverse this change occurs
- control of *body temperature, blood sugar levels* and *water balance* are all examples of homeostasis.

The concept of negative feedback is very important

Homeostasis is the process by which an organism controls its internal environment. This is necessary if it is to be independent of its surroundings, and if its body processes are to function efficiently. Most homeostatic mechanisms work by *negative feedback,* i.e. if there is a change away from the optimum value, a process which will reverse this change automatically occurs. There must be a *sensor* to detect the change, and an *effector* to reverse the change.

Maintenance of body temperature

Humans are endothermic, i.e. they maintain a constant body temperature of 37°C, irrespective of the surrounding temperature. This is very important because body temperature affects *metabolic rate,* i.e. the rate of chemical reactions within the body. There is an efficient homeostatic mechanism which regulates body temperature by causing changes in the skin (Fig. 4.147).

Optimum value: 37°C
Sensor: cells in hypothalamus of brain
Effectors: capillaries in skin, hair erector muscles in skin, sweat glands in skin.

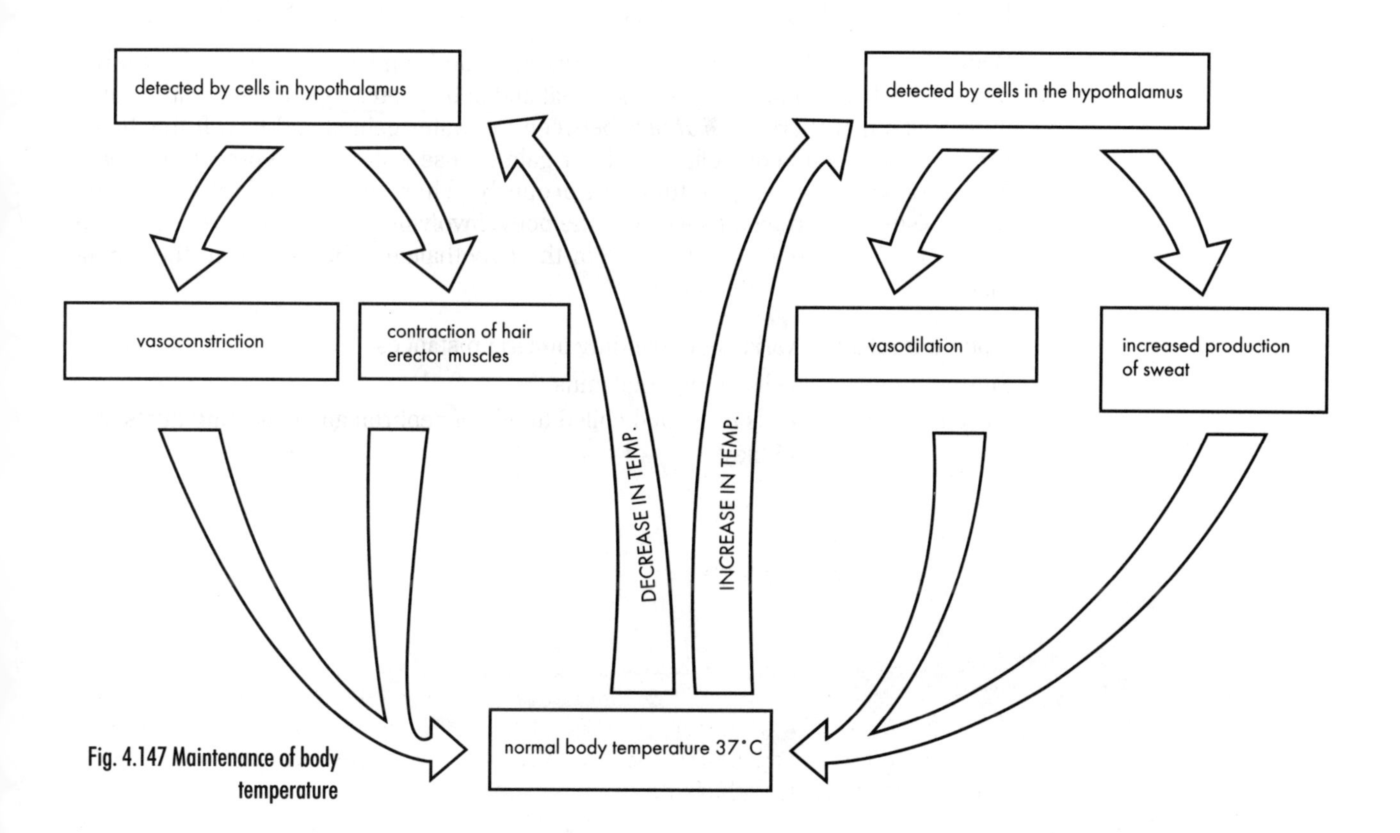

Fig. 4.147 Maintenance of body temperature

Regulation of blood sugar levels

Most body cells need a regular supply of glucose for respiration. If there is not enough glucose in the blood, cells cannot function properly and the person may go into a coma. If there is too much glucose in the blood, some will be lost in urine. There is an efficient homeostatic mechanism which regulates blood sugar level, involving 2 hormones made by the pancreas, *glucagon* and *insulin* (Fig. 4.148).

Optimum value: 90 mg of glucose per 100 ml blood
Sensor: cells in pancreas
Effector: cells in liver and muscle

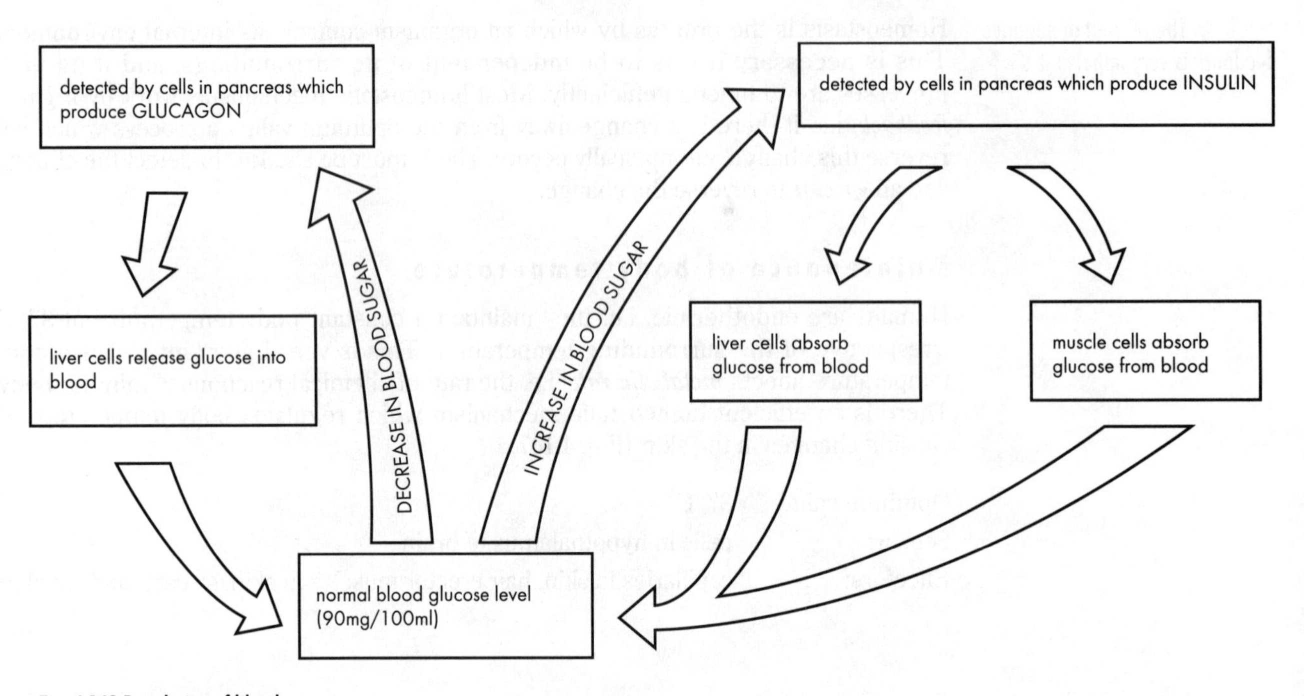

Fig. 4.148 Regulation of blood sugar level

Osmoregulation (regulation of water within the body)

Water is gained by the body in drinks and food, and made within the body by respiration. It is lost in urine, faeces, sweat and in the water vapour we breathe out.

It is vital that there is a *balance* between the water gained and lost. If this balance breaks down, then body cells will either gain or lose water by the process of *osmosis*, and they would no longer function properly. There is an efficient homeostatic mechanism which regulates water in the body, involving *ADH* (anti-diuretic hormone) (Fig. 4.149). This hormone is made in the hypothalamus but stored in the pituitary gland.

Optimum value: variable, depending on circumstances
Sensor: cells in hypothalamus
Effector: cells in second coiled tubule of nephron and collecting ducts of kidney

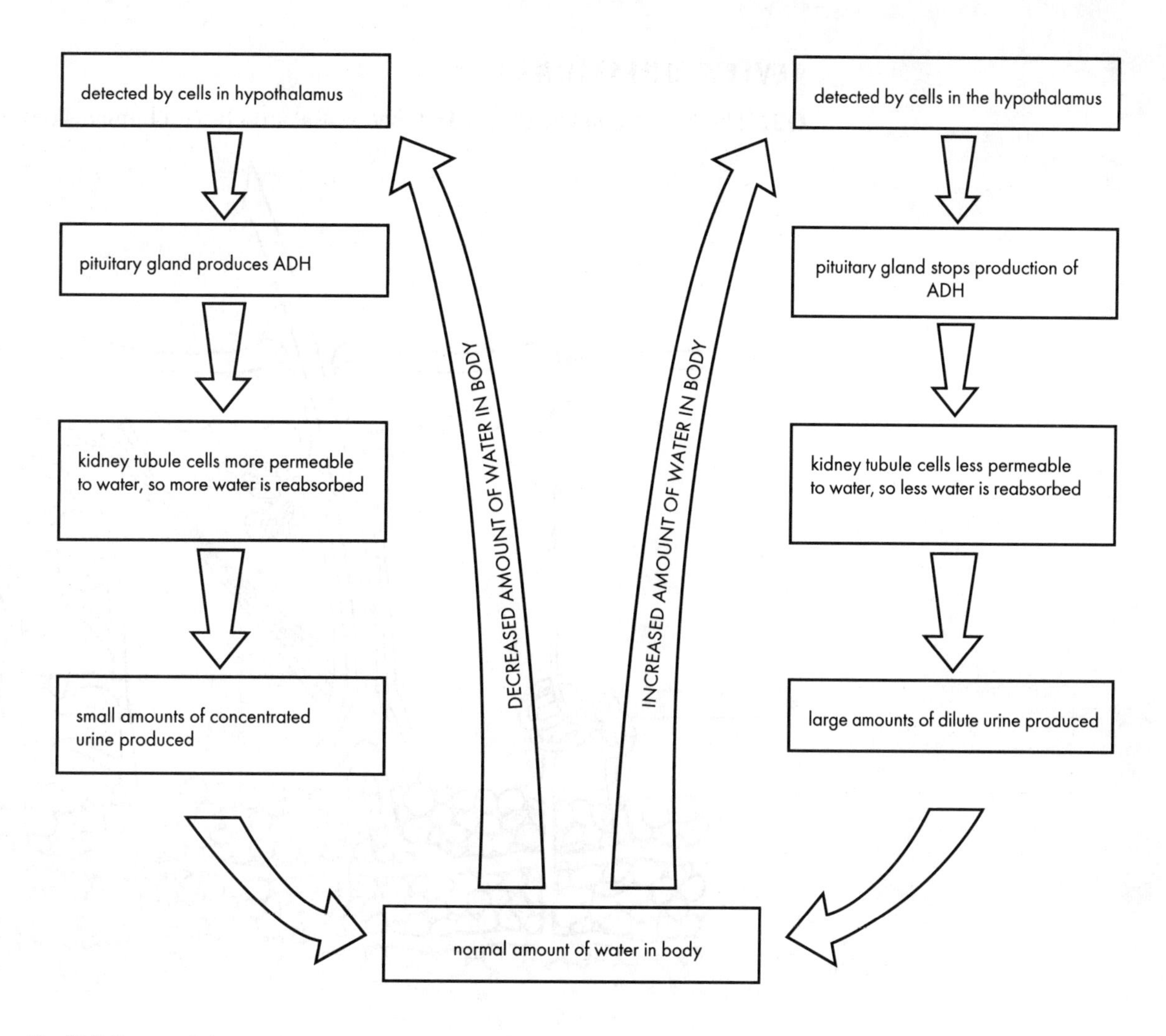

Fig. 4.149 Osmoregulation

Links with other topics

Related topics elsewhere in this book are as follows:

Topic	Chapter and Topic number	Page number
Osmosis	3.5	26
Cells	4.3	36
Survival in a natural habitat	4.4	43
Reproduction	4.5	67
Excretion	4.5	76
Photosynthesis	4.6	105
Hormonal system	4.8	152

REVIEW QUESTIONS

Q22 The diagram below (Fig. 4.150) shows the structure of human skin.

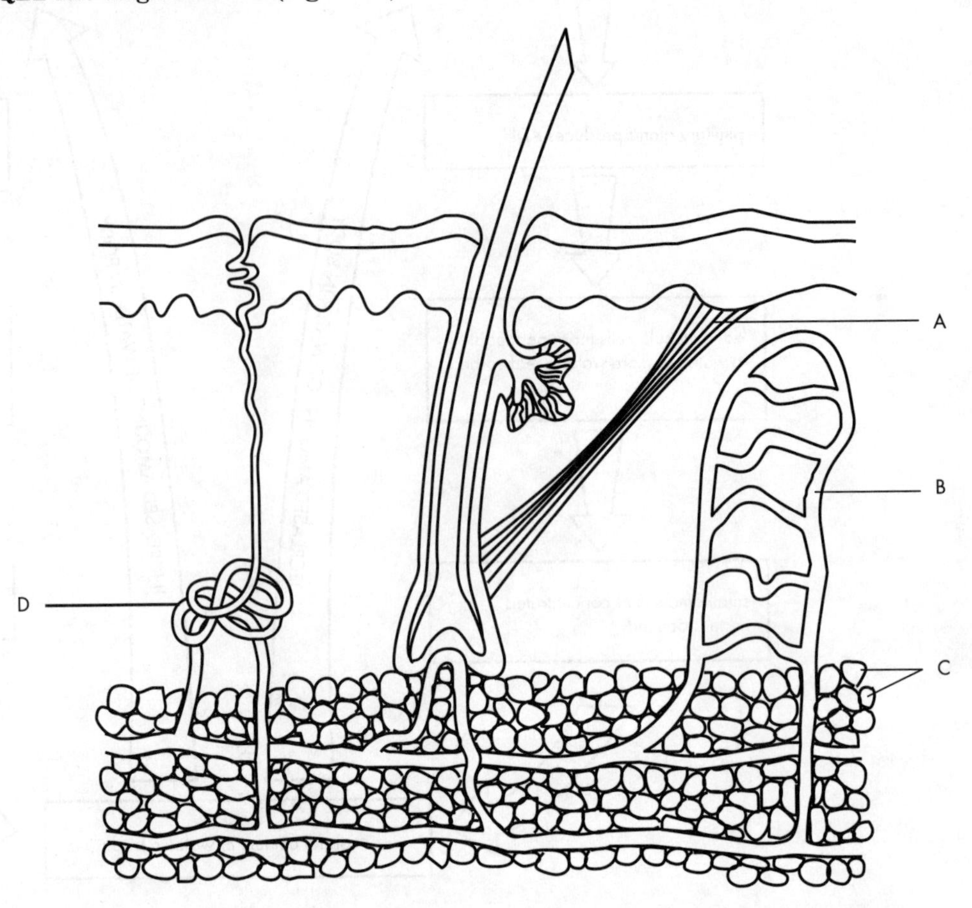

Fig. 4.150

(a) (i) Complete the table below by naming the parts **A**, **B** and **C** and then say how each part helps us keep warm in cold conditions.

	Name of part	How it helps us keep warm
A		
B		
C		

(ii) Name part **D**.

(1)

(iii) How does part **D** react to a fall in body temperature?

(1)

(b) The skin helps with homeostasis. What is homeostasis?

(1)

(c) (i) What is hypothermia?

___

___

(1)

(ii) Give one biological reason why hypothermia mainly affects old people.

___

___

(1)

(iii) State one thing which should be done to treat someone suffering from hypothermia.

___

___

(1)

Q23 The diagram below (Fig. 4.151) shows a human embryo developing in the uterus.

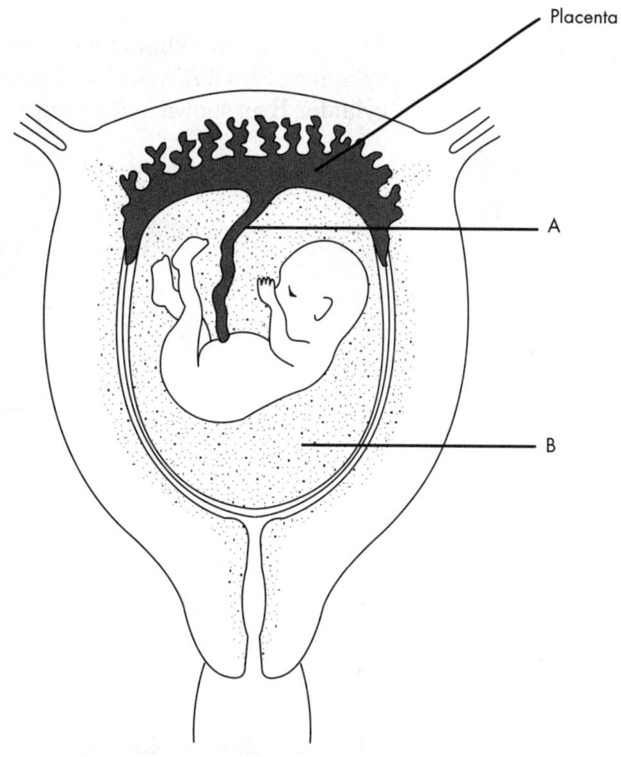

Fig. 4.151

(a) Name the parts labelled **A** and **B**.

A ___

B ___

(2)

(b) (i) Name *two* substances needed by the embryo which pass from the mother through the placenta.

1. ___

2. ___

(2)

(ii) Name one substance excreted by the embryo which passes through the placenta to the mother.

(1)

(c) Blood vessels in structure **A** carry blood to and from the placenta. In which direction is the blood flowing at a higher pressure? Give one reason for your answer.

(1)

Q24 A student wanted to find out which surface of a leaf lost water by transpiration more quickly. The student cut two similar shoots, **A** and **B**, from the same plant. The upper surface of the leaves on shoot **A** and the lower surface of the leaves on shoot **B** were covered with vaseline.

The ends of the shoots were each placed in 100 cm^3 of water in measuring cylinders. Shoot **A** was placed in measuring cylinder **A** and shoot **B** in measuring cylinder **B** as shown in the diagram below.

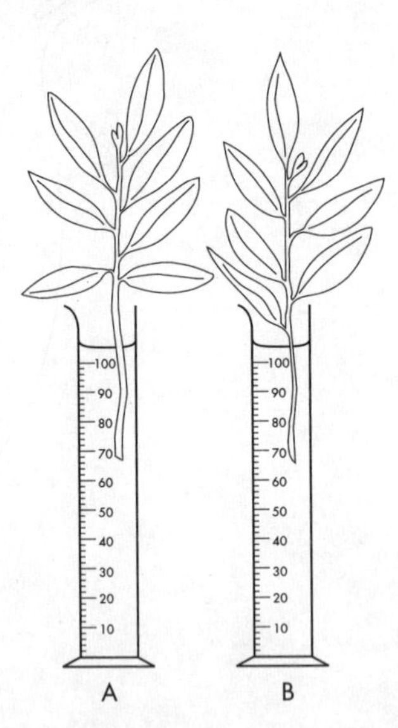

Fig. 4.152

After standing for 48 hours on a laboratory bench the student removed the shoots from the measuring cylinders which are shown below.

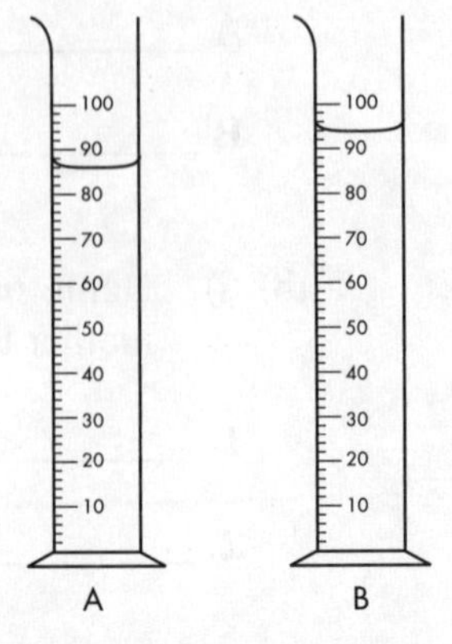

Fig. 4.153

(a) Complete the table below using information from the diagrams.

Volume of water in cm^3	Measuring cylinder	
	A	B
Left in cylinder		
Removed by plant		

(4)

(b) (i) The student used two shoots from the same plant. This means that the leaves were of the same type (species).

Write down one other way in which the shoots should have been similar.

____

____

(1)

(ii) Why were the shoots removed from the cylinders before the final volumes were measured?

____

____

(1)

(iii) Which surface of the leaves lost more water?

____

(1)

(iv) It was suggested that evaporation had taken place from the water surface in the cylinders. How might this have been prevented in the investigation?

____

____

(1)

(v) Write down one assumption which has been made in the above investigation.

____

____

(1)

(c) A number of different environmental factors can affect the rate of transpiration. Suggest a simple way of altering this investigation to test the effect of the dampness of the air on the rate of water loss in a similar shoot.

____

____

____

____

(1)

Q25 Three strips of plant epidermal cells were placed on microscope slides. Each was covered by one drop of a solution, as shown in the table blow.

Strip 1	Strip 2	Strip 3
Distilled water	Dilute sugar solution	Concentrated sugar solution

The diagrams below show the appearance of some of the cells after one hour. Two of the strips had cells as shown in diagram **A**. One strip had cells as shown in diagram **B**.

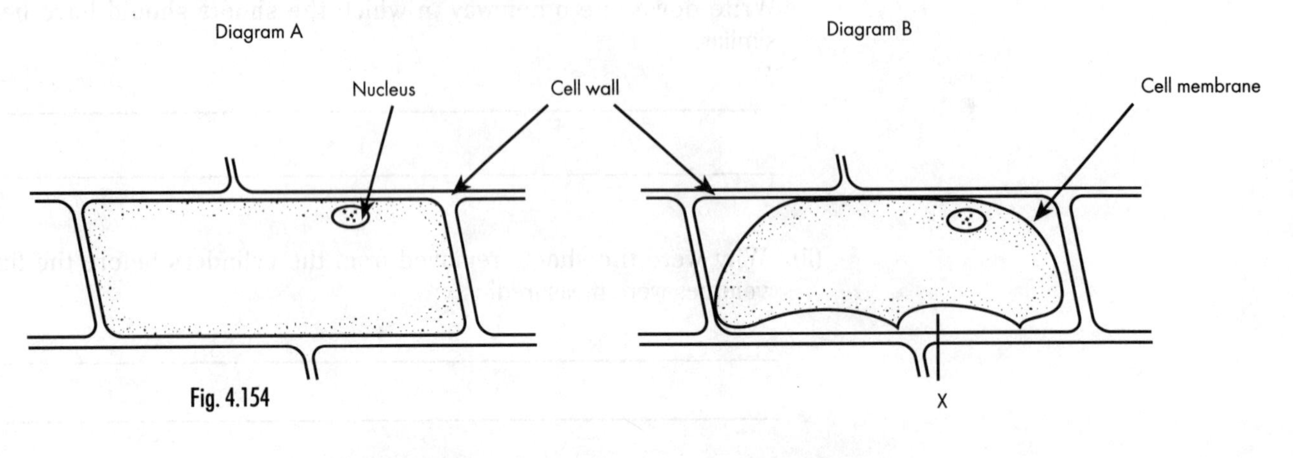

Fig. 4.154

(a) (i) Which *one* of the strips 1, 2 or 3 would have cells that look like diagram **B**?

____________________ (1)

(ii) Explain what happens to a cell to make it look like diagram **B**.

____

____

____________________ (1)

(b) Which one of the following would be found at the position labelled **X** in cell **B**? Tick the correct answer.

Water alone ☐ Sugar solution ☐ Cell sap ☐

Air ☐ Cytoplasm ☐

(1)

(c) What would happen to a leaf if most of its cells became like diagram **B**?

____________________ (1)

Q26 An experiment was carried out to investigate the movement of water in a plant. A stick of celery was placed in a beaker of water containing a dye. After two hours, the celery was washed. Sections were cut at different distances from the base of the celery. The parts of the celery containing dye show the path taken by water. This experiment is shown in diagram form below.

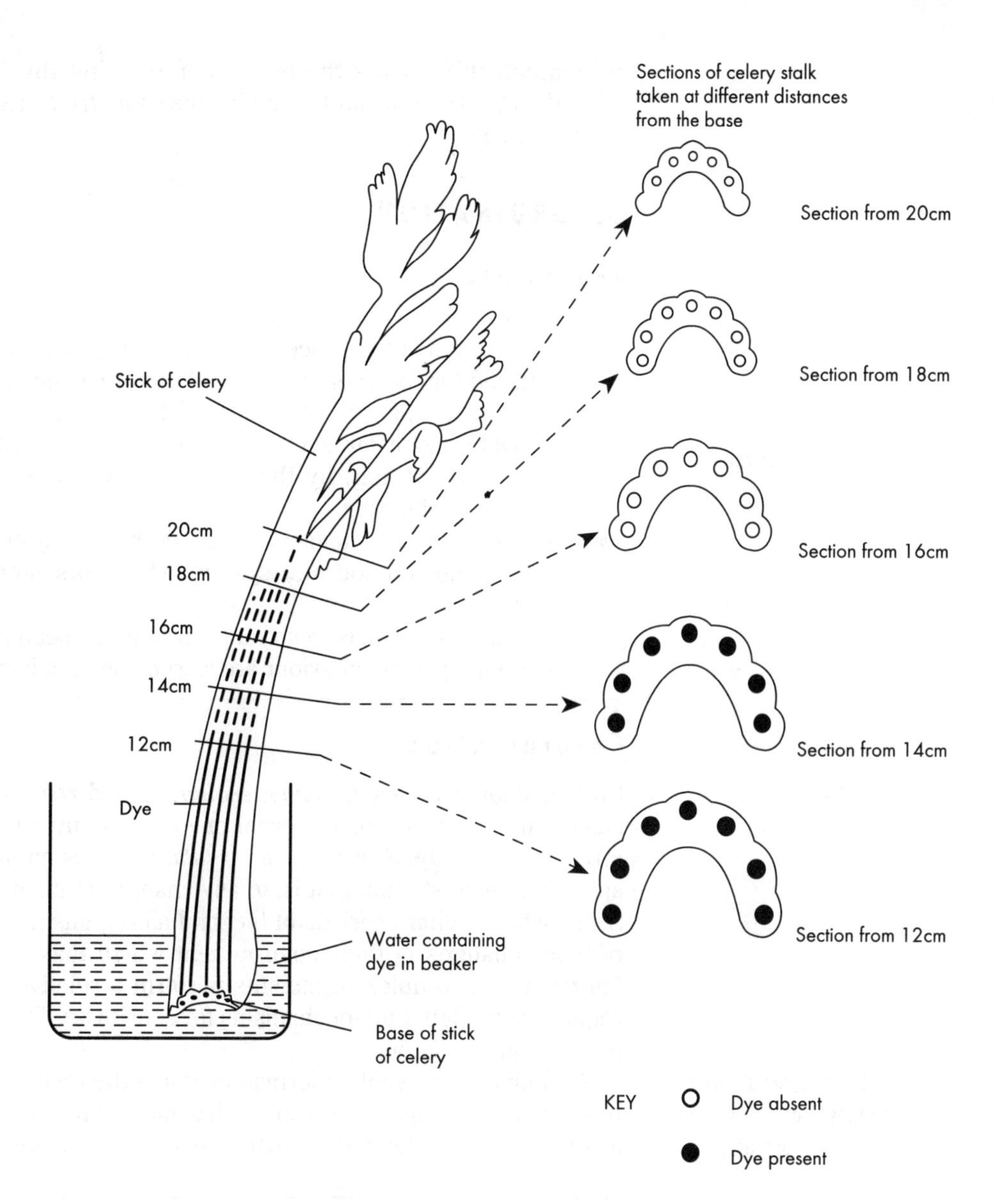

Fig. 4.155

(i) How far had the dye travelled up the celery stalk in this experiment?

____

(ii) Calculate the speed of movement of the dye in centimetres per hour. Show your working.

Answer _______________

8 CO-ORDINATION IN MAMMALS

Co-ordination is the process by which an organism's activities are organised to increase its chances of survival. This is especially important in large, complex organisms which are capable of carrying out many simultaneous processes in specialised cells and tissues. An individual process in one part of the organism cannot occur independently of other processes in the same body. These various activities need to be co-ordinated so that they remain appropriate for the organism's overall needs.

In this Topic Group, coordination is described for the two main 'networks' in a mammal's body, the ***nervous system*** and the ***hormonal (endocrine) system***. Both systems involve *detection* and *response* to a stimulus, from either outside or inside the animal. The type of response which is produced is controlled by the central nervous system, which includes the brain. Before using this Topic group, you may find it helpful to look again at Topic Groups *Sensitivity* (Section 4.5) and *Homeostasis* (Section 4.7.)

CO-ORDINATION

Key Points

Level 9

- Co-ordination is the process by which the different activities of an organism are organised for the overall benefit of the organism. Co-ordination involves several linked processes. Firstly, information about the changing environment is *detected* by a *receptor*, such as sensory cells, or a sense organ. The information is then *processed*, e.g. by the central nervous system. Lastly, an appropriate response is triggered.
- Co-ordination in humans is organised by two systems; the nervous system and the hormone (endocrine) system. The two systems operate in very different ways.
- Responses of animals to stimuli can often be seen as changes in behaviour. The two main types of behaviour are learned and instinctive.

Co-ordination

Co-ordination involves *detecting, evaluating* and *responding* to stimuli. A *stimulus* is a change in the organism's internal or external environment. Being 'aware' of stimuli involves sensitivity. *Sensitivity* or *irritability* allows an organism to detect and respond appropriately to stimuli, that is, to any changes in its internal or external environment. Sensitivity is a characteristic of life, and all organisms are capable of some awareness of those changes in their environment which may affect their chances of survival. Larger, more complex organisms perform a relatively wide range of activities and require more elaborate mechanisms for sensitivity. The basic mechanism of sensitivity is shown in Fig. 4.156.

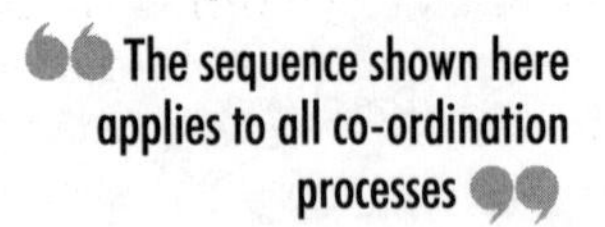

A stimulus represents information which the receptor converts into a form that can be relayed within the organism, for instance by the release of a chemical message, i.e. a hormone, or an electrochemical message, i.e. a nerve impulse.

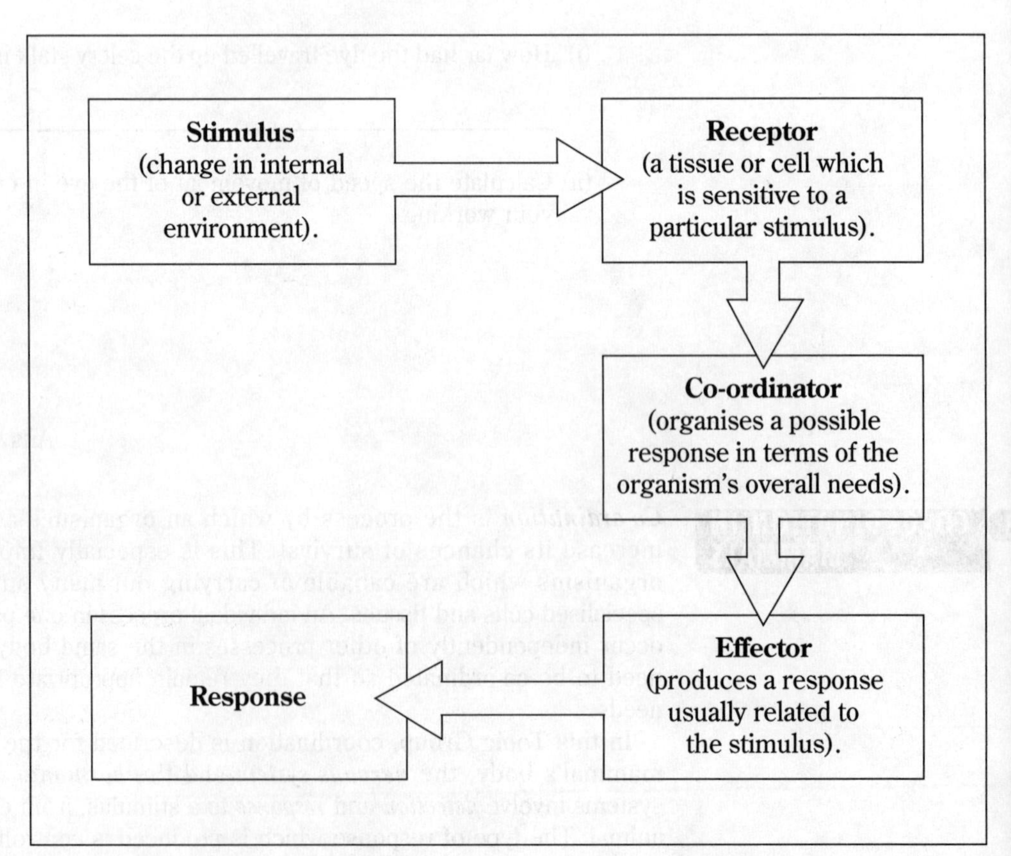

Fig. 4.156 The mechanism of sensitivity

Co-ordination is especially important in *multicellular*, complex organisms, within which different tissues can communicate with each other.

Simpler animals respond to relatively few stimuli and the range of possible responses is fairly limited. More complex animals tend to have specialised tissues for sensitivity, for instance including sense organs and a relatively developed co-ordinating system, for example involving a brain.

The two main systems which have a direct role in sensitivity in higher animals are the *nervous system* and the *endocrine (hormone) system*. These operate together but perform their overall function of sensitivity in different ways. The nervous and endocrine systems are compared in Fig. 4.157.

Comparison	*Nervous system*	*Endocrine system*
Message	Nerve impulse	Hormones
Route	Nervous system	Blood system
Transmission rate	Rapid	Slow, depends on circulation
Origin of message	Receptor (sense organ)	Endocrine gland
Destination of message	Effector (muscle)	Target organ or organs
Speed and duration of effect	Immediate, brief	Delayed, prolonged

Fig. 4.157 Comparison of nervous and endocrine systems

Behaviour

The observable outcome of all the co-ordinating processes of an organism is called *behaviour*. The range of behaviour that is expressed by an organism depends on the range of stimuli that it can respond to. The type of behaviour may depend on the direction of the stimulus, and the response may be by part of, or by the whole organism. The outcome of all behaviour is often to increase an organism's chances of survival. Patterns of behaviour are characteristic of different species but also vary between individuals within a species. Plants show much less obvious behaviour than animals; animals generally show relatively elaborate behaviour. Behaviour is especially complex in social animals, including insects such as ants and bees, and also in more advanced animals such as birds and mammals.

There are two main types of behaviour, *instinctive (innate)* and *learned (conditioned)*. Instinctive behaviour is genetically determined and therefore is inherited. This increases an organism's chances of survival without the need to learn behaviour during the early, critical period of development. Examples of instinctive behaviour include the simple reflex, courtship, mating and territorial behaviour and tactic response. *Tactic response (taxes)* involve a movement of the whole organism in response to a directional stimulus. For example, woodlice may move towards regions of high moisture or low light intensity within a choice chamber (Fig. 4.158).

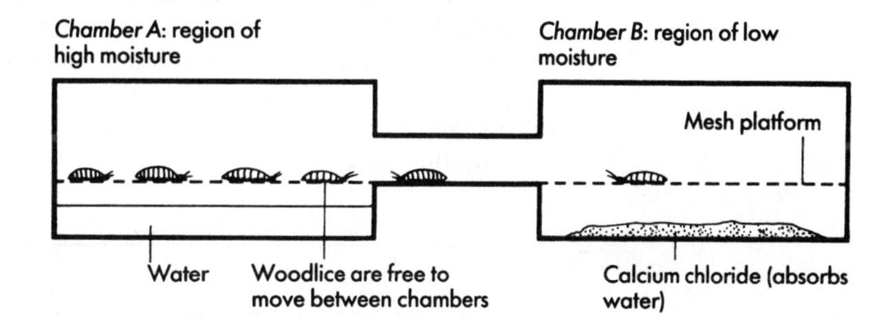

Fig. 4.158 An example of instinctive (innate) behaviour: the tactic response of woodlice (*Oniscus*) to a moisture gradient within a choice chamber

Links with other topics

Links with related topics in this book are as follows:

Topic	Chapter and Topic number	Page number
Sensitivity	4.5	90
The nervous system	4.8	149
The hormonal system	4.8	152

REVIEW QUESTIONS

Q27 Humans are sensitive to a variety of conditions. A change in the conditions surrounding the human body produces responses from the hormonal and nervous systems.

(a) How are hormones transported from where they are made to the place where they act?

(b) In the table (Fig. 4.159) below, describe three differences between the transmission of messages by the nervous system and the transmission of messages by the hormonal system.

	nervous system	hormonal system
1.		
2.		
3.		

Fig. 4.159

When a pupil accidentally touched a hot object, she took her hand away quickly. This is an example of a spinal reflex. The diagram (Fig. 4.160) shows the components of a simple reflex arc.

(c) Explain how this reflex arc is able to produce a quick response to the contact with the hot surface.

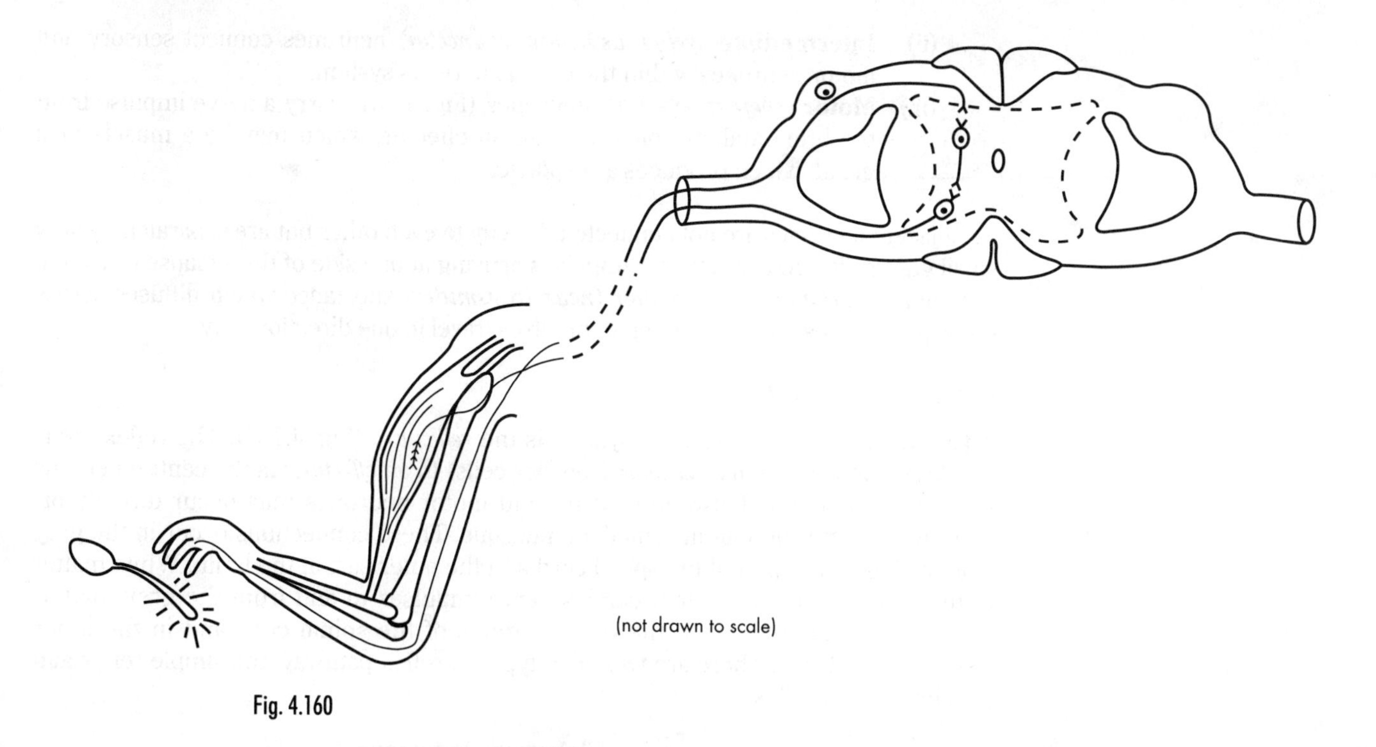

Fig. 4.160

THE NERVOUS SYSTEM

Key Points

Level 9

- The nervous system is composed of millions of *nerve cells*, or *neurones*. They are responsible for: (a) transmitting nerve impulses from sensory tissue in response to a *stimulus*, (b) relaying the information through the *central nervous system*, and (c) passing the nerve impulse to tissues which produce a *response*. This sequence of neurones is called the *reflex arc*.
- A *simple reflex arc* does not involve the brain in 'approving' a response to a stimulus. A *conditioned reflex* is one in which the brain decides on the type of response, if any, which is produced.

Nerve cells

The *nervous system* in higher animals consists of two main parts: the *central nervous system* (CNS), which is composed of the brain and spinal cord, and the *peripheral nervous system* (PNS), which includes all other nerves.

The structural unit of the nervous system is the nerve cell, or *neurone* (see Fig. 4.161). Neurones are adapted for the transmission of nerve impulses, for instance by being elongated. Groups of neurones are arranged together in bundles, known as nerves.

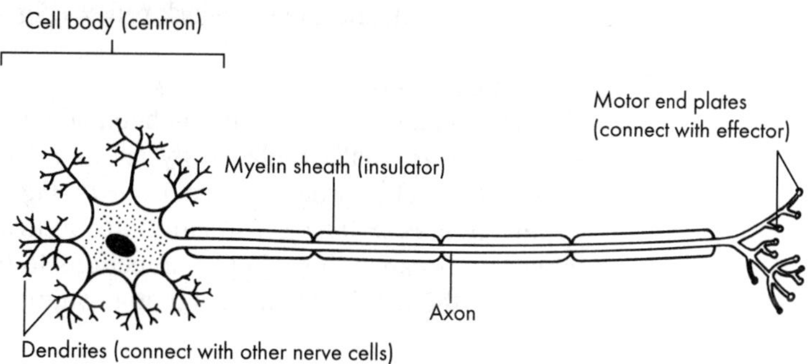

Fig. 4.161 Structure of a motor neurone

There are three main types of neurones:

(i) **Sensory** (*afferent*) neurones carry a nerve impulse from a receptor (sensory cells) to the spinal cord and/or brain.

(ii) **Intermediate** (*relay, associate, connector*) neurones connect sensory and motor neurones within the central nervous system.

(iii) **Motor** (*efferent, effector*), neurones (Fig. 4.161) carry a nerve impulse from the brain and/or spinal cord to an effector, which may be a muscle or a gland, which produces a response.

Adjacent neurones are not connected directly to each other but are separated by very small gaps called *synapses*. Nerve impulses arriving at one side of the synapse cause the secretion of a *chemical transmitter* (*neurotransmitter*) substance which diffuses across the gap. Synapses ensure that nerve impulses travel in one direction only.

Nerve pathways

A functional unit of the nervous system is the *reflex arc* (Fig. 4.162). The reflex arc is the direct pathway from a *receptor* (sensory cells) to an *effector*, via the central nervous system. Connections between sensory and motor neurones may occur directly or, more usually, through an intermediate neurone. These connections occur in the grey matter of the inner part of the spinal cord and the outer part of the brain. White matter in the central nervous system carries nerve impulses to and from the grey matter. White matter is situated in the outer regions of the spinal cord and in the inner regions of the brain. There are two main types of reflex pathway; the simple reflex and the conditioned reflex.

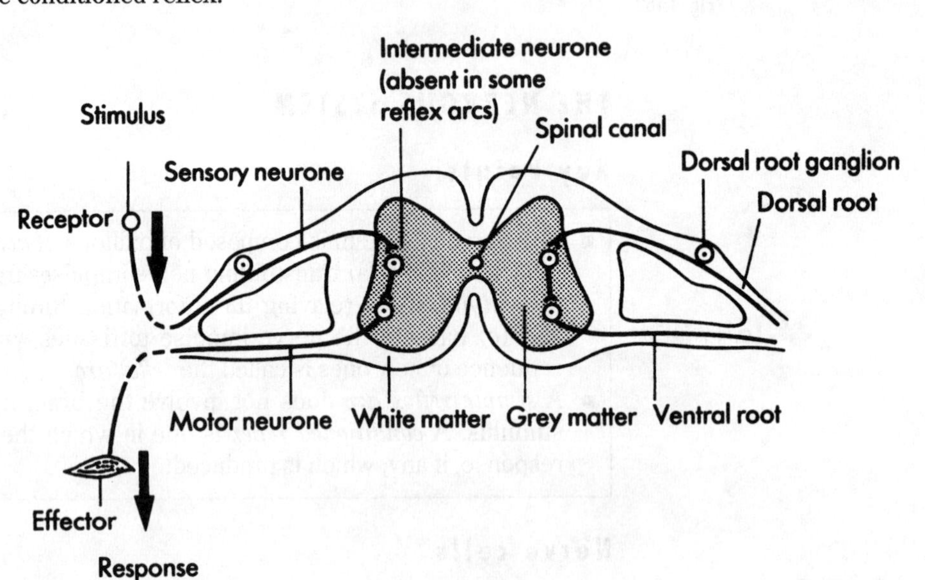

Fig. 4.162 The reflex arc: summary of main components

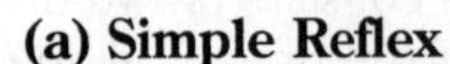

(a) Simple Reflex

This involves a pathway consisting of the reflex arc only. The pathway may involve the brain (cranial reflex) or the spinal cord (spinal reflex). A simple reflex results in a very rapid, involuntary (automatic) response to a stimulus. Simple reflexes are instinctive and often have a survival value, for instance in allowing the animal to escape danger. Examples of reflexes include eye blinking and eye focusing, withdrawing a limb from a source of pain, and the knee-jerk reflex (Fig. 4.163).

“Exam questions on the reflex frequency occur.”

(b) Conditioned Reflex

This is a simple reflex which has been adapted by experience. The original (primary) stimulus is gradually replaced by another (secondary) stimulus which produces the same effect. This process is called *learning* and involves the brain which can store experience as *memory*. Intermediate neurones in the brain and spinal cord are important because they allow connections to be made with many other neurones; this increases the range of activities that an animal can perform.

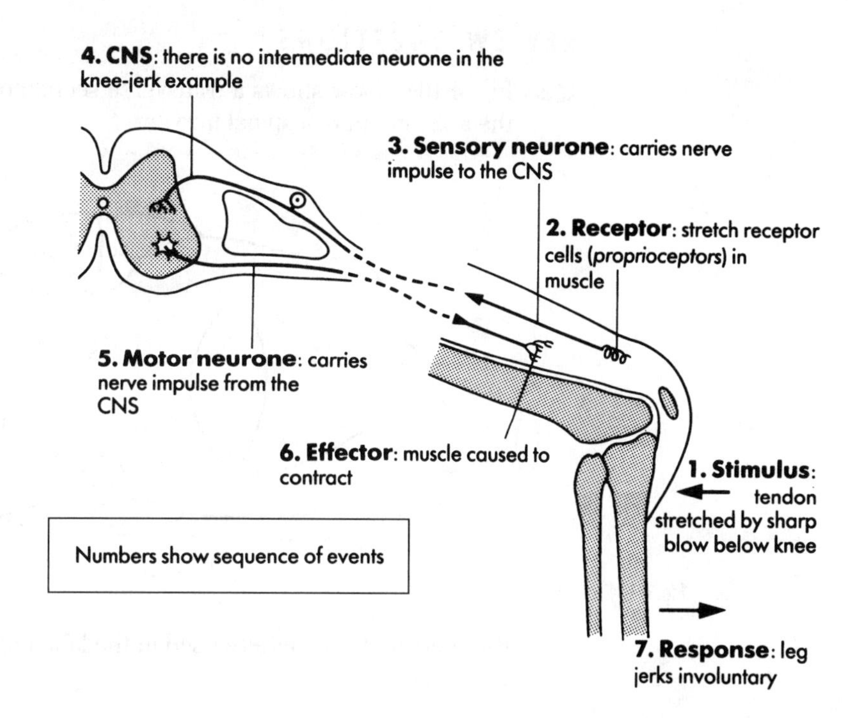

Fig. 4.163 Simple reflex: the knee jerk

Practical: reaction time

A 1 metre ruler is held by the experimenter, vertically above and in front of the subject, who should be ready to catch the ruler between finger and thumb when it is released by the experimenter. The experimenter releases the ruler without warning. The subject should detect a movement in the ruler either (i) by sight or (ii) by touch. The distance (in cm) the ruler falls before it is caught can be used as an indication of 'reaction time'. If the subject's hand was at 100 cm at the start of the experiment, a high distance value indicates a fast reaction time.

The experiment can be repeated to investigate any of the following conditions:

- the difference in reaction time for an individual subject when using sight and touch
- the difference in reaction time for an individual subject over several attempts
- the difference in reaction time for different individuals using the same experimental conditions.

Links with other topics

Links with related topics elsewhere in the book are as follows:

Topic	Chapter and Topic number	Page number
Sensitivity	4.5	90
Coordination	4.8	149
The hormonal system	4.8	152

REVIEW QUESTIONS

Q28 Fig. 4.164 below shows a transverse section of the spinal cord of a mammal and the attachment of a spinal nerve.

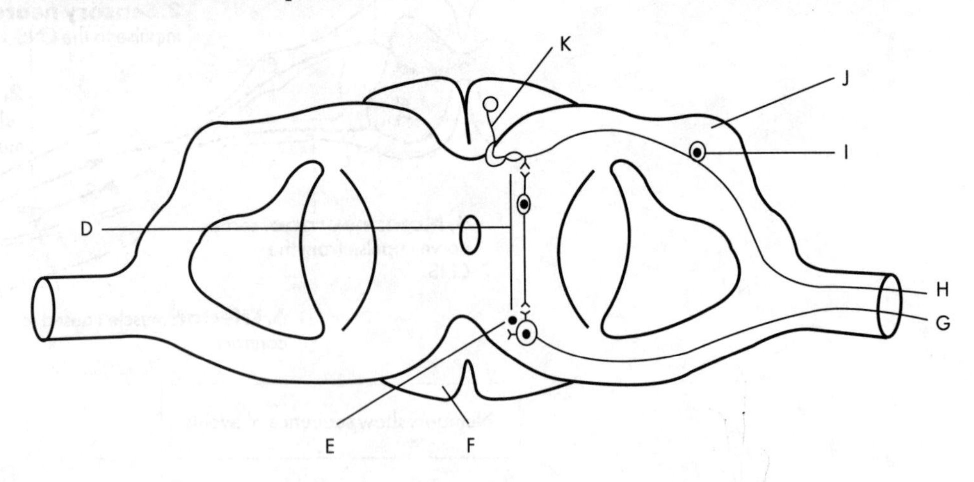

Fig. 4.164

By referring to the letters used in the labelling of Fig. 4.164 answer the following questions:

(a) Give the name of cell **D**. ____________________

(b) What name is given to **E**, where nerve cells meet? ____________________

(c) What is region **F** called? ____________________

(d) What is the destination of fibre **G**? ____________________

(e) To what type of neurone does fibre **H** belong? ____________________

(f) What is **I**? ____________________

(g) Name the structure **J**. ____________________

(h) What is the function of cell **K**? ____________________

THE HORMONAL SYSTEM

Key Points

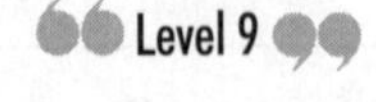
Level 9

- The *hormonal (endocrine) system* coordinates the body's activities by using 'chemical messengers', or *hormones*, which are carried in the blood system. Hormones are produced by *endocrine glands* and have their effects on *target tissues*.
- The action of hormones is illustrated by *adrenaline, oestrogens* and *progesterone*. Each of these hormones has several functions, including affecting the production of other hormones.
- Hormones are used artificially for a range of purposes, including *fertility control* in humans and *growth production* in crop plants and domestic animals.

The hormonal system

The *hormonal (endocrione) system* coordinates the activities of the animal by 'chemical messengers', or *hormones*. Some hormones are proteins and all are secreted by *endocrine glands* (sometimes called 'ductless glands'). These glands produce and store particular hormones, which are released directly into the bloodstream when required. The *activities of endocrine glands* are controlled by other hormones, called *trophic hormones*, by the *nervous system* and by *negative feedback* (see Section 4.7).

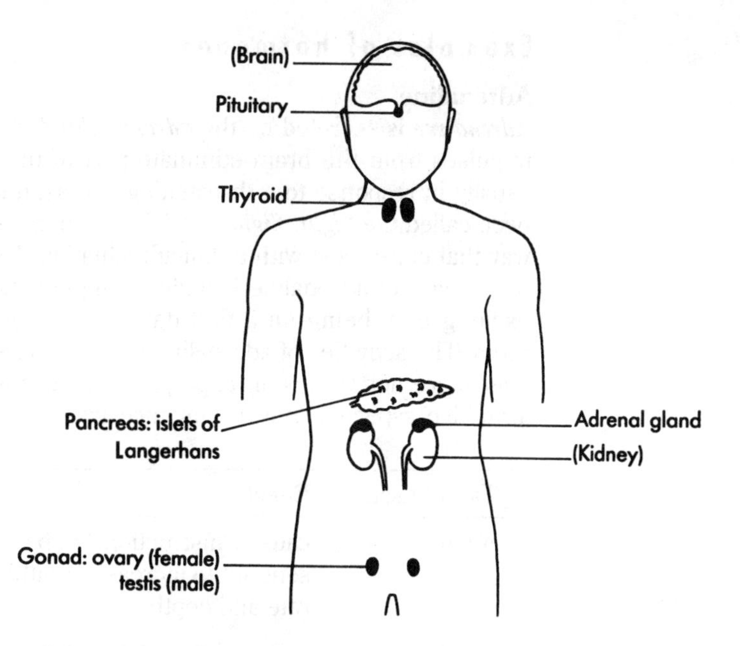

Fig. 4.165 Position of the main endocrine glands in humans (the relative position of other organs is also shown)

“You should be familiar with this comparison”

The overall function of the endocrine system, like that of the nervous system, is to co-ordinate the body's activities; a comparison of the two systems is given earlier in this section.

The position of the main endocrine glands is shown in Fig. 4.165; some of these glands have other functions which may not necessarily be related to that of the endocrine system. Some of the functions of the main endocrine glands are summarised in Fig. 4.166. Hormones have their effects on target tissues which are particularly sensitive to the hormone. Whilst hormones may have a long term effect, such as in growth and development, they do not permanently remain in the blood; they are broken down in the liver. The presence of a hormone in the body may cause the endocrine gland which produced it to slow down the rate of secretion; this is an example of negative feedback.

Gland	Hormone	Effects
Pituitary	Trophic hormones	Cause other endocrine glands, e.g. thyroid, adrenal and gonads, to release their hormones.
	ADH (vasopressin)	Increases water reabsorption in **nephrons** of the **kidney**.
	Oxytocin	Causes contraction of the **uterus** during birth.
	Prolactin	Stimulates milk production from **breasts**.
Thyroid	Thyroxine	Increases the general rate of **metabolism** (chemical reactions in the body) and stimulates **growth**.
Adrenal gland	Adrenaline	Sometimes called the **fight, flight or fright hormone**; prepares the body for potentially difficult or dangerous situations, for instance by increasing **heart rate**, efficiency of **muscles** and **breathing rate**. Adrenaline also raises the *blood glucose* level (see below).
Pancreas (islets of Langerhans)	Insulin	Causes the conversion of glucose to glycogen.
	Glucagon	Causes the conversion of glycogen to glucose.
Gonads: Ovary	Oestrogen	Promotes the development of female secondary sexual characteristics.
	Progesterone	Maintains the uterus lining during pregnancy.
Testes	Testosterone	Promotes the development of male secondary sexual characteristics.

Fig. 4.166 Summary of some of the main hormones in humans

Examples of hormones

Adrenaline

Adrenaline is secreted by the *adrenal glands* (situated just above the kidneys). Nerve impulses from the brain stimulate part of the adrenal glands to release adrenaline, usually in response to a threatening, stressful or dangerous situation. Adrenaline has been called the '*fight, flight or fright*' hormone, since it prepares the body to react in a way that copes best with a difficult situation. Examples of situations which may cause the release of adrenaline include: being confronted by a violent person, watching an exciting film, being on a first date(!), playing in a recital, entering an examination room. The activities of adrenaline are summarised in Fig 4.167. Note that some of the effects are not always an appropriate or useful response to some stimuli, but reveal 'primitive' aspects of our distant ancestry.

Target tissue	Effect	Purpose
brain	causes 'instructions' to be sent out to increase breathing rate and depth	increased supply of oxygen, and removal of carbon dioxide
heart	pacemaker makes heart beat faster	more rapid delivery of essential substances, e.g. glucose, oxygen, adrenaline itself.
blood supply to skin	reduced	blood to skin is not urgently needed, so can be diverted
body muscles	causes muscles to tense	body muscles need an increased supply of essential substances
blood supply to gut	reduced	blood to gut is not urgently needed, so can be diverted
gut muscles	causes muscles to relax	digestion is not urgent, and can be temporally stopped
liver	increased conversion of glycogen to glucose	glucose is needed for respiration in body muscles
fat deposits	increased conversion of fats to fatty acids	fatty acids are needed for body muscle contraction

Fig. 4.167 Summary of the action of the hormone adrenaline

Oestrogen and Progestrone

Oestrogens (a group of related hormones) and *progesterone* are hormones involved in the *menstrual cycle* and *pregnancy* in females (see Section 4.5). Following menstruation (the 'period'), oestrogens are produced within the ovaries by follicles, which surround the ovum (egg). Oestrogens cause the uterus (womb) lining to thicken and develop, and increase the blood supply, in preparation for a possible implantation by a fertilised egg. After the release of the ovum (during ovulation), the production of oestrogens declines, and the remains of the follicle develop into a so-called 'yellow body' (corpus luteum). This produces progesterone, which maintains the uterus lining during the remainder of the menstrual cycle, and also during pregnancy if fertilisation takes place. If fertilisation does not occur, progesterone levels drop, and the lining of the uterus is lost in menstruation; the cycle is then repeated again. These processes are summarised in Fig. 4.168.

"You should also be able to understand graphs based on this theme"

Hormones in preventing or promoting pregnancy

During pregnancy, progesterone has several effects on the woman's body, including the prevention of egg release (ovulation). The *contraceptive pill* contains chemicals which produce similar effects to the oestrogens and progesterone. They prevent ovulation, so eggs are not released and therefore cannot be fertilised. The woman stops taking the contraceptive pill for a few days each month to allow menstruation to occur as usual.

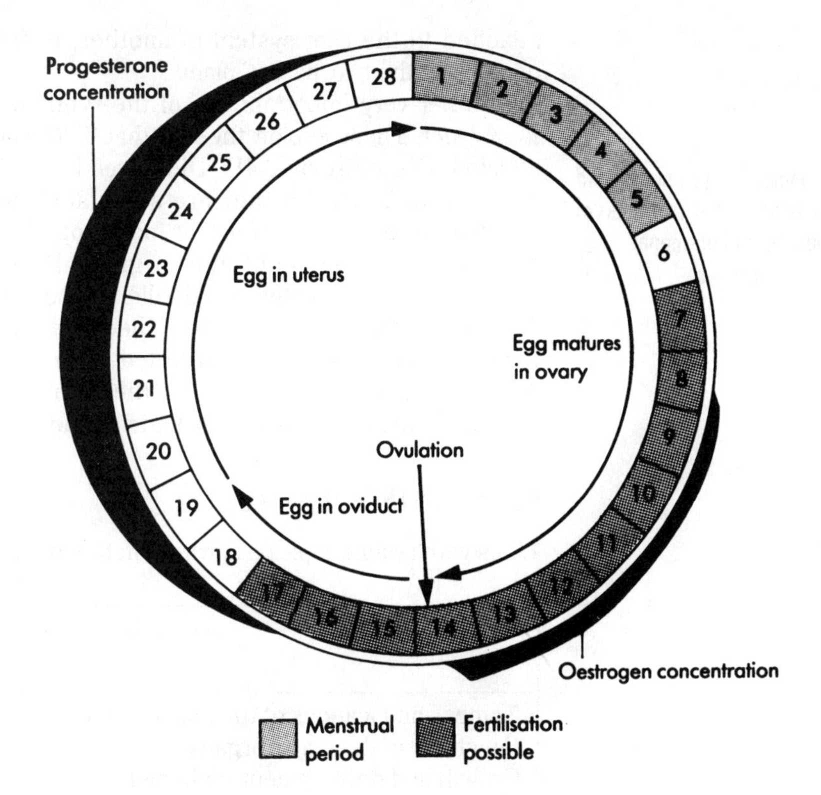

Fig. 4.168 Diagram to show the main events of the menstrual cycle, including relative hormone concentrations (shown by different thickness of line)

In contrast, hormones are also used to increase the chances of pregnancy in women who are normally unable to produce sufficient eggs for successful fertilisation. Injections of *follicle stimulating hormone* (FSH) and *luteinising hormone* (LH) are given. These cause her ovaries to release several mature eggs simultaneously, so increasing the chances of fertilisation taking place. Although this technique has proved very effective in many cases, one potential problem is the increased incidence of 'multiple births', e.g. twins, triplets, quadruplets, etc. Individual babies may be unusually small, and require considerable medical care immediately after they are born. Also, multiple births may impose a difficult burden for parents, who need to care for several children of the same age.

The use of hormones to increase egg production is also used as part of the procedure of *in vitro fertilisation* ('test tube babies'). The numerous eggs produced can be collected from the mother, and then fertilised in the laboratory using the father's sperm. If an egg is successfully fertilised, it can be implanted in the mother's uterus, and pregnancy can continue as normal.

Hormones in agriculture and horticulture

Synthetic plant and animal hormones, which work in a similar way to certain natural hormones, can often be easily made by industrial processes. They are therefore relatively inexpensive and are widely used for a variety of purposes.

One use of **synthetic animal hormones** is to control insect pests by interrupting their normal life cycle. Other hormones act as artificial 'sex attractants'. The effect is either to kill the insect directly, or prevent them from reproducing. One problem is that they often need to be applied in large quantities to control a pest insect, and other animals (e.g. aquatic crustaceans) may also be affected. Another use for animal hormones is to promote growth and production in farm animals. Some poultry are treated with oestrogen-type hormones to increase egg production. However, residual amounts of hormone may be present in poultry meat, and can then accumulate in human consumers. There have even been reports of men developing breasts, following long-term consumption of contaminated chicken meat.

Synthetic plant hormones have been put to many uses in horticulture and agriculture. One example which has many different uses is the synthetic hormone 2:4D. For example, it is used as 'hormone rooting powder', which stimulates the formation of so-called *adventitious roots* from the shoots of 'cuttings'. Synthetic hormones can also be used to promote *grafting*, when the shoot of one plant is attached to the root system of another, to form a new plant which combines the qualities of the two 'parent' plants.

Another very important use of the synthetic plant hormone 2:4D is as a *selective weed killer*. These exploit the fact that dicot and monocot plants (see Chapter 5) are affected differently by 2:4D. Dicot weeds, such as dandelions and thistles are very sensitive to 2:4D, and they are stimulated to grow 'too fast', and die, e.g. due to blocked phloem tubes. Monocot food crops, such as wheat and maize are much less sensitive, and they continue to grow. However, there are possible ecological disadvantages of selective weed killers. One is that 'weeds' may be killed that provide food for pollinating insects (e.g. bees), which die because of lack of food. Some herbivorous insects are also killed directly by the weed killer, which therefore disrupts normal food chains. Another problem is that over-use of weed killers has resulted in the evolution of resistant varieties of weeds.

Make sure you understand the benefits and drawbacks of these uses of hormones

Links with other topics

Links with related topics elsewhere in this book are as follows:

Topic	Chapter and Topic number	Page number
Names and locations of the major organs	4.1	30
Functions of the major organs	4.1	31
Growth and development (animals)	4.5	63
Reproduction (animals)	4.5	67
Homeostasis	4.7	137
Co-ordination	4.8	149
The nervous system	4.8	152

REVIEW QUESTIONS

Q29 Fig. 4.169 below shows some of the organs in the body which produce hormones.

(a) On Fig. 4.169, label the organs **A**, **B** and **C**.

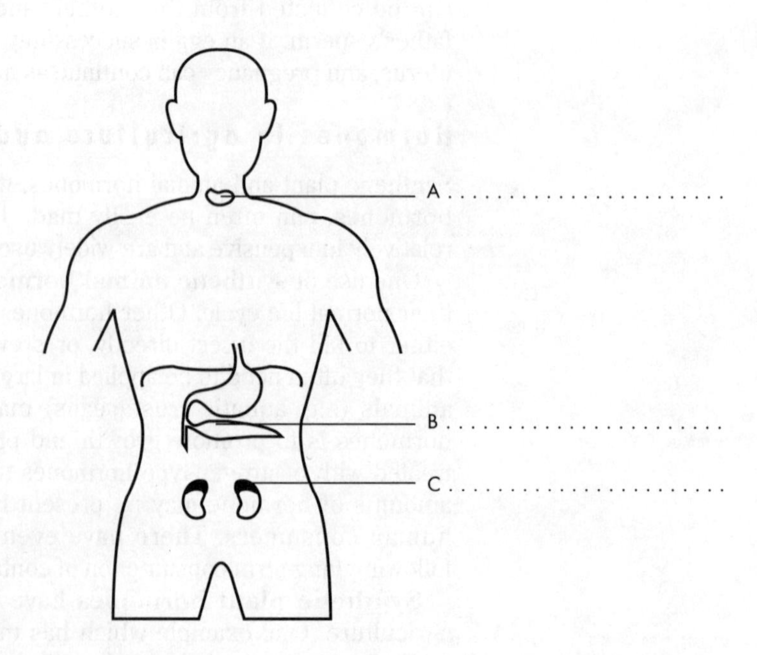

Fig. 4.169

(b) For **each** of the 3 organs, name the hormone secreted and state **one** way in which **each** hormone affects the body.

ANSWERS TO REVIEW QUESTIONS

A1 (a) **A** Appendix **C** Liver **E** Salivary gland
F Stomach **J** Large intestine

(b) See Fig. 4.3

A2 (a) See Fig. 4.5
(b) Stem – transports substances around the plant.
Leaf – to carry out photosynthesis
Roots – anchor plant to ground
Petal – attract insects for pollination

A3 Stem = asparagus; fruit = pepper;
Flower = broccoli; seed = pea;
leaf = watercress; root = parsnip.

A4 (a) **B** 3, **D** 1, **A** 4, **C** 2.

(b) All contain a cell membrane, all contain a nucleus.

(c) Cellulose cell wall

(d) 1. White Blood Cell 2. Sperm Cell

A5 (a) **A, B, D, E, C.**

(b) 1. This cell has a cellulose cell wall; a cheek cell does not.
2. This cell has chloroplasts; a cheek cell does not.

A6 1. thick cuticle: reduces evaporation of water
2. hairy surface: slows down diffusion of water vapour
3. stomata at bottom of pits: longer diffusion path.

A7 (a) small spiny leaves help to reduce water loss (by transpiration); Fat fleshy stem stores water; deep root helps water collection

(b) asexual reproduction gives genetically identical clones; important when parent already well adapted to environment. Also important when the reproducing organism is isolated from other members of species, as with cactus.

A8 Flat profile; less resistance to water flow; hooks on end of legs; to cling to rock; well developed legs; for swimming against current.

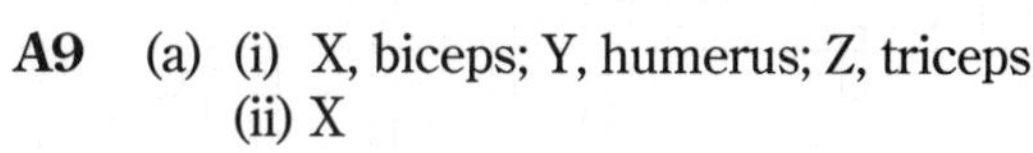

A9 (a) (i) X, biceps; Y, humerus; Z, triceps
(ii) X

(b) Nerve fibres; Blood vessels

(c) See Figure

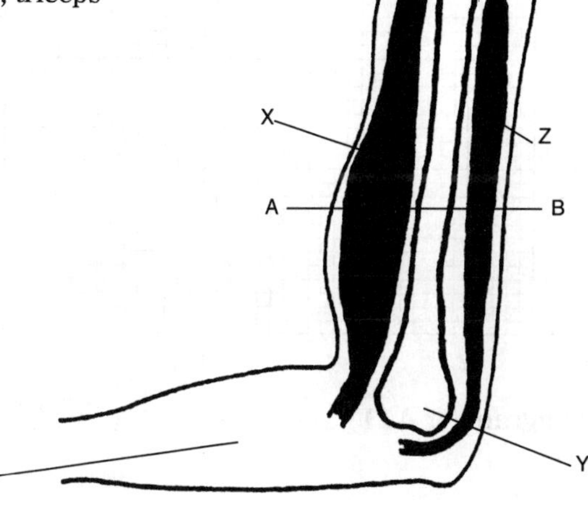

Comments

You may be used to seeing diagrams of structures from a particular 'viewpoint', e.g. diagram of the human arm from the side, in section. Diagrams in the page of a book have 'two dimensions' only. However, most structures in biology, actually have 'three dimensions'! If you understand diagrams fully, you should have no difficulty in interpreting diagrams of the same structure, from another viewpoint.

(a) (ii) Remember that muscles get shorter and 'fatter' when they contract. There are several possible answers to (b).

A10 (a) (i) Tar, (ii) Alveoli, (iii) Cigarette smoke contains carbon monoxide, which 'competes' with oxygen to combine with the oxygen-carrying molecule haemoglobin. This reduces the oxygen-carrying capacity of the blood by up to 15 per cent. (iv) It can cause lung cancer and chronic bronchitis.

(b) Toxic substances from smoke circulate in mother's blood, and can cross the placenta to enter the blood circulation of the embryo. Nicotine narrows blood vessels, and carbon monoxide displaces oxygen from haemoglobin. The embryo is deprived of sufficient oxygen, possibly resulting in retarded growth.

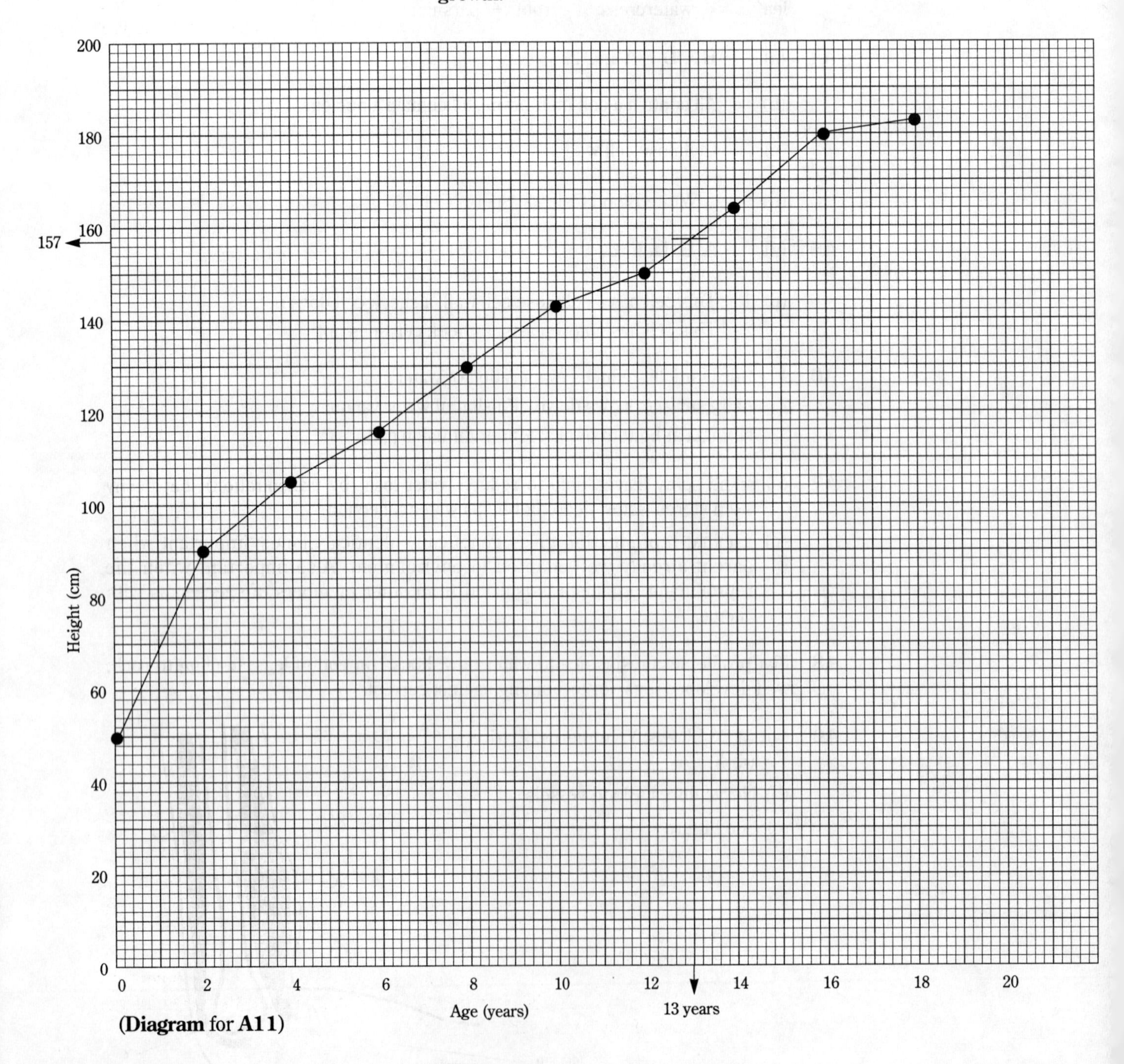

(Diagram for A11)

Comments

You should thoroughly revise the effects of smoking health on health – questions on this tend to be popular with examiners! In this particular Review Sheet, an understanding of blood circulation and also aspects of human reproduction are needed.

A11 (a) (i) 40 cm, (ii) 13 cm, (iii) 16 cm.
(b) (i) 157 cm, (ii) 0–2 years (See diagram on page 158)

A12 (a) (i) 0–4 years, (ii) 12–16 years
(b) The nervous system is needed from an early stage of development, because the baby needs to respond to stimuli even before birth, and the baby/young child needs to use information about the environment to learn.

Comments

This question tests your ability to draw and interpret graphs – a very important skill in biology! Not that you should also be able to perform simple calculations (have a calculator and ruler handy in all biology exams! In an examination, you will get marks if your answers are within the correct range; e.g. for (ii), 12–15 would also be acceptable.

A13 (a) (i) Depends on mother's nutrition, (ii) Depends on characteristics inherited from parents.
(b) (i) They have inherited very similar genes controlling growth and development, (ii) They are not genetically identical.
(c) Smoking may cause the baby to be deprived of sufficient oxygen.

Comments

This question tests your understanding of reproduction, and also the genetic and environmental influences on development. Revise these topics if you had difficulty with the Review Question.

(a) The mother's health would be another factor.

(b) Also, they both experienced similar conditions during pregnancy, but (c) conditions in the womb may not have been the same for each baby, e.g. because of different positions of the placentas. Also for question (c), you could answer that smoking reduces the mother's appetite, so less nutrients may have been available. Smoking may also result in premature birth, but this would not answer this question about differences between the two babies.

A14 (a) (i) Excretion of metabolic wastes; control of the body's water content by removing excess water or mineral salts (osmoregulation). (ii) Renal artery.
(b) (i) It must be selectively permeable, allowing small molecules only to pass through, (ii) To increase its length, and therefore the surface area for exchange.
(c) (i) Blood proteins, (ii) Glucose, urea, (iii) To maintain a concentration similar to that in the blood, to prevent essential salts being 'lost' from the blood by diffusion.

Comments

(a) (i) another function is to control blood pH. (b) If this question is confusing, it might help to revise the topic on diffusion and osmosis. A 'large surface area' is also characteristic of all exchange surfaces in the body itself (e.g. see the topic on breathing). (c) But not blood cells, which are not a constituent of the blood plasma itself. (iii) may take some thinking about! Revise diffusion and osmosis if you're not sure.

A15

(a) (i) In this meal, which food gave the most protein?

baked beans

"Correct answers."

(ii) Name *one* other food not eaten in this meal which is rich in protein.

cheese

(iii) Why does the girl need protein?

for energy

"Incorrect answer; the main function of protein is for growth and repair."

(b) (i) The total energy value of the meal is 6600 kJ.
In one day the girl needs 9600 kJ.
If she ate this meal, now many *more* kJ would she need in that day?

3000

"Correct answer, though units should also be given."

(ii) What would happen if she eats much more than 9600 kJ of food every day?

she would get fat

"More precisely, excess food would be stored, resulting in obesity."

(iii) A lot of energy comes from fat.
Name the *two* foods in this meal which gave her most energy.

sausages

apple pie

"Correct answers."

(c) The girl needs 14mg of iron and 25mg of Vitamin C each day to keep healthy.

(i) How much of her daily iron needs did this meal give her?

7mg

"Correct."

(ii) How much of her daily Vitamin C needs did this meal give her?

24mg

"Incorrect; the answer should be 25 mg."

(iii) What will happen if she does not have enough iron and Vitamin C?

Not enough iron may cause anaemia

Not enough Vitimin C may cause scurvy

(d) Why should she eat fibre (roughage) every day?

to stop constipation

"Correct answers."

A16 (a) (i) See diagram

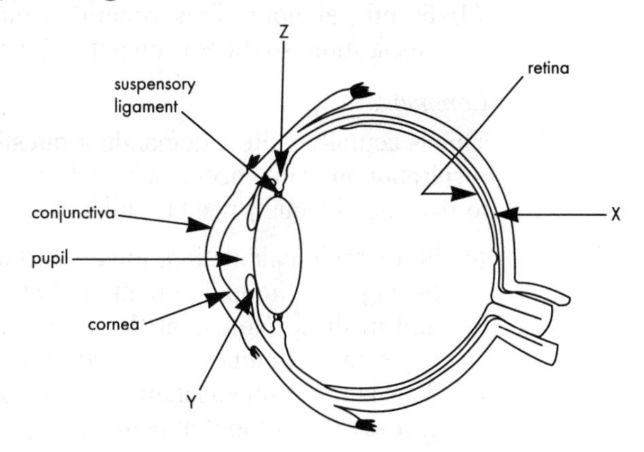

(b) (i) Choroid, (ii) It contains blood vessels, which supplies food and oxygen to, and removes wastes from, the retina. It prevents reflections inside the eye.

(c) (i) Accommodation, (ii) The lens is made more convex ('fatter') by contraction of the circular ciliary muscles (at 'Z'), which reduces tension of the suspensory ligaments, allowing the lens to assume its 'normal' thickened shape.

Comments

This is a fairly straight-forward question, which tests your understanding of both structure and function of the eye.

(a) Label these parts carefully, since any inaccuracy or lack of clarity will lose marks.

(c) (ii) This reduces the focal length of the lens. Use your understanding of physics to help you here!

A17 (a) (i) Plasma, (ii) Tissue fluid (lymph), (iii) Nucleus of phagocyte, (iv) Capillary.

(b) Thin walls (one cell layer thick), 'Leaky' walls, allowing an exchange of substances between blood and surrounding tissues.

(c) There are two main methods, both involving white blood cells which move from the blood to the infected area. Phagocytes engulf bacteria. Lymphocytes produce antibodies, which attack the bacteria.

Comments

Revise this topic again if you had difficulty with any part of the Review Sheet.

(a) Another clue is the fact that blood cells are 'squeezing' through the narrow blood vessel.

(c) Marks would be given for any brief answer that included reference to phagocytes and lymphocytes.

A18 (a) Oxygen is given out to air, Water is taken from the soil, Carbon dioxide is taken from the air.

(b) Glucose or sugar, (c) Chlorophyll, (d) To 'trap' light energy, (e) There is more chlorophyll in the upper part of the leaf (palisade layer) than the lower part (spongy mesophyll). Chlorophyll inside the leaf is visible through the transparent epidermis on each side of the leaf.

Comments

(a) Don't refer to 'minerals... taken from the soil', as the question is on photosynthesis only.

(d) Make sure you refer to light in your answer. The light energy is used to split water during photosynthesis.

(e) Revise leaf structure if this part is unclear.

A19 (a) 9 hours, (b) 21 hours, (c) 50 ml × 21 = 1 050 ml = 1.05 litres.
(d) 50 ml per hour. This amount is equal to the amount being produced by respiration, so there is no net gain or loss of oxygen at 'X'.

Comments

This is actually quite a demanding question! The key thing is to understand the respiration in plants occurs *all the time*. So, if you gave '12 hours' as your answer to (b), you obviously need to think more carefully about this.

(c) is a simple calculation, once you have understood that respiration occurs throughout the experiment, and at the same rate (50 ml per hour) in light and in dark. Note that the temperature is kept constant throughout the experiment (temperature affects the rate of respiration).
(d) This is the compensation point. So, oxygen gained (by respiration) = oxygen lost (by respiration) at point 'X'. If you answered '0' ml per hour, think again!

A20 (a) B, D. (b) (i) To remove dissolved oxygen. (ii) Using freshly de-ionized water.
(c) D. The conditions in 'D' are the same as in 'E', except for temperature.
(d) (i) 'B', (iii) Water, warmth, oxygen.
(e) Seeds in 'E' would be larger, since they have absorbed water.

Comments

(d) (i) The definition of a 'control' is that it has all the factors being tested in the experiment.
(e) Also, the outer testa might be split. Other changes which occur during the early stages of germination might be apparent (e.g. emergence of radicle or plumule), but question (c) suggests that no germination has not occurred.

A21 (a) (i) To provide water for germination. (ii) A control for comparison with ungerminating seeds. (iii) To kill seeds, to prevent germination.

(b) (i) Activity of microbes inside the flask. (ii) Use of disinfectant to sterilise the seeds.

(c) Place thermometer in direct contact with seeds. Place insulation (e.g. cotton wool) around vacuum flask neck. Seeds soaked for longer, to ensure all seeds are germinating. More peas in flasks

Comments

This question requires a knowledge of conditions necessary for seed germination, respiration (in plants), and also experimental design.

(a) (ii) If necessary, revise the meaning of the term 'control' in experiments.

(b) (i) 'Decomposition of seeds' would also be a correct answer.

(c) There are three marks for this question; this means that the examiner is looking for three distinct parts to your answer. Question (c) tests your understanding of experimental method. It is a good idea to be asking yourself 'why is the experiment designed like this?' whenever you conduct an experiment. Even better, design your own experiments!

A22 (a) (i) **A** – Muscle — Raise hairs to trap air
B – blood capillary — Constriction reduces blood flow
C – Fat/adipose — Insulates so keeps body heat in

(ii) Sweat gland (iii) Reduces or stops production of sweat

(b) Control or regulation of internal environment of body

(c) (i) Rapid drop in temperature so heat loss exceeds heat produced

(ii) Move about less or have less body fat for insulation

(iii) Wrap in a blanket to warm them up or give them a warm drink.

A23 (a) **A** Umbilical cord **B** Amniotic fluid

(b) (i) Glucose/mineral salts/amino acids: oxygen

(ii) Urea/carbon dioxide

(c) To the placenta. Blood flowing in this direction has recently been pumped by the embryo's heart (it is in the umbilical artery).

A24 (a)

A	B
86 cm^3	94 cm^3
14 cm^3	6 cm^3

(b) (i) Same size with same number of leaves

(ii) Level rises when shoots are in the water

(iii) Lower surface

(iv) Cover water with layer of oil or vaseline

(v) No loss of water takes place through the stem or any other part of the plant

(c) Dry the air around one plant by using a hair dryer or enclose one plant in a polythene bag to trap moisture.

A25 (a) (i) Strip 3

(ii) Water leaves cell vacuole; from region of high concentration of water to region of lower water concentration; by osmosis.

(b) Air

(c) It would wilt.

A26 (i) 14 cms (units must be stated)

(ii) $\frac{14}{2} = 7$ cm/hr

A27 (a) In the bloodstream.

(b)

	nervous system	hormonal system
1.	rapid, brief response	relatively slow prolongeol response
2.	message sent to specific target tissues only	message sent throughout the body including target tissues
3.	transmission is chemical and electrical	transmission is chemical only

(c) A nerve message is transmitted from the sensory neurone to the motor neurone by a connector (relay) neurone, which acts like a 'short cut'. The brain is not involved in a time-consuming decision-making process.

Comments

Comparisons between the nervous and hormonal systems are frequently asked for in examinations! When you revise this topic, make sure you are familiar with both systems.

(b) Another pair of answers is: transmitted by nerves (direct pathway)/ transmitted by bloodstream (indirect pathway).

(c) Note that there are several alternative names for each type of neurone; use which ever ones you are familiar with. The main point to get across is that the brain is not involved (the action is not conscious – until afterwards), and that the route is as short as possible.

A28 (a) Connector (relay, intermediate, associate) neurone. (b) Synapse. (c) White matter. (d) Effector (e.g. muscle). (e) Sensory. (f) Cell body. (g) Ganglion. (h) To conduct nerve impulses to the brain, allowing conscious decisions to be made.

Comments

This is a straightforward question about the central nervous system (CNS). Make sure you are familiar with the names and functions of all the main structures. (c) White matter consists mostly of thousands of axons – the elongated part of neurones. Grey matter (region 'D' in the diagram) consists mainly of cell bodies and synapses. (d) The clue 'destination' in the question should tell you that this is a motor neurone, not a sensory neurone. Also, motor neurones do not have the swellings (i.e. the 'dorsal root ganglion', 'J' in the diagram) which are characteristic for collections of sensory neurones. (h) Note that there is another neurone which carries impulses from the brain, to the motor neurone.

A29 (a) A = thyroid, B = pancreas, C = adrenal glands

(b) Thyroid: Thyroxine (growth-promoting hormone), which stimulates growth during development. Pancreas: insulin, causes the conversion of glucose to its stored form of glycogen. Adrenal glands: adrenaline, which stimulates the pacemaker to increase the heart rate.

Comments

(a) 'Islets of Langerhans' would also be acceptable.

(b) Each of these hormone-producing (endocrine) glands produces more than one hormone. The hormones given in the answer are those mostly commonly associated with these glands – others may not be included on your syllabus.

EXAMINATION QUESTIONS

Here is a selection of examination questions based on the topic groups in this chapter. You will already have gained practice in answering examination questions when you tried the *Review Questions* at the end of the topic groups. You can find answers to these examination questions at the end of the book (p351).

Question 1

The drawing shows a young plant.

(a) Plants make their own food. Which part of the plant does this?

____

(b) Give **one** function of the root hairs.

____

(c) Roots grow downwards and sideways.

Give **two** advantages of sideways growth.

1. ______________________________

____

2. ______________________________

____

Fig 4.170

Question 2

The diagrams below show the roots of two different plants, **A** and **B**.

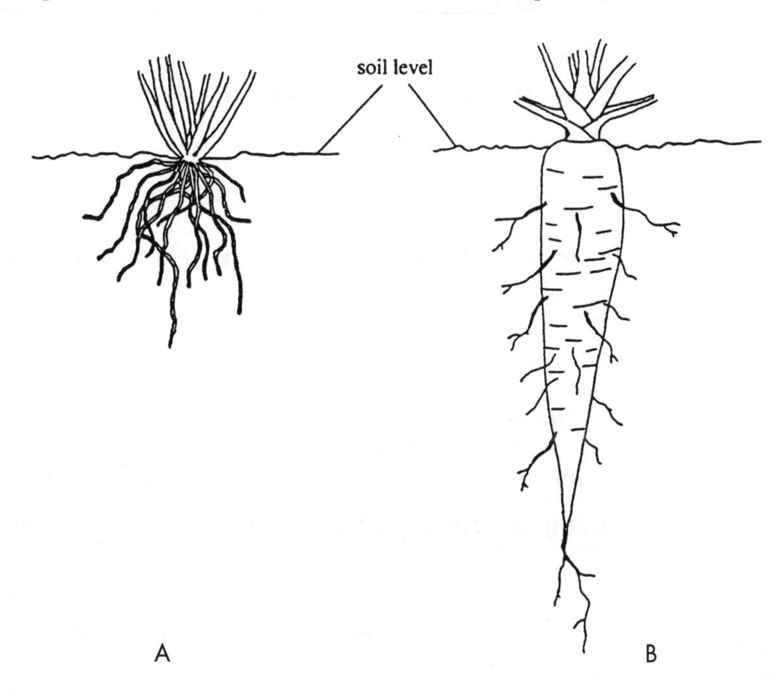

Fig 4.171

(a) Describe *two* ways in which the roots of **A** differ from the roots of **B**.

1. ______________________________

____

2. ______________________________ *(2)*

(b) (i) Give **three** functions which are carried out by both roots.

1. ______________________________

2. ______________________________

3. ______________________________ *(3)*

(ii) Suggest **one** function which is carried out by root **B** but not by root **A**.

______________________________ *(1)*

Question 3
The diagrams below show two different plant cells.

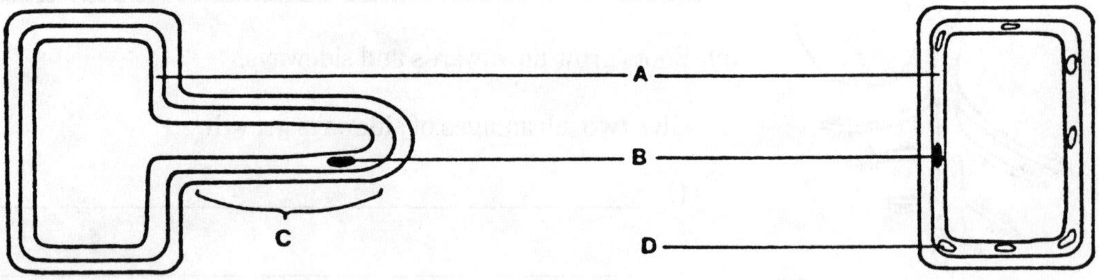

Fig 4.172

(a) Name the parts labelled **A** and **B**.

A ______________________________

B ______________________________ *(2)*

(b) How does the part labelled **C** help the root to carry out one of its functions?

______________________________ *(2)*

(c) How does the part labelled **D** help the leaf to carry out its main function?

______________________________ *(2)*

Question 4
The diagram shows a cell.
(a) Name the part labelled **P** on the line provided. *(1)*

(b) (i) The cell is from either a plant or an animal. Write down which one you think it is.

______________________________ *(1)*

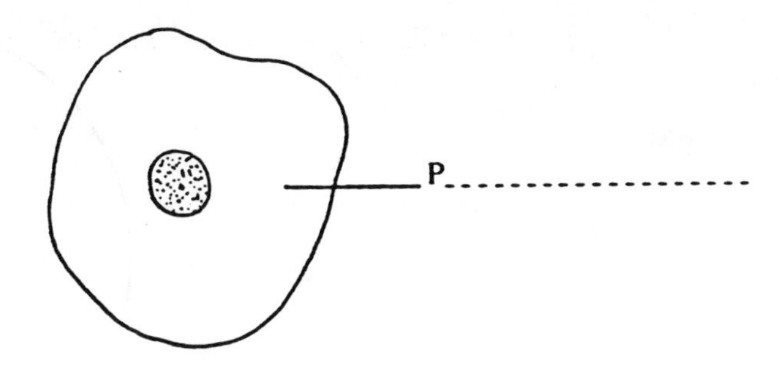

Fig 4.173

(ii) Give a reason for your answer in (i).

(1)

(c) The width across the centre of the live cell is 0.03 mm. Explain how you could work out the magnification of the drawing. Do not do the actual calculation.

(2)

Question 5

Figure 1 below represents a leaf cell.

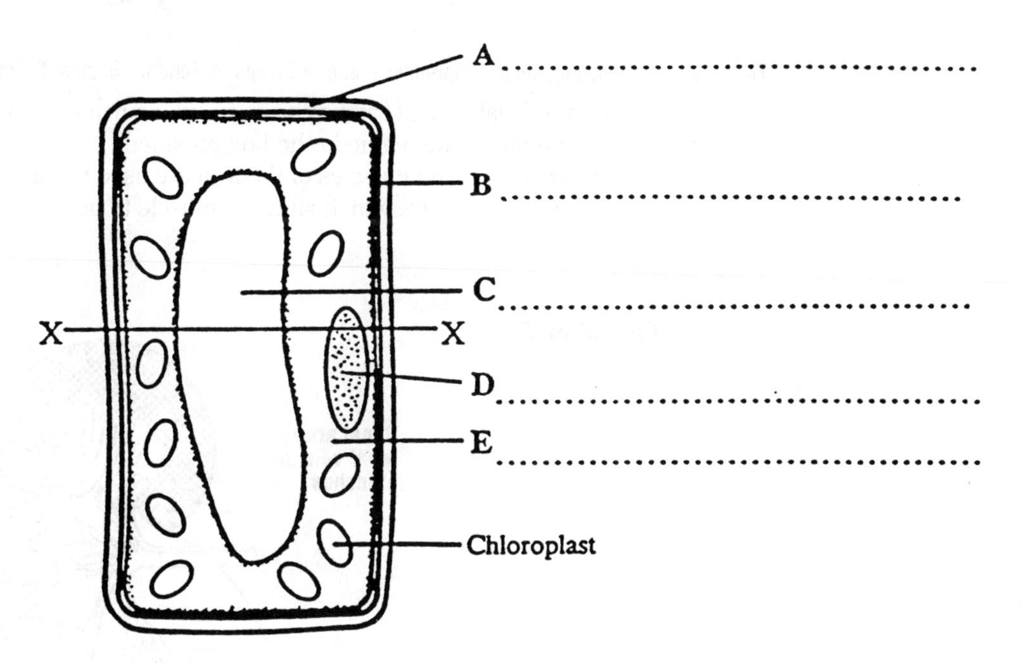

Fig 4.174

(a) Name the structures labelled **A** to **E** on the lines provided *(5)*

This leaf cell was cut across the line **X** – **X** on Figure 1.
Figure 4.175 below represents the outer edge of the cross-section. Complete the drawing of the cross-section in Figure 4.175. Label any structures you draw with the correct letter from Figure 4.174.

(4)

Cross-section

Fig 4.175

Question 6

The diagram below shows certain bones and muscles in the arm.

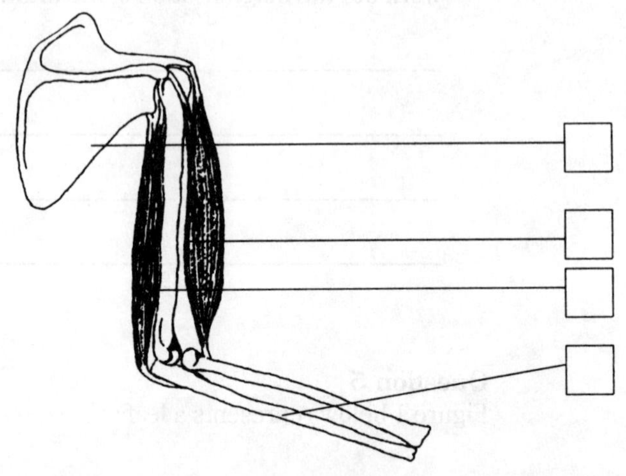

Fig 4.176 **1.** Ligament **2.**Shoulder Blade **3.**Biceps **4.**Tendon **5.**Ulna **6.**Triceps **7.**Humerus **8.**Radius

(a) Correctly label the structures shown, using names chosen from the list. Write only the number of the name in the box provided. *(2)*

(b) Describe how the muscles of the arm are used to raise it. *(2)*

(c) Name the structure which attaches muscle to bone. *(1)*

Total 5 marks (WJEC)

Question 7

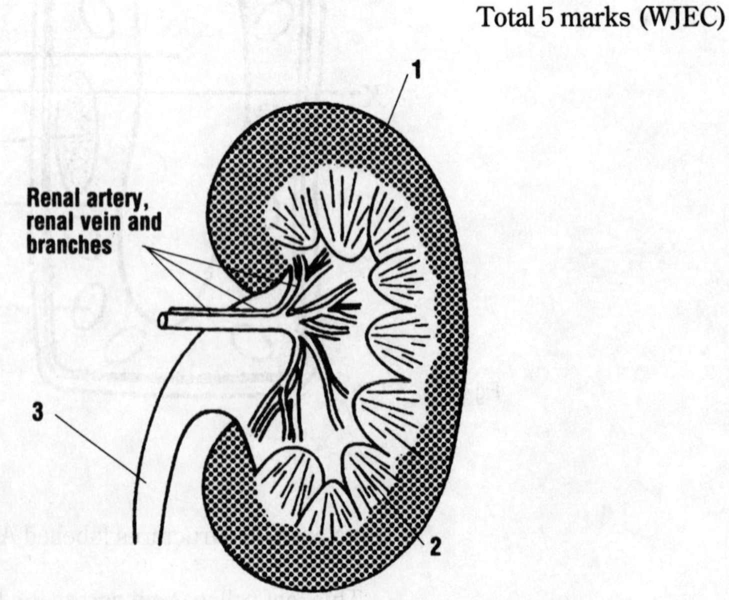

(a) The diagram shows a section through a mammal's kidney. Name the parts 1, 2 and 3.

Rate of blood flow in kidneys	Rate of filtration into kidney tubules (nephrons)	Rate of urine passing out of kidneys
1.2 dm^3 per minute	0.12 dm^3 per minute	1.5 dm^3 per day

(b) The table gives information about the human kidney.
   (i) What percentage of blood passing into the kidney is filtered into the kidney tubules? *(1)*
   (ii) Where in the kidney (part 1, 2 or 3 in the diagram) does filtration take place? *(1)*
   (iii) About 17.2 dm^3 are filtered from the blood into the kidney tubules per day, yet only 1.5 dm^3 of urine are excreted. What happens to the other 15.7 dm^3? *(2)*

	Urine lost per day (dm^3)	Sweat lost per day (dm^3)	Salt (sodium chloride) lost per day	
			In urine (g)	In sweat (g)
Normal day	1.5	0.5	18.0	1.5
Cold day	2.0	0.0	19.5	0.0
Hot day	0.375	2.0	13.5	6.0

(c) The table shows the average amounts of urine, sweat and salt (sodium chloride) lost on a normal day, a cold day and a hot day. (Assume that food and drink are the same on all days.)

   (i) Why is more urine lost on a cold day than a normal day? *(2)*
   (ii) Why do you think the total amount of salt lost on each of the three days is the same? *(2)*
   (iii) The minimum amount of urine excreted in a day is 0.375 dm^3. Why do you think the kidneys always produced some urine? *(2)*
   (iv) What *must* someone losing more than 7 dm^3 of sweat in a day do in order to remain healthy? *(2)*

Total 15 marks (SEG)

Question 8

(a) Explain the terms ***homeostasis*** and ***negative feedback*** in relation to body temperature of a mammal. (A detailed account of temperature regulation is **not** required.) *(7)*

(b) (i) Describe the homeostatic control of the water content of the body. *(7)*

   (ii) Describe the influence of the liver, pancreas and adrenal glands on the sugar content of the blood. *(6)*

Total 20 marks (IGCSE)

Question 9

A 22-year-old woman kept a precise record of her food and drink intake for one day. The quantities of some major nutrients were calculated, and the totals compared with the average daily requirements of a woman of that age. The figures are summarised in the table below (Fig. 4.177).

Meal	Item	Quantity	Energy (kJ)	Protein (g)	Fat (g)	Carbo-hydrate (g)	Calcium (mg)	Iron (mg)	Vitamin C (mg)
Breakfast	White bread	90 g	950	7	2	50	90	1	0
	Butter	15 g	450	0	12	0	2	0	0
	Jam	30 g	330	0	0	19	5	0	1
	Black coffee	1 cup	20	0	0	1	4	0	0
Lunch	Hamburger	150 g	1560	30	15	30	50	4	0
	Ice cream	100 g	800	4	12	20	130	0	1
	Fizzy drink	1 can	550	0	0	30	0	0	0
Evening meal	Sausages	75 g	1150	9	24	10	30	0.5	0
	Chips	200 g	2100	8	20	70	25	2	20
	Baked beans	220 g	600	10	1	20	100	2.5	4
	Apple pie	150 g	1800	5	25	60	60	1	1
	Cream	30 g	550	0.5	15	1	20	0	0
	Tea with milk	2 cups	200	2	4	6	100	0	0
Snacks	Chocolate	50 g	1200	5	20	25	120	1	0
	Peanuts	50 g	1200	15	25	5	30	1	0
TOTAL INTAKE FOR DAY			13460	95.5	175	347	766	13	27
AVERAGE DAILY REQUIREMENT			9400	58	*	*	600	14	30

*Amounts variable

Fig 4.177

(a) (i) By how much did the energy content of the day's diet exceed the average daily energy requirements of the 22-year-old woman? *(1)*

(ii) What would be the probable effect of this difference on an average woman of this age over a long time? *(1)*

(iii) This woman could have a daily energy requirement much greater than average. Suggest *one* reason why this would be so. *(2)*

(b) The recommended carbohydrate : fat ratio in a balanced diet is 5:1 by weight.

(i) Which individual meal in the day given in the table had a carbohydrate:fat ratio of exactly 5:1? *(1)*

(ii) To the nearest whole number, what is the carbohydrate:fat ratio for the whole day's intake? *(1)*

(iii) What effects may this proportion of fat in the daily diet have on the woman's circulatory system? *(2)*

(c) *Excluding* coffee, tea and fizzy drinks, which of the items eaten during the day had the highest level of calcium per gram of food? Give the calcium content of this item in mg per g of food. *(2)*

(d) Identify ***one*** nutrient in which the day's food intake is ***deficient*** and name a deficiency disease that may result if the woman's diet continues to provide too little of the nutrient. *(1)*

(e) A sample of the woman's urine taken the morning *after* the day described above contained a high concentration of urea. How might this have been predicted from the information in the table? *(1)*

Total 12 marks (ULEAC)

Question 10

The figure below shows a diagrammatic representation of an alveolus and its blood supply.

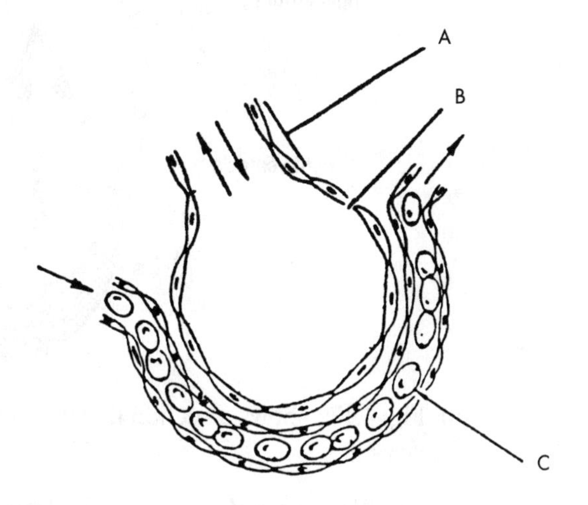

Fig 4.178

(a) Name the structure labelled **A, B** and **C**.

____

(b) Name the large blood vessel along which blood travels from the heart to the lung.

____

(c) (i) Name the process by which oxygen and carbon dioxide enter and leave the blood at an alveolus.

____

(ii) For the process to work effectively there must be differences in concentration between the gases in the blood and in the alveolus. Give two ways in which these differences are maintained.

____

____

(d) State three adaptations, shown in the diagram, which enable the alveolus to function efficiently.

____

____

____

(e) In which parts of the blood, and in what forms, are the two gases carried to and from the lungs?

(i) Carbon dioxide is carried mainly in the ____________ as ____________.

(ii) Oxygen is carried mainly in the ____________ as ____________.

Question 11

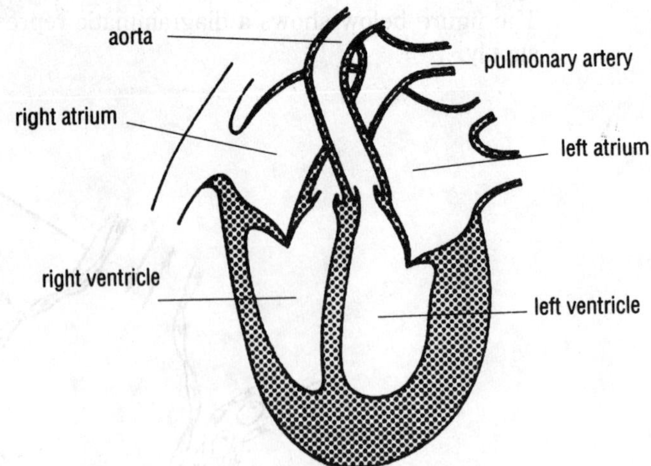

Fig 4.179

(a) Fig. 4.179 shows a section through a mammal's heart and some of its major blood vessels.

(i) What route does blood take from the lungs to the aorta? *(3)*

(ii) Explain how the valve between the left ventricle and the aorta works. *(4)*

(b) Fig. 4.180 shows blood pressure in the left ventricle and the aorta throughout two cycles of the heart. Using the graph:

(i) How long is a heart cycle (the time interval between the start of one heartbeat and the start of the next)? *(1)*

(ii) For how long during each heart cycle is the valve between the aorta and the left ventricle closed? How did you work out your answer? *(2)*

(iii) What is the difference between the maximum and minimum pressure in the aorta? *(2)*

(c) Explain how blood pressure is increased in the left ventricle. *(3)*

Total 15 marks (SEG)

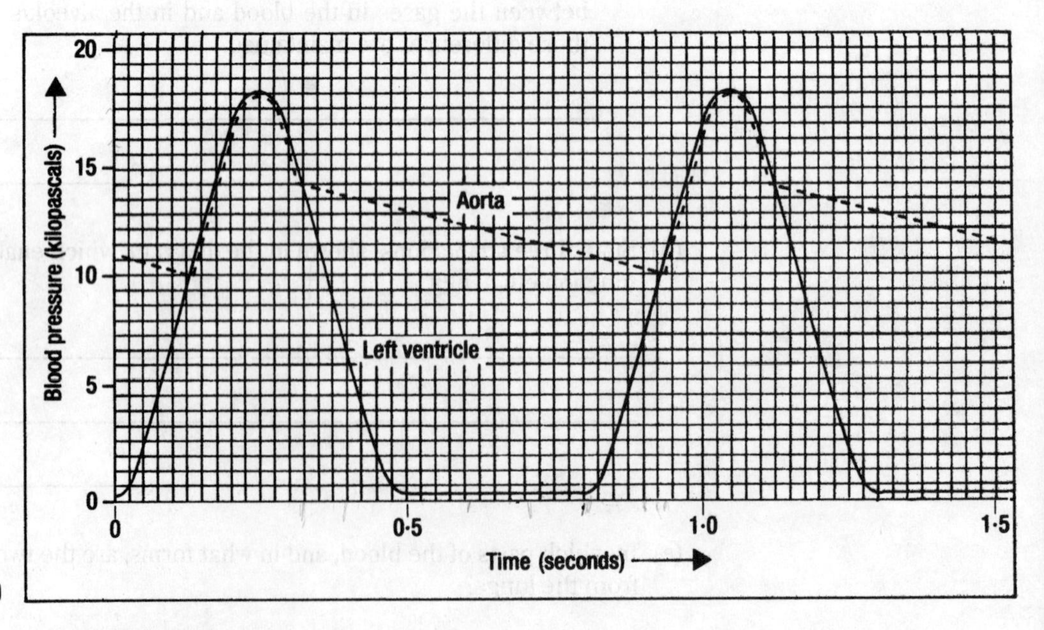

Fig 4.180

Question 12

In what circumstances is adrenaline secreted? What are the effects of adrenaline on the body and how might it help the person concerned? *(10)*

STUDENT'S ANSWERS WITH EXAMINER'S COMMENTS

Questions 13

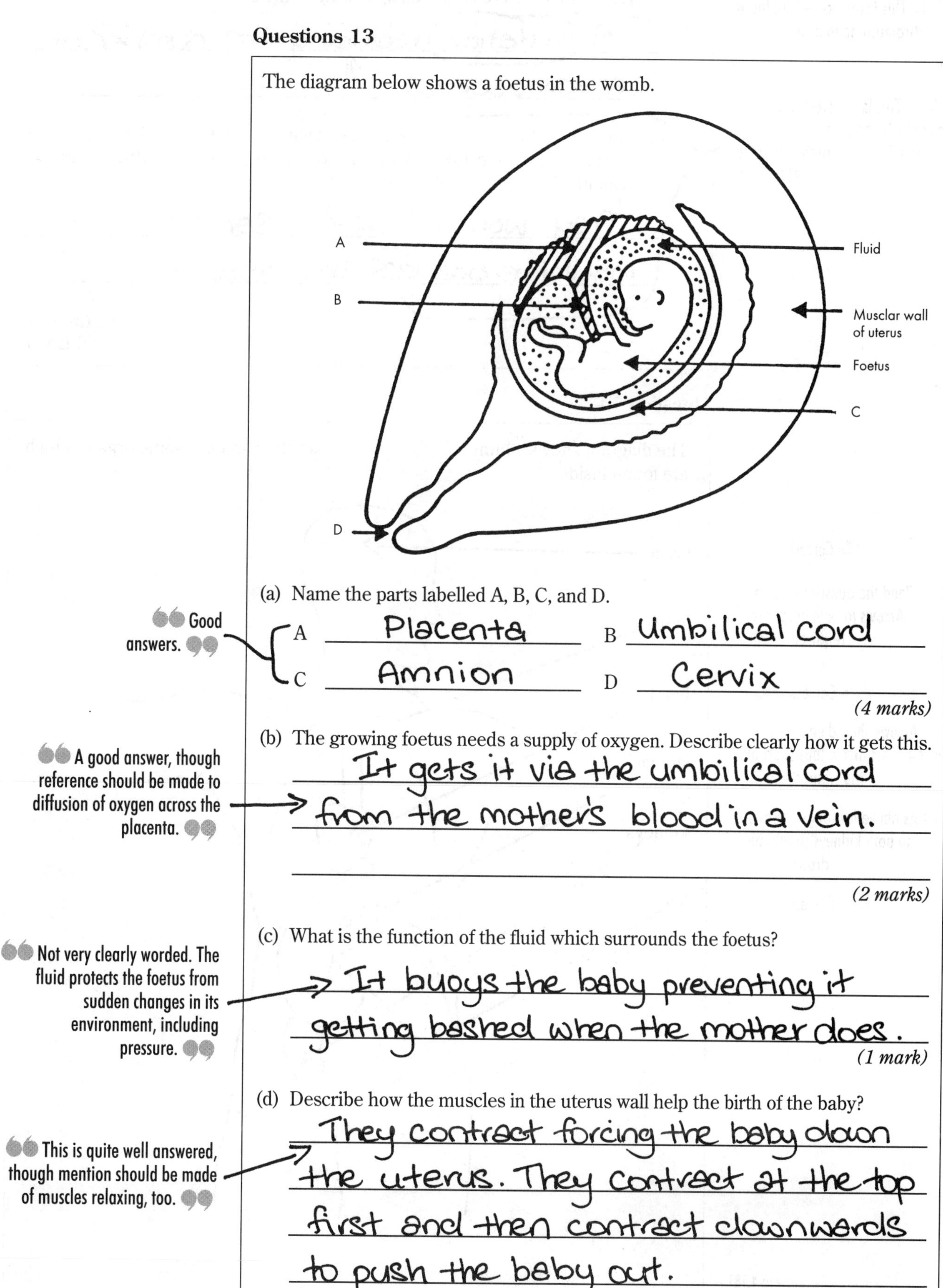

The diagram below shows a foetus in the womb.

(a) Name the parts labelled A, B, C, and D.

A Placenta B Umbilical cord

C Amnion D Cervix

(4 marks)

Good answers.

(b) The growing foetus needs a supply of oxygen. Describe clearly how it gets this.

It gets it via the umbilical cord from the mother's blood in a vein.

(2 marks)

A good answer, though reference should be made to diffusion of oxygen across the placenta.

(c) What is the function of the fluid which surrounds the foetus?

It buoys the baby preventing it getting bashed when the mother does.

(1 mark)

Not very clearly worded. The fluid protects the foetus from sudden changes in its environment, including pressure.

(d) Describe how the muscles in the uterus wall help the birth of the baby?

They contract forcing the baby down the uterus. They contract at the top first and then contract downwards to push the baby out.

(2 marks)

This is quite well answered, though mention should be made of muscles relaxing, too.

(e) Concern has been expressed at the number of children suffering from Down's Syndrome being born to families living along the east coast of Northern Ireland.

(i) What may have happened in the mother's body which could have been responsible for the condition appearing in the child?

A mutation reducing the number of chromosomes

(1 mark)

> This is correct; it is in fact a chromosome mutation.

(ii) How would the chromosones in a cell taken from a Down's Syndrome baby differ from those taken from a baby who does not suffer from the condition?

They would look the same but there would be one less chromosome

(1 mark)

> This is incorrect; there will actually be an extra chromosome (i.e. 2n = 47 rather than the 46).

(Total 11 marks)

(NISEAC)

Question 14

The diagram shows a human body. Beside it are the names of some organs which are found inside.

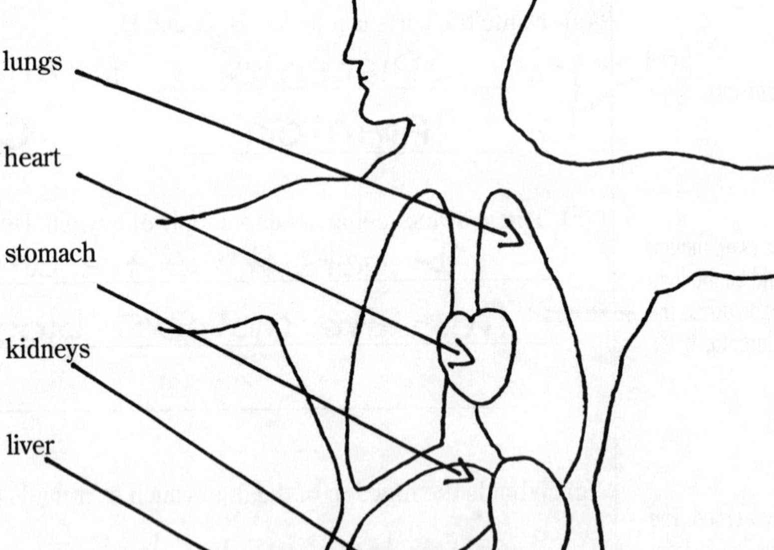

> Correct (brain)

> Read the questions again. Arrows to *both* lungs are required (lungs)

> Correct (heart)

> Arrows should end on or in the organ. This arrow falls short of the stomach (stomach)

> As above for lungs—arrows to both kidneys should be drawn (kidneys)

> Correct (liver)

Fig 4.181

(a) Draw arrows from each of the names to the shapes in the body where the organ should be. (If there are two similar organs, draw an arrow to **both**). *(6)*

(b) What is the main job of

(i) the heart?

pump blood around the body.

(2)

❝ Good ❞

(ii) the kidneys?

remove waste from the blood.

(2)

❝ Filter blood is required for the second mark ❞

(iii) the lungs?

take oxygen into the body and remove carbon dioxide.

(2)

❝ Good ❞

Question 15

(a) The table shows the approximate composition and volume of the fluid entering the kidneys by the renal arteries, passing into the nephrons and then leaving the kidney as urine.

percentage composition	entering kidneys	entering nephrons	leaving as urine
proteins	8.0	0	0
salt	0.7	0.7	1.2
urea	0.03	0.03	2.0
sugar	0.1	0.1	0
Fluid volume in cm^3 per minute	750	125	1

(i) From the table,

(a) Which substances are unable to pass through the membrane of the Bowman's capsule?

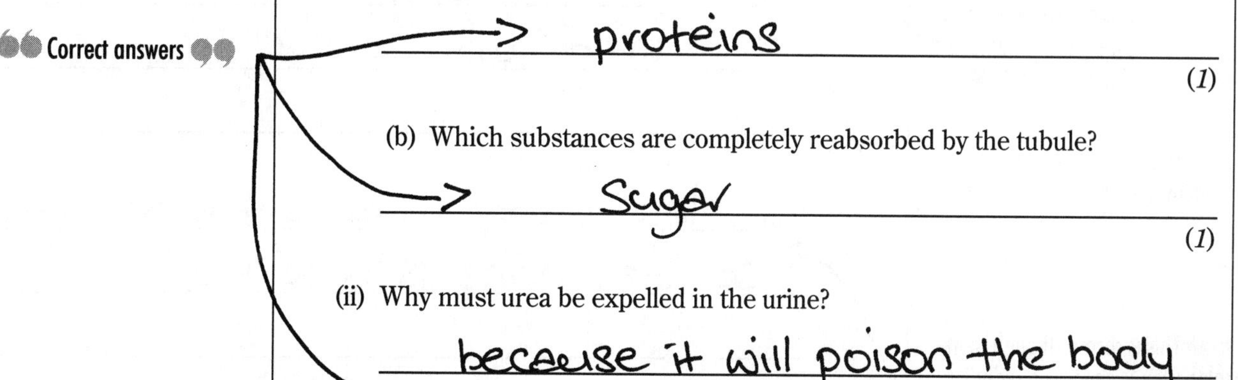

(1)

(1)

(ii) Why must urea be expelled in the urine?

because it will poison the body if it builds up.

(1)

(iii) What differences in composition of urine would you expect in

(a) a starving refugee living on starch and water.

it would contain less urea

(*1*)

“This is correct; the water content (and the volume) of the urine would increase.”

(b) a weight lifter eating lots of steak,

more protein would enter the kidney

(*1*)

“Protein does not enter the urine so this answer is not relevant. There *would* be an increase in urea concentration in urine, from the breakdown of excess protein in the liver.”

(c) a jockey trying to keep his mass down by rationing his fluid intake?

a more concentrated urine, less urine is produced.

(*1*)

“Correct.”

(b) Some young children pass through a developmental stage in which they sometimes urinate whilst asleep rather than awakening before this happens (they wet the bed), possible treatments involve

(i) not having an evening drink,

(ii) using a special pad in the bed which, as soon as it is dampened by the first urine, rings a bell to wake the child,

(iii) giving a medicine in the evening to reduce the blood supply to the kidney during the night.

In each of the three treatments, give a brief explanation of the body mechanism which is being altered to prevent bed-wetting.

(i) less water is filtered by the kidney so there is less water in the urine and less urine.

(*2*)

“This is correct; it is worth using the word ‘volume’ in relation to amount.”

(ii) The bell tells the child to wake up.

(*2*)

“For two marks another point needs to be made, e.g. that this produces a conditioned reflex.”

(iii) less water reaches the kidney, so less urine is produced.

(*2*)

“This is correct, though some explanation is needed; a reduction in the volume of blood reaching the kidney decreases the volume of urine produced.”

(*Total 12 marks*)

(WJEC)

Question 16

The diagram represents a section through the human eye.

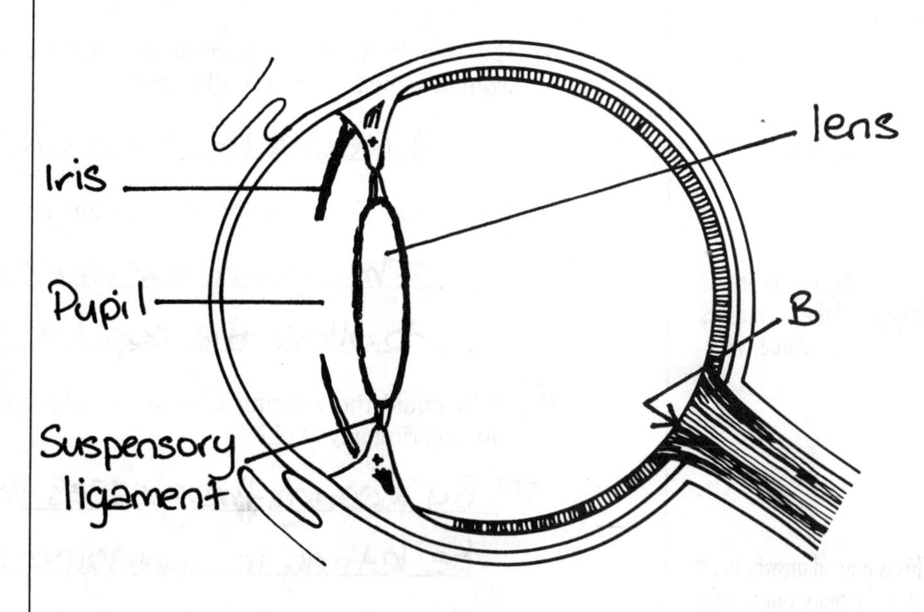

❝ A good answer; the structures are correctly shown. ❞

(a) Complete the diagram by drawing in the correct position of the lens, the suspensory ligament, the iris and the pupil. Label each clearly.

(4)

(b) Use the letter B to label the area of the retina which does not contain sensitive cells.

(1)

(c) State two ways in which a person's sight would be affected by the loss of one eye.

1. Objects would look flat ie not 3-dimensional

2. We would not be able to judge distances very well

❝ Correct; the term *'stereoscopic vision'* is an alternative to '3-dimensional'. ❞

❝ Correct answer. ❞

(d) A football spectator was seated in the grandstand reading the match programme when the players ran on to the pitch. What changes would have to occur in the focusing mechanism of the spectator's eyes to enable the players to be seen clearly?

To enable the players to be seen clearly the lens would become thinner as the ciliary muscles relax. The pupil would become smaller as more light would flood into the eye. The iris muscles will contract.

(3)

❝ Correct, though the term *'less convex'* is more precise than 'thinner'. ❞

❝ The student seems to understand the mechanism involved but needs to state *which* iris muscles (radial or circular) contract or relax. ❞

(e) Some evidence indicates that many sunglasses were not doing their job of cutting out the harmful ultra violet rays in the sunlight. The lenses may look dark but wearing these glasses may be even more damaging than wearing none at all.

(i) How would the appearance of the pupil behind the dark lenses differ from its appearance in bright light?

It would be much bigger.

" Correct answer. "

(ii) How do the muscles in the iris bring about the change noted above?

The sphincter muscles relax to allow the pupil to enlarge.

" Again, the precise functioning of the iris muscles needs to be stated. "

(iii) Why could the wearing of defective glasses be more damaging than not wearing any at all?

By wearing the glasses the pupil would be letting in damaging rays that if the glasses were not worn would not be let in as the eye would be able to sense that the light was so intense.
By not wearing them the eye senses the intensive light and so the pupil becomes smaller allowing less light to enter.

" This is a good answer but it needs to be more concise; the student should use the space and mark allocation as a guide for answer length. "

(Total 15 marks)
(NISEAC)

Note: Answers to the examination questions can be found at the end of the book.

STRAND (II) VARIATION AND THE MECHANISMS OF INHERITANCE AND EVOLUTION

CLASSIFICATION OF LIVING THINGS

SORTING LIVING THINGS INTO GROUPS

FUNGI

PLANTS

ANIMALS

FEATURES OF PLANTS & ANIMALS

IDENTIFYING PLANTS & ANIMALS

GETTING STARTED

This chapter and the next cover the materials involved in the core area of the Biology Key Stage 4 Syllabus 'Variation, Inheritance and Evolution'. By the end of Chapters 5 and 6 you should have met all the strand (ii) statements of the National Curriculum outlined below.

Pupils should use keys to assign organisms to their major groups and have opportunities to measure the differences between individuals. This is covered in Chapter 5. The other strand (ii) statements outlined below are covered in Chapter 6.

- They should consider the interaction of genetic and environmental factors (including radiation) in variation.
- They should be introduced to the gene as a section of a DNA molecule and study how DNA is able to replicate itself and control protein synthesis by means of a base code.
- Using the concept of the gene, they should explore the basic principles of inheritance in plants and animals and their application in the understanding of how sex is determined in human beings and how some diseases can be inherited.
- Using sources which give a range of perspectives, they should have the opportunity to consider the basic principle of genetic engineering, for example, *in relation to drug and hormone production*, as well as being aware of any ethical considerations that such production involves.
- They should consider the evidence for evolution and explore the ideas of variability and selection leading to evolution and selective breeding.
- They should consider the social, economic and ethical aspects of cloning and selective breeding.

Answers to the 'Examination Questions' later in this chapter can be found at the end of the book.

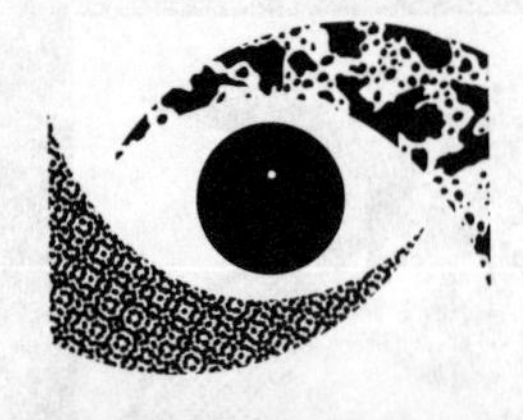

ESSENTIAL PRINCIPLES

1 SORTING LIVING THINGS INTO GROUPS

There are thousands of different types of living things on Earth. Scientists who have studied them have sorted them into groups of similar organisms: this is classification. Scientists have classified all living things into one of five **kingdoms**:

Fungi
Plants
Animals
Bacteria
Protoctists (simple, single-celled organisms)

We will look at 3 of these kingdoms in detail.

It is useful for scientists to sort out living things into groups of similar individuals – this is CLASSIFICATION. When scientists have classified organisms, they give them a BINOMIAL NAME (this is in Latin).

e.g. oak tree – *Quercus petraea*
daisy – *Bellis perennis*
human – *Homo sapiens*

This is useful because:

- it tells other scientists which organisms belong to the same group, e.g. *Panthera leo* (lion), *Panthera tigris* (tiger) and *Panthera pardus* (leopard) are all part of the same group.
- all scientists use these names, no matter which country they come from, so they are able to understand each other.

We can use **keys** to identify unfamiliar organisms. A key is a set of questions about the appearance of the organism.

In this rather short chapter instead of *Review Questions* at the end of each section there will be a broad coverage of *Examination Questions* at the end of the chapter.

Key Points

"Level 4"

- all organisms have been classified so that they belong to one of the five *kingdoms*
- all organisms have been given a Latin Binomial name, e.g. daisy is *Bellis perennis*
- branching keys or number *keys* can be used to identify organisms
- fungi have cell walls made of *chitin*, they never contain chlorophyll and they reproduce by making *spores*
- plants have cell walls made of *cellulose* and they can *photosynthesise* to make food
- animals have *no cell walls*, and cells never contain chlorophyll

2 FUNGI

The **Fungi** kingdom includes a large variety of organisms. Some of them are *single-celled,* e.g. Yeasts, and others are *multicellular,* e.g. pin mould (*Mucor*). Multicellular fungi consists of threads called *hyphae* which grow to make a tangled mass called a *mycelium* (Fig. 5.1).

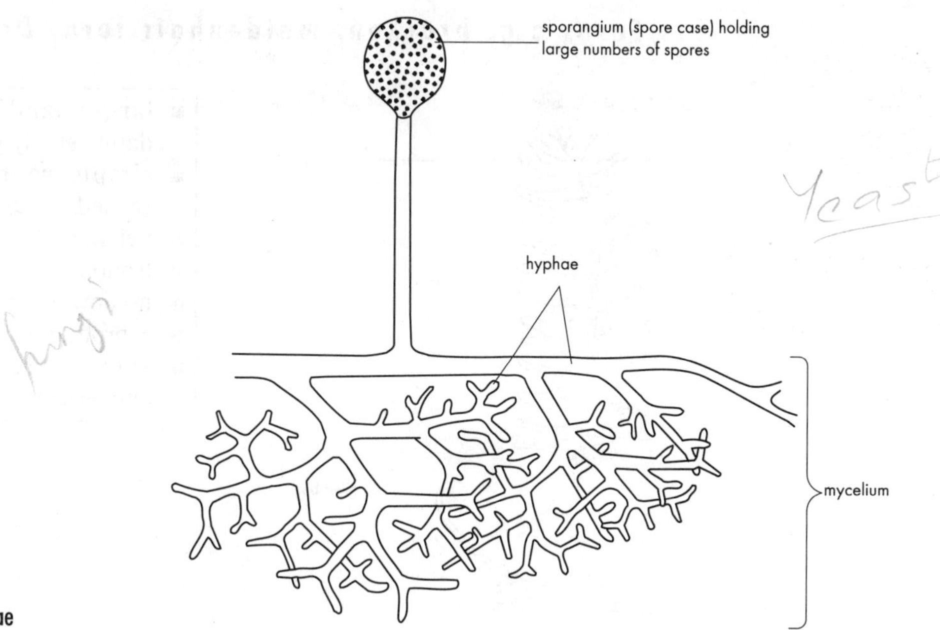

Fig 5.1 Fungal hyphae

All **fungi** have three things in common

- they have cell walls made of *chitin*
- they do not contain chlorophyll, so they can't photosynthesise
- they reproduce by making *spores*.

3 PLANTS

All **plants** have two things in common

- they have cell walls made of *cellulose*
- they *photosynthesise*

We can divide the **plant kingdom** into 4 *groups*:

mosses
ferns
conifers
flowering plants

You should know the main features of these groups

1. Mosses (Bryophytes) e.g. *Sphagnum*

Fig 5.2

- small land plants which live in damp places
- no proper roots
- simple stem and leaves
- no xylem or phloem tubes
- no flowers or seeds
- reproduce by making *spores*
- spores are released from a capsule

2. Ferns e.g. bracken, maidenhair fern, *Dryopteris*

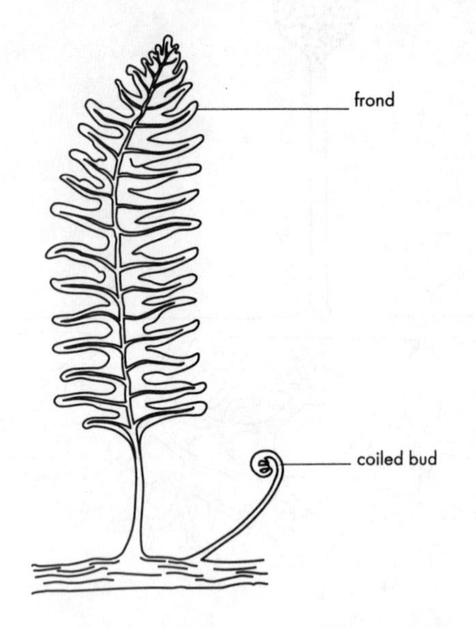

Fig 5.3

- larger land plants which live in damp, shady places
- simple roots, stems and leaves (called *fronds*)
- xylem and phloem tubes present
- fronds are coiled inside the bud
- no flowers or seeds
- reproduces by making *spores*
- spores are usually grouped in clusters on the underside of fronds.

3. Conifers e.g. pine (*Pinus sylvestris*), larch

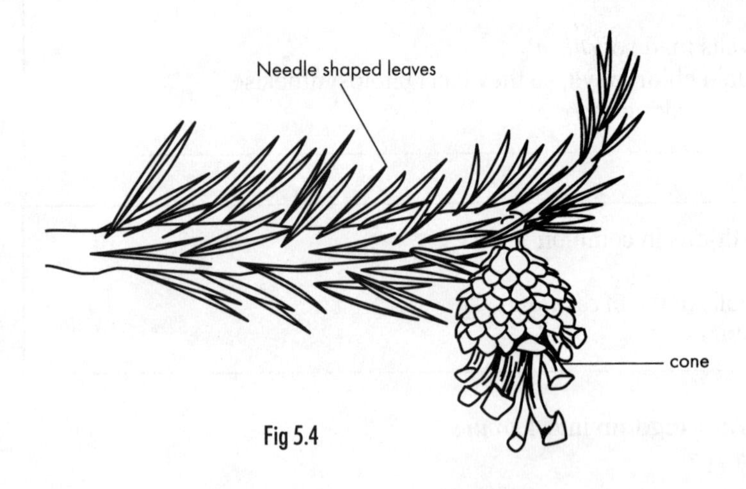

Fig 5.4

- trees or shrubs which often live in dry areas
- proper roots, stems and leaves
- simple xylem and phloem tubes
- leaves are usually needle shaped
- no flowers
- produces seeds inside *cones*

4. Flowering plants

- live on land or in water, very variable in size
- proper roots, stems and leaves
- xylem and phloem tubes present
- have *flowers* which contain reproductive organs
- produce seeds which are enclosed in a *fruit*
- can be divided into two main types: monocots and dicots

Monocots, e.g. grass, daffodil, iris, barley, maize

- have narrow leaves with *parallel veins*
- have one seed leaf (cotyledon) inside seeds

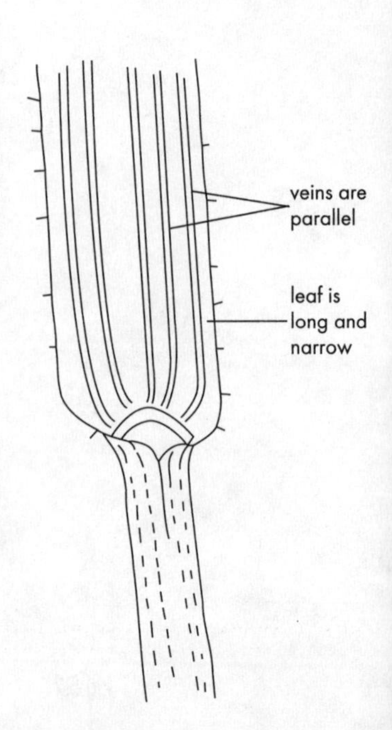

Fig 5.5 Part of a monocot leaf

Dicots, e.g. daisy, ivy, rose, oak, sycamore

- have broad leaves with a *network of veins*
- have two seed leaves (cotyledons) inside seeds.

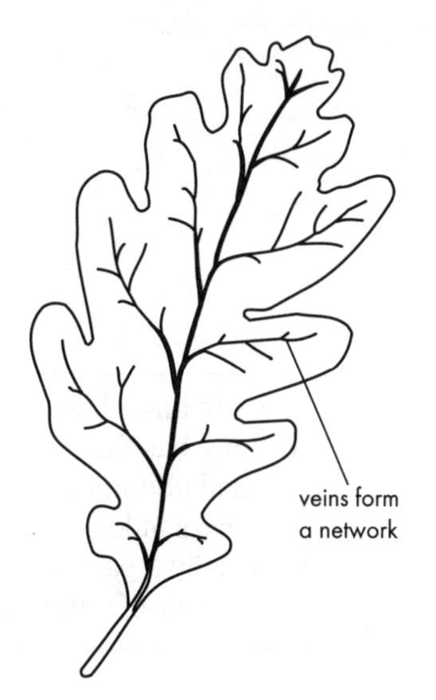

Fig 5.6 Dicot leaf

4 ANIMALS

All **animals** have two things in common
- cells never have a cell wall
- they feed on organic material, e.g. proteins, carbohydrates

"You should know the main features of these groups"

Animals can be divided into two main groups:
- invertebrates
- vertebrates

1. Invertebrates have no bones or skeleton inside their body

There are many groups of invertebrates, and we will consider six in detail:

annelids	crustaceans
molluscs	arachnids
insects	myriapods

Annelids (ringed worms) e.g. earthworms

Fig 5.7

- body is divided into *segments*
- no legs
- most have bristles (chaetae) to anchor them in the soil
- damp body surace (covered in mucus)

Molluscs e.g. snails, slugs, limpets, mussels

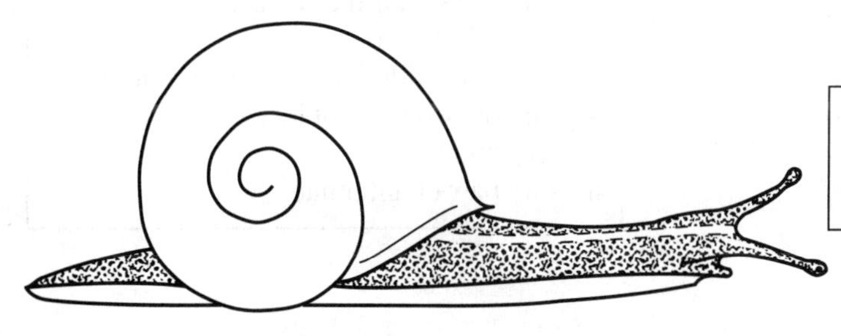

Fig 5.8

- soft body
- no legs
- body may be protected by one or two *shells*

Insects e.g. housefly, locust, dragonfly

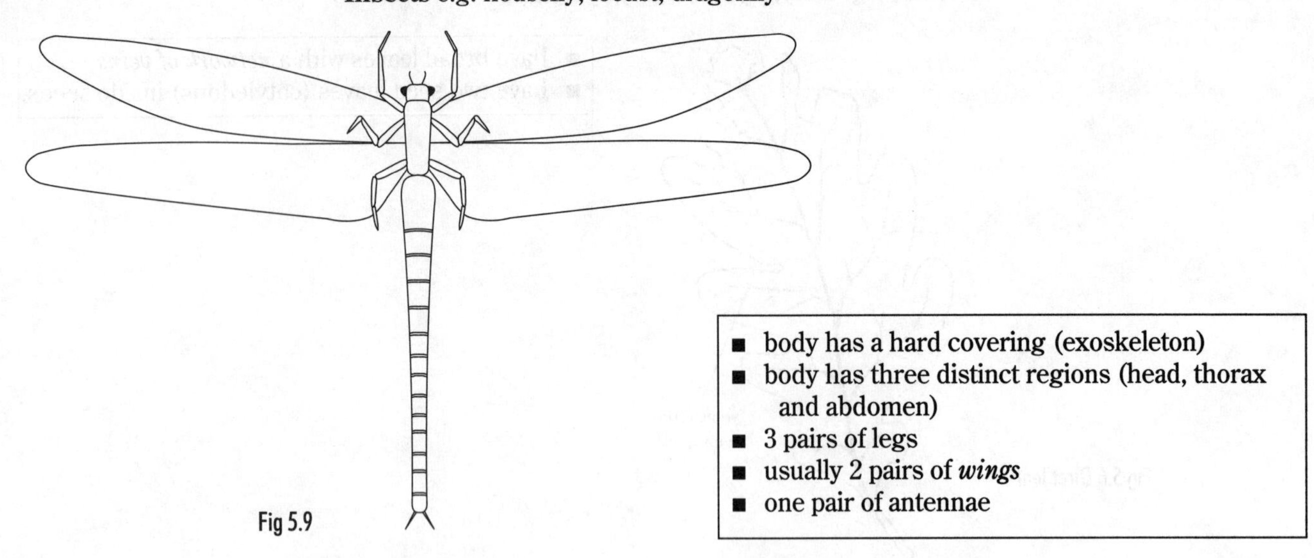

Fig 5.9

- body has a hard covering (exoskeleton)
- body has three distinct regions (head, thorax and abdomen)
- 3 pairs of legs
- usually 2 pairs of *wings*
- one pair of antennae

Crustaceans e.g. woodlice, crabs, shrimps

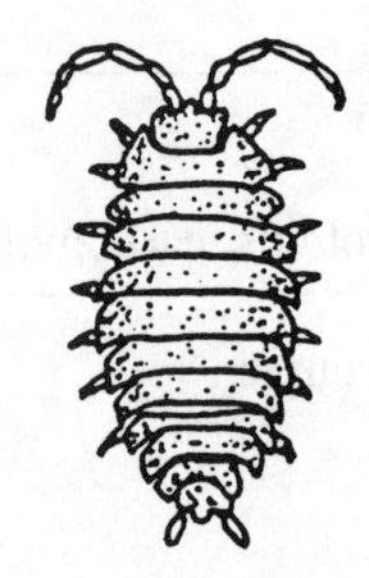

Fig 5.10

- body has a hard covering (exoskeleton)
- many legs or 'leg-like structures'
- 2 pairs of antennae

Arachnids e.g. garden spider, harvestman, mites

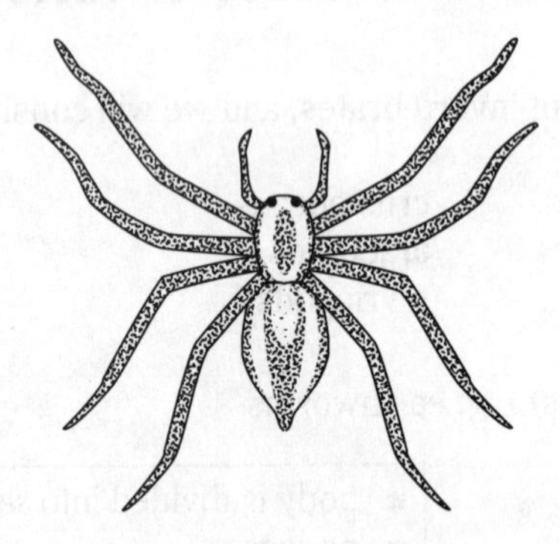

Fig 15.11

- body has a hard covering (exoskeleton)
- 4 pairs of legs
- no antennae

Myriapods e.g. centipedes, millipedes

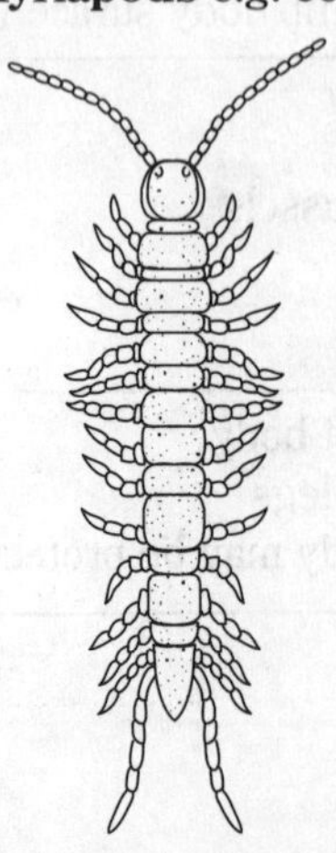

Fig 5.12

- body has a hard covering (exoskeleton)
- body is divided into many segments
- one or two pairs of legs on each segment
- one pair of antennae

2 Vertebrates have a skeleton inside the body, including a backbone

There are five groups of vertebrates:
fish
amphibians
reptiles
birds
mammals

Fish e.g. trout, carp, cod, pike

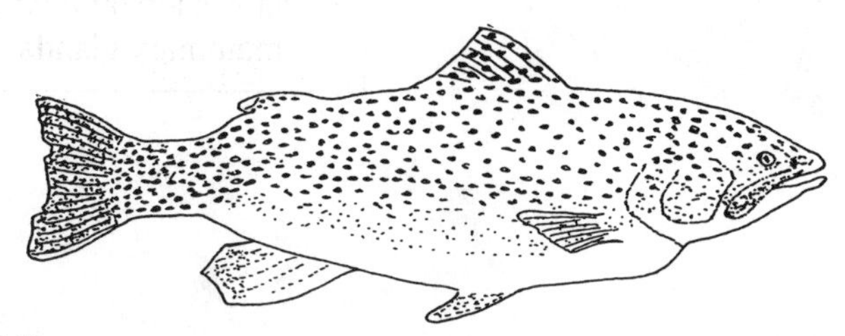

Fig 5.13

- body covered in *damp scales*
- no legs
- *fins* to move through water
- breathe using *gills*
- cold blooded
- always live in water
- lay eggs with no shells in water

Amphibians e.g. frogs, toads, newts

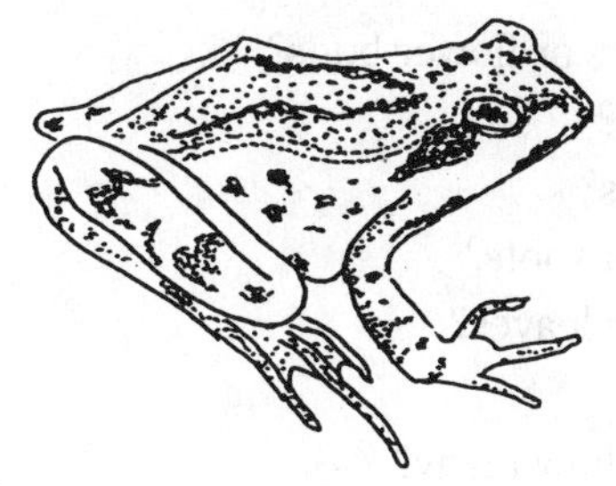

Fig 5.14

- moist body covering (no scales)
- 4 legs
- breathe using lungs (when adult)
- cold blooded
- adults live on land, but lay eggs in water
- lay eggs with no shells

Reptiles e.g. crocodile, lizard, tortoise, snake

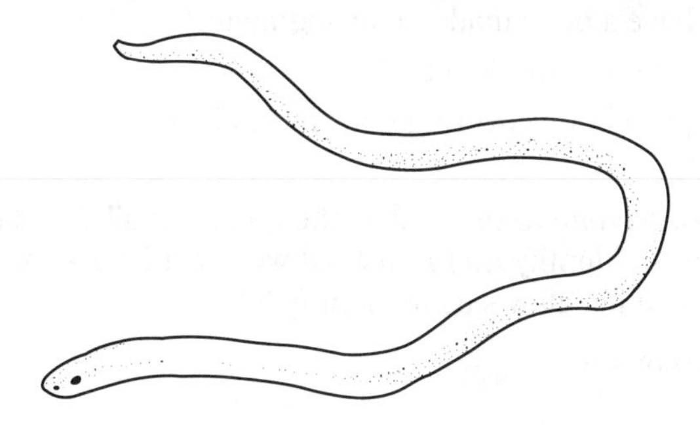

Fig 5.15

- body covered with *dry scales*
- most have 4 legs
- breathe using lungs
- cold blooded
- mostly live on land, some feed in water
- lay eggs with soft, leathery shells on land

Birds e.g. starling, crow, owl, gull

Fig 5.16

- body covered with *feathers*
- 2 legs
- one pair of wings
- breathe using lungs
- warm blooded
- most live on land
- lay hard shelled eggs on land

Mammals e.g. humans, dogs, squirrels, whales

Fig 5.17

- skin covered in ***hair or fur***
- most have 4 legs
- breathe using lungs
- warm blooded
- most live on land
- eggs develop ***inside*** the mother's body so babies are born live
- babies fed on ***milk*** from the mother's mammary glands

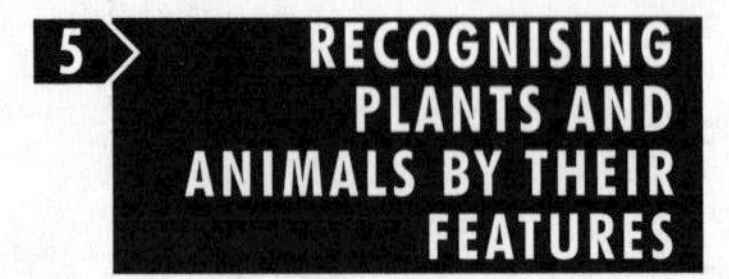

5 RECOGNISING PLANTS AND ANIMALS BY THEIR FEATURES

When you try to *identify* plants or animals, it is important to observe them very carefully. These are some of the main features to look for:

Plants: does it have flowers or flower buds?
does it have cones?
does it have spores?
does it have proper roots?
what shape are the leaves?

Animals: does it have legs? how many?
does it have wings?
does it have antennae? how many?
does it have a body made up of segments?
does it have a bony skeleton?
what type of body covering does it have?

6 USING KEYS TO IDENTIFY PLANTS AND ANIMALS

It is not possible for anyone to remember the names of all the plants and animals that exist, and to be able to identify them. Instead we use a KEY – this is a list of questions about the organism which allows us to identify it.

There are two types of key:

(a) Branching key

This is set out as a flowchart (Fig. 5.18). For each question there is a simple choice and only one of the answers can be correct.

This type of key is easy to use, but is only suitable for a small number of organisms because it takes up a lot of space.

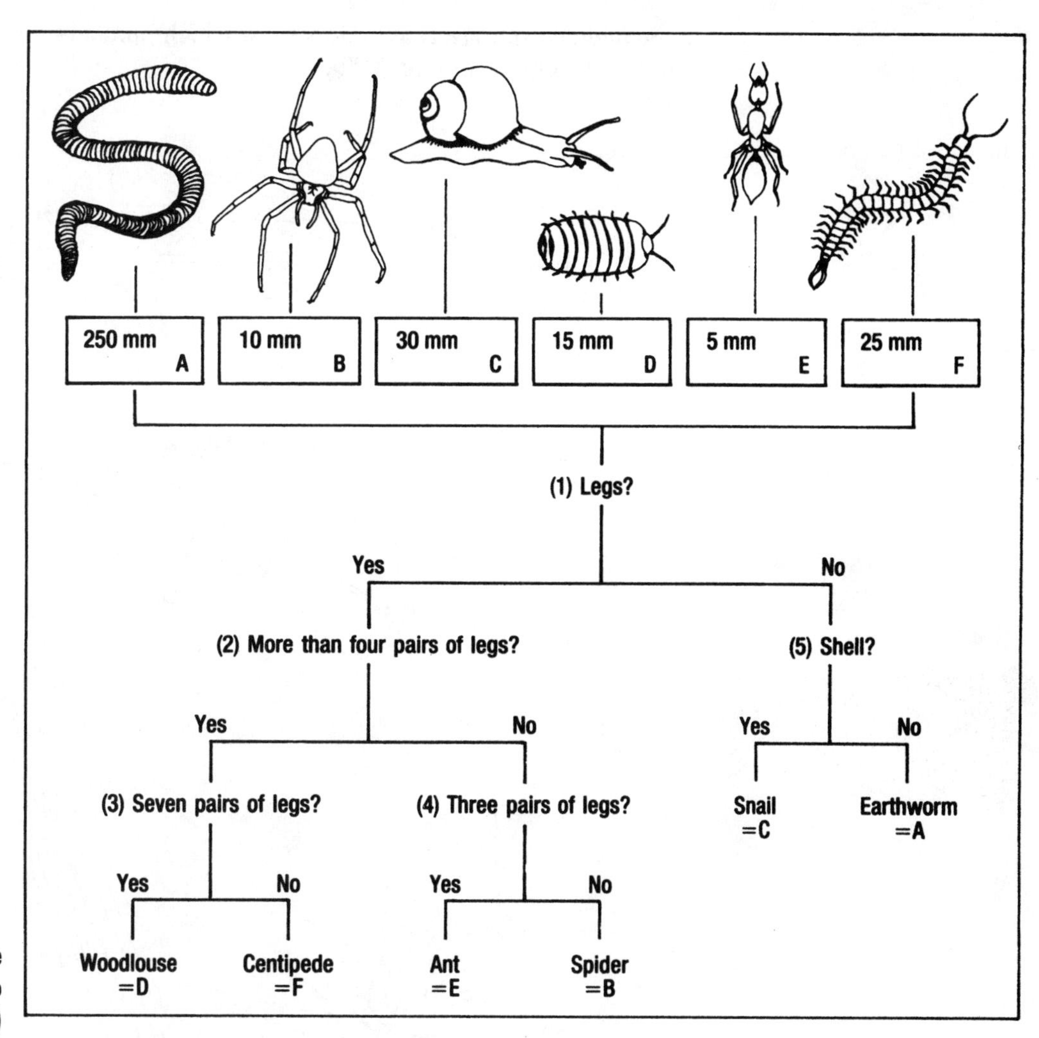

Fig 5.18 Branching key for some invertebrates (Drawings not to scale)

(b) Number Key

This is set out as a list of questions (Fig. 5.19). For each question there is a simple choice and an instruction about what to do next, e.g. go to 2.

1.	Legs Present?	Yes:	2	(B,D,E,F)
	....	No:	5	(A,C)
2.	More than Four pairs of legs?	Yes:	3	(D,F)
	....	No:	4	(B,E)
3.	Seven pairs of legs present?	Yes:	woodlouse	(D)
	....	No:	centipede	(F)
4.	Three pairs of legs present?	Yes:	ant	(E)
	....	No:	spider	(B)
5.	Shell present?	Yes:	snail	(C)
	....	No:	earthworm	(A)

Fig 5.19

This type of key can hold much more information in a small space.

Sometimes you are required to show which numbers you have used to reach the name of the organism (Fig. 5.20).

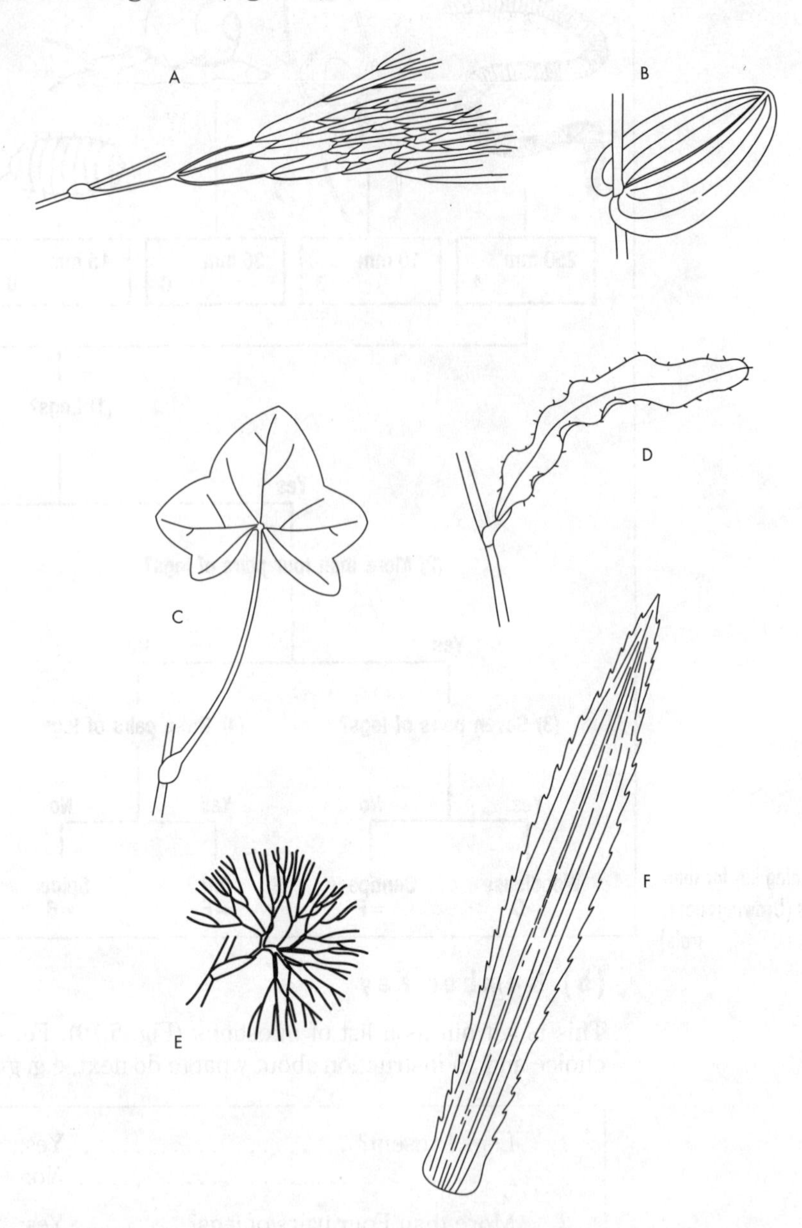

Fig 5.20

Use the key to name each of the leaves. As you work through the key for each leaf, tick the boxes in the table to show how you got our answer.
Leaf A has been done for you.

KEY

			Name of plant
1	(*a*)	Leaf of many fine strands — go to number 2	
	(*b*)	Leaf formed of a flat blade — go to number 3	
2	(*a*)	Strands branch into threes	**River crowfoot**
	(*b*)	Strands branch irregularly	**Round-leaved crowfoot**
3	(*a*)	Leaf has a toothed edge — go to number 4	
	(*b*)	Edge of leaf smooth — go to number 5	
4	(*a*)	Flat leaf sharply toothed	**Water soldier**
	(*b*)	Wavy leaf with many small teeth	**Curled pondweed**
5	(*a*)	Oblong leaf without lobes	**Perforated pondweed**
	(*b*)	Leaf has five main lobes	**Ivy-leaved crowfoot**

Key questions used

❝ You will get marks for filling in the table correctly, as well as for the right answer ❞

Leaf	1(*a*)	1(*b*)	2(*a*)	2(*b*)	3(*a*)	3(*b*)	4(*a*)	4(*b*)	5(*a*)	5(*b*)	Name of plant
A	✔		✔								**River crowfoot**
B		✓				✓			✓		perforated pondweed
C		✓				✓				✓	Ivy leaved crowfoot
D		✓				✓		✓			curled pondweed
E	✓			✓							Round leaved Crowfoot
F					✓		✓				water solder.

Links with other topic groups

Related topics elsewhere in this book are as follows:

Topic	Chapter and Topic number	Page no.
Investigating ecosystems:	7.1	275
Rocky shore ecosystem	7.1	280
Pond ecosystem	7.1	283
Woodland ecosystem	7.1	286
Sequence of decay	8.2	318
Role of living things in decay and cycling of nutrients	8.2	323

EXAMINATION QUESTIONS

Question 1

The diagrams below show leaves of six different trees.

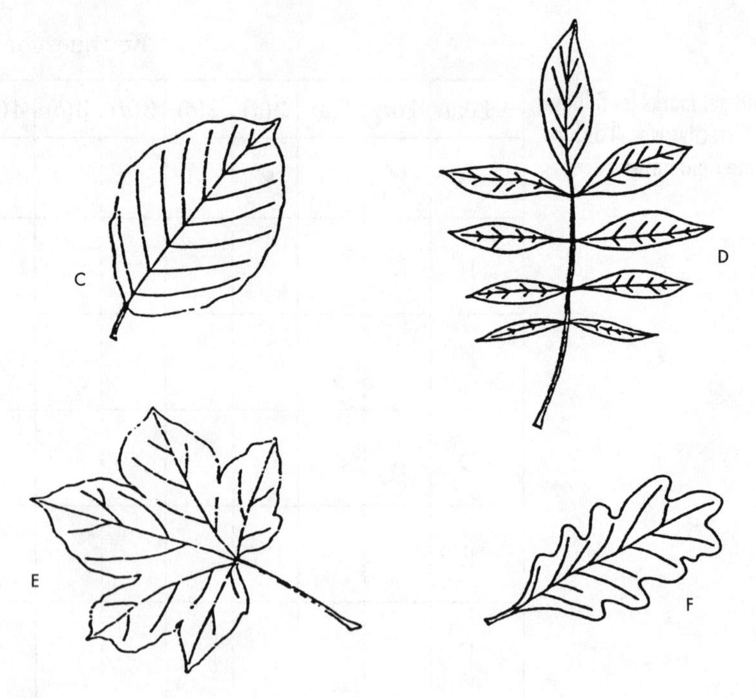

Fig 5.21

Use the key below to name each of the leaves.

	Key	Name of tree
1	(a) Leaf divided into leaflets – go to number 2	
	(b) Leaf not divided into leaflets – go to number 3	
2	(a) Leaflets form a fan on the leaf stalk	**Horse chestnut**
	(b) Leaflets are in pairs on the leaf stalk	**Ash**
3	(a) Edge of leaf has spikes	**Holly**
	(b) Edge of leaf has no spikes – go to number 4	
4	(a) Leaf outline smooth	**Beech**
	(b) Leaf outline lobed – go to number 5	
5	(a) Leaf outline has more than five lobes	**Oak**
	(b) Leaf outline has five main lobes	**Sycamore**

(*Total 6 marks*)
(ULEAC)

Leaf	Name of tree
A	
B	
C	
D	
E	
F	

Question 2

(a) The drawings below show a centipede and a millipede.
Find *three* differences between these two animals. Design a table and list the differences between a centipede and millipede in it.

(b) (i) Name the major animal group of which the centipede and millipede are examples.

(ii) List 3 features of this group visible from the diagrams.

(8)
(ULEAC)

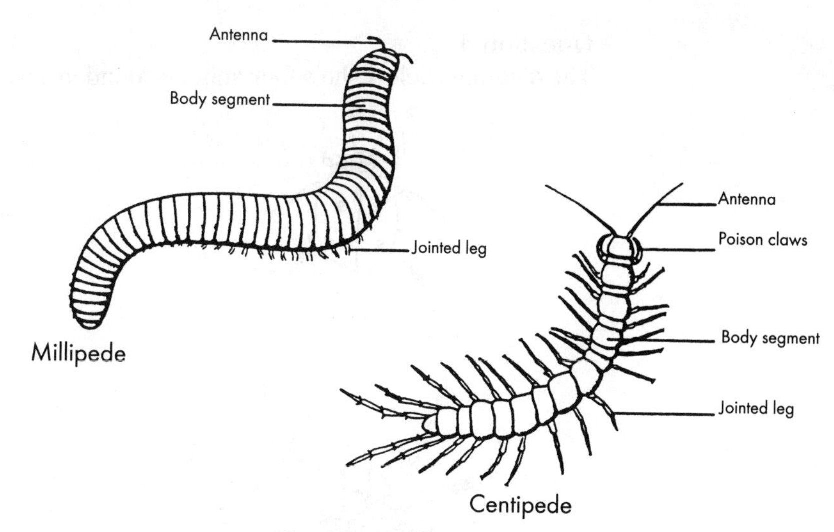

Fig 5.22

Question 3

A sample of soil and decaying leaves was collected from a local wood.

(a) The diagram shows some animals found in this sample.

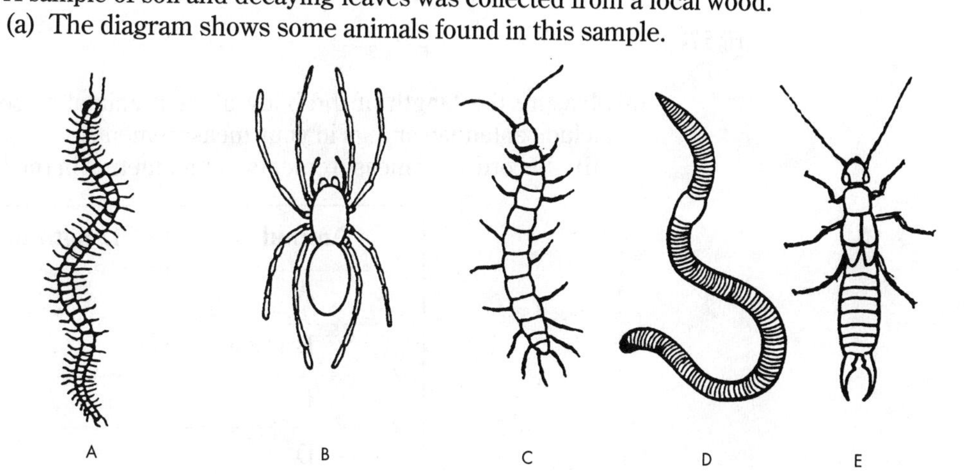

Fig 5.23

Use the key below to identify the animals. Write the letters **A** to **D** in the box below. Animal **E** has been identified for you.

1. 8 or more walking legs present 2
   Fewer than 8 walking legs present 4

2. Legs shorter than the body 3
   Legs longer than the body .... *Lycosa*

3. All body segments have one pair of legs .... *Geophilus*
   Some body segments without legs .... *Scutigerella*

4. 'Pincers' on tip of abdomen .... *Forficula*
   No 'pincers' on tip of abdomen .... *Enchytraeid*

Name of animal	Letter
Lycosa	
Geophilus	
Scutigerella	
Forficula	**E**
Enchytraeid	

(4)

(ULEAC)

Question 4
The diagrams below show four animals found in a wood.

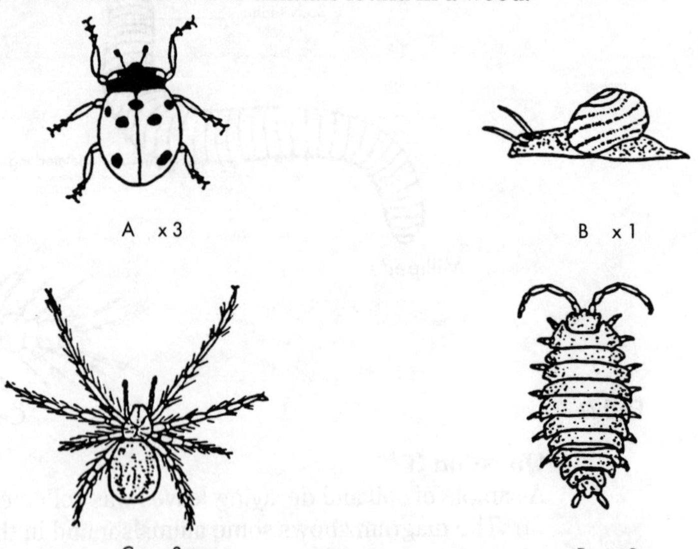

Fig 5.24

(a) Measure the length of the body of each animal as seen in the diagrams. Do not include antennae or legs in your measurements.

(i) Record your measurements in millimetres in the table below.

Animal	Length in mm
A	
B	
C	
D	

(2)

(ii) Use these lengths and the scale beside each animal to calculate their real lengths.

Animal	Real length in mm
A	
B	
C	
D	

(2)

(iii) Put the animals in the order of their real lengths, starting with the shortest.

Order	Animal
Shortest 1	
2	
3	
Longest 4	

(1)

(b) (i) Write down *two* features, which you can see in the diagram, which show that animal **B** is a snail.

1. ______________________________

____

2. ______________________________

____

(2)

(ii) How many of the animals in the diagrams are insects?

____

(1)

(Total 8 marks)

(ULEAC)

Question 5

Match each letter with the **most** suitable number.

(a)	**A**	Earthworm	1	Warm blooded
	B	Bird	2	Tentacles
	C	Mollusc	3	Scales
	D	Fish	4	Saddle

A____ **B**____ **C**____ **D**____

(b)	**A**	Moss	1	Lenticel
	B	Fungus	2	Rhizome
	C	Fern	3	Mycelium
	D	Woody flowering plant	4	Rhizoid

A____ **B**____ **C**____ **D**____

(WJEC)

Question 6

Look at these diagrams of plants and animals.

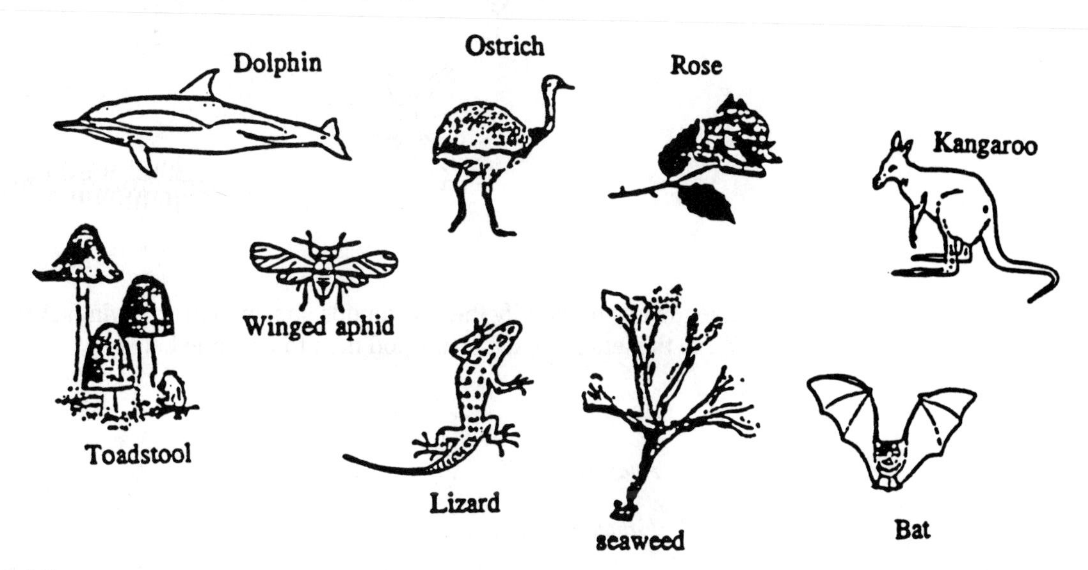

Fig 5.25

(a) (i) Name one animal shown which lives in water. *(1)*

____

(ii) Write down **one** way in which **this** animal is suited to life in water. *(1)*

___

(b) Name one plant shown which grows under water. *(1)*

___

(c) (i) Write down **one** way in which the bat is suited to flying. *(1)*

___

(ii) Name **one** other animal shown in the diagrams which can fly. *(1)*

___

(d) The table shows the groups to which some of the plants and animals in the diagrams belong. Fill in the spaces. *(5)*

Name of plant or animal	**Group**
Ostrich	
	Flowering plant
	Reptile
	Mammal
	Insect

(SEG)

Question 7

The drawings below show five different arthropods, **A**, **B**, **C**, **D** and **E** which are found in woods. The key which follows can be used to identify them.

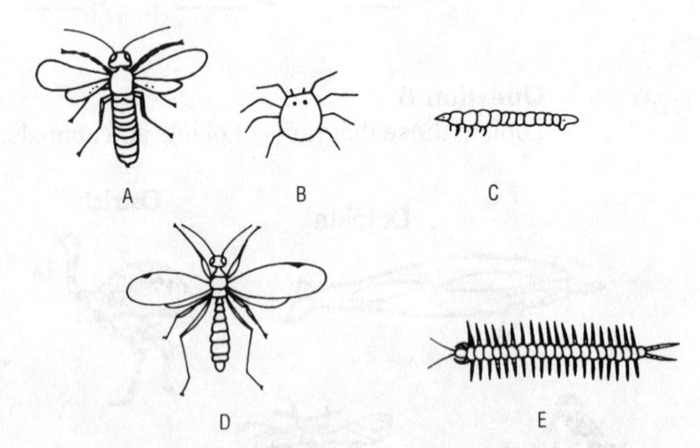

Use the key to identify the arthropods shown in the drawings **A** to **E**. Write the letter of each arthropod next to its name below.

Agriotes ....

Apanteles ....

Lithobius ....

Pergamasus ....

Tenthredo ....

Total 5 marks (ULEAC)

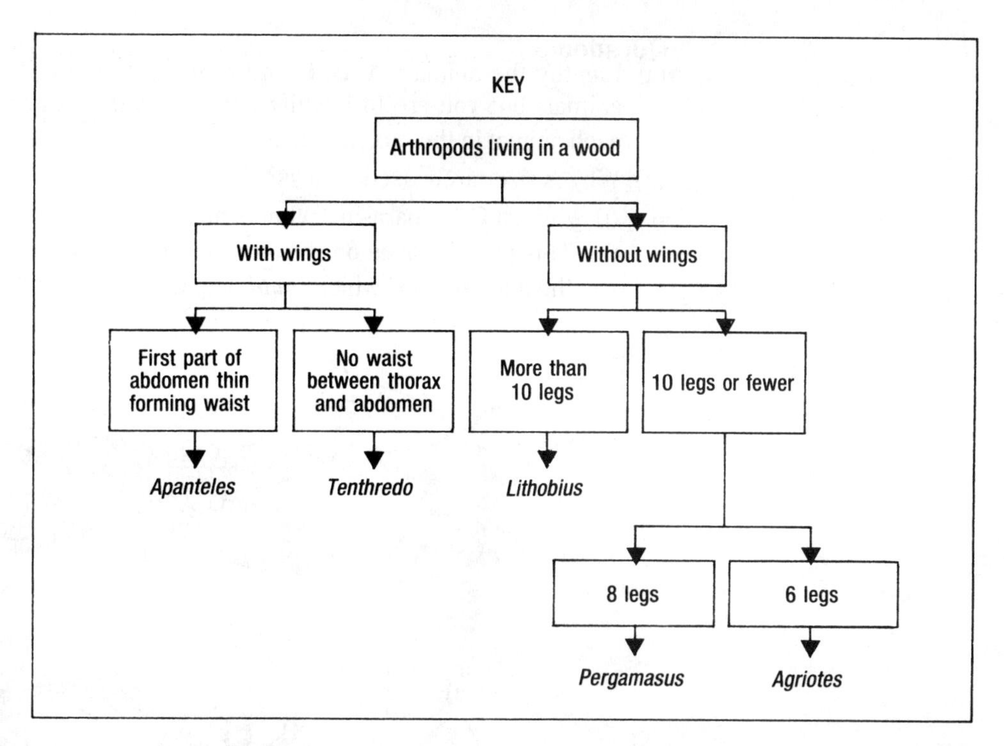

Fig 5.26

Question 8

The animals drawn below were all found in a pitfall trap set up in a hedgerow. They are not drawn to the same scale.

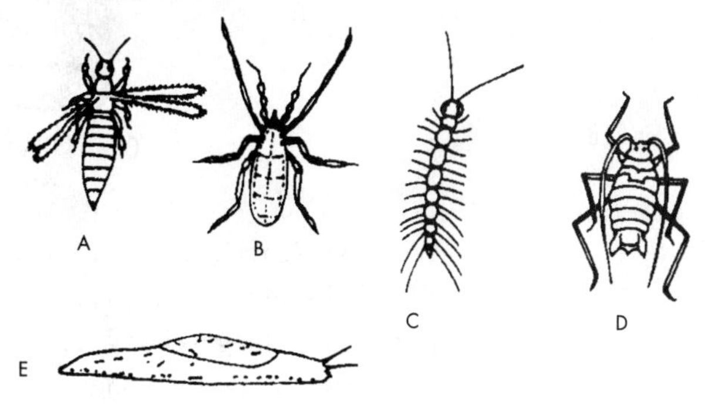

Fig 5.27

(a) What is animal **E**? *(1)*

(b) Use the key below to identify animals **A** and **B** *(2)*

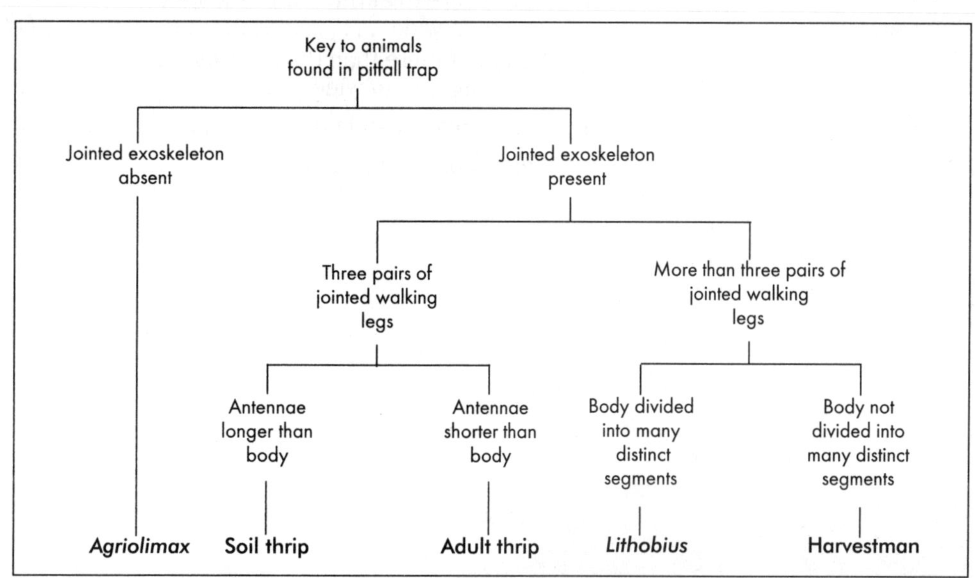

(c) Describe the appearance of *Lithobius* using the information given in the key *(2)*

Total 5 marks (ULEAC)

Question 9

(a) Identify the animals **A**, **B**, **C** and **D** using the key. The key can be used for six animals but you are to identify only the four that are shown. Write the letter of each animal in the box next to its name. *(4)*

(b) Why is size rarely used in keys? *(1)*

(c) (i) Animal **C** is a parasite. What is meant by the term **parasite**? *(2)*

(ii) This parasite lives on the skin of a mammal. Describe **one** feature shown in the diagram of **C** which *could* help it to live as a parasite. *(1)*

Total 8 marks

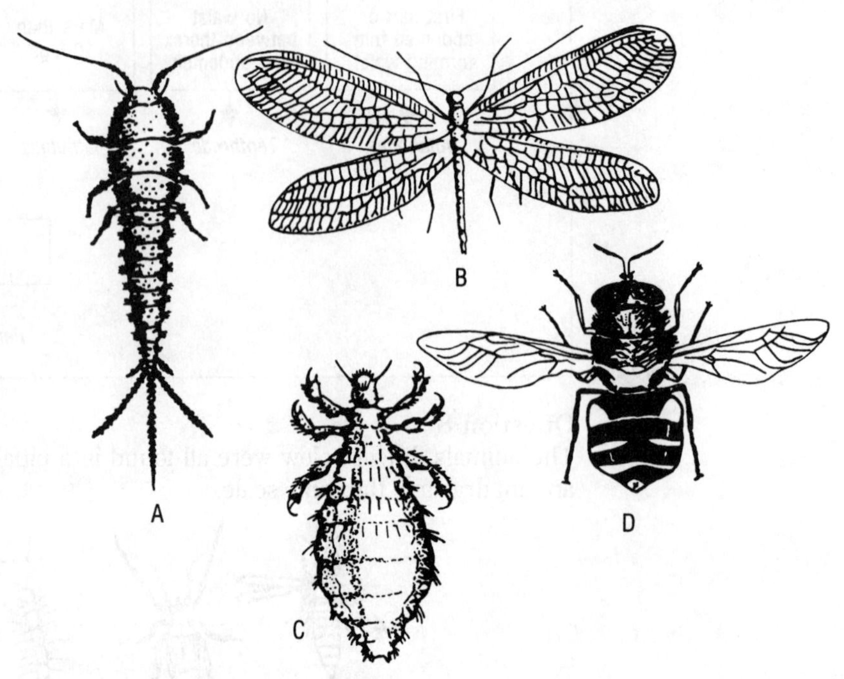

Fig 5.28

1. Wings present 2
   Wings absent 4
2. Antennae as long as body .... *Chrysopa septempunctata* ☐
   Antennae shorter than body 3
3. One pair of wings .... *Stratiomyia potamida* ☐
   Two pairs of wings .... *Aphis rumicis* ☐
4. Body more than three times longer than wide .... *Lepisma saccharina* ☐
   Body not more than three times longer than wide 5
5. Legs longer than body .... *Tegenaria domestica* ☐
   Legs shorter than body .... *Pediculus humanus* ☐

Question 10

STUDENT'S ANSWER – EXAMINER'S COMMENTS

(a) Name TWO major groups of vertebrates which maintain a constant body temperature.

1. dog
2. whale

"These are good examples of endothermic animals, though they do not answer the question as set."

"These are not 'major groups of vertebrates', though they are examples of vertebrates. The correct answer is mammals and birds"

(b) Describe one way in which you would distinguish between

(i) a bird and a reptile,

wings and feathers

"Good answer, though only one structure is needed here. Does this refer to bird or reptile"

(ii) a reptile and an amphibian,

Scaly skin
they live in water

"A good distinguishing feature, but does this refer to reptile or amphibian?"

(iii) an amphibian and a fish.

fish can swim

"It is best to avoid behavioural differences; refer instead to structures."

(5 marks)
(Total 5 marks)

Note: Answers to the Examination Questions 1–9 can be found at the end of the book

CHAPTER

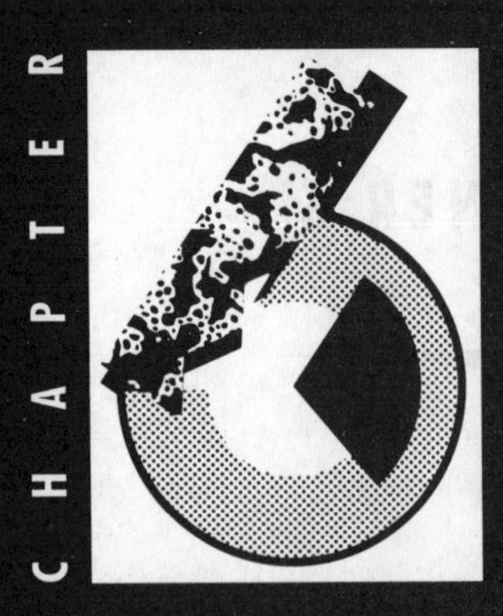

VARIATION, INHERITANCE AND EVOLUTION

INFORMATION FROM GENES

SELECTIVE BREEDING

GENETICS AND CELLS

PATTERNS OF INHERITANCE

HUMAN GENETICS

CAUSES OF GENETIC VARIATION

VARIATION, NATURAL SELECTION AND EVOLUTION

DNA AND PROTEIN SYNTHESIS

GENETIC ENGINEERING

GETTING STARTED

Heredity. Organisms often share certain characteristics with their parents. These characteristics are determined by genetic information transferred during reproduction. Individuals produced by sexual reproduction tend to be more varied than those produced by asexual reproduction. This is because sexual reproduction usually involves the combination of genetic information from two rather than one parent. Genetic information is carried within the nucleus of cells within thread-like **chromosomes**. Particular sections of each chromosome, called **genes**, determine certain characteristics within the organism. Characteristics are controlled by one gene, or by several genes operating together.

Variation. There is both considerable variation and uniformity amongst living things. The way in which individual organisms are similar or different to each other is determined by the genetic information they contain (their **genotypes**) and also the effect of the environment in which they live. During an organism's growth and development there is an interaction between the genotype and its environment. The outcome of this interaction is the **phenotype**, which is the set of characteristics which make up the organism.

The genotype is what the organism *could* become. The phenotype is what the organism *does* become as a result of its genotype and its environment acting together.

Even in organisms of the same **species**, the phenotype can be quite varied for certain characteristics. There are two types of variation, continuous and discontinuous. **Continuous variation** tends to occur when characteristics are determined by several genes, or because growth and development are readily affected by the environment. **Discontinuous variation** tends to occur when characteristics are determined by one gene, or when the environment does not directly affect that particular characteristic.

Evolution is the process by which living things gradually develop from earlier forms by progressive inheritable changes.

There are two important outcomes of evolution:

1. **Increase in complexity.** The change from one form to another usually involves the production of more complex organisms from simpler *ancestral* forms.
2. **Increase in diversity.** The formation of *new species* is a significant part of the process of evolution.

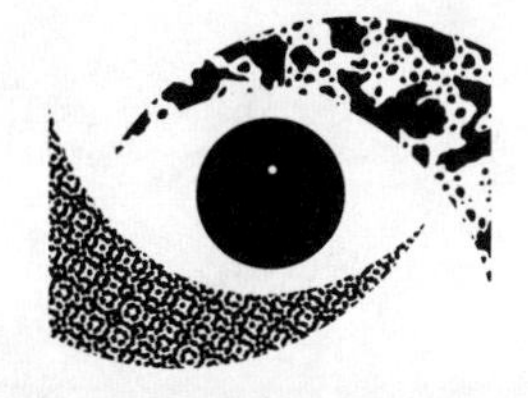

ESSENTIAL PRINCIPLES

1 INFORMATION FROM GENES

The Topic Group 'Information from genes' describes the way in which genetic information is inherited from one generation to the next. The 'unit of inheritance' is the *gene*. In general, each gene is responsible for 'instructing' the cell to perform a particular function. Genes are carried on *chromosomes* within the nuclei of all cells. Chromosomes are arranged in pairs in most cells. The gene for each pair exists as two possible forms, called *alleles*. Alleles are really the two alternative ways in which genes can be expressed. Having different alleles increases the variety of genetic information can be organised, since even simple organisms contain thousands of genes, and there is an infinite number of all allele combinations!

If you are new to genetics, you may need to spend time becoming familiar with the terms used. You will also need to practice drawing or interpreting *genetic diagrams*. When you're revising this part of the syllabus, be sure to refer to other related topics, such as cell division and reproduction.

GENES IN INHERITANCE

Key Points

- *Genes* are inherited units of information, contained in *chromosomes*, within each cell nucleus. Genes are passed on to offspring during reproduction. In sexual reproduction, genes are contained in male and female sex cells. These fuse during fertilisation, forming a cell which contains genes from both parents. The cell then develops into a new organism consisting of many cells. Copies of the original set of genes are copied each time a new cell is produced.
- In most cases, each gene is responsible for controlling a particular characteristic in an organism. Each gene can exist in two alternative forms, called *alleles*. These occur in pairs normal body cells, but become separated when sex cells are made. Each offspring inherits one allele of each gene from each parent. Sexual reproduction allows a 'mixing' of alleles, so that offspring from the same parents are not the same as each other, or their parents.

"Level 5."

1 Heredity and genes

"This is an important difference between asexual and sexual reproduction."

Organisms often share certain characteristics with their parents. These characteristics are determined by genetic information transferred during reproduction. Individuals produced by sexual reproduction tend to be more varied than those produced by asexual reproduction. This is because sexual reproduction usually involves the combination of genetic information from two rather than one parent. Genetic information is carried within the nucleus of cells within thread-like *chromosomes*.

Particular sections of each chromosome, called *genes* determine certain characteristics within the organism. Characteristics are controlled by one gene, or by several genes operating together.

Genes are particular regions of chromosome which are responsible for determining a particular inherited characteristic in organisms. Genes are 'units of heredity' which are transferred from one generation to the next during reproduction. Offspring inherit their genes from their parents.

Chromosomes occur in pairs, and each chromosome of the pair carries the same set of genes. However, the genes on the two chromosomes are not necessarily a 'matching set'. This is because genes can exist in different forms, or *alleles*. Alleles are alternative forms of genes, occupying the same relative position on paired chromosomes. For example, the gene controlling fur colour in mice may have two alleles, one for black fur, the other for brown fur (Fig. 6.1). In studies of inheritance, or *genetics*, alleles are given symbols, for convenience. In this example, the alleles are labelled with the letters 'B' and 'b'. This is explained below.

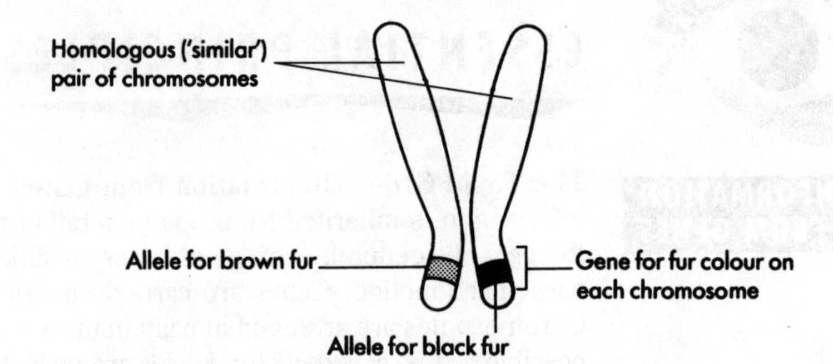

Fig. 6.1 Example of alternative alleles for a single gene: fur colour in mice

2 Inheritance of genes and alleles

During sexual reproduction, sex cells (gametes) are formed by a special type of cell division called *meiosis*. This causes the pairs of chromosomes to separate. Alleles are also separated during this process (Fig. 6.2).

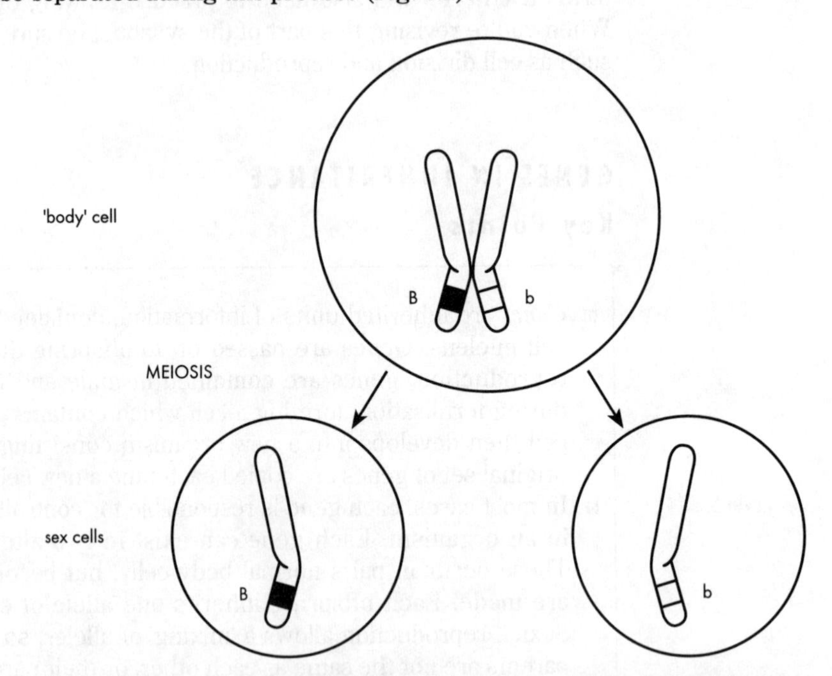

Fig. 6.2 Separation of alleles 'B' and 'b' during meiosis

Sexual reproduction involves fertilisation, which brings pairs of chromosomes back together, although they are not the original pairs, since one chromsome is contributed by each parent. This results in new combinations of alleles (Fig. 6.3). The offspring, like the parents, now has two alleles for each gene. However, the allele combination may be different from either or both parents.

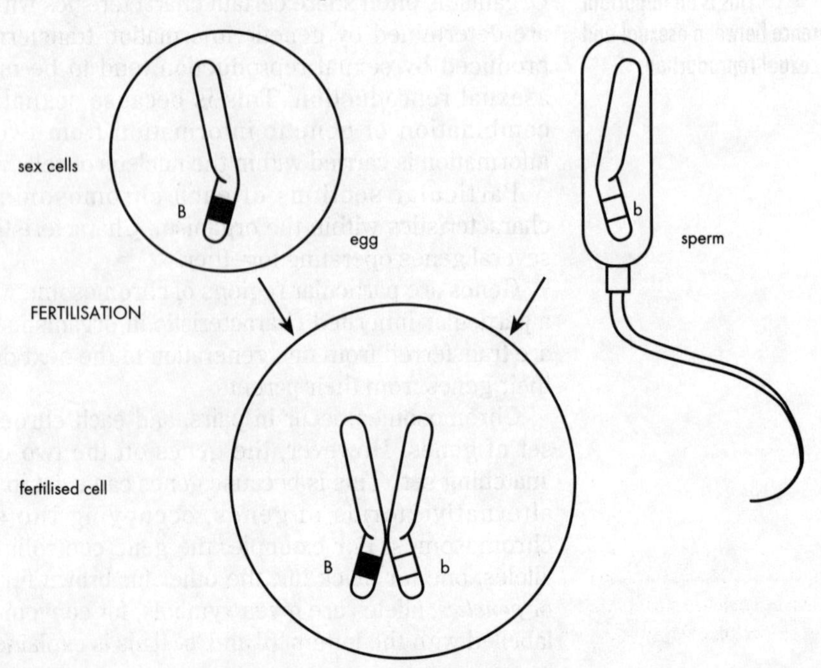

Fig. 6.3 Combination of alleles 'B' and 'b' during fertilisation

The way in which alleles are actually 'expressed' in the offspring depends on whether one allele is ***dominant*** over the other. Dominant alleles are always expressed if they are present in the cell. The alternative type of allele is called ***recessive.*** Recessive alleles will only be expressed if no dominant allele is present. In our example, black is the dominant allele, and brown is the recessive allele.

We can now combine Figs. 6.2 and 6.3 to show the possible fur colour in the offspring (Fig. 6.4). In 'genetic diagrams' such as Fig. 6.4, letters are used for convenience. The letter used is normally based on the dominant allele and is written in upper case, i.e. dominant black = 'B'. The recessive allele is written in the lower case, i.e. brown = 'b'. In this example, 'b' means 'recessive of the dominant allele 'B', rather than brown. If the recessive fur colour was cream, the symbol would still be 'b'!

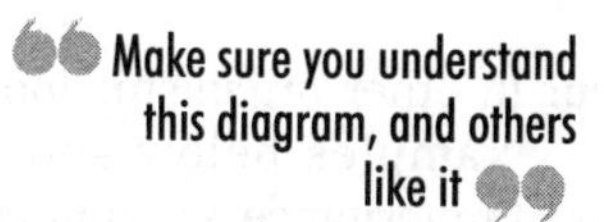

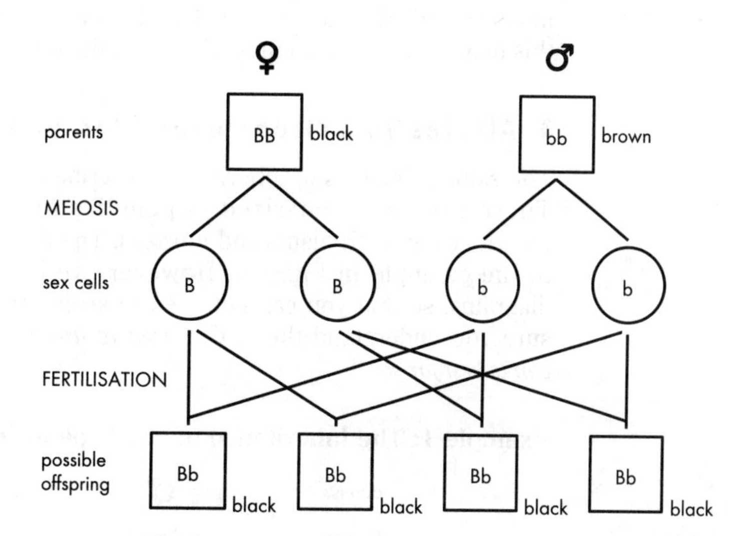

Fig. 6.4 The possible offspring produced from a black mouse and a brown mouse; parents both true-breeding

The offspring are all black, since the black allele is present in all offspring (Bb), and black is the dominant allele. Note that the parents of the fertilisation (or '*genetic cross*') shown Fig. 6.4 each contain *either* alleles for black fur (BB), *or* alleles for brown fur (bb), but not both. If two black mice were parents, all the offspring would be black. If two brown mice were the parents, all the offspring would be brown. Such parents are said to be ***true-breeding***, or ***pure breeding***. The offspring shown in Fig. 6.4 are ***not true breeding***. This can be shown in another genetics diagram (Fig. 6.5). This can be regarded as the '***second generation***' (= 'F_2'), since fertilisation by true-breeding parents is often regarded as the '***first generation***' (= 'F_1') in a genetics example.

The possible offspring from fertilisation involving two not true breeding mice are three black mice and one brown mouse. Although three allele combinations contain

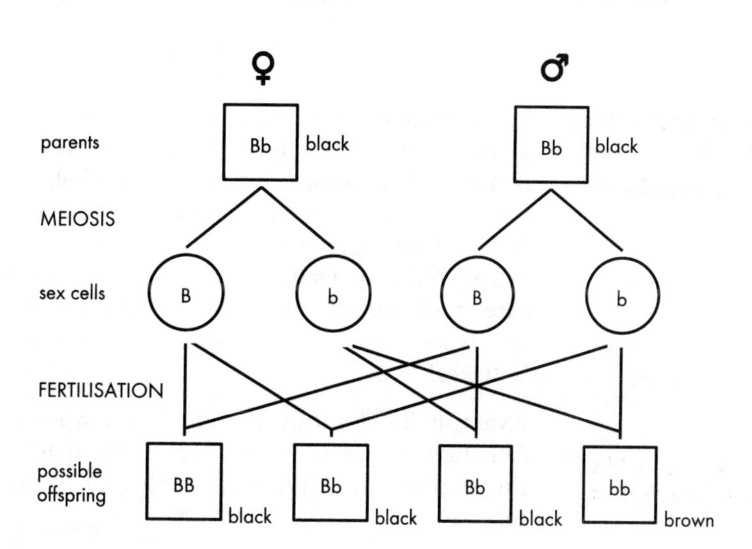

Fig. 6.5 The possible offspring produced from two black mice; parents both not true breeding

the recessive allele (i.e. for brown fur), only one actually appears in the offspring; this is a *true breeding* brown mouse (bb). The brown allele in the other two offspring is 'hidden' by the dominant allele (i.e. for black fur) (Bb). The remaining mouse is *true-breeding* back (BB). In genetics, there are terms to describe the combination of alleles which are either true breeding or not true breeding. *Homozygous* combinations consist of the same type of alleles, e.g. 'BB' or 'bb'. The two types of homozygous combinations are called *homozygous dominant* (e.g. 'BB') and *homozygous recessive* (e.g. 'bb'). *Heterozygous* involves both types of alleles (e.g. 'Bb').

It is important for you to learn the meaning of these terms

Notice in Fig. 6.5 the ratio of dominant : recessive offspring is 3:1. This *3:1 ratio* is characteristic of a genetic cross involving two heterozygous parents. It should become familiar to you as you study genetics. This ratio (and others, too) can be used to *predict the probability* of a certain type of allele combination occuring amongst the offspring. In our example shown in Fig. 6.5, the theoretical probability of a brown mouse occurring would be 1 in 4. However, because of the chance events involved, this may not actually occur, especially if the number of offspring is small.

3 Alleles in action: examples of inheritance

The same principles of inheritance described for mice occur in other organisms too. These can be summarised in genetics diagrams. The examples below show inheritance in both plants and humans. The basic principles of inheritance are similar to the example in Fig. 6.4. However, we have used different types of genetics diagrams, so that you can get used to seeing them. Study them carefully, and make sure you understand them. *Get used to drawing the type of genetics diagram which works best for you*!!

Example 1: The inheritance of petal colour in garden pea flowers

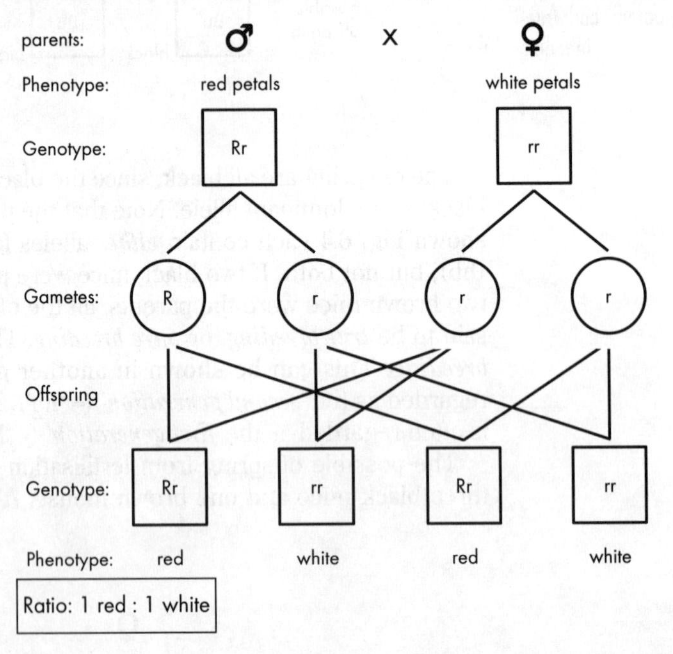

Fig. 6.6 Inheritance of petal colour in pea flowers

These examples should help you become familiar with genetics diagrams.

In this example, red (R) is dominant, and white (r) is recessive. The 'genetic cross' shown in this example (Fig. 6.6) involves one heterozygous (Rr) parent and one homozygous recessive (rr) parent. Symbols are used for male ♂ and female ♀, but these are optional. In fact, in this example the identity of the parents doesn't matter. Note that the gametes (= sex cells) are shown in circles; this helps them 'stand out'. When you're drawing this type of genetic diagram, *be very careful to link each gamete from each parent with each gamete from the other parent*. There should be four genotypes (some, or all, may be the same!), and *gametes from the same parent must not be linked*!

Example 2: The inheritance of hair wavyness in humans

Although this example may look unfamiliar, in fact it is very similar to Fig. 6.5; only the symbols are different! Curly (C) is the dominant allele, straight (c) is the recessive.

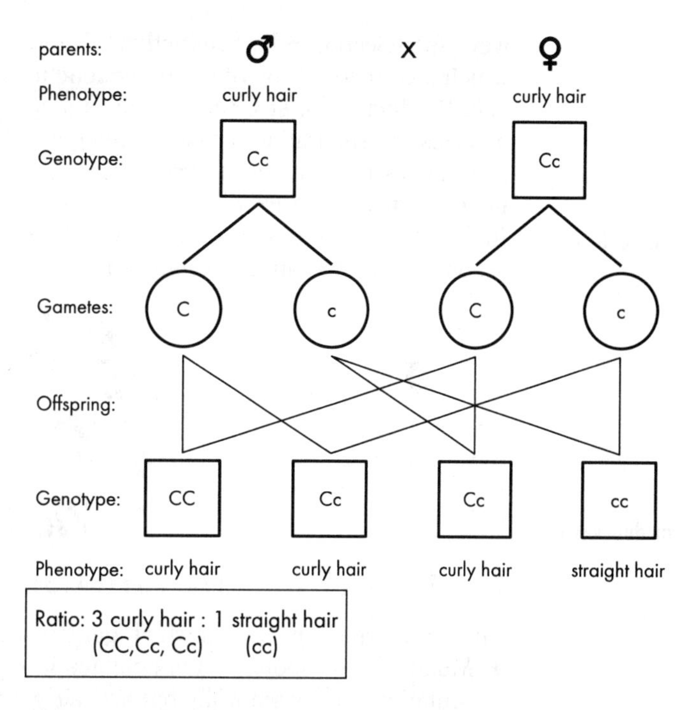

Fig. 6.7 Inheritance of hair wavyness in humans

GENETIC AND ENVIRONMENTAL SOURCES OF VARIATION

Key Points

- *Body (somatic) cells* contain a complete set of genetic material in their nuclei. The genetic information is organised into *chromosomes*, each of which carries a set of *genes*. Chromosomes occur in pairs, so that each pair carries two of each gene. Two different 'versions', or *alleles*, exist for each gene. Paired chromosomes may carry different alleles of the same genes.
- During the production of *sex cells (gametes)*, paired chromosomes separate. Each sex cell contains half the amount of genetic material of body cells. Different sex cells may contain different alleles for each gene.
- During fertilisation in *sexual reproduction*, a sex cell from each parent become fused. The resulting fertilised cell contains a full set of genetic material, half contributed by each parent. Sexual reproduction can produce genetic variability because it allows a re-mixing of alleles, resulting in a unique combination in most offspring.
- During development, organisms are subject to more variation, caused by the effects of the *environment* on the body's functioning, and even on the genetic material itself. In general, the growth and development of higher animals tend to be influenced less by environmental factors compared with lower animals and plants.

1 The sources of variation: genetics and the environment

There is both considerable variation and uniformity amongst living things. The way in which individual organisms are similar or different to each other is determined by the genetic information they contain, their *genotype*, and also the effect of the environment in which they live. During an organism's growth and development there is an interaction between the genotype and its environment. The outcome of this interaction is the *phenotype*, which is the set of characteristics which make up the organism.

The genotype is what the organism *could* become. The phenotype is what the organism *does* become as a result of its genotype and its environment acting together.

"Identical twins have the same genotype. Why are they not always alike?"

Genetic information is contained within almost all living cells. In each body cell (sex cells are an exception), a full set of genetic information is carried. The information is contained within paired *homologous* (= 'same') *chromosomes*, which carry *genes*, responsible for particular functions. Each gene can exist in two alternative forms, called *alleles*. Different alleles can cause the gene to be expressed in different ways (see Section 6.1). Production of body cells involves cell division (by mitosis) in which a complete copy of all the genetic material is passed on to each new 'daughter' cell. Production of sex cells involves a special type of cell division (*meiosis*) which involves the separation of the homologous chromosome pairs. Each part (half) of the pair carries its own set of alleles. This means that sex cells may have quite different genetic information from each other, though they both have a complete set of genes.

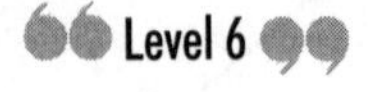

In the case of humans, normal body cells contain 46 chromosomes (Fig. 6.9). Sex cells (eggs, sperm) contain 23 chromosomes.

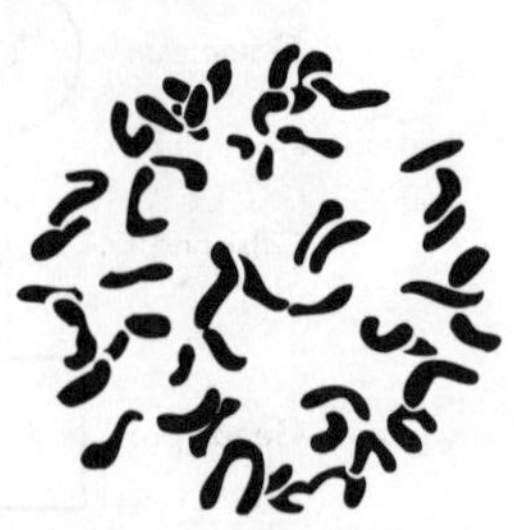

Fig 6.9 Human chromosomes

2 Effects of Genetic Variation

There are two main ways in which genetic variation can occur:

- Mutations, or spontaneous changes in the structure of chromosomes and genes. Mutations occur naturally but are fairly rare. However, the rate at which mutations occur is increased by environmental factors. This is described in more detail below.
- Re-assortment of chromosomes and alleles during sexual reproduction. Sexual reproduction involves a separation of paired chromosomes when sex cells are formed (by meiosis). Fertilisation brings genetic material together from two parents. The actual way in which genetic material is combined involves chance events. The examples which follow should illustrate this idea. You should refer back to Section 6.1 if you have difficulty understanding these examples.

(a) Variations in chromosomes

Level 8

A good example of genetic variation involving whole chromosomes is the inheritance of sex in humans. Sex chromosomes determine the *gender* (male or female) of an individual. Sex chromosomes carry genes which provide genetic information relating to gender, such as *primary* and *secondary development.*

The sex chromosomes also carry genes which are not directly related to gender. All other genetic information is carried within the rest of the set of chromosomes which contain an individual's genotype. The non-sex chromosomes are sometimes called *autosomes.*

In humans, the sex chromosomes are either 'X' or 'Y', and the alternative genotypes are XX = female and XY = male (note these are respectively homozygous and heterozygous genotypes). The pattern of inheritance of the sex chromosomes is shown in Fig. 6.10. There is an expected 1:1 ratio of males:females in the offspring, but the chances of this happening are reduced when parents have fewer children. However, there is a male:female ratio of approximately 1:1 in the human population as a whole.

Sex-linked genes are carried on the sex chromosomes but are not directly involved with sexual development. Examples of those causing disease include red-green colour blindness and haemophilia (See topic group 6.5).

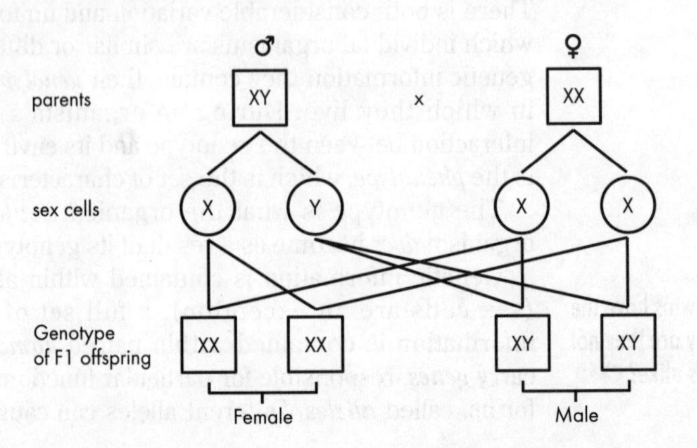

Fig. 6.10 Inheritance of sex in humans

(b) Variations in alleles

A classic example of this is based on experiments conducted by Gregor Mendel (1822–1884). Mendel used the garden pea which has a hermaphrodite flower and is normally self-pollinated. For this reason the plants were usually true breeding before they were artificially cross-pollinated.

Mendel carefully observed patterns of inheritance in the phenotypes of many crosses. One classic example of his work is briefly presented here. (Modern terminology is given in brackets):

(i) When tall or dwarf pea plants were self-pollinated ('selfed') repeatedly, no variation in height occurred in the offspring. Conclusion: each parent was true-breeding (homozygous).

(ii) When tall and dwarf were cross-pollinated ('crossed', or hybridised), the offspring from the first generation were all tall. Conclusion: the 'germinal unit' (allele) for tallness was dominant. The germinal unit for dwarfness was recessive.

(iii) Hybrid (heterozygous) offspring from the first generation (F1) were either self-pollinated, or else cross-pollinated with each other. The offspring in the next (F2) generation were tall and dwarf in the ratio 3:1. (This is known as the *monohybrid* ratio). Conclusion: a pair of contrasting characteristics (alleles) are carried separately within individual gametes. This example is shown as a genetics diagram in Fig. 6.11.

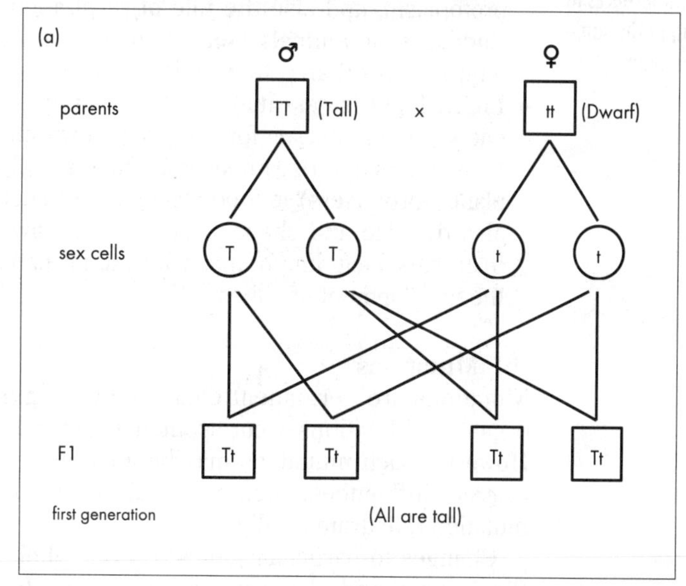

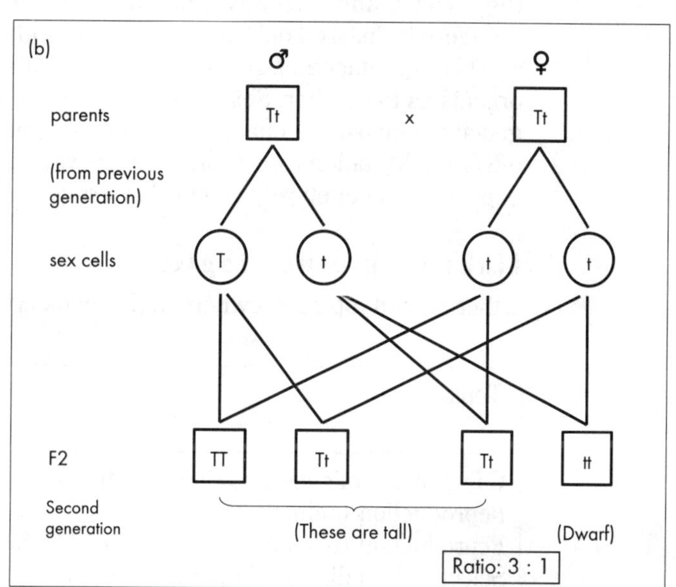

Fig. 6.11 Genetic variation in the height of peas. The genetic diagrams show (a) the first generation, and (b) the second generation.

The monohybrid (3:1) ratio shown in Fig. 6.11 is characteristic of a cross involving two heterozygous (e.g. *Tt*) parents, or a selfing of one heterozygous hermaphrodite

Can you think what advantage a pea plant might have by being tall?

individual. You should note that the 3:1 ratio assumes random fertilisation of gametes and is only approximate in a small number of samples. In Mendel's original experiment, he obtained 787 tall (i.e. about 200 cm high) and 277 dwarf (i.e. about 30 cm high) plants; this ratio is actually 2.84:1.

The 3:1 ratio (and other such ratios in genetics) can be used to *predict* the results of crosses. For example, the chances of a recessive fruit appearing as a result of a monohybrid cross would be 1 in 4.

3 Effects of environmental variation

The environment can modify the way in which genetic information is expressed, and it can even alter the genetic material itself, in mutations.

(a) Variations in the growth of plants and animals

Both plants and, to a lesser extent, animals respond to numerous factors.

Level 6

- **Nutrients.** The availability of nutrients, including water, influences growth because they provide the raw materials for the synthesis of protoplasm. Nutrients are also needed for the production of extracellular substances, e.g. cellulose and lignin in plants, chitin and bone in animals.

Devise an experiment to show either (a) environmental influences on genetically identical plants, or (b) genetic differences in plants growing in the same environment.

- **Temperature.** Growth tends to occur more rapidly with increased temperature, within certain limits and provided other factors are favourable. Temperature influences the rate of metabolic reactions, including the synthesis of new protoplasm, and also the rate of respiration, which provides energy for synthesis. Endothermic animals (see Chapter 4) are less directly dependent on favourable temperatures than ectothermic animals.
- **Light.** Light is essential for photosynthesis in green plants. This process provides energy and materials for the plants' own metabolism and also for other organisms (consumers and decomposers) which are directly or indirectly dependent on green plants (producers) in food chains. Light affects the development of chlorophyll and also the size and shape of stems and leaves. Light is directly important in some mammals, including humans, for the formation of vitamin D, which indirectly affects the development of bones.

(b) Mutations

Mutations are permanent changes in the genetic composition, or genotype, of an organism. Mutations occur spontaneously in dividing cells, because of *internal* factors. However, such mutations may be a fairly rare event. Mutations can also result from *external* influences, such as certain radiation and chemicals; these may raise the mutation rate dramatically.

Changes to particular genes or even whole chromosomes affect the cell in which they occur, and also any cells produced from it by cell division. Mutations can therefore be 'passed on' from one cell to another.

This inheritance of genetic change can also be passed on from one generation of organisms to another. Sex cells or other cells involved in reproduction may have their genetic composition changed by mutation, and this may have a significant effect on offspring. Mutations are really the only way in which variation can be introduced into a population of genetically identical *clones*, produced by asexual reproduction.

Links with other topics

Other related topics elsewhere in this book are as follows:

Topic	Chapter and Topic number	Page number
Components of cells, and their function	4.3	36
Reproduction (animals)	4.5	65
Reproduction (plants)	4.6	117
Selective breeding (influences)	6.2	209
Chromosomes	6.3	214
Meiosis	6.3	219
The monohybrid cross	6.4	223
Protein synthesis	6.8	253

REVIEW QUESTIONS

Q1 An eye disease called 'glaucoma' can be inherited.

The diagram (Fig. 6.8) shows a family in which some of the people have glaucoma.

(a) What does 'inherited' mean?

____

____ *(1)*

(b) Use the diagram to explain why Jill has glaucoma.

____

____ *(1)*

(c) Why does John not have glaucoma?

____

____ *(1)*

(d) One of Jane's grandfathers had glaucoma.

Tick the box to show which of her two grandfathers this was more likely to be.

Jean's father **or** David's father

Give a reason *for* your answer.

____

____ *(1)* (4 marks)

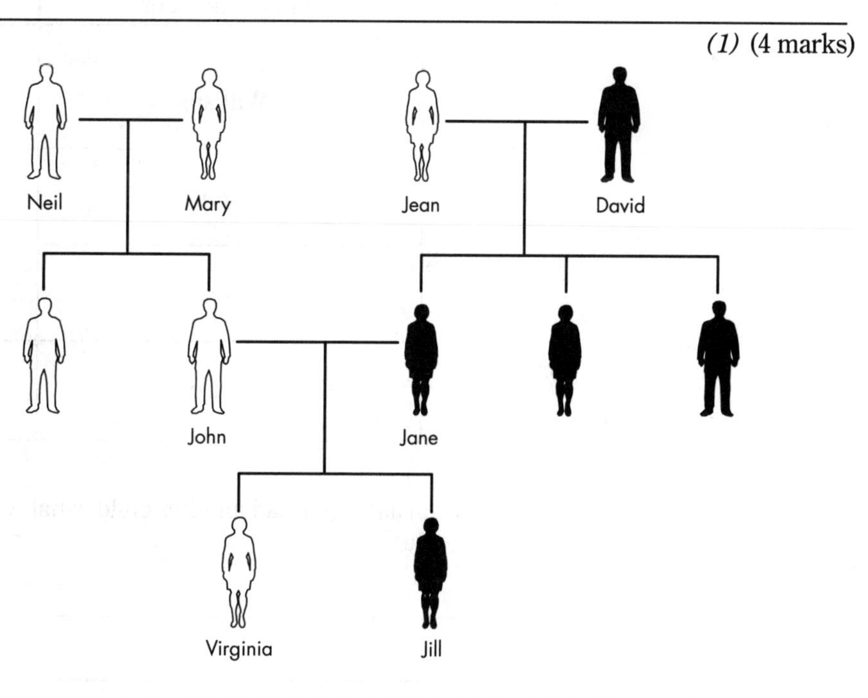

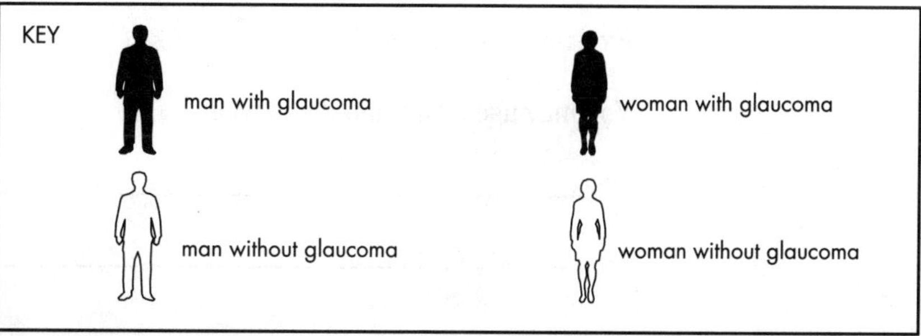

Fig. 6.8

Q2 Freckles are small patches of coloured skin.

The allele for freckles (a) is recessive to the allele for unfreckled skin (A).

The diagram (Fig. 6.12) below shows a family where some of the people have freckles.

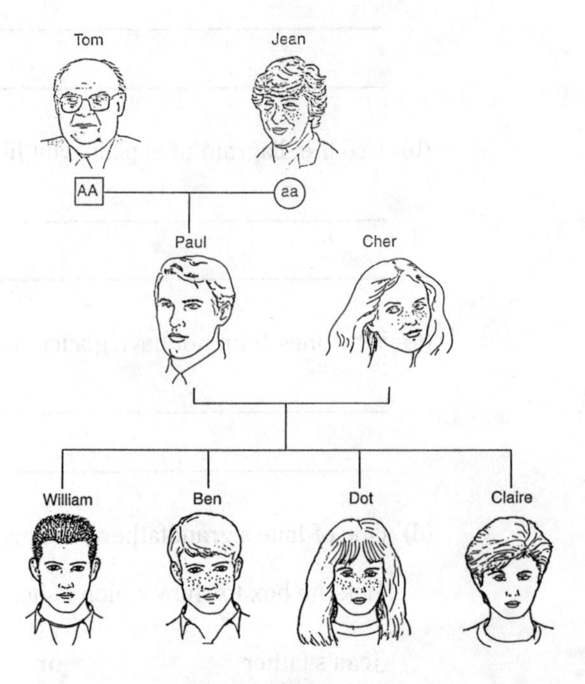

Fig. 6.12

(a) Look carefully at the diagram (Fig. 6.13) and complete the table to show the genetic make-up (genotype) of William, Ben, Dot and Claire.

family member	genetic make-up
William	
Ben	
Dot	
Claire	

Fig. 6.13

(b) If Paul and Cher had another child, what would be the chance of it having freckles?

____

____

Explain your answer.

You may use a diagram if you wish.

____

____

____

Q3 Fig. 6.14 represents a genetic experiment involving black animals and white animals. The parents 1 and 2 had many more offspring than are shown and all of them were black.

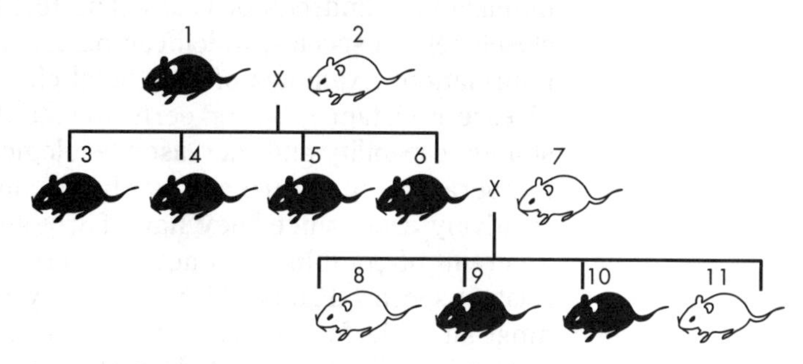

Fig. 6.14

(a) If the genotype of animal number 5 is Bb, state the genotypes of animals number 1, number 7 and number 9.

____

____

(b) In this experiment, what is the phenotype of an animal which is homozygous recessive?

____

(c) List the numbers of all the animals shown in the diagram which are homozygous.

____

____

(d) State the ratio of the phenotypes of the offspring produced as a result of crossing animals 9 and 10.

____

2 SELECTIVE BREEDING

Selective breeding involves the application of genetic principles for the improvement of crops and domestic animals, by human intervention. Human communities are completely dependent on crops and domestic animals for food, clothing materials, building materials, and many other uses. However, humans may need to modify the range of 'natural' characteristics of crops and domestic animals, to make them more appropriate for human needs. This can be done by repeatedly selecting offspring with desirable characteristics over several generations.

Selective breeding, also known as *artificial selection* is really a kind of 'experiment in evolution', because desirable characteristics are methodically selected, and undesirable characteristics are rejected. At this stage, you may need to revise previous topics on genetics and variation.

Key Points

Level 7

- The purpose of *selective breeding* of *crops* is to increase desirable characteristics such as high yield and disease resistance. This is achieved by 'crossing' parents which can each contribute a useful characteristic. The result of a successful cross are plants which combine the characteristics of the parent plants.
- *Selective breeding* has been used over many years to change and improve characteristics of *domestic animals*. Individual animals with different desirable traits have been bred together, to produce offspring which combine the characteristics of their parents. Examples of domestic animals which have been selectively bred include cattle, poultry, pigs, horses and dogs.

SELECTIVE BREEDING-CROPS

Selective breeding, or *artificial selection*, is a technique which has been used by humans for hundreds of years. The technique exploits natural variation between closely-related species, which can be manipulated to combine their beneficial genetic information. Examples of beneficial characteristics include high yield, increased disease resistance, better performance during mechanical harvesting, improved storage capability and increased 'ecological tolerance' to a range of environmental temperature, moisture and daylight conditions. Selective breeding in plants is relatively easy, since they have fairly short generation times, and are generally 'tolerant' of combining genetic material from plants from other (usually closely related) species. Plants which are usually *self-pollinating* (see Fig. 6.11) tend to be the most suitable for selective breeding programmes, since they do not become 'contaminated' with unwanted genetic material from other plants.

The result of many successful breeding programmes, particularly recently, has been a dramatic improvement in the quality and quantity of crop products. This has been called the 'green revolution'. Using this approach, for example, different strains of maize have been bred to produce maize with either high protein and low oil content, or low protein and high fat. Different varieties produced in this way are available for different markets and different products.

The procedure of selective breeding basically involves repeated selection of useful characteristics, and rejection of undesirable characteristics, over many generations. For example, if a tomato plant produces large fruit, a cultivator can remove the seeds and germinate them to produce more offspring. This process may need to be repeated many times, especially if a particular characteristic is controlled by more than one gene. Also, genes from other plants can be introduced, e.g. a tomato plant with large fruits can be *cross-pollinated* (*'crossed'*, or *hybridised*) with a plant which has fruit with excellent storage properties. However, once a desirable combination of characteristics have been achieved, selective breeding is also needed to chose those which are *true-breeding* (*pure-breeding*), i.e. those which are *homozygous*.

Revise these terms if you're not sure what they mean

The principle of selective breeding in crop plants can be illustrated by a further example.

Example of selective breeding: bread wheat

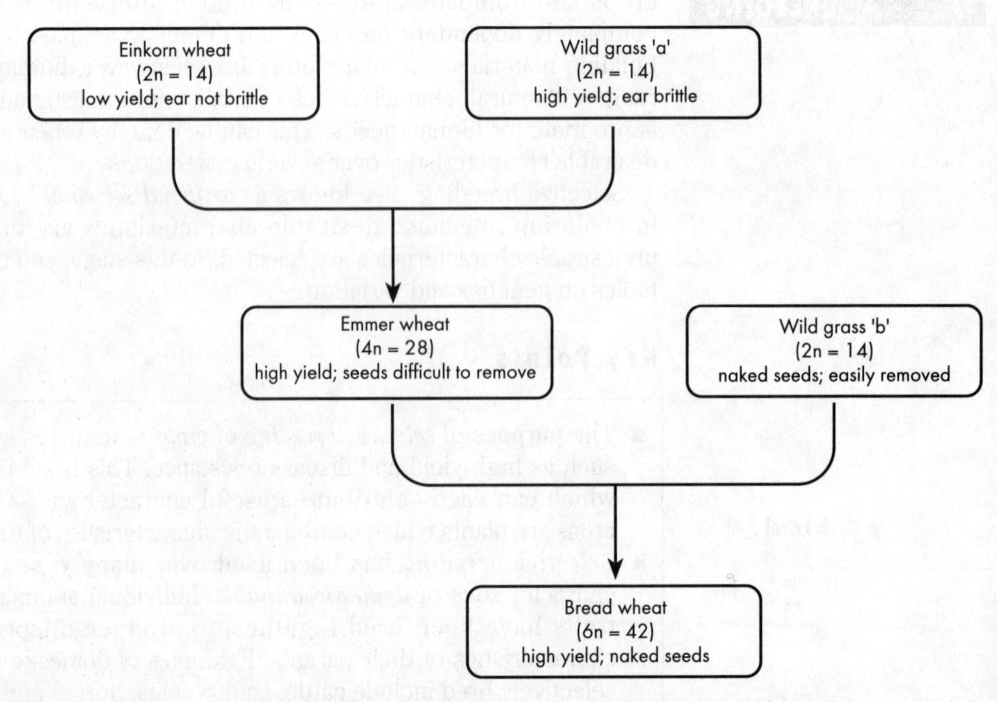

Fig. 6.15 Summary of the selective breeding which resulted in bread wheat (numbers of chromosomes shown in brackets)

Selective breeding of wheat for bread has been going on for about 10 000 years. Much of the early selective breeding of wheat took place in the Middle East. The original wild wheat was not suitable for cultivation because the 'ear' (i.e. the collection of seeds in the fruit) was brittle, and broke off during harvesting. This was replaced by another variety, called ***Einkorn*** wheat, which was non-brittle. However, Einkorn wheat gave a low yield, and was replaced by another variety, ***Emmer*** wheat. This was selectively bred from Einkorn wheat and from a wild grass relative of wheat. During this selective breeding, the chromosome number was increased, by a process known as *polyploidy*. Emmer wheat is still grown in parts of India and Russia. However, the grain (i.e. seeds) of Emmer wheat are difficult to remove from the surrounding chaff during threshing. Modern bread wheat has 'naked' seeds which are easily removed during mechanised threshing. This has resulted from selective breeding of Emmer wheat with a wild grass. The selective breeding processes involved in the production of modern wheat are summarised in Fig. 6.15. Notice the increase of chromosome number which accompanied this process.

SELECTIVE BREEDING – DOMESTIC ANIMALS

Domestic animals are those which live as part of human communities, and which are specifically bred for human use. They are very important in human societies as a source of food, clothing material, transport and for pulling machinery. Rural communities in particular have been dependent on domestic animals for thousands of years. Some animals, particularly dogs and horses, have traditionally been closely associated with humans for centuries; in rural situations they continue to be used for hunting and for livestock control. In urban situations, dogs are used for companionship and security, and they also have specialised uses, e.g. in police and military use for detecting drugs and explosives, and as guide dogs for the blind.

All domestic animals have wild ancestors, and in some cases animals which have 'ancestral characteristics' still exist, e.g. wild pigs and wolves. However, the ancestors of most modern domestic animals are likely to have been very different from their recent descendants. For example, ancestors of modern food animals from the wild would have been difficult to control, and might not have provided food of the same quality and quantity as their modern counterparts.

Present-day cattle, for example, have been selectively bred to be docile, disease resistant, high yielding and true-breeding. Different 'breeds' of cattle have been specifically bred to provide high milk yield (e.g. Fresian, Jersey), or high beef yield (e.g. Hereford). *True-breeding, pure-breeding*, or *pedigree* animals from a given 'stock' are those which, when bred together, usually produce offspring with the same genetically-determined characteristics. Individual animals which breed well and which have all the desired characteristics may be specifically used for breeding purposes. One common technique in selective breeding of farm animals is *artificial insemination*. Semen is collected from the male, and 'injected' into the female. This method is an advantage for farmers who do not have suitable male animals for breeding, and also avoids the need for animals to be moved over large distances for selective breeding.

You may be asked to interpret data, e.g. on milk yields

There is a vast number of different breeds of dog, all produced by selective breeding. Like many breeds of domestic animal, all dogs belong to one species. In theory, they should be able to interbreed to produce an infinite variety of offspring. However, unless selective breeding is done carefully, offspring will not be true-breeding (i.e. they will be 'crosses', or 'mongrels'). This means that desirable characteristics may not be retained in the offspring, or that undesirable characteristics are introduced. Also, certain breeds might be 'technically incapable' of inter-breeding because of size differences, e.g. St. Bernard and Chihuahua breeds of dog.

Selective breeding in domestic animals can be a slow process compared with that of crops, because generation times are long, and relatively few offspring are produced from which to choose for further selective breeding.

Links with other topics

Other related topics elsewhere in this book are as follows:

Topic	Chapter and Topic number	Page number
Sources of variation	6.6	233
Mixing of genes	6.6	234
Variation	6.7	238
Natural selection	6.7	241
Evolution and species formation	6.7	245

REVIEW QUESTIONS

Q4 Two different varieties of tomato plant **A** and **B** were bred together. This produced a new variety.

The diagrams (Fig. 6.16) show all three varieties of tomato plant bearing ripe fruit.

(a) Give **two** ways, shown in the diagrams, in which the new variety is different from the parent varieties.

1. ______________________________

2. ______________________________ (*2*)

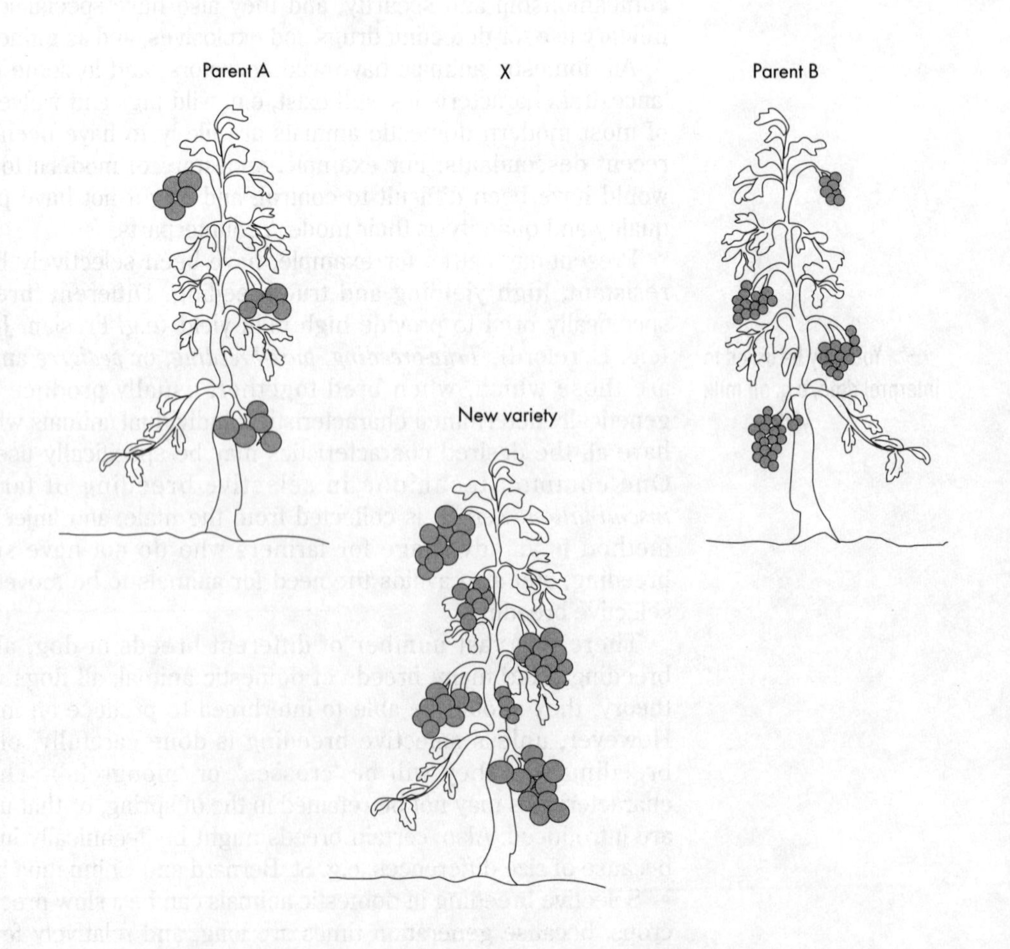

Fig. 6.16

(b) Suggest **one** other characteristic of the tomatoes from the new variety which might be different from the tomatoes of either parent.

____ *(2)*

(c) Researchers at a research farm wish to try and improve the new variety further. They start with the tomato seeds from the new variety. Describe steps they should take in this new selective breeding experiment. *(2)*

____

____

____

____

____

____ *(2)*

Q5 Fig. 6.17 shows F1 and F2 generations produced as a result of crossing two varieties of pure-bred dogs.

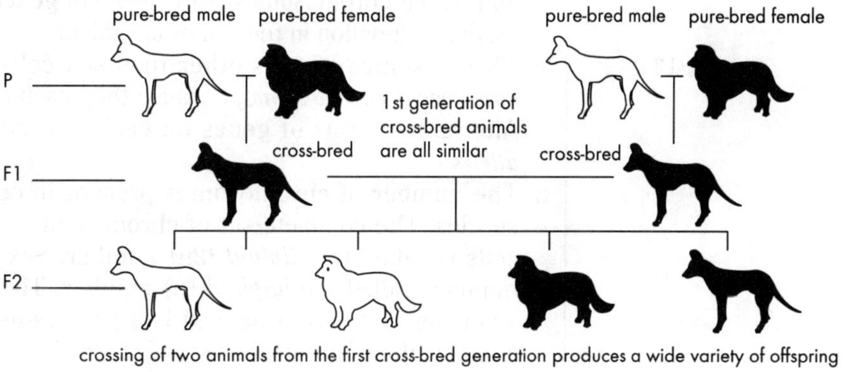

Fig. 6.17

(a) Which two pairs of observable phenotypes are present in the parental generation and subsequent generations?

____

____ *(2)*

(b) State the dominant forms of these phenotypes observable in the F1 generation.

(i) ____

(ii) ____ *(2)*

(c) Give a simple explanation why the second generation (F2) shows more variation than the members of the F1 generation.

____

____ *(2)*

3 GENETICS AND CELLS

Genetic material, contained within the nuclei of cells, is highly organised. *Genes*, which are the 'units of inheritance', are carried by *chromosomes*. In body cells, chromosomes are arranged in so-called *homologous pairs*. The same set of genes is carried by each chromosome in the pair. Chromosomes from different homologous pairs carry different sets of genes.

During *cell division*, each new cell receives at least one homologous chromosome. Normal body cells receive a complete set of homologous chromosomes. Sex cells receive only one homologous chromosome from each pair. When sex cells fuse during fertilisation, the number of chromosomes is restored to the full number again.

Understanding the behaviour of chromosomes during cell division is important in explaining how genes or alleles (different forms of a particular gene) are inherited. It also helps us to understand how a whole set of characteristics can be inherited together, in the case of *sex chromosomes*. These determine whether an individual is male or female.

This Topic Group should provide a useful preparation to understanding later sections on inheritance and genetic variation.

CHROMOSOMES

Key Points

- *Chromosomes* are structures, contained within cell nuclei, which carry genetic information. The genetic information is held as a chemical code. The functional unit within chromosomes is the *gene*. In general, each gene is responsible for a particular function in the cell or organism.

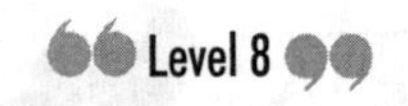

- Chromosomes of cells other than sex cells occur in pairs. They are called *homologous chromosomes*, because they each carry the same sequence of genes. Alternative forms of genes on each chromosome within the pair are called *alleles*.
- The number of chromosomes present in cells is normally fixed for a given species. The complete set of chromosomes contained in most body (somatic) cells is called the *diploid* (2n) number. Sex cells (gametes) contain half this number, called the *haploid* (n) number. The halving of chromosome number when sex cells are produced is a preparation for fertilisation, when sex cells combine their chromosomes, and the diploid number is restored.

1 What are chromosomes?

Chromosomes are thread-like structures, contained within the nucleus of a cell, which carry genetic information. Genetic information is held as a 'chemical code' by the molecules *DNA* and *RNA* (see Section 2.8). The code provides instructions for the cell to make particular proteins. DNA is packaged as tight coils around proteins (*histones*) to form the chromosomes. In cells which are not dividing, chromosomes are long and thin. In this state, the genetic information can be 'read', so that the cell has instructions for making proteins. During cell division, chromosomes become short and compact, so that they can be moved from one place to another more easily. They become much easier to see in this state.

Each chromosome carries a set of *genes*. A gene is a 'unit of inheritance', and provides the genetic information for the cell to make a particular protein. Each body (somatic) cell normally contains a full set of chromosomes. These are arranged in pairs, each of which carries a similar set of genes. For this reason, paired chromosomes are known as *homologous chromosomes*. The full set of chromosomes in a cell is called the *diploid* number, and is given the symbol *2n*.

"These are important concepts – make sure you understand them."

Sex cells (gametes) are produced by a special type of cell division called *meiosis*. This process involves the separation of homologous chromosomes, and results in a halving of the chromosome number. Cells in this condition are said to have a *haploid* number of chromosomes, and this is given the symbol *n*. This is a preparation for fertilisation during sexual reproduction; male and female sex cells combine, and this restores the chromosome number of the 'normal' diploid. If sex cells were also diploid, instead of haploid, each fertilization would double the number of chromosomes in the offspring!

2 Chromosome number

The chromosome number of cells within an organism is characteristic of the species it belongs to. Some examples of diploid chromosome numbers in different species are given in Fig. 6.18. Humans have a diploid number of 46, and therefore a haploid number of 23. Note that chromosome number does not necessarily reflect the complexity of the organism. One reason for this is chromosome length; some species have many short chromosomes, whilst another may have a few long chromosomes, though they may be carrying more genes.

Species	Chromosome Number
PLANT	
Maize *(Zea mays)*	10
Pea *(Pisum sativum)*	7 or 16
Onion *(Allium spp.)*	6
ANIMAL	
Fruit-fly *(Drosophila spp.)*	6 or 12
Human *(Homo sapiens)*	46
Frog *(Rana pipiens)*	26
Goldfish	100

Fig. 6.18 Diploid chromosome numbers in various species

The number of chromosomes in an organism or species can vary, however. One way in which this can happen is due to errors during meiosis. If this affects only one chromosome, the condition is known as *polysomy*. An example of this in humans is Down's syndrome; individuals with this condition have an additional type-21 chromosome; their diploid chromosome number is therefore 47 rather than the normal 46 (Fig. 6.18). Sometimes a whole set of chromosomes fail to separate during meiosis. The result is *polyploidy*. This is more common in plants than animals. In fact, many crop plants and fruits that we eat are polyploid. For example, some types of wheat have 28 chromosomes, whilst 'original' forms had 14. These two types of wheat have been bred together (hybridised) to obtain wheat with 42 (i.e. 28 + 14) chromosomes. Polyploid plants are often larger and more disease resistant.

3 Chromosomes in the inheritance of sex

“Exam questions on this come up fairly often!”

Individuals are either male or female. Their sex is determined, not by a particular gene, but by one homologous pair of chromosomes, called the *sex chromosomes*. Other chromosomes are sometimes called *autosomes*.

Each of the sex chromosomes is either 'X' or 'Y'. The alternative genotypes are XX = female and XY = male. The pattern of inheritance of sex in chromosomes is shown in Fig. 6.19. There is an expected 1:1 ratio of males:females in the offspring, but the

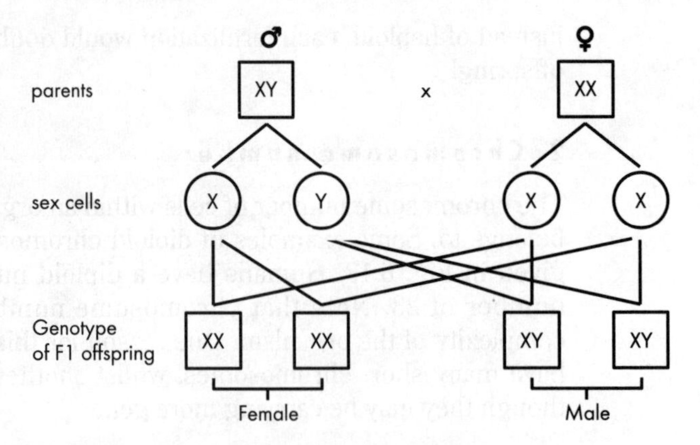

Fig. 6.19

chances of this happening are reduced when parents have fewer offspring. Note that the female genotype is homozygous (XX) whilst the male is heterozygous (XY) for this characteristic.

The sex chromosomes obviously carry genes which determine primary and secondary sexual development. *Sex-linked genes* are carried on the sex chromosomes but are not directly involved with sexual development. Examples of those causing disease include colour blindness and haemophilia.

Links with other topics

Other related topics elsewhere in this book are as follows:

Topic	Chapter and Topic number	Page number
Components of cells, and their function	4.3	36
Genes in variation	6.1	203
Sources of variation	6.6	233
Mixing of genes	6.6	234
Mutations	6.6	235
DNA and RNA	6.8	249
The genetic code	6.8	252
Protein synthesis	6.8	253

REVIEW QUESTIONS

Q6 (a) The technique called *amniocentesis* involves the sampling, by a doctor, of the amniotic fluid which surrounds a human embryo. A long needle is inserted into the mother's womb, and some of the fluid is sucked out. The fluid contains cells which have become detached from the embryo. Examination of the cell nuclei provides information about the chromosomes and genes of the embryo.

(i) Explain why female embryos are found to have slightly more chromosome material than males.

____

____

(ii) Amniocentesis can be used to find out whether an embryo is male or female, before the baby is born. What is the probability of the embryo being a boy? Explain your answer with a genetic diagram if you wish.

____

____

(b) A species of plant has a chromosome number of 8 in the nucleus of each its root cells.

(i) What chromosome number would you expect to find in the sex cells contained in its pollen grains?

(ii) What name is given to describe the number of chromosomes contained in the sex cells?

(c) A normal human body cell contains 46 chromosomes.

(i) How many pairs of autosomes are present in such cells?

(ii) How many pairs of sex chromosomes would be present in such cells?

(iii) How many autosomes are present in human sex cells?

MITOSIS

Key Points

- Cell division is necessary for *growth, development* and *repair*, and also during reproduction. *Asexual reproduction* in simple organisms consists basically of cell division in which the new cells become separated from the 'parent'.
- *Mitosis* is a type of cell division used by *body cells* (*somatic cells*). During mitosis, exact copies of all the genetic material in the cell nucleus are made. The genetic material then divides, resulting in two nuclei. Each nucleus contains a full set of chromosomes.
- Mitosis can most easily be seen in parts of an organism where rapid growth is occurring. Examples include root tips in plants.

1 The purposes of cell division

Cells and their contents, including chromosomes, divide to produce new cells. Cell division is important in growth, development and reproduction:

- **Growth and maintenance**
  Cell division allows an organism to grow larger; numerous cells have a greater combined surface area than a single cell of the same volume. The surface area is important for the exchange of substances with the cell's environment. Cell division also allows damaged tissues to be repaired.
- **Development**
  Cell division allows cells to become *specialised* for different purposes. This allows a range of functions to be performed within a single organism.

■ **Reproduction**
Cell division allows asexual reproduction to take place, and is necessary for the production of gametes ('sex cells') for sexual reproduction. Meiosis is also necessary for sexual reproduction, and is described in the next section.

Asexual reproduction in simple one-celled organisms consists of cell division in which the new cells become partially or completely separated from the 'parent'. The offspring usually contain the same genetic information as their 'parent'. Some examples of the main types of asexual reproduction are given here:

(a) **Binary fission**
Binary fission involves a splitting of a single cell into two. This involves an equal division of the nucleus, followed by organelles and cytoplasm. Binary fission occurs in bacteria, protozoa, e.g. *Amoeba*, and simple algae, e.g. *Pleurococcus*.

The total number of cells in the population can, in theory, increase in an *exponential* way, e.g. 1, 2, 4, 8, 16, 32. However, environmental factors limit such growth in number. Individual cells cease to exist when they divide; they do not necessarily die, so perhaps they are immortal!

(b) **Budding**
Budding occurs in the unicellular fungi called yeasts and also in *Hydra* (animal) and *Bryophyllum* (plant). This involves the formation of a swelling or *bud* which then develops into a complete organism. In yeast, since the buds may not separate when formed, chains of cells result. *Hydra* is a multicellular animal and buds are formed only from undifferentiated cells.

"There are many different ways in which mitosis is used by organisms"

(c) **Fragmentation**
Part of the parent organism breaks away and develops separately. This occurs in *Spirogyra* (plant) and *Planaria* (animal). The multicellular, filamentous algae such as *Spirogyra* produce additional cells by growing extra cross walls (*septation*). A new filament is formed when it breaks away from an existing strand.

(d) **Spore formation**
Spores are small particles, consisting of one or more cells, which are covered by a tough outer covering and are resistant to harsh conditions, such as desiccation. Spores are easily dispersed by air currents and germinate in favourable conditions. Large quantities of spores are produced to offset possible losses.

Examples of organisms forming spores include protozoa, mosses and ferns and also fungi such as *Mucor* (pin mould) (Fig. 6.20).

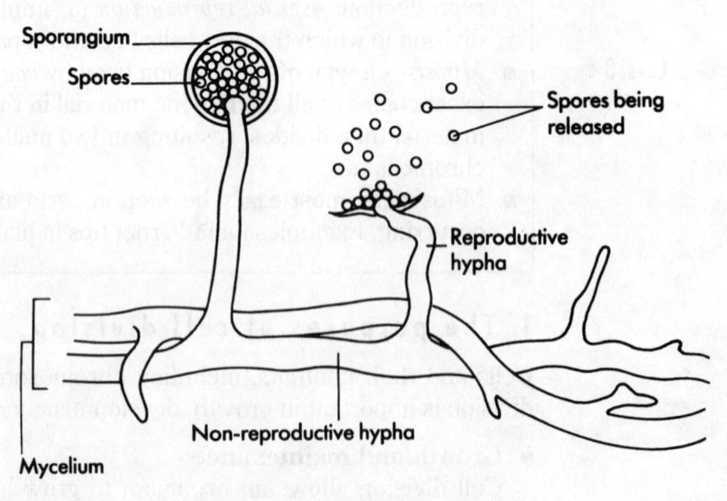

Fig. 6.20 Spore formation in the fungus *Mucor*

2 Cell division by mitosis

There are two main types of cell division. The difference between them depends on whether the number of chromosomes is maintained or reduced in the new cells; *mitosis* and *meiosis*. Mitosis is a type of cell division which *maintains the chromosome number*. In other words, the 'normal' *diploid* (2n) number of chromosomes is maintained in the new cells. These new cells are sometimes called *somatic* ('body') cells.

Mitosis cell division is used by all multicellular organisms to increase cell number, mostly for growth. In many such organisms, mitosis occurs only in certain specialised tissues, e.g. root and shoot tips (meristem) in plants (Fig. 6.21).

Mitosis is also used by ***unicellular*** organisms for asexual reproduction, by binary fission.

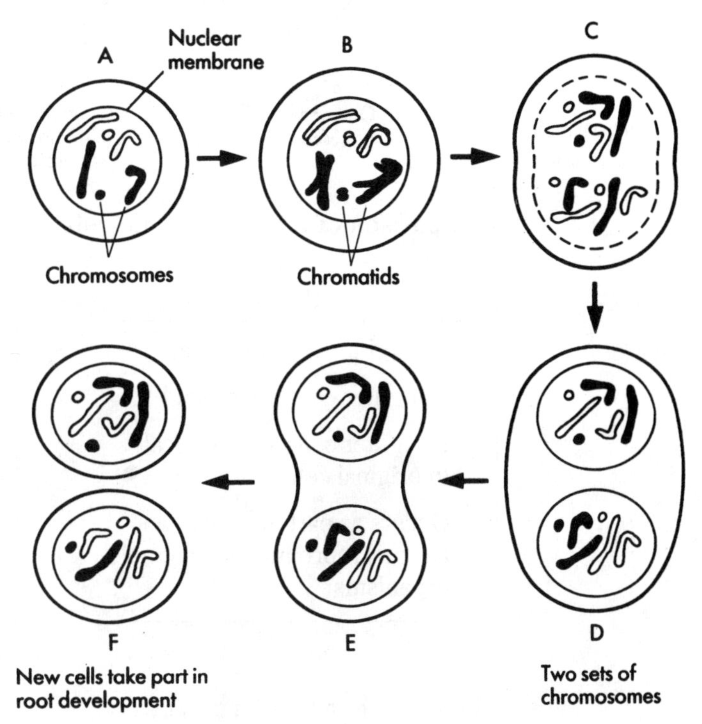

Compare this Figure with Fig. 6.22

Fig. 6.21 Mitosis in a root tip (crocus; *Crocus balansae*)

■ **Summary of events in mitosis**

(A) cell before cell division. It contains 6 chromosomes (diploid number).

(B) each chromosomes makes a copy of itself so it is made up of two chromatids.

(C) the chromatids separate and move to opposite ends (poles) of the cell.

(D) a nuclear membrane forms around each group of chromosomes.

(E) the cell begins to divide.

(F) cell division has occurred. There are two daughter cells, and each contains 6 chromosomes (diploid number).

MEIOSIS

Key Points

Level 8

■ *Meiosis* is a special type of cell division, which occurs during the formation of *sex cells* (*gametes*). During the last stages of meiosis, pairs of *homologous chromosomes* separate. This results in sex cells which contain half the normal number of chromosomes. For this reason, meiosis is referred to as a reduction division.

Meiosis is a type of cell division which ***reduces the chromosome number***. In other words, the 'normal' ***diploid*** (2n) number of chromosomes is halved to a ***haploid*** (n) number in the new cells. These new cells are 'sex cells', or gametes, for sexual reproduction. Gametes combine during *fertilisation* to form a single cell (a fertilised egg), and the chromosome number is restored to diploid (Fig. 6.22).

Meiosis is sometimes called a 'reduction division' because it consists of mitosis followed by a reduction of chromosome number; each chromosome of a ***homologous pair*** separates.

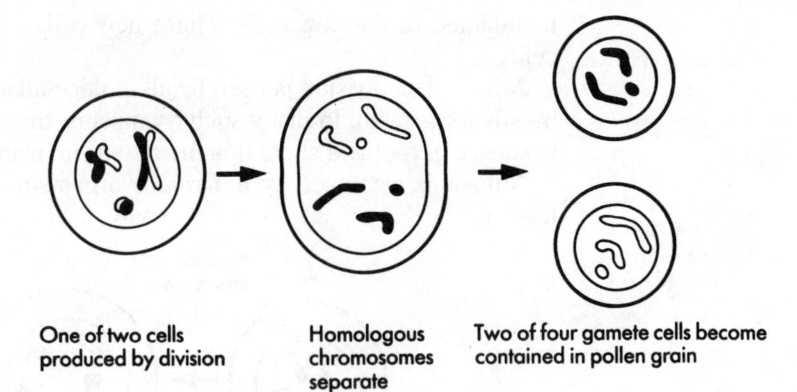

Fig. 6.22 Meiosis in young anther of crocus *(Crocus balansae)*

Comparison of mitosis and meiosis

Mitosis	Meiosis
2 daughter cells are formed Daughter cells are diploid Daughter cells are identical to original cell Occurs as part of growth and asexual reproduction in simple organisms	4 daughter cells are formed Daughter cells are haploid Daughter cells are not identical to original cell Occurs in gamete formation (sexual reproduction) in plants and animals

Link with other topics

Other related topics elsewhere in this book are as follows:

Topic	Chapter and Topic number	Page number
Components of cells, and their function	4.3	36
Reproduction (animal)	4.5	65
Reproduction (plant)	4.6	117
Chromosomes	6.3	214
Sources of variation	6.6	233
Mixing of genes	6.6	234

REVIEW QUESTIONS

Q7 The drawings (Fig. 6.23) show stages in cell division (mitosis).

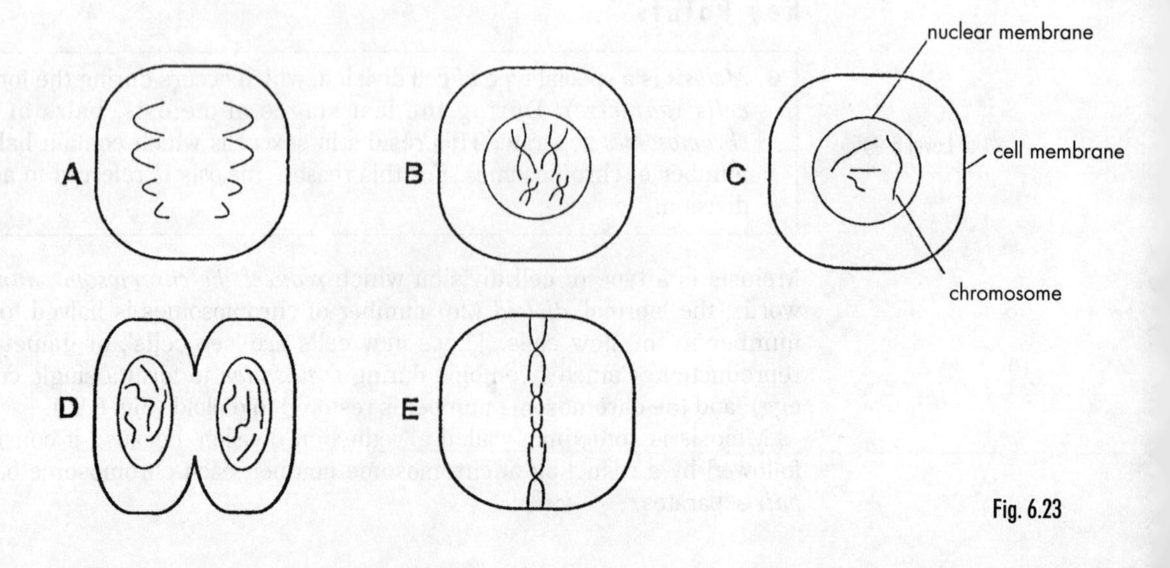

Fig. 6.23

(a) Using the letters **A–E**, put these stages in the correct order.

(first) _____ _____ _____ _____ _____ (last)

(b) A human skin cell contains 46 chromosomes.

If the cell divides once, how many chromosomes would each daughter skin cell contain?

_____

(*1*)

(c) Each body cell of a mouse contains 40 chromosomes.

How many chromosomes would you expect to find in a mouse sperm cell?

_____

(*1*)

(d) A sperm cell with one chromosome missing fertilises an egg.

Suggest one reason why the embryo is unlikely to develop in the normal way when some genes are absent.

_____

_____

(*1*)

Q8 The diagram below (Fig. 6.24) shows a summary of the life cycle in humans.

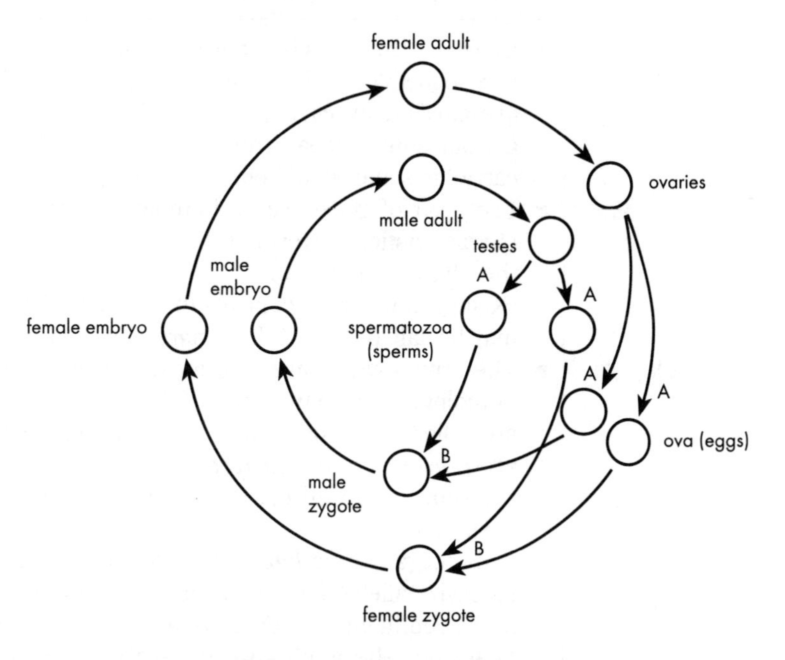

Fig. 6.24

(a) Fill in all the circles in the diagram by writing the symbols for the sex chromosomes. Use the following list of four alternatives to help you **XY XX X Y.**

(b) Name the type of cell division which is occurring at each of the following parts of the diagram:

(i) region **A** _____

(ii) region **B** _____

(c) What changes, if any, are happening to the *total amount of genetic material* in each of the regions '**A**' and '**B**'?

(i) changes in region '**A**' ______________________________

(ii) changes in region '**B**' ______________________________

4 PATTERNS OF INHERITANCE

There are some fundamental principles of inheritance, which are most easily understood in examples involving single characteristics (i.e. in so-called *monohybrid crosses*). In the simplest cases, there is a single gene for a particular characteristic, and the gene can have two 'versions', or alleles. These alleles occur in pairs, and can be either *dominant* or *recessive*, depending on how they are expressed within the pair. However the inheritance of some characteristics is more complex, for instance involving *incomplete dominance* or *co-dominance*.

The study of inheritance (genetics) deserves careful revision for several reasons. Firstly, it is a part of the syllabus which causes some candidates difficulties. One reason for this is the terminology which is used. However, terms used in genetics are well worth understanding, partly because they are often used in exam questions, and partly because, once you are confident with them, you will be able to use them yourself in your answers.

Secondly, genetics helps us to understand how certain important human conditions, including disease are inherited. Genetics is also important as a basis for understanding *variation* and *evolution*, so you may need to revise genetics when you are also revising these other topics.

Finally, genetics questions occur quite often in biology exams! Some questions are easier than others, so you should get a chance to show what you know!

INHERITANCE OF ALLELES

Key Points

- *Genes* are the 'units of inheritance' in all organisms. In general, each type of gene is responsible for determining a particular characteristic in the organism. Genes are carried by *chromosomes* which, in 'normal' body cells, exist in pairs. Chromosomes in pairs are known as *homologous chromosomes*, since they usually carry the same set of genes as each other.
- Each pair of genes within homologous chromosomes is responsible for a given characteristic. However, for most genes, there are two possible ways in which the characteristic can be expressed. These alternatives are called *alleles*. For example if the gene controls the characteristic 'height', the two alternative alleles may be 'tall' or 'short'. This increases the variability in inheritance.

- The way in which alleles are expressed depends on how they are 'paired up' in homologous chromosomes. Alleles which are always expressed when present are called *dominant*. Those which are only expressed when no dominant alleles are present are called *recessive*. In some genes, neither allele is dominant in relation to the other; this results from either *codominance* or *incomplete dominance*.
- If the organism is *homozygous* for a given gene, it has both dominant or both recessive alleles (i.e. the same). If the organism is *heterozygous* for a given gene, it has dominant and recessive alleles (i.e. different). The *genotype* of an organism describes the alleles which are present for a given gene. The organism's *phenotype* describes how the alleles actually have their effect.
- The principles of inheritance can be illustrated using a *monohybrid cross*. This involves a study of inheritance of a *single characteristic* (e.g. 'tallness').

1 The inheritance of genes

In the simplest patterns of inheritance, each gene determines a particular 'type' of characteristic. However, individual genes can be expressed in different ways because they may have different *alleles*. Each allele determines a particular 'version' of a characteristic. This means that individual genes can be expressed (i.e. have their effect) in different ways.

Genes operate, directly or indirectly, through *protein synthesis*, i.e. genes affect which, and when, certain proteins are made. Proteins affect characteristics in the organism by determining both structure and function. Individual genes may have their effect by controlling the assembly of a particular protein (this is the 'one gene, one protein' theory). Alternatively, several genes may act together to influence a single characteristic. In an organism consisting of many different specialised cells, each type of cell is not likely to use all its genes; however, each cell inherits a full set of genes during cell division.

The usual number of chromosomes contained within each cell tends to be characteristic for each species. These normal, or *diploid* numbers occur in all body (*somatic*) cells but not in *gametes* (sex cells).

All chromosomes within any individual are precise copies of the original chromosomes inherited from the parents. In sexually reproducing organisms, chromosomes, contained in the gametes, are contributed by each parent. Gametes contain half the normal chromosome number, i.e. they are *haploid*.

When they combine during *fertilisation*, the normal diploid number is restored. Fertilisation tends to be called 'crossing' in genetics; individuals are crossed with other organisms. Hermaphrodite organisms ('two-sexed') can be self-fertilised, or 'selfed'.

A diploid cell contains two complete sets of chromosomes, consisting of pairs of similar, or *homologous chromosomes*. A haploid cell has only one set of chromosomes, one from each homologous pair. Each pair of chromosomes carries pairs of alleles. These pairs of alleles control the same characteristic, e.g. fur colour in mice, height in peas. However, each allele may be expressed in different ways; for example black or brown fur colour in mice, and tallness or shortness in peas. These are examples of *single factor* inheritance, in which a particular gene (or pair of alleles) determines a single characteristic.

2 Single factor inheritance

Studies of inheritance are made simpler by using examples involving characteristics determined by a single pair of alleles.

"There are several words in genetics which may be unfamiliar. However, once their meanings are understood they can be used to give very precise answers to genetics questions."

During sexual reproduction each parent contributes gametes, each of which contains a haploid set of chromosomes. At fertilisation, homologous chromosomes pair up so that similar genes from each chromosome are positioned next to each other. Each allele is therefore contributed by one of the parents.

In simpler patterns of inheritance, there are various predictable ways in which parental alleles can be passed on to offspring. How these alleles are actually expressed depends on an interaction between them; different alleles may be dominant or recessive:

dominant alleles will always be expressed if at least one is present;

recessive alleles will normally only be expressed if a corresponding dominant allele is not also present, i.e. both are recessive.

The way in which alleles are expressed can be illustrated by two examples.

Example 1: Inheritance of fur colour in mice

One possible way in which fur colour is inherited in mice is shown in Fig. 6.25.

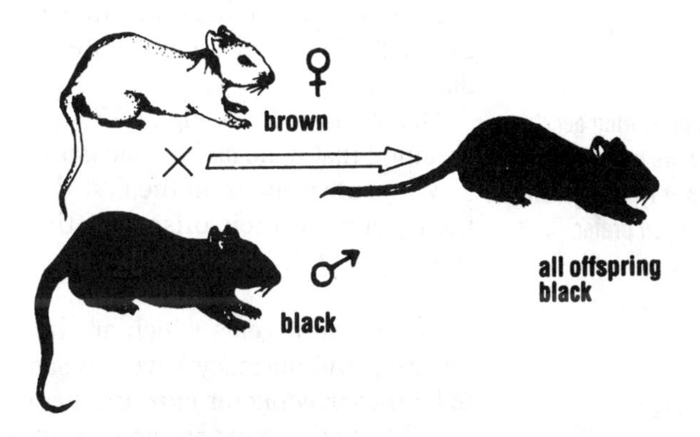

Fig. 6.25 Inheritance of fur colour in mice.

In the example given in Fig. 6.25 the allele for black fur is clearly dominant because it is expressed in all the offspring. The allele for brown fur is obviously recessive, because it is not expressed in the offspring.

The inheritance of different alleles can be shown by symbols, often determined by the dominant allele. The appropriate symbols in this example are 'B' = dominant (black) and 'b' = recessive (brown). Organisms are normally diploid, so for a particular gene both alleles are shown. Three possible combinations are possible in each case; the different types of combinations are:

homozygous dominant, i.e. pair of identical dominant alleles;
homozygous recessive, i.e. pair of identical recessive alleles;
heterozygous, i.e. pair of different alleles, one dominant and one recessive.

The various allele combinations form part of the individual's *genotype*, or genetic content. The *phenotype* is the genetic (and environmental) expression. An example of genotype and phenotype is given in Fig. 6.26. You can see that, for any such gene consisting of dominant and recessive alleles, there will be two possible phenotypes, determined by three possible genotypes.

Fig. 6.26 Possible genotypes and phenotypes for fur colour in mice

GENOTYPE	PHENOTYPE	DESCRIPTION
BB	Black	Homozygous dominant
Bb	Black	Heterozygous
bb	Brown	Homozygous recessive

Patterns of inheritance can best be studied by using genetic diagrams. One such diagram is shown in Fig. 6.27; this is based on the situation in Fig. 6.25.

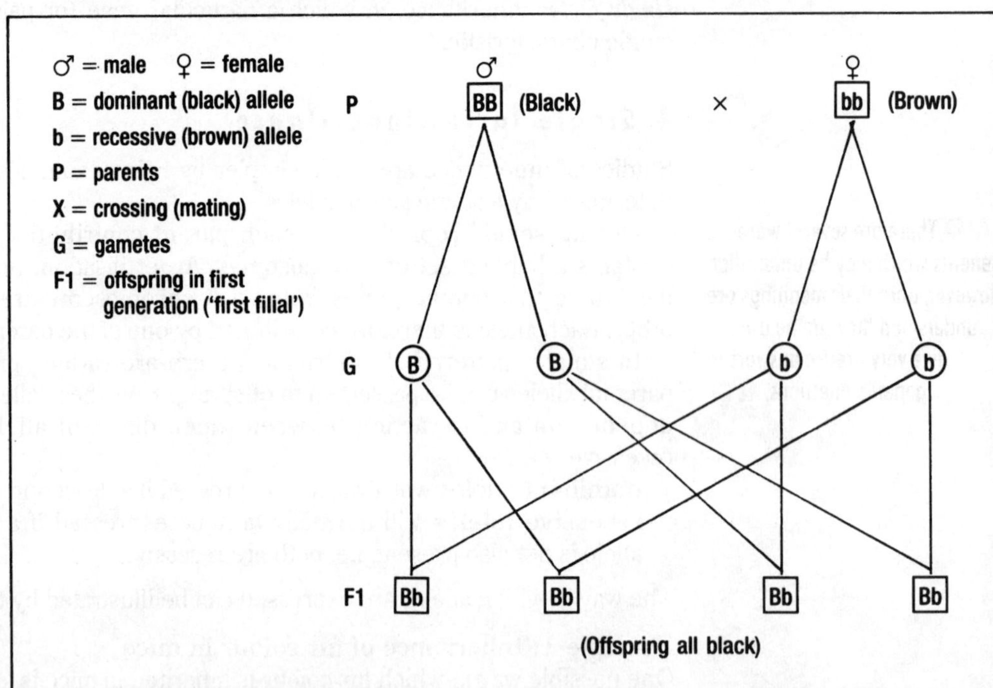

Fig. 6.27 First (F1) generation inheritance of fur colour in mice (*Mus domestica*)

Another type of diagram (the ***punnett square***) showing this inheritance is shown in Fig. 6.28. The alleles contributed by the male are usually shown on the left of the diagram.

" In answering genetics questions, you are usually free to use whichever type of genetics diagram you prefer. "

Diagrams such as Figs 6.27 and 6.28 show all possible gamete combinations; it is assumed that male and female gametes fuse in a random way.

When offspring from the first (F1) generation are sexually mature, they can pass on their genes to their offspring, the second (F2) generation. One possible cross is shown in Fig. 6.29.

When two parents which are homozygous for the same allele are crossed, all the offspring will normally have the same genotype as the parents. Such parents are said to be ***true-breeding*** (or ***pure-breeding***).

When two parents are crossed which are homozygous for different alleles (see Fig. 6.26), or are heterozygous, the offspring will not necessarily have the same genotype as the parents. Parents which are homozygous for different alleles produce ***hybrid*** offspring; this process is called ***hybridisation***.

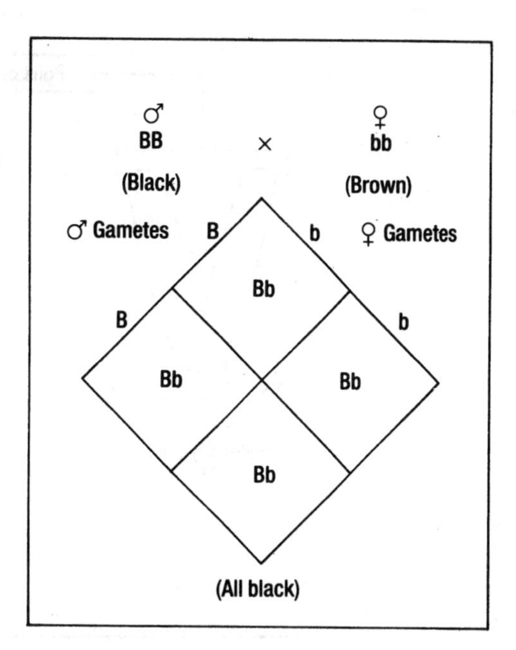

Fig. 6.28 First (F1) generation inheritance of fur colour in mice (*Mus domestica*): Punnett Square

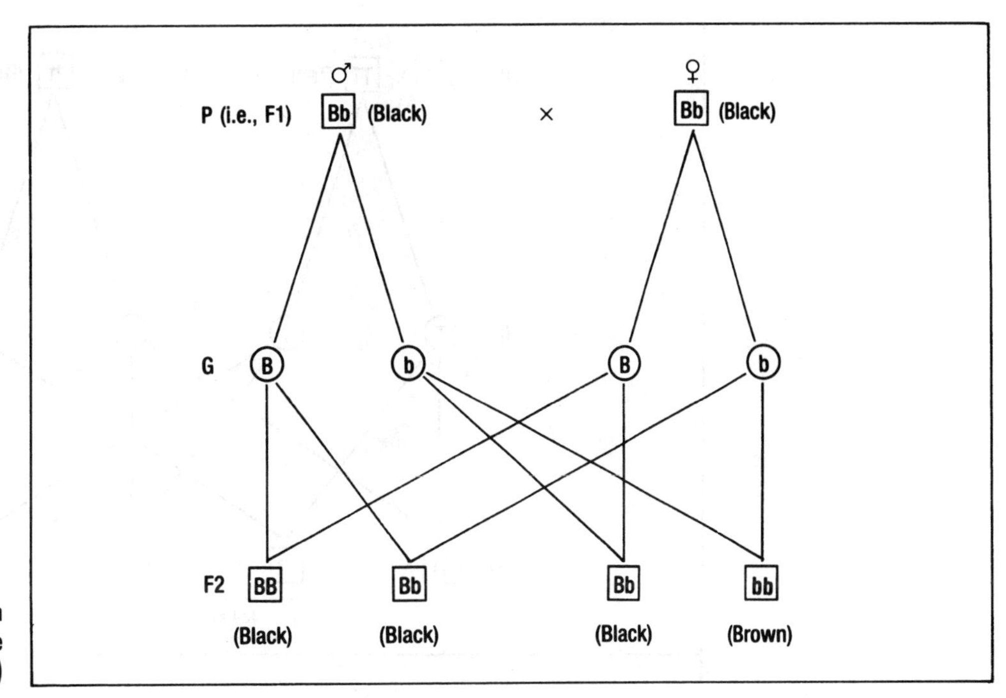

Fig. 6.29 Second (F2) generation inheritance of fur colour in mice (Mus domestica)

Example 2: Inheritance of height in peas

The principles of inheritance of fur colour in mice apply to many other organisms, too. In some classic early experiments in genetics, Gregor Mendel used the garden pea to study the inheritance of a variety of characteristics. For each characteristic, he used pure-breeding (i.e. homozgous) plants. He was able to transfer the pollen from one plant to another by using a small paint brush. In one experiment, he used the character for 'tallness'. The two alleles for the 'tallness' gene are 'tall' and 'short'; the 'tall' allele is dominant. The three alternative genotypes are shown diagramatically in Fig. 6.30.

The results of a cross during two generations, starting with pure-bred parents, are shown in Fig. 6.31. Note the monohydrid ratio (3:1) in the second (F2) generation.

The genotype of an individual which is homozygous for a recessive allele will be obvious from its phenotype. For example, the genotype of a dwarf pea plant can only be *tt*. However, the genotype of a tall plant could be homozygous dominant (e.g. *TT*) or heterozygous (e.g. *Tt*); there are two genotypes for one phenotype.

It is not possible to see genes directly, even using a very powerful microscope. To determine whether an individual with a dominant phenotype is homozygous or heterozygous, it is necessary to perform a *backcross* (or '*test-cross*'). This involves a homozygous recessive individual, whose genotype is therefore dwarf.

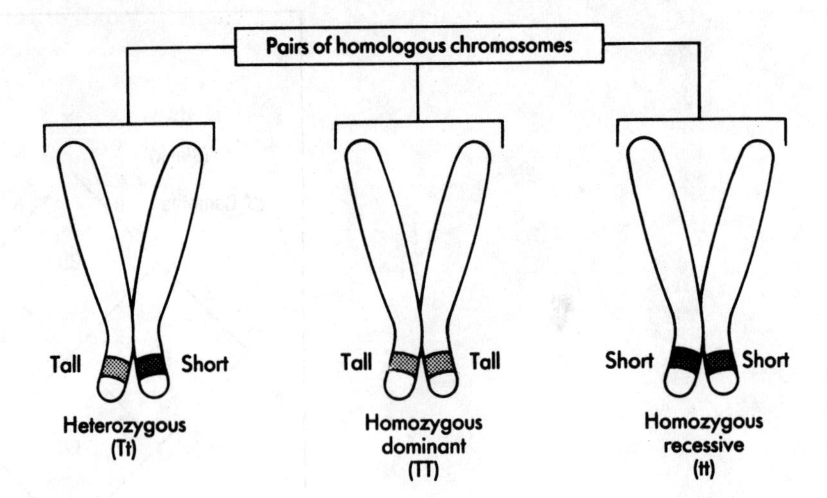

Fig. 6.30

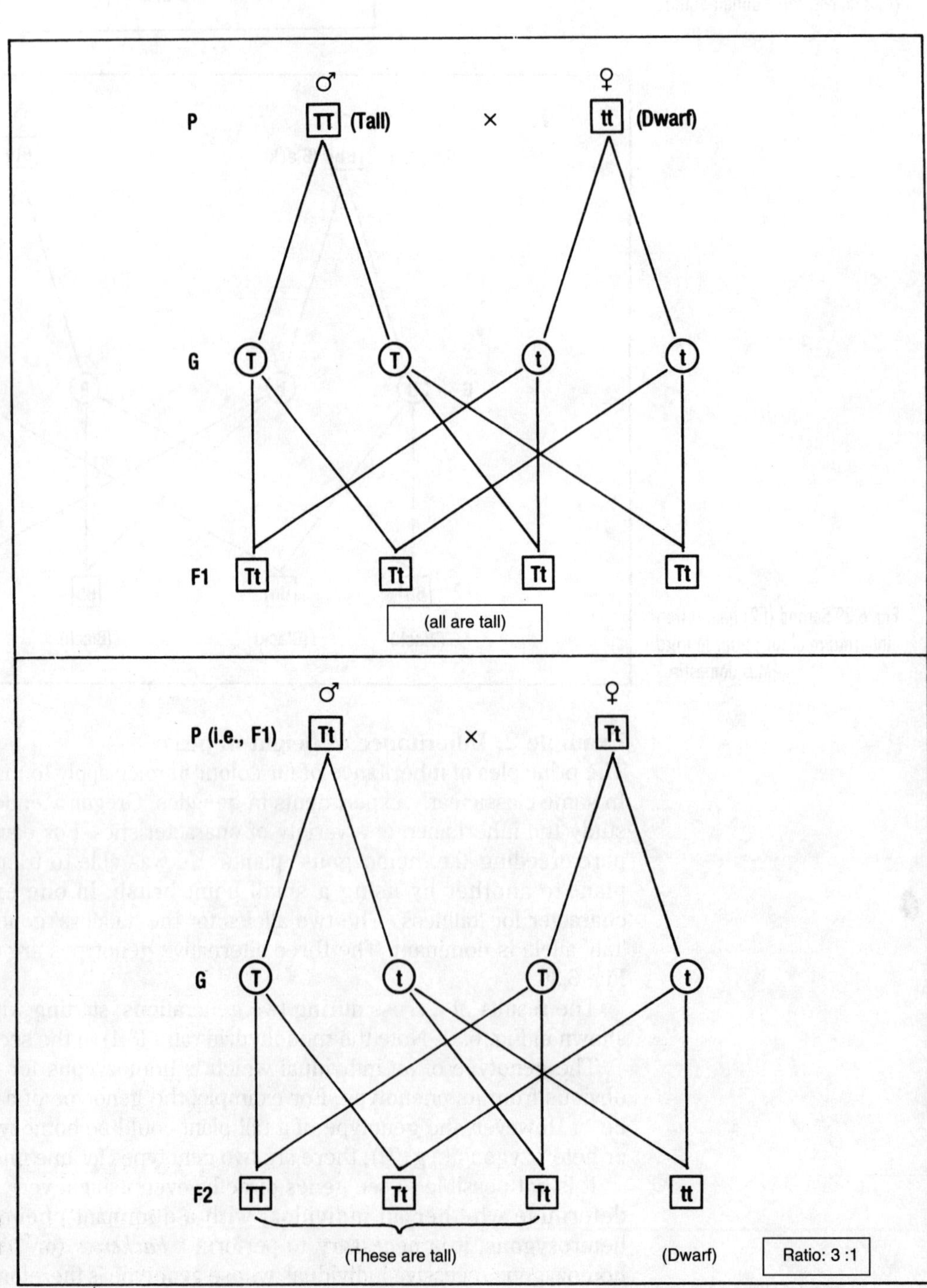

Fig. 6.31 Hybridisation and the monohybrid ratio in pea (*Pisum sativum*)

There are two possible outcomes to a backcross, these are shown in Fig. 6.32.

(a) *All offspring are tall*; this is characteristic of a cross involving a homozygous dominant individual (see Fig. 6.31).
(b) *Half of the offspring are tall*, the other half are dwarf; this is characteristic of a cross involving a heterozygous individual.

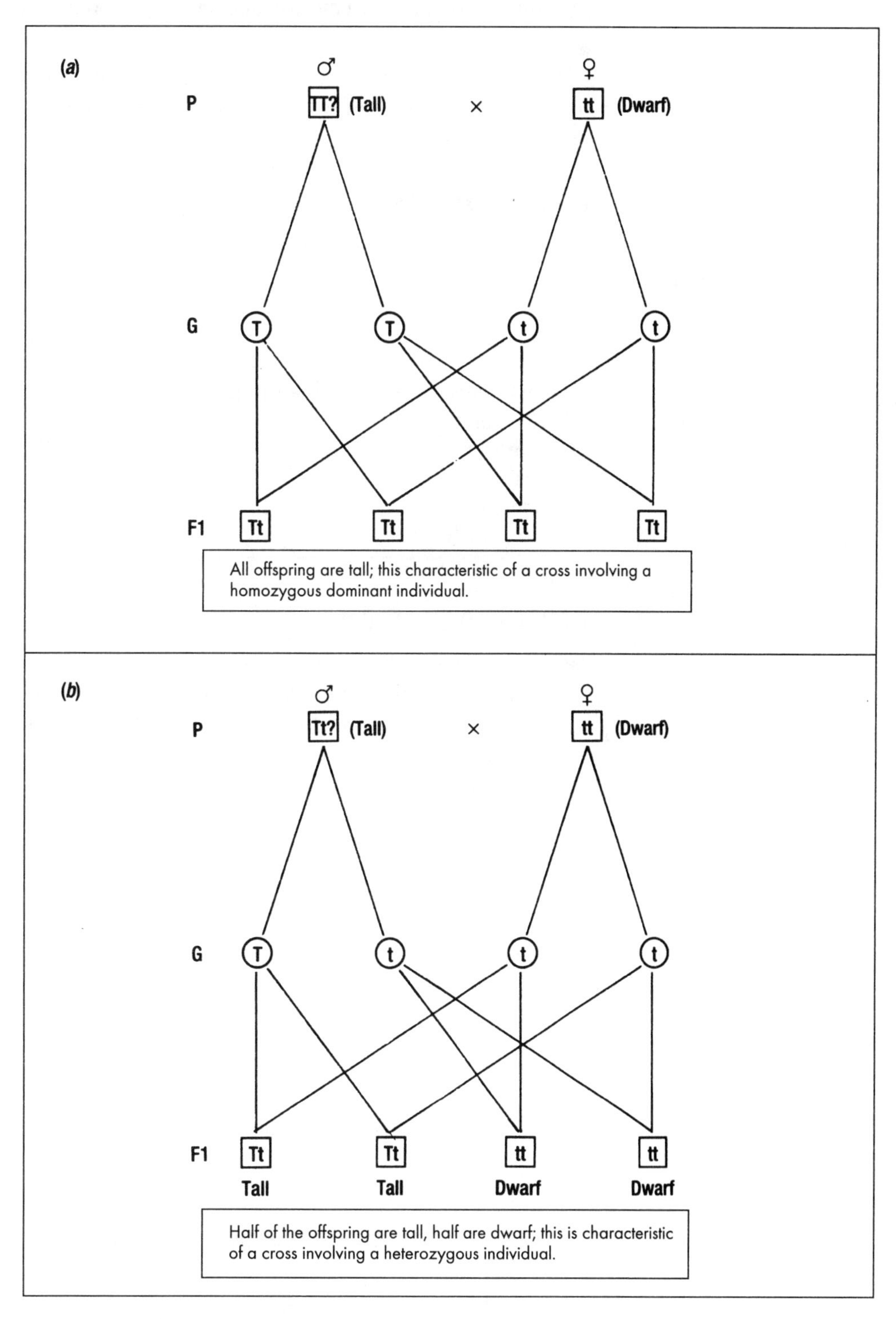

Fig. 6.32 Backcross in the pea (*Pisum sativum*)

3 Incomplete dominance and co-dominance

Not all alleles are either dominant or recessive in their effects. There are two ways in which a pair of alleles in the genotype can have 'mixed' effect in the phenotype. There are examples of each of these in the next section.

■ **Incomplete dominance**

Heterozygous (hybrid) individuals sometimes show a third, possibly intermediate, 'blended' phenotype. One way in which this occurs is for the recessive allele to be only partially 'masked' by the dominant allele in the heterozygous condition. This is known as *incomplete dominance* and an example of this is sickle-cell anaemia (see topic group 6.5). New symbols often need to be invented for genetics diagrams showing incomplete dominance. This also applies for co-dominance.

■ **Co-dominance**

Genes may have alternative alleles which are neither dominant nor recessive to each other. This is called *co-dominance* and, like incomplete dominance, results in a third phenotype. An example of this is blood groups in humans (see section 6.5).

Links with other topics

Other related topics elsewhere in this book are as follows:

Topic	Chapter and Topic number	Page number
Components of cells, and their functions	4.3	36
Sources of variation	6.6	233
Chromosomes	6.3	214
Meiosis	6.3	219
Human genetics	6.5	229
Mixing of alleles	6.6	234

REVIEW QUESTIONS

Q9 The genetic diagram Fig. 6.33 **A** represents a cross between parent plants produced from round (R) seeds or wrinkled (r) seeds. Fig. 6.33 **B** represents a cross between parent plants produced from round seeds.

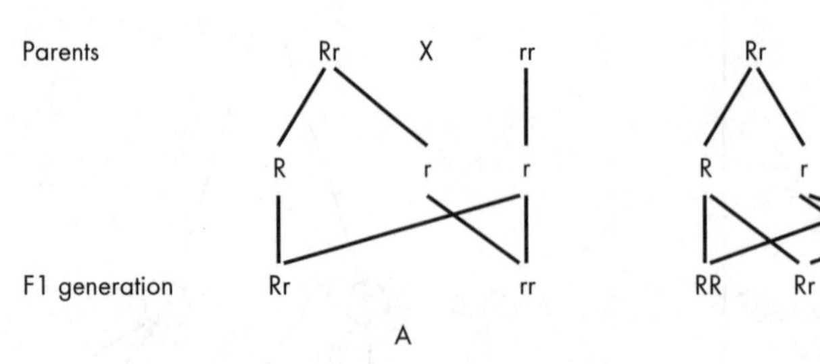

Fig. 6.33

From these diagrams state:

(a) a homozygous genotype, ______

(b) the different gametes shown, ______

(c) the ratios of the phenotypes in the F1 generation in the following crosses:

(i) Rr × rr ______

(ii) Rr × Rr ______

(d) the recessive phenotype

______

(e) Draw a genetic diagram to show what would happen if a cross was made between the homozygous parent from the first cross and one of the homozygous F1 plants showing the dominant phenotype.

5 HUMAN GENETICS

Key Points

Level 8

- Human inheritance is similar in principle to that of simpler organisms. However, there are relatively few characteristics controlled by a single gene. Also, alleles are not always simply dominant or recessive. Some alleles are neither fully dominant nor fully recessive, in situations involving *incomplete dominance* and *co-dominance*. Examples include the inheritance of sickle-cell anaemia and blood groups.
- Human gender (male or female) is controlled by a pair of *sex chromosomes*. These control factors which are characteristic of males or females. Sex chromosomes also carry so-called *sex-linked genes*, some of which cause inherited diseases. The way in which diseases are inherited can be shown using 'family trees'.

1 Human genetics

The principles of genetics established by Mendel and others are mostly based on work with fairly simple organisms. Examples of such organisms include pea (*Pisum sativum*), mouse (*Mus domestica*), maize (*Zea mays*) and fruit fly (*Drosophila melanogaster*). These species have relatively few chromosomes or genes, their generation time is short and breeding experiments present perhaps fewer ethical problems.

An understanding of human genetics is derived mostly from studies of patterns of inheritance in closely related individuals, for instance within families. Many human characteristics are controlled by several genes (*polygenic inheritance*) and alleles are not necessarily dominant or recessive. However, some human features are controlled by single genes, for example tongue-rolling ability, the capacity to taste PTC (*phenylthiocarbamide*) and even which thumb (left or right) is placed over the other when the hands are clasped!

Human inheritance is often shown as '*family trees*' instead of the sort of genetic diagrams we have seen in previous sections. Family trees occur in exam. questions, so it is worth becoming familiar with them. For example, Fig. 6.34 shows part of a family tree for the gene controlling the ability to taste PTC. The ability to taste is a dominant allele (T), so 'tasters' are either TT or Tt. 'Non-tasters' are homozygous recessive (tt).

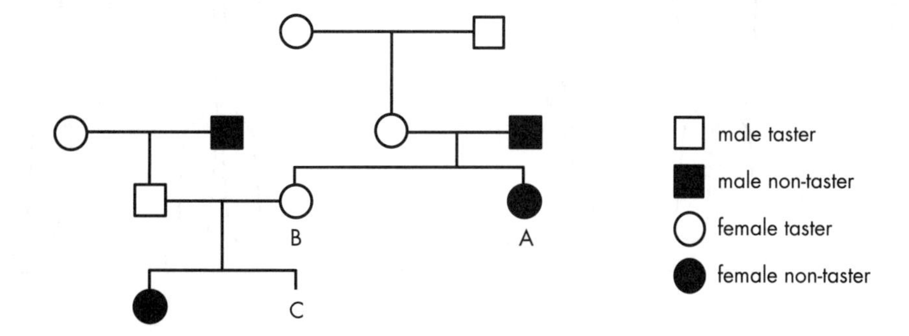

Fig. 6.34 An example of a 'family tree' in genetics

Take care to understand diagrams like this which occur in exam. questions – all the information you need will be there!

Note that, although males and females are distinguished in this diagram, the gene involved is not sex-linked (see below), since both males and females have the 'taster' and 'non-taster' phenotypes. From this diagram, it is possible to determine genotypes of some of the offspring or their ancestors. For example, the genotype of female A must be 'tt' since all non-tasters are homozygous recessive. Female B must be 'Tt'; she is not 'TT' (the only alternative) because only one of her parents was a taster. You can also use diagrams like this to predict the likelihood of a certain allele being inherited. In this example, child C (male or female) would have a 75% (i.e. 3 in 4) chance of being a taster; since both Child C's parents are heterozygous, this gives a monohybrid ratio (3:1, tasters:non-tasters).

In the remainder of this topic, human inheritance will be illustrated using examples of the inheritance of sex and sex-linked genes (e.g. colour blindness, haemophilia), incomplete dominance (e.g. sickle-cell anaemia), and co-dominance (e.g. blood groups).

2 Inheritance of sex and sex-linked diseases

Individuals are either male or female. Their sex is determined, not by a particular gene, but by one homologous pair of chromosomes, called the *sex chromosomes*. Other chromosomes are sometimes called *autosomes*.

Each of the sex chromosomes is either 'X' or 'Y'. The alternative genotypes are XX = female and XY = male. The pattern of inheritance of sex in chromosomes is shown in Fig. 6.19. There is an expected 1:1 ratio of males:females in the offspring, but the chances of this happening are reduced when parents have fewer offspring. Note that the female genotype is homozygous (XX) whilst the male is heterozygous (XY) for this characteristic.

The sex chromosomes obviously carry genes which determine primary and secondary sexual development *Sex-linked genes* are carried on the sex chromosomes but are not directly involved with sexual development. Examples include colour blindness and haemophilia.

Haemophilia is a potentially lethal condition in which the ability to clot blood is much reduced, so that even a small injury can result in copious bleeding. **Red-green colour blindness** is a non-lethal condition in which red and green colours cannot be distinguished.

Both colour blindness and haemophilia are caused by alleles on the X chromosome. More precisely, the alleles are situated on the part of the X chromosome for which there is no corresponding part of the Y chromosome (Fig. 6.36). This leaves this part of the X chromosome 'exposed' in males, so that even a recessive allele can be expressed, since there is no corresponding dominant allele to 'mask' it. This is why these sex-linked alleles tend to affect the males (XY) more frequently than females (XX). They will normally only become apparent in a female if she is homozygous recessive. Females who are heterozygous for this allele are said to be *carriers*, since the condition is present in the genotype but not apparent in the phenotype.

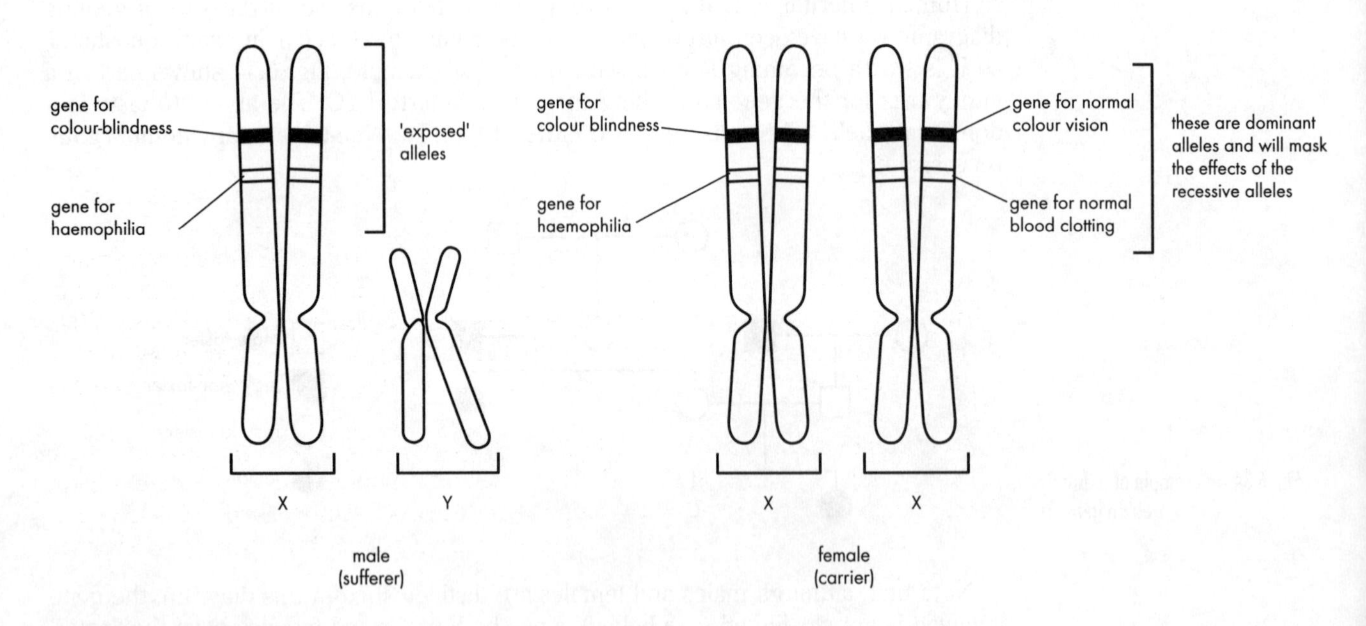

Fig. 6.36 Sex chromosomes showing how sex-linked genes can be expressed in a male (sufferer) but not necessarily in a female carrier (heterozygote)

Malaria

3 Incomplete dominance

Incomplete dominance sometimes occurs for certain genes when a recessive allele is expressed even though a dominant allele is also present. Incomplete dominance therefore occurs in a heterozygous individual (i.e. when the alleles are mixed), in which the recessive allele is not completely 'masked' by the dominant allele. The result is a third phenotype (i.e. genetic 'outcome'); in more usual situations, only two phenotypes result from different combinations of dominant and recessive alleles for a given gene.

An example of incomplete dominance in humans is **sickle-cell anaemia**. This disease results from a recurrent ***mutation*** within certain human populations which affects red blood cells. The dominant allele (*N*) is for normal cells. The recessive allele (*n*) may result in distorted (sickle-shaped) cells which are unable to carry oxygen properly and may also cause a blockage of fine blood vessels. However, these cells confer resistance to the disease malaria. Incomplete dominance allows an intermediate 'blended' phenotype called ***sickle-cell trait*** to occur. Individuals having the trait generally suffer a fairly mild form of the anaemia while having a resistance to malaria (Fig. 6.37).

"In regions where malaria is common, it is a real advantage to be heterozygous."

GENOTYPE	PHENOTYPE	ANAEMIA	MALARIAL RESISTANCE	OCCURRENCE IN POPULATIONS IN MALARIAL REGIONS
NN Homozygous dominant	Normal	Not anaemic	Not resistant	25% surviving
Nn Heterozygous (incomplete dominance)	Sickle cell trait	Not anaemic	Resistant	50% surviving
nn Homozygous recessive	Sickle cell anaemia	Anaemic	Resistant	25% dying

Fig. 6.37 Summary of the genetics of sickle cell anaemia

4 Co-dominance

Co-dominance is a condition in which two *different* alleles of a particular gene may be expressed (i.e. have their effect) together. Generally, each gene is responsible for a different 'type' of characteristic. Alleles result in different 'versions' of a characteristic. Different combinations of alleles are expressed as a different phenotype. Alleles occur in pairs. If either one or both dominant alleles is present, they will be expressed as the 'dominant' phenotype. If neither dominant allele is present, i.e. if only recessive alleles are present, the result will be the 'recessive' phenotype. Co-dominant alleles create the possibility of a third phenotype.

An example of co-dominance is human blood groups. The composition of blood includes red cells. These have an inherited chemical 'identity' and this is the basis for blood groups. The identifying protein on the surface of certain types of red blood cells is called an ***antigen***. There are three alleles determining the presence or absence of these antigens:

This is a complicated bit of biology; you may need to re-read this section carefully.

allele A – causes antigen A to be formed
allele B – causes antigen B to be formed
allele O – causes no antigen to be formed

Alleles A and B are co-dominant with each other. Allele O is recessive to both alleles A and B. Inheritance of these alleles occurs in a Mendelian way. Fig. 6.38 summarises possible genotypes with corresponding phenotypes. The relative number of these phenotypes in the population varies throughout the world.

PHENOTYPE (BLOOD GROUP)	GENOTYPE	OCCURRENCE IN UK POPULATION (%)
A dominant	AA or AO	40
B dominant	BB or BO	10
AB co-dominant	AB	3
O recessive	OO	47

Fig. 6.38 Co-dominance in humans: blood groups

Blood plasma may contain special sorts of *antibodies*. These antibodies are called *agglutinogens* because they cause 'foreign' blood to coagulate or agglutinate. The antibodies are produced in response to 'foreign' antigens, i.e. A or B, or both. There are two types of antibody, anti-A and anti-B. One or both of these are produced by individuals with blood groups A, B or O.

antigens.

In blood transfusions, care has to be taken to ensure that blood from the *donor* is compatible (matched) with that of the *recipient*. If not, coagulation (*agglutination*) may occur. This is summarised in Fig. 6.39. Note that blood group AB is a universal recipient, whilst blood group O is a universal donor.

BLOOD GROUP	ANTIGEN PRESENT ON RED CELLS	ANTIBODY IN PLASMA	BLOOD GROUPS WHICH CAN BE RECEIVED	BLOOD GROUPS WHICH CAN RECEIVE
A	A	anti-B	A and O	A and AB
B	B	anti-A	B and O	B and AB
AB	A and B	none	any	AB
O	none	anti-A and anti-B	O	any

Fig. 6.39 Compatibility in human blood groups

Links with other topics

Other related topics elsewhere in this book are as follows:

Topic	Chapter and Topic number	Page number
Reproduction (animals)	4.5	65
Sources of variation	6.6	233
Chromosomes	6.3	214
Meiosis	6.3	219
Mixing of alleles	6.6	234

REVIEW QUESTIONS

Q10 I^A, I^B and I^o are the alleles of the gene which controls human blood groups.

(a) Using these symbols, what are the possible genotypes of people with the following blood group?

(i) Group AB ______________________

(ii) Group O ______________________

(iii) Group B ______________________

(b) Which blood group is the result of co-dominance? ______________

(c) Complete the genetic diagram below to show the possible blood groups amongst the offspring of a father who is heterozygous for blood group A and a mother who is heterozygous for blood group B.

	father		*mother*	
phenotypes	Blood group A		Blood group B	
genotypes	______		______	
gametes	______	______	______	______
genotypes of offspring	______	______	______	______
phenotypes of offspring	______	______	______	______

(4)

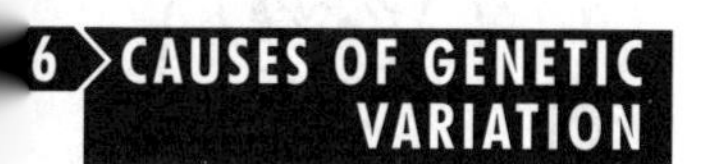

6 CAUSES OF GENETIC VARIATION

Genetic variation arises for two main reasons. In sexual reproduction, genetic material contained in sex cells from the parents is combined. This will almost certainly create a unique and new arrangement of genes in the offspring. Part of the reason for this is due to differences which already exist between the parents. It is also because the actual process by which sex cells are produced allows further genetic changes. Within the cells of any individual, another source of genetic change is from mutations. Mutations consist of changes to genes or chromosomes.

The idea of genetic change is important in understanding the basis for inherited variation and, consequently, for evolution. You should revise this part of the syllabus in conjunction with the relevant Topic Groups which follow on evolution and DNA.

MIXING OF ALLELES

Key Points

> **Level 9**

- *Genes* are the so-called 'units of inheritance' in cells, and are carried by *chromosomes*. Chromosomes occur in *homologous* pairs, each chromosome in the pair usually carrying the same set of genes. However, different 'versions' of genes, or *alleles*, exist, which allow genes to be expressed in alternative ways.
- During *meiosis*, the type of cell division which produces sex cells, alleles become separated *between* different chromosomes of the homologous pair. The chromosomes also break and rejoin during meiosis, resulting in alleles being separated *within* chromosomes. These processes result in a mixing of alleles.

Sexual reproduction contributes to genetic variety for two main reasons, *fertilisation* and *meiosis*. Fertilisation involves the merging of genetic information from (usually) two organisms which are already likely to be genetically different. For any given characteristic, there are often at least three different forms of the genetic information (the *genotype*) which controls them. The complete genotype for even the simplest organism consists of thousands of genes, controlling thousands of characteristics. The number of different combinations of genes in each individual's genotype is almost infinite. Even though certain genes can only be expressed in one way (because any alternative would be lethal), that still leaves a vast amount of genetic variation. The way in which offspring inherit genetic information from their parents has been covered in the previous Topic Group (2.5).

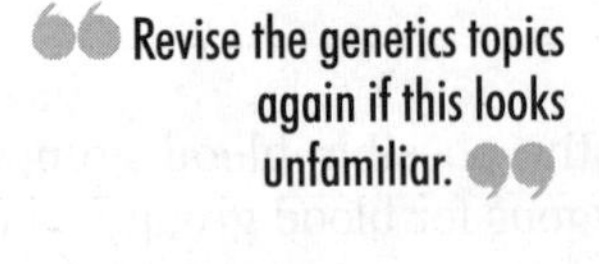

In this section, we will describe the role of meiosis in increasing genetic variability. Meiosis is a special type of cell division only involved in the production of *gametes* (sex cells) for sexual reproduction. During meiosis, the normal *diploid* (2n) number of chromosomes is reduced by a half, resulting in *haploid* (n) gametes. Meiosis includes a 'reduction division' in which *homologous chromosomes* (which carry the same set of genes) separate from their normal pair arrangement. The gametes only receive one homologous chromosome from each pair. In fact, this type of cell division begins with a normal type of cell divisioon, *mitosis* (see 2.4.2) which makes a copy of all the genetic material in the 'original' cell. This is then used in the reduction division, resulting in four sex cells. This process is summarised in Fig. 6.40.

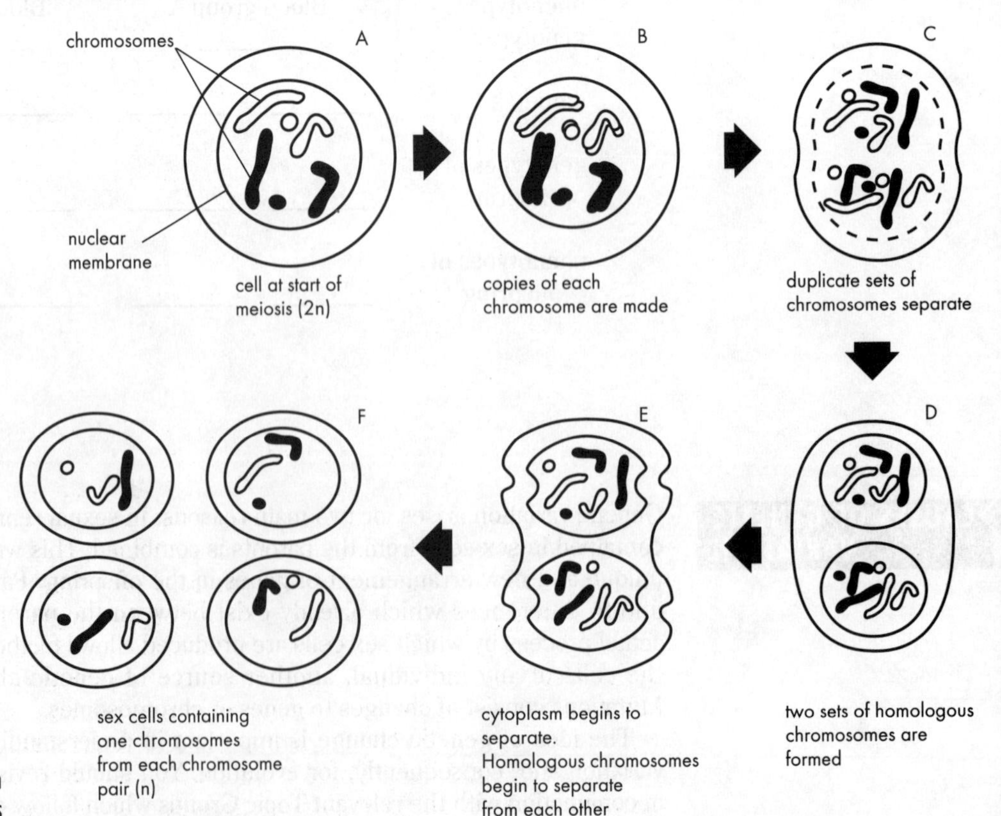

Fig. 6.40 Simplified diagram showing the main stages in meiosis

During the separation of homologous chromosomes, each chromosome is broken and rejoined. Fragments of chromosome are 'swapped' with each other, during *crossing over* (Fig. 6.41). This process occurs simultaneously in all homologous chromosome pairs, and each cross-over event in one pair is independent of those in other chromosome pairs. This is known as *independent assortment.*

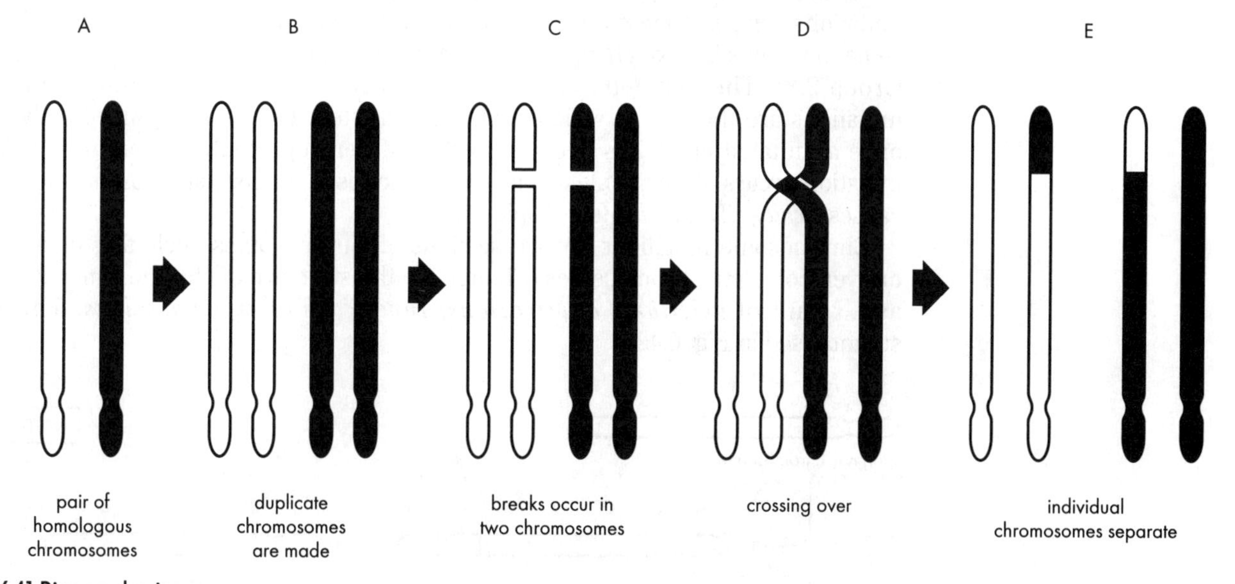

Fig. 6.41 Diagram showing cross-over between homologous chromosomes

Note: Use the Review Questions in the next section to check your understanding of causes of genetic variation.

MUTATIONS

Key Points

- *Mutations* are changes in genetic material. Some mutations can be inherited. To be inherited, mutations must affect the cells which are directly involved with reproduction. Mutations occur spontaneously, but the rate at which they occur can be increased by ***mutagens*** – mutation-causing chemicals and radiation.
- There are two main types of mutation; ***gene mutations*** and ***chromosome mutations***. Gene mutations involve an error in copying the gene during cell division. Chromosome mutations involve deletions or additions to individual chromosomes, or changes in the number of chromosomes.
- Mutations are a source of genetic variation. Many are harmful, some are harmless, and a very few are beneficial.

1 Mutations and their causes

A ***mutation*** is a change in genetic material. The type of cell affected determines whether mutations can be inherited. Mutations in *body (somatic) cells* may only be inherited if the organism reproduces asexually, i.e. if body cells carrying the change become part of the new organism. However, mutations in body cells can be serious for the individual organism; skin cancer is an example of this. In organisms using sexual reproduction, mutations which affect the ***gametes*** *(sex cells)* are inherited.

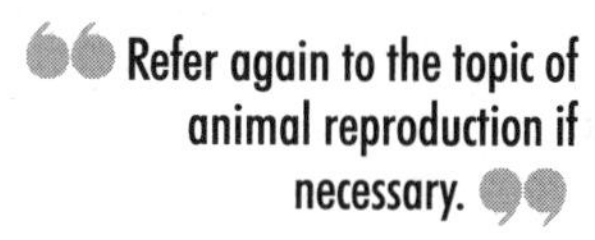

Mutations are generally harmful, although some are harmless. The reason for the low occurrence of useful mutations is that existing genes have mostly been 'tried and tested' during long evolution, and organisms are in general already well-adapted to their environment (otherwise they would not survive). However, if there was a change of environment, a 'new' gene produced by mutation might be useful.

Mutations occur spontaneously in cells. However, mutation rates can be increased by so-called ***mutagens*** in the organism's external environment. Examples of mutagens include chemicals such as hydrogen peroxide, nitrous acid and 'mustard gas'-type agents. High-energy ionising radiation can also act as mutagens, e.g. ultra-violet light, X-rays and gamma radiation. Humans who work with these agents must protect themselves from direct exposure. However, some agents affect humans more generally. The proportion of ultra-violet radiation in sunlight has been increasing recently, due to 'holes' in the normally protective ozone layer in the atmosphere.

2 Types of mutation

There are two main type of mutation; *gene mutations* and *chromosome mutations*. Gene mutations (or *point mutations*) result from an error during chromosome replication. Gene mutations involve changes to the *base pairs* within the DNA molecule (see Topic Group 2.8). The mutated gene is effectively an allele of the original, and in fact mutations are the way in which alleles are created. Unless it is immediately lethal, once a mutated allele has been formed it is then repeatedly copied until a further mutation occurs. An example of a gene mutation is in the formation of the allele which causes *sickle-cell anaemia* (see Topic 6.5).

Chromosome mutations involve sections of chromosomes, including many genes, or even complete chromosomes. Changes in the structure of chromosomes can occur as a result of *deletions, duplications, inversions* or *translocations*. These are summarised in Fig. 6.42.

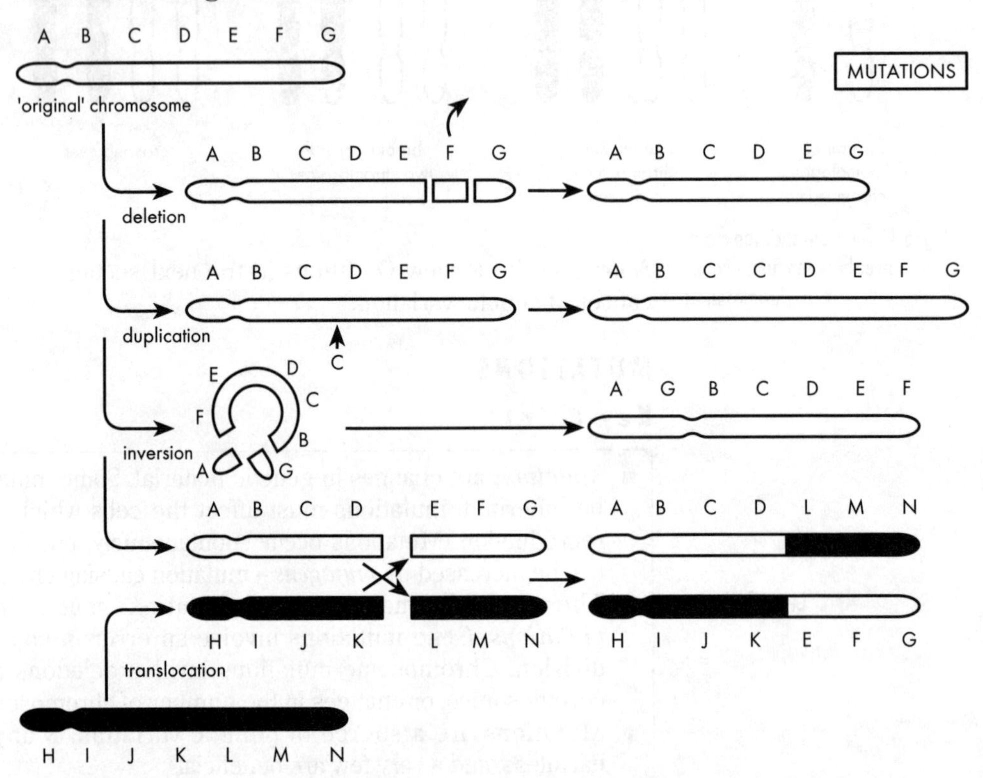

Fig. 6.42 Diagram showing the main types of chromosome mutation

Changes in *chromosome number* involves loss or gain of individual chromosomes (e.g. in Down's syndrome in humans) or even a whole set of chromosomes (e.g. polyploidy in plants).

Links with other topics

Other related topics elsewhere in this book are as follows:

REVIEW QUESTIONS

Q11 (a) Write brief descriptions or definitions for each of the following terms:

1. Mutation ____________________

____

2. Mutagen ____________________

____

3. Homologous chromosomes ____________________

____

4. Meiosis ____________________

____

5. Crossing-over (in meiosis) ____________________

____

(b) Briefly explain why mutations may be:

(i) a possible *disadvantage* for the organism

____

____

(ii) a possible *advantage* for the organism

____

____

(iii) *neutral* for the organism, having no distinct advantage or disadvantage

____

____

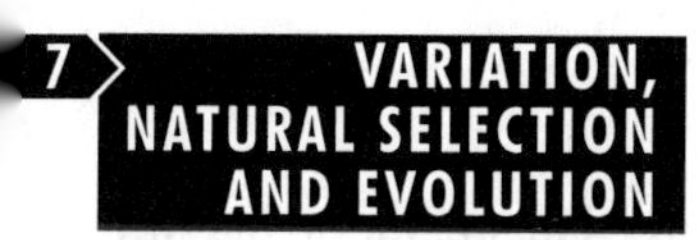

7 VARIATION, NATURAL SELECTION AND EVOLUTION

In this Topic Group, we look at the reasons why organisms are different from each other, both within species and between species. Variation of organisms is produced in different ways. Those types of variation which result in a change in genetic material which can be inherited are particularly important. They form the basis for the process of natural selection, in which 'new' characteristics provide an improved adaptation of the organism to its environment, and so enable the organism to compete more effectively. This process of variation-adaptation-survival is thought to be the mechanism by which new species evolve.

This Topic Group draws on many other parts of the syllabus, which should ideally be revised together. Important sections to refer back to at this stage include Animal and Plant Groups, Selective Breeding, Patterns of Inheritance and Causes of Variation.

VARIATION

Key Points

Level 9

- *Variation* of characteristics between organisms results from two main sources; *genetic differences* and from *environmental influences*. For any given characteristic, there are two main types of variation; *continuous* and *discontinuous*. Continuous variation (e.g. height in humans) usually results from the influence of several genes, and also the effects of the environment. Discontinuous variation (e.g. 'tallness' in peas, human blood groups) tend to result from a single genes, with little or no environmental effect.
- Variation which can be inherited is the 'raw material' for *natural selection* and, hence evolution.

Variation consists of differences between individuals belonging to the same species. Variation is caused by the separate or combined effects of *genetic* (internal) and *environmental* (external) factors. In other words, the way in which individual organisms are different from (or similar to) to each other is determined by the genetic information that they inherit and by the environment in which they live.

1 The interaction of genotype and environment

The total amount of genetic information present within an organism's cells is called its *genotype*. During an organism's growth and development there is an *interaction* between the genotype and the environment. There is also a sort of interaction 'within' the genotype; for instance, in the way in which certain *alleles* (different versions of *genes*) are expressed ('have their effect'), and in the combined effects of more than one gene acting together (this is *polygenic inheritance*).

Revise previous topics on genetics if you are unsure of this.

The outcome of this interaction is the *phenotype*, the set of characteristics which make up the organism. In a sense, the genotype is what the organism *could* become. The phenotype is what the organism *does* become as a result of the genotype and environment acting together.

Organisms within a species are likely to have a very similar or identical genotype if they have been produced by asexual reproduction. This is because the genetic information is supplied by just one parent. Organisms produced by sexual reproduction will be genetically different from each other (and their parents) because genetic information is supplied by two parents and is 'mixed' in a fairly random way.

2 Inherited variation

Characteristics determined by genes are inherited, but the way in which these are expressed in the phenotype is determined by:

(i) the interaction of alleles and genes in the genotype, for instance dominant and recessive alleles, polygenic inheritance
(ii) the interaction of the genotype with the environment.

Inherited characteristics are significant because they arise through evolution and may ensure that offspring are 'pre-adapted' to their environment. Some characteristics cannot be expressed in an alternative way without being lethal; variation will not occur in such characteristics.

Changes to the Genotype
There are three main ways in which a genotype may be changed in an inheritable way:

- **Meiosis.** During the production of gametes by meiosis, separating chromosomes may acquire different alleles. This is known as 'independent assortment'. Errors in meiosis may, for instance, cause an additional chromosome to be inherited, resulting in Down's syndrome.
- **Fertilization.** During sexual reproduction, new combinations of homologous chromosomes will result. This is a 'random' process.

- **Mutation.** Mutations are permanent changes in the genetic composition, or genotype, of an organism. Changes to particular genes or even whole chromosomes affect the cell in which they occur, and also any cells produced from it by cell division. Mutations can therefore be 'passed on' from one cell to another.

3 Types of variation

There are two main types of genetically determined variations, *discontinuous* and *continuous*. There is also variation caused by *acquired characteristics* which is not genetically determined:

Discontinuous variation

Discontinuous variation occurs when characteristics are determined by one gene, or when the environment does not directly affect that characteristic. The result of discontinuous variation will be a *limited number of phenotypes, with no intermediate forms*. Examples of this include blood groups in humans (Fig. 6.43) and height in pea plants (Fig. 6.44). Discontinuous variation is more common in less complex organisms for general characteristics, and in all organisms for 'vital' characteristics.

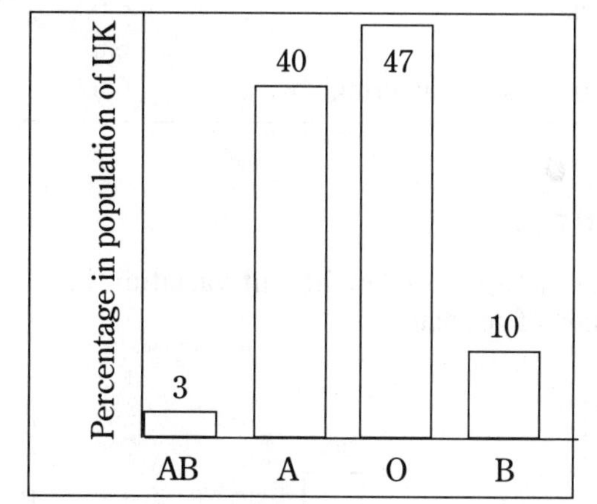

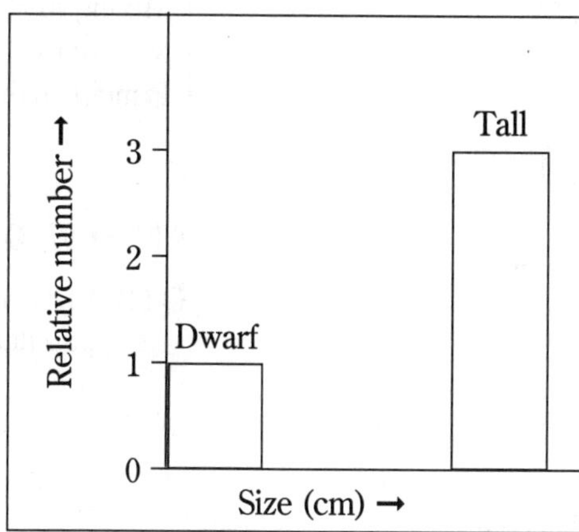

Fig. 6.43 (left) Discontinuous variation in humans: blood groups

Fig. 6.44 (right) Number and size of tall and dwarf varieties of pea plant

Continuous variation

Continuous variation occurs when the characteristic is determined by several genes, or because growth and development are readily affected by the environment. Examples include the size of insect larvae (Fig. 6.45) and height in humans (Fig. 6.46). Note that human height is less variable than mass in genetically similar individuals (e.g. identical twins). This suggests that the genotype for human height is more strongly controlled than that for mass.

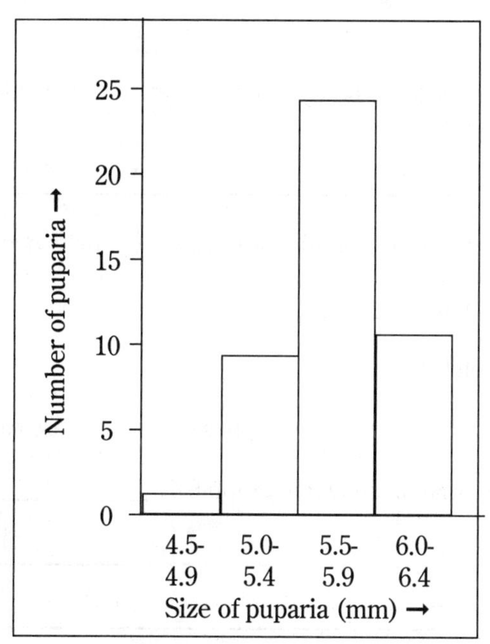

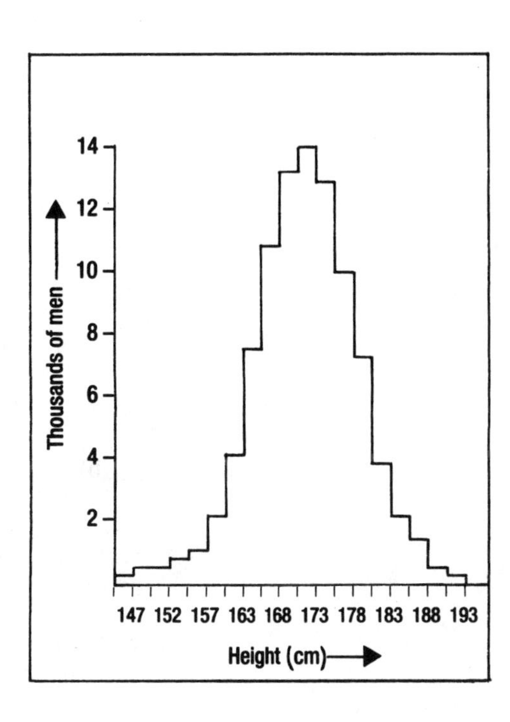

Fig. 6.45 (left) Number and size of pupa of house fly

Fig. 6.46 (right) Continuous variation in human males: height

Acquired characteristics

Acquired characteristics arise from the effects of the environment and do not result in a change in the genotype. Variation caused in this way is adaptive; it is a means of modifying the organism in response to a changing environment. Examples include increased human fitness through exercise, and tanning of the skin.

Variation within a population creates possibilities for ***natural selection***. This is because some organisms may, as part of their variation, have *adaptations* which allow them to *compete* more effectively with each other. For this reason, variation has been called the 'raw material of evolution'.

Links with other topics

Topic	Chapter and Topic number	Page number
Adaptations for survival	4.4	44
Sources of variation	6.6	233
Mixing of alleles	6.6	234
Mutations	6.6	235
Principles of genetic engineering	6.9	254

REVIEW QUESTIONS

Q12 Figs 6.47 and 6.48 show height variation in a human population and in a population of pea plants.

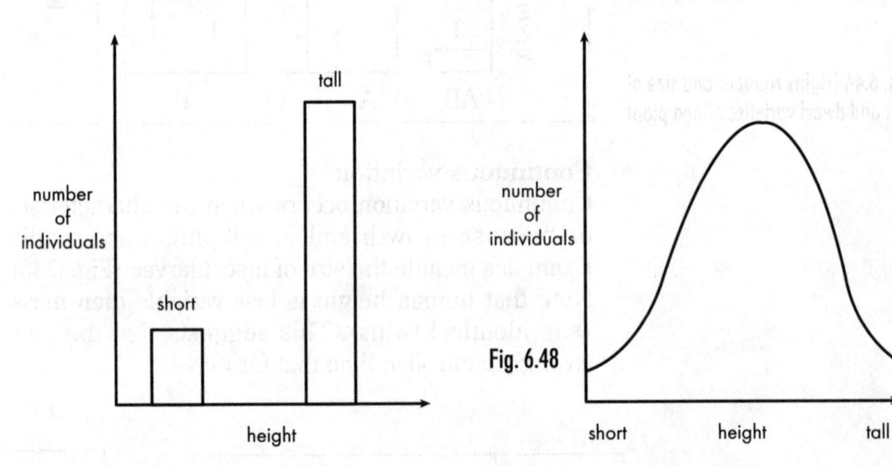

Fig. 6.47

Fig. 6.48

(a) (i) Which graph Fig. 6.47 or Fig. 6.48, shows height variation in human beings?

____

(ii) What types of variation are shown by the two graphs?

Graph 6.47 ______________

Graph 6.48 ______________

(b) Explain briefly how the two types of variation are caused.

Type of variation shown in graph 6.47 ______________________________

____

____

Type of variation shown in graph 6.48 ______________________

____

____

(c) Whatever the normal skin colour of a population, occasionally an albino child is born. An albino has no skin pigment.

(i) What name is given to the unexpected occurrence of such a characteristic?

____

(ii) Suggest what processes may have taken place leading to the birth of an albino rather than to the birth of a normally-pigmented child.

____

____

____

____

NATURAL SELECTION

Key Points

- *Natural selection* has been proposed as the most likely mechanism for *evolution*. Natural selection is the process by which organisms which have better *heritable adaptations* to their environment tend to have an increased chance of survival. 'Heritable adaptations' are those which can be inherited. Because of genetic variation, it is likely that some organisms will have a certain adaptation and others will not. If an organism has a greater chance of survival because of an adaptation, then it is more likely to reproduce, and so 'pass on' the adaptation to its offspring.
- Natural populations tend to stay fairly constant, but organisms tend to be capable of producing more than enough offspring to replace themselves. This suggests that there is *competition* for survival, or 'struggle for existence'. In competition, natural selection favours those organisms which are best adapted to their environment; this is called the 'survival of the fittest'.
- An example of natural selection is the inheritance of *industrial melanism* in the peppered moth.

1 What is natural selection?

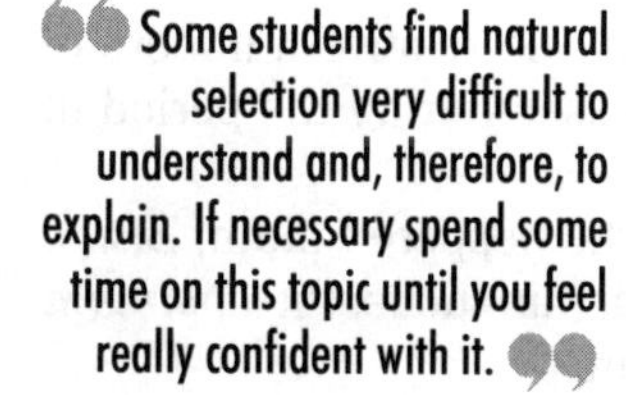

Natural selection is a process by which organisms which are better adapted to their environment tend to have an increased chance of survival; they therefore have a greater opportunity to reproduce, and pass on to their offspring those genetically determined characteristics which cause them to be better adapted.

The theory of natural selection is based on the following related observations:

1. **Reproductive Potential.** All organisms are capable of producing more than enough offspring to replace themselves. The potential for a very rapid increase in population was noted, amongst others, by Thomas Malthus in 1778. However, the number of surviving organisms in a population tends to remain, on average, *fairly constant*.

2. **Competition.** Competition occurs when organisms require the same limited opportunities, such as access to food, oxygen, warmth and space and the avoidance of predators and disease. Competition is often more intense between members of the same species because their requirements tend to be similar. This leads to a 'struggle for existence'.

3. **Variation.** Within any species there are variations caused by the *inheritance of different characteristics*. Variation occurs particularly in those organisms which reproduce sexually, because new genetic combinations are created. Whilst offspring are often different from their parents they also share many important similarities, too.

4. **Adaptation.** Certain variations may allow particular individuals within a species to compete more effectively because they are better adapted. This increases their chances of survival in a given environment, and they are more likely to have the opportunity to reproduce. In this way, genetically determined characteristics may be 'selected' because they provide an advantage in comparison with other variations. This has been called 'survival of the fittest' i.e. survival of those best fitted to their environment.

2 How natural selection works

Natural selection occurs as the result of an *interaction of organisms with their environment*. Organisms are not always fully adapted to their surroundings, for example because their environment is *changing* or because they have *colonised* a new environment. In such circumstances any organism that becomes better adapted through a change in its inherited characteristics has a *selective advantage.*

bacteria

Genetically-determined ('heritable') variation is an essential pre-requisite for natural selection. This is because any variation which makes members of a species better adapted than others to survive can be passed on to subsequent generations. In other words, if the variation gives organisms a selective advantage it will tend to spread through the population.

This idea can be demonstrated by a relatively simple example:

The disease-causing bacterium *Neisseria gonorrhoeae* became resistant to the *antibiotic* penicillin because of a new heritable variation. This variation was caused by a *mutation,* a change in the chromosome of one bacterium. The resistant cell was better-adapted because it had an *allele* which could prevent the cell from being killed by penicillin. The offspring of the mutant bacterium inherited the resistance (Fig. 6.49) by natural selection; the allele providing resistance gave a selective advantage to the bacteria carrying it and so spread quickly through the population.

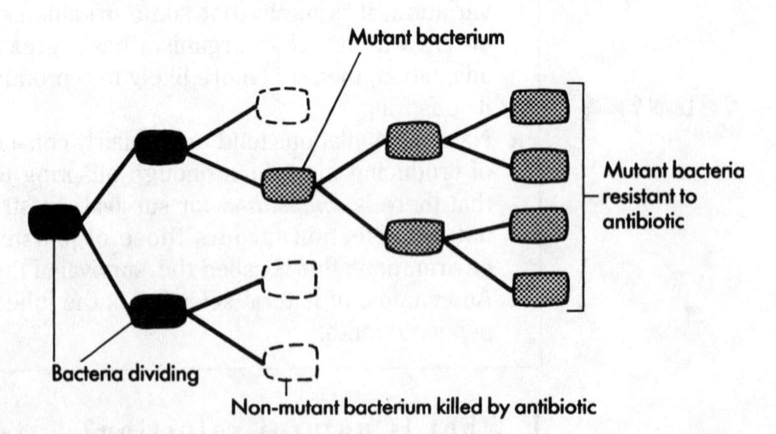

Fig. 6.49 Resistance to antibiotics: in this example, the disease-causing bacterium *Neisseria gonorrhoeae* has become resistant to the antibiotic penicillin because a new mutant strain of the bacterium is not affected

3 Natural selection in action: industrial melanism

Dark (melanic) varieties occur in nearly 80 species of moth; the relative numbers of the dark variety in relation to the 'normal' paler forms increased during the period of industrialization in the UK from about 1850 to 1950.

One species which has been particularly well studied is the peppered moth, *Biston betularia*. The moth shows two genetically-determined colour variations; the *dominant* dark (melanic) form and the *recessive* pale (non-melanic) form.

The relative number of each type of the (nocturnal) peppered moth eaten by birds may depend on how well camouflaged they are whilst resting on surfaces such as tree trunks during daytime (Fig. 6.50).

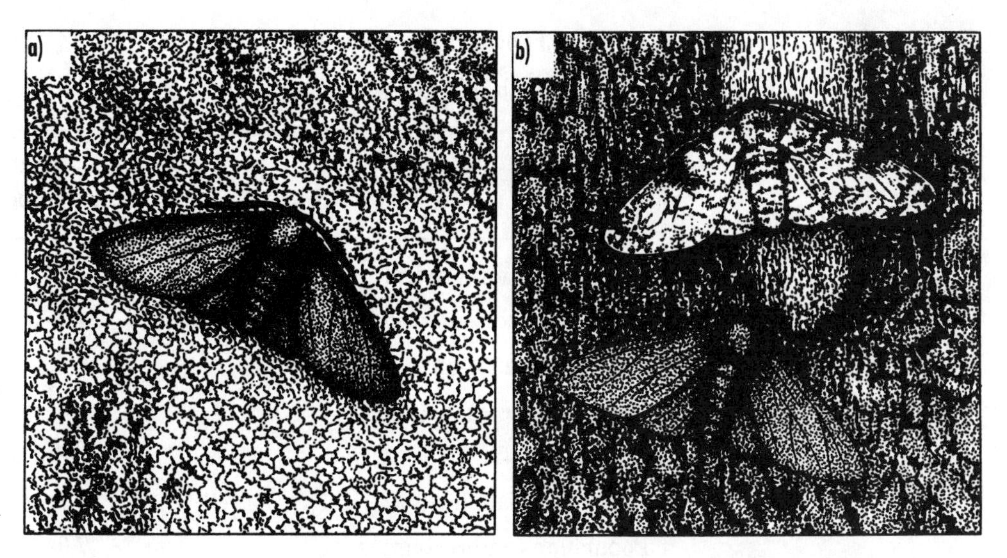

Fig. 6.50 Pale and dark forms of the peppered moth:
a) Lichen covered tree in unpolluted area.
b) blackened tree in polluted area.

The increase in numbers of the dark form in the Manchester, UK, area during nearly 50 years of industrialisation occurred very quickly as is shown in Fig. 6.51. This increase corresponded with the blackening of tree trunks by smoke. At the same time, the growth of lichen, which is sensitive to atmospheric pollution was seriously reduced. The relative distribution of dark and pale forms in 1958 (Fig. 6.52) shows that the number of pale forms remained relatively high in the non-industrialised areas of Britain.

	PERCENTAGE OF EACH FORM	
YEAR	DARK	PALE
1848	1	99
1894	99	1

Fig. 6.51 Increase in relative numbers of the dark form of the peppered moth in an industrial area

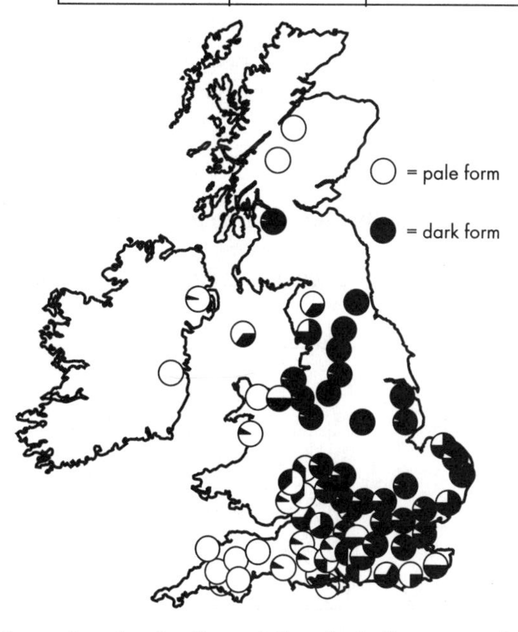

Fig. 6.52 Distribution of pale and dark forms of the peppered moth (*Biston betularia*) in Britain in 1958. (Republished from the journal *Heredity*, with permission)

In this example, natural selection of the dark form may have occurred during a period of increasing pollution because they were better camouflaged on blackened tree trunks than the pale form. Being dark gave moths a selective advantage in industrial areas because they were better adapted to their environment. Since the Clean Air Act of 1959 air pollution has declined and so also have the numbers of the dark form of the peppered moth.

These observations have been confirmed by capturing and marking moths, releasing them and then recapturing them. This allows estimates to be made of the relative numbers of each type of moth in a given area.

Links with other topics

Other related topics elsewhere in this book are as follows:

REVIEW QUESTIONS

Q13 Fig. 6.53 below shows the range of wing sizes in populations of a species of small bird on different islands (**A–F**) to the north of the United Kingdom.

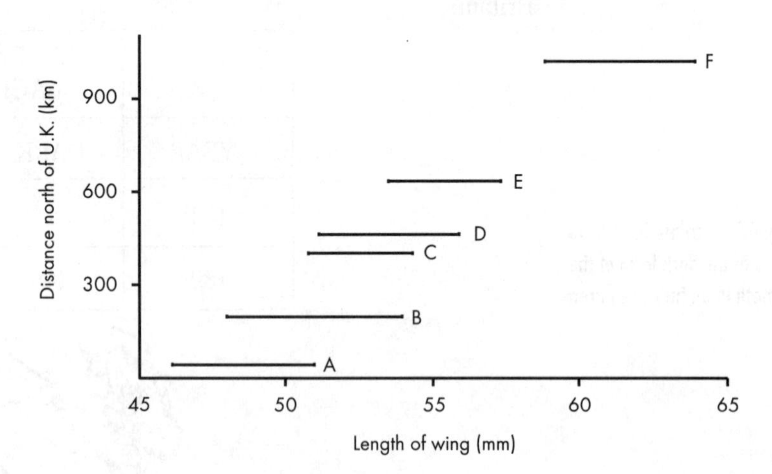

Fig. 6.53

Using the information given in Fig. 6.53, answer the following questions:

(a) A bird with wings 53 mm long might have lived on which island(s)?

___

(b) Which island has the smallest range of wing sizes?

___

(c) What general trend is seen to be taking place (that is, how are the populations changing) as the samples are taken from further north?

___

___

(d) Assuming that the birds with longer wings have larger bodies, how does this change in body size help the birds to survive?

___

___

___

(e) Draw a line connecting the mid-points of **E** and **F** and from this find the mid-point of the range of wing length which would be present on an island situated 800 km north of the United Kingdom.

(f) Explain how mutation and natural selection could have led to the distribution of wing lengths shown.

____

____

____

____

EVOLUTION AND SPECIES FORMATION

Key Points

- *Evolution* is the process by which new *species* are thought to have arisen from pre-existing species by gradual changes in *inheritable characteristics*. All species are therefore related to each other by *ancestry*. Species are different from each other because of different *adaptations* to their mode of life. These adaptations arise from *modifications* in their genetic composition.
- Various theories exist for the *mechanism for evolution*, including *natural selection*. *Evidence* for evolution includes taxonomy (classification), fossil evidence, anatomical similarities and artificial selection.

1 The history of life

The way in which life began and has subsequently developed is clearly an important aspect of biology. There are several explanations for the origins of life, including:

- **Special Creation.** Many people throughout the world believe that life was created at a particular time by God and that, once formed, species do not change.
- **Evolution.** Most biologists believe that life was created over a very long period of time by gradual changes, involving natural selection. New species are formed from pre-existing species as part of a continuing process. The origins of life through evolution involved the assembly of biologically important molecules during *chemical evolution*, followed by gradual development of increasingly complex organisms from simpler forms during *organic evolution*.

Evolution can explain both *similarities* and *differences* between organisms. Organisms are *similar* because they are descended from a common ancestor; each generation is related to previous ones by the inheritance of genetic information. Organisms are *different* because *variation* occurs between parents and offspring.

2 Formation of species

Speciation is the formation of a new species and involves natural selection. In a population which can freely interbreed similar genetic information is, in a sense, shared by all individuals.

The flow of *genes* within the population is prevented if members of the species are not free to interbreed, for instance if they are separated by a physical barrier such as a mass of land or water. If separation occurs for a sufficient period of time, separated groups of organisms will gradually become genetically different. This is because natural selection will be acting independently on each group. The cumulative differences may eventually be so great that the groups will not successfully interbreed even if they become free to do so (Fig. 6.54).

Barriers which prevent interbreeding and which therefore promote speciation include:

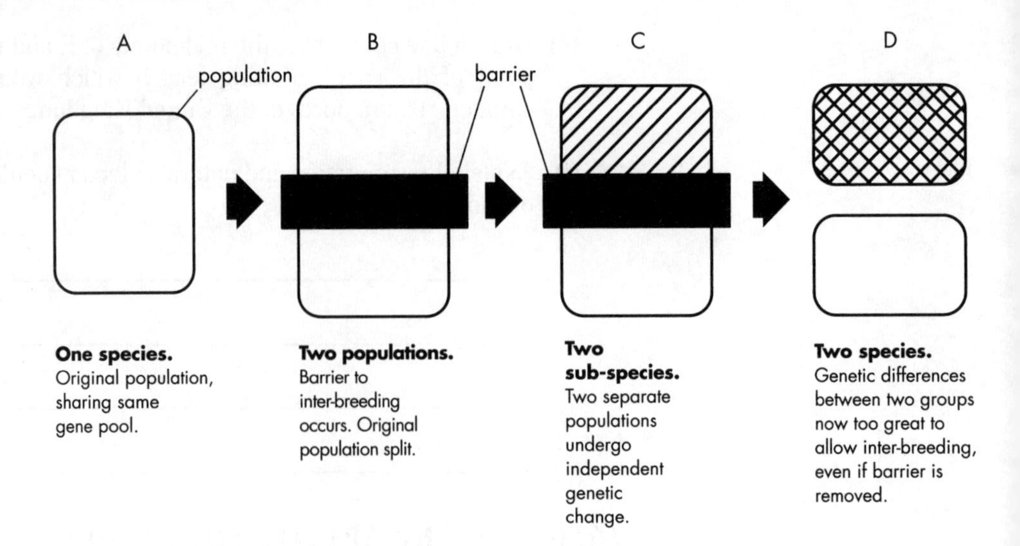

Fig 6.54 Simplified diagram, showing species formation

- **Geographical isolation**

Populations divided by the formation of oceans, rivers, mountains.

Example: different species of fruitfly (*Drosophila*) on the islands of Hawaii.

- **Behavioural isolation**

Populations adopt different breeding cycles or courtship patterns.

Example: different species of grasshopper having different mating calls.

Even if interbreeding does occur, genetic differences may be such that the *hybrid* is infertile; for example in the mule. However some hybrids, particularly in plants, may be fertile and may be considered to be a new species; for example, bread wheat (see topic 6.2).

3 Evidence for evolution

There are various forms of evidence for evolution. This includes *examples of natural selection* which have occurred, or are occurring, over a short enough time scale to be observed by scientists. Two examples are presented elsewhere in this book; industrial melanism in the peppered moth and sickle-cell anaemia. *Taxonomy* (the study of the way in which species are related) shows that species are related in a systematic way. It also reveals a general hierarchical trend of increasing complexity in more recently-evolved species.

Other types of evidence for evolution are from fossils, comparative anatomy and artificial selection.

Fossil evidence

Fossils are the preserved remains or traces of organisms which were living in the past. Many fossils were formed in *sedimentary rock* as particles of sand, silt or mud settled around dead organisms and then became compressed into solid rock. The impression embedded in the rock reveals the shape of the original organism. Other fossils have been formed by organisms being preserved in substances such as amber or peat.

The age of fossils can be estimated from the position where they are found; older fossils normally occupy a deeper position in a rock formation. Fossils provide a record of change of one form of organism into another; for example in the evolution of the modern horse from its ancestral forms.

Fossils provide a record of those organisms that have become extinct; they probably failed to adapt to a changing environment. However, the fossil record is not continuous or complete. This may be because soft-bodied organisms do not readily form fossils and because other conditions may have been unsuitable.

Anatomical evidence

Organisms which are thought to have evolved in a similar way often have comparable structures within their body. *Homologous structures* are thought to have been derived from a *common ancestor* but do not necessarily perform the same function. Although they are thought to share common origins, homologous structures may have become adapted to different functions; this is known as *divergent* or *adaptive radiation* in evolution.

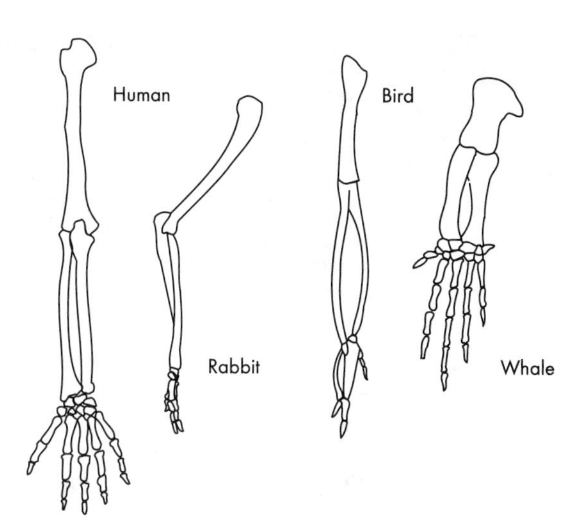

Fig. 6.55 The pentadactyl limb as a homologous structure

A good example of such homologous structures is the *pentadactyl limb* in vertebrates (Fig. 6.55). This may perform different functions in locomotion so is adapted in different ways. However, similarities in bone structure indicate that the pentadactyl limb may have been derived from a common ancestor.

Analagous structures are similar in function but do *not* indicate close evolutionary relationships between certain organisms. For example, the wings of the bat, bee and a sycamore fruit perform a similar function but have been derived very differently.

Artificial selection

Artificial selection is the development of 'improved' types of domesticated organisms useful to humans by *selective breeding*. The main purpose of artificial selection is to provide useful varieties, but the process also shows how natural selection could operate. In the case of artificial selection, the selection of adaptations is by humans rather than by purely 'natural' means.

For example, farm animals and plant crops are selectively bred for increased yield and resistance to disease. Variation naturally occurs, for instance because of mutations; organisms showing variations which are desirable are selectively bred with each other.

Links with other topics

Other related topics elsewhere in this book are as follows:

Topic	Chapter and Topic number	Page number
Adaptations for survival	4.4	44
The main plant and animal groups	5.1	181
Selective breeding – crops	6.2	210
Selective breeding – domestic animals	6.2	211
Types of variation	6.7	238

REVIEW QUESTIONS

Q14 The Galapagos Islands are situated in the Pacific Ocean about 1000 km west of Ecuador in South America.

When Charles Darwin visited the Galapagos islands in 1835, he noticed **six** different varieties of beak amongst the finches on the islands.

The diagram (Fig. 6.56) shows these beak shapes and the main food eaten by the finches.

Darwin suggested that, originally, the large ground finches came to these islands from South America and that all the other types evolved from these.

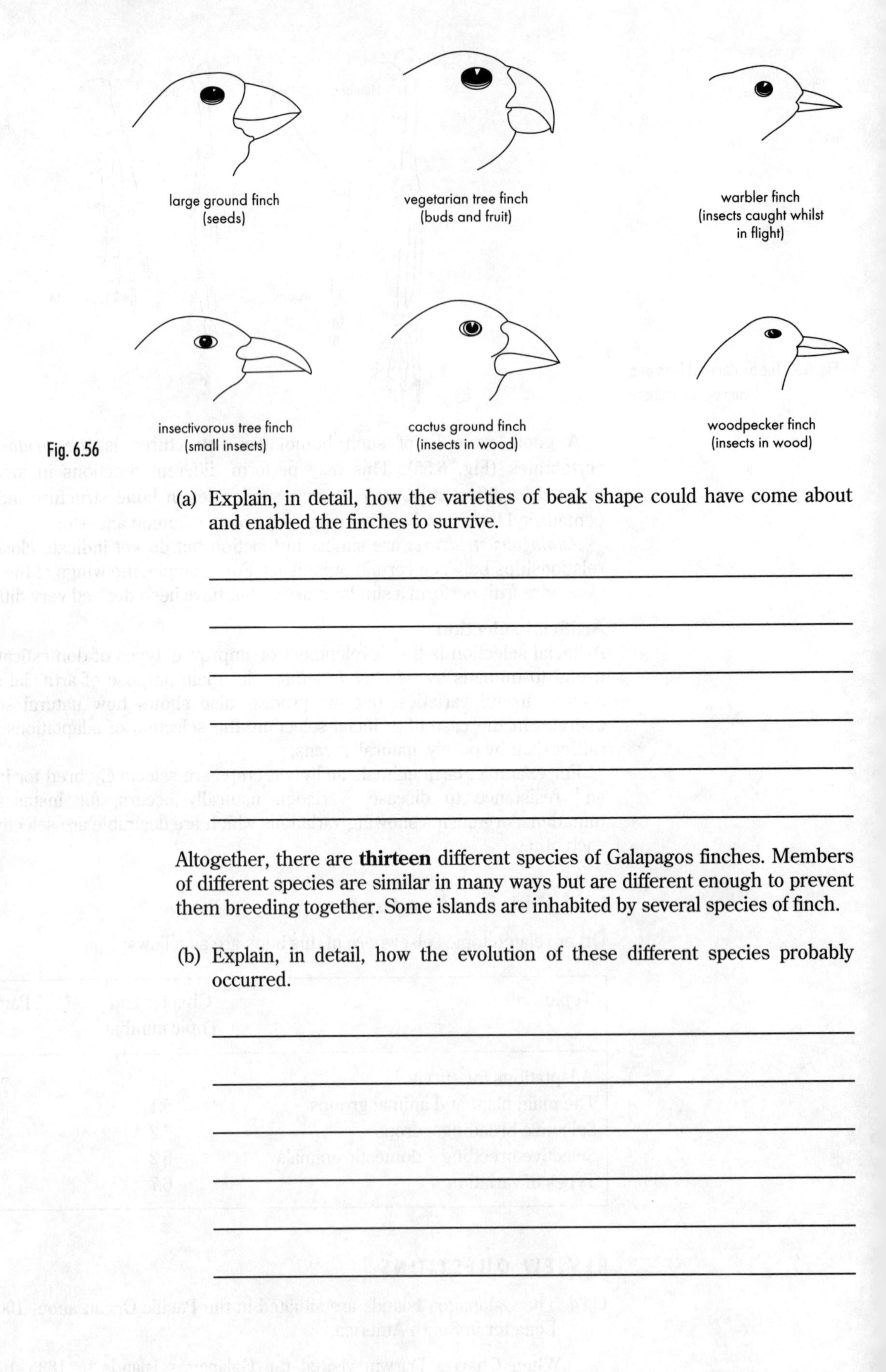

Fig. 6.56

(a) Explain, in detail, how the varieties of beak shape could have come about and enabled the finches to survive.

Altogether, there are **thirteen** different species of Galapagos finches. Members of different species are similar in many ways but are different enough to prevent them breeding together. Some islands are inhabited by several species of finch.

(b) Explain, in detail, how the evolution of these different species probably occurred.

8 DNA AND PROTEIN SYNTHESIS

Inside the nucleus of all plant and animal cells there are long threads of DNA called chromosomes. Each chromosome carries hundreds of genes, and each gene codes for a single protein.

Level 10

The DNA inside the cell will replicate (make a copy of itself) before the cell divides, so that each daughter cell will carry the genetic information coded on the chromosomes.

The genetic information is in the form of the genetic code; this depends on the order of bases in the DNA. Three adjacent bases will code for each amino acid.

When proteins are being made, a copy of one gene, called the RNA template, is made inside the nucleus, then this moves out into the cytoplasm. Amino acids are joined together according to the genetic code on the RNA.

Key Points

Level 10

- DNA is made up of repeating units called *nucleotides*
- DNA is a *double helix* i.e. two strands of nucleotides are twisted together in a spiral shape
- *base-pairing* holds the two strands together
- when DNA replicates, *identical* strands of DNA are formed
- an RNA copy of part of the DNA moves from the nucleus to the cytoplasm during protein synthesis
- amino acids line-up on the RNA according to the order of the bases (*genetic code*) and are joined together to make a protein.

Structure of DNA

DNA is short for *deoxyribonucleic acid.* The structure of DNA was worked out by a British scientist, Francis Crick, and and American scientist, James Watson, in 1953. It is made up of a large number of repeating units called *nucleotides.* One nucleotide looks like this (Fig. 6.57).

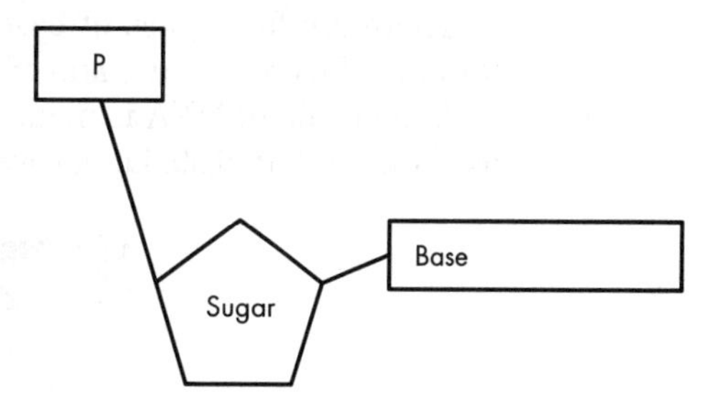

P = phosphate

Fig. 6.57 Structure of a nucleotide

The nucleotides are joined together to make a long chain, (Fig. 6.58).

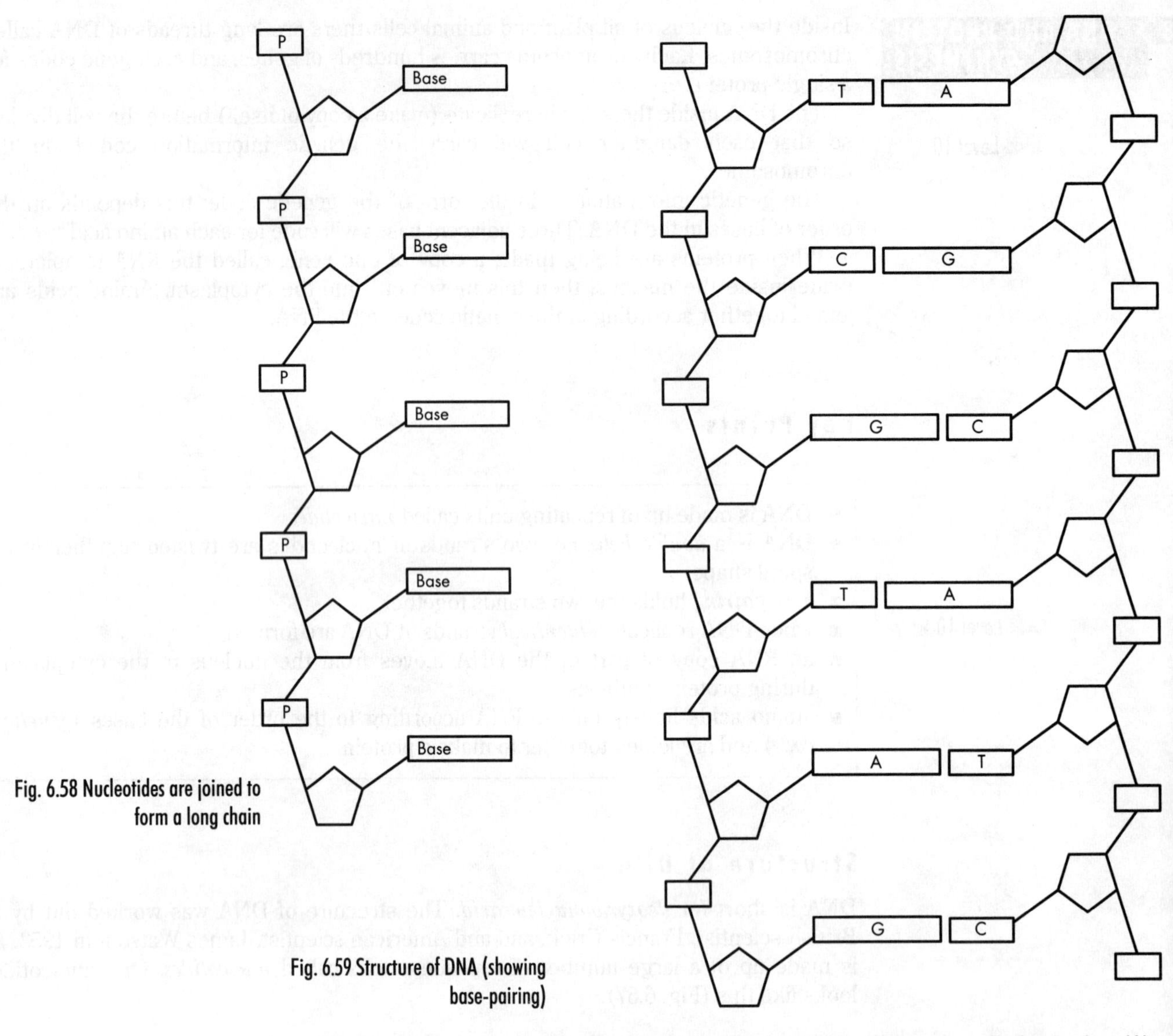

Fig. 6.58 Nucleotides are joined to form a long chain

Fig. 6.59 Structure of DNA (showing base-pairing)

There are four types of base which can be included. They are called *adenine* (A), *thymine* (T), *cytosine* (C) and *guanine* (G).

A molecule of DNA is made up of two strands of nucleotides joined by their bases, as shown in Fig. 6.59. Look carefully at how they are joined together.

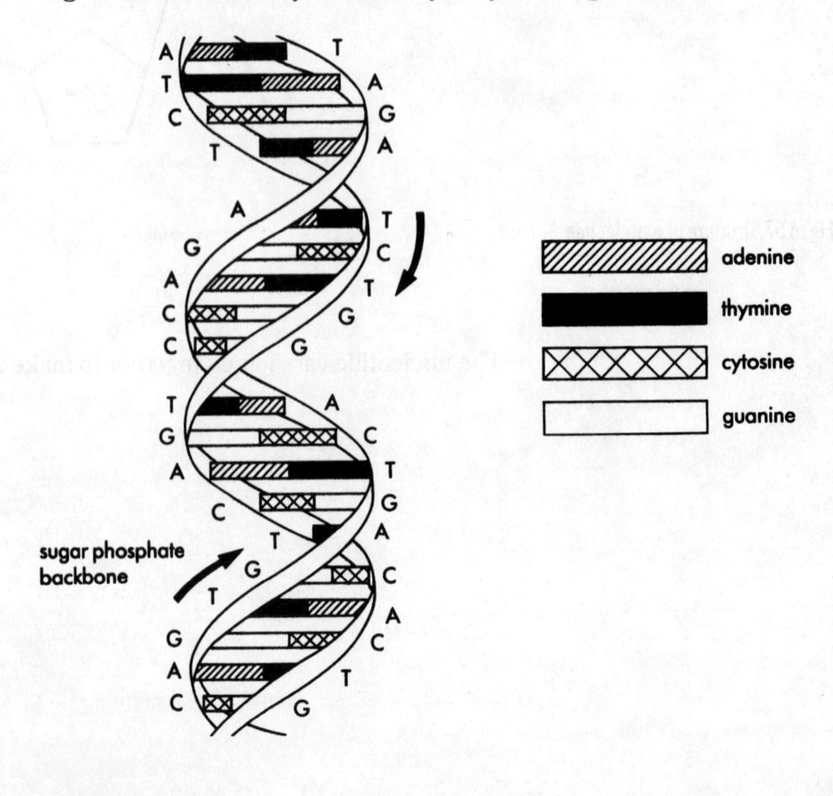

Fig. 6.60 Structure of a DNA molecule

Adenine always bonds to thymine, and cytosine always bonds to guanine: this is called *base-pairing*.

The shape of the strands causes the molecule to twist into a ***helix*** (spiral) shape. It looks like a twisted ladder, where the rungs are the pairs of bases (Fig. 6.60).

Replication of DNA

A typical human body cell e.g. skin cell, bone cell, has 46 chromosomes. When that cell divides by mitosis, two daughter cells are formed, each with 46 chromosomes. (Fig. 6.61)

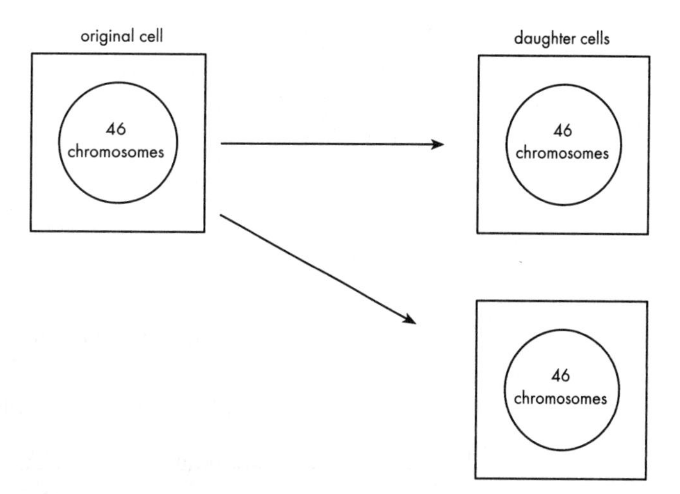

Fig. 6.61 Mitosis

This means that new chromosomes must have been formed. In fact, the DNA makes a copy of itself – the scientific term for this is *REPLICATION.*

There are 4 main stages in replication (Fig. 6.62)

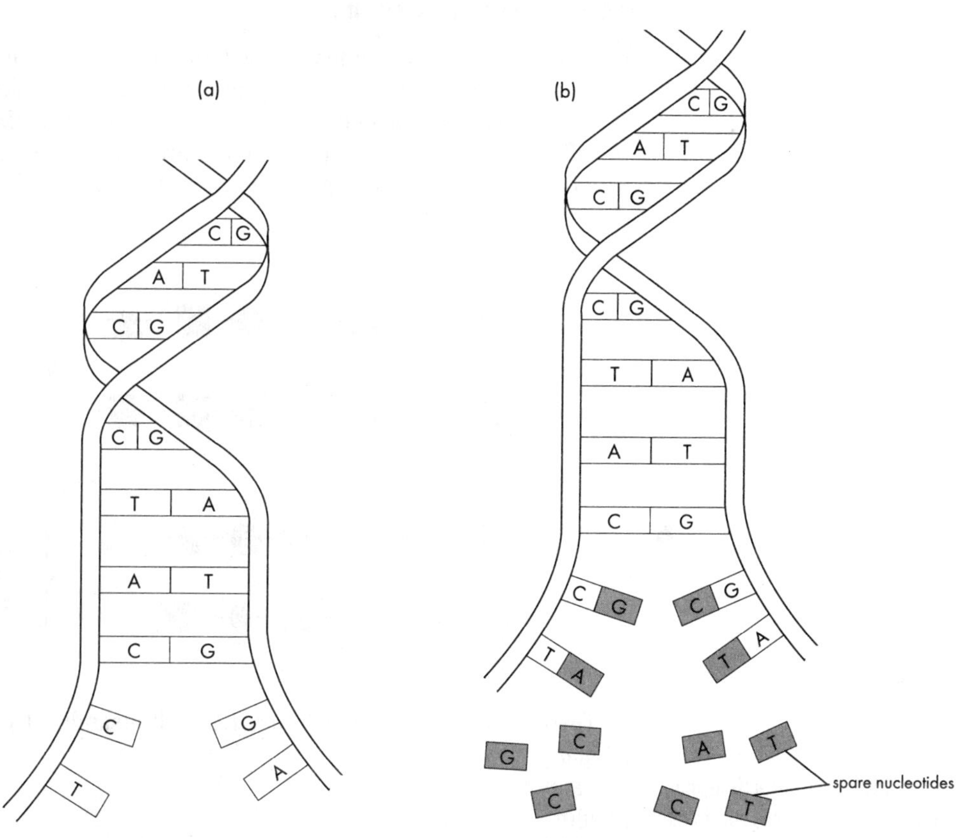

Fig. 6.62 Stages in the replication of DNA

(a) The DNA molecule is double-stranded and helix shaped.
    It untwists and begins to separate into two strands.
    The bonds between the bases are broken.

(b) Spare nucleotides inside the nucleus pair up with the bases on the separated strands.

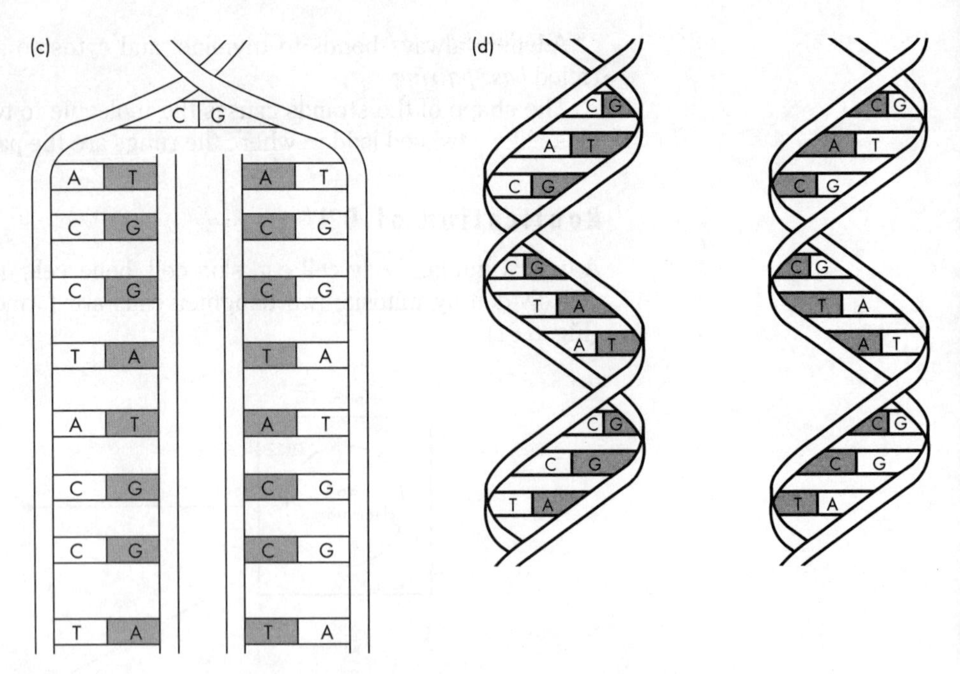

Fig. 6.62 (Cont.)

(c) As the helix unwinds, more nucleotides pair up with the bases on the separated strands.

Eventually the whole DNA molecule will separate in this way.

(d) There are now two molecules of DNA, each double-stranded.

Each one twists up to form a helix shape.

The two DNA molecules are *identical* to each other.

Genes and proteins

The DNA in chromosomes carries hundreds of genes. Each gene *codes* for one protein; that means that it carries the instructions to make one protein.

Remember, each protein is made up of a chain of *amino acids*. Altogether, there are 20 different amino acids, and proteins are different to each other due to the *type* of amino acids they contain, and the *order* of the amino acids. (Fig. 6.63)

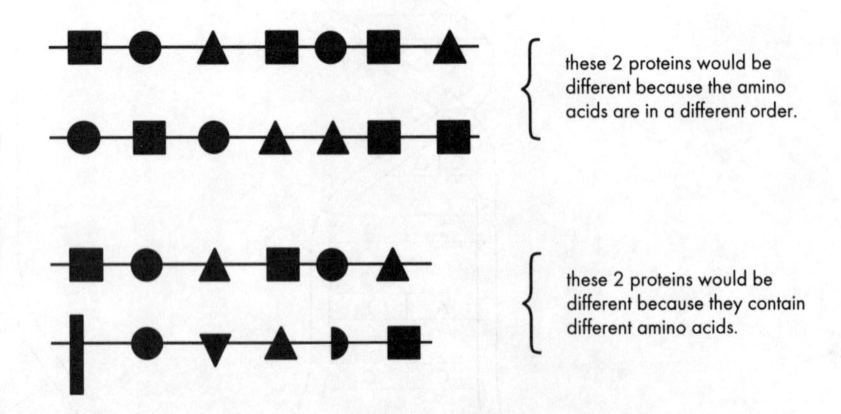

Fig. 6.63

To be healthy, a human must have thousands of different proteins, all working properly.

e.g. amylase to digest starch
insulin to control blood sugar level
growth hormone to regulate growth

Proteins control thousands of important reactions within the body.

So, a gene must carry instructions for the type of amino acids, and the order of amino acids, to make a protein. It does this through the *genetic code* i.e. the order of the bases on the DNA molecule. Each set of 3 bases is called a *triplet*, and it will code for a single amino acid.

Protein synthesis

DNA (in chromosomes) carries the instructions for making a particular protein, and the chromosomes are inside the nucleus.

Protein synthesis (putting together the right amino acids in the right order) occurs in the cytoplasm of the cell. The chromosomes are too big to move out of the nucleus, so a copy is made of part of the DNA (one gene), and this moves out into the cytoplasm. This copy acts as a pattern, or ***template***, to make the protein, and is made of a substance called *RNA* (ribonucleic acid).

The RNA is much shorter than the DNA it is copied from; it is exactly the length of a single gene. It is single stranded, i.e. made up of one chain of nucleotides, and contains organic bases. However, it never contains ***thymine*** – this is replaced with ***uracil*** (U).

RNA is made by base-pairing with the DNA in a process called TRANSCRIPTION. The RNA is complementary to one of the DNA strands – means that the bases match it.

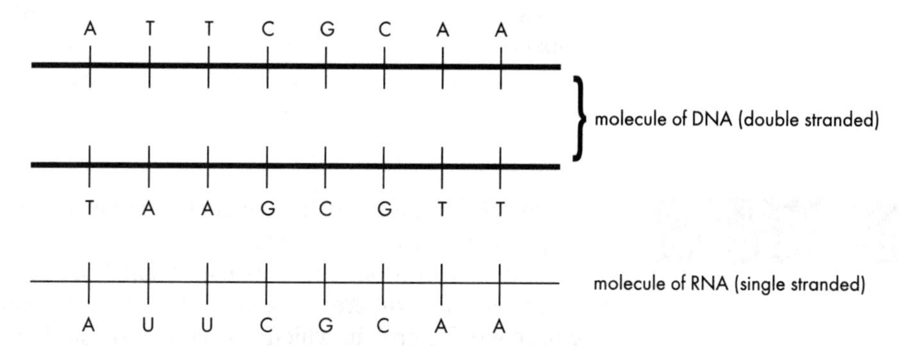

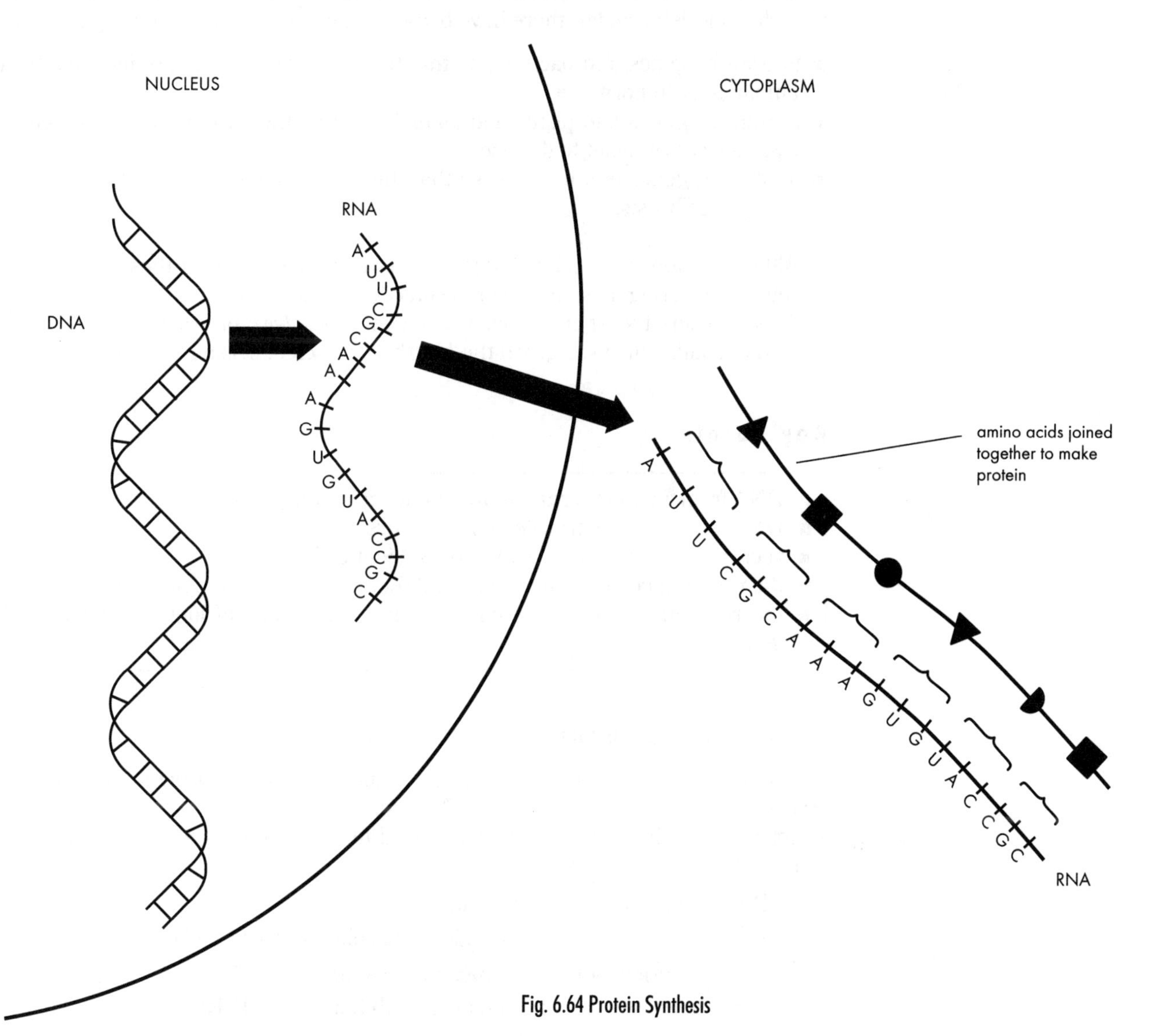

Fig. 6.64 Protein Synthesis

“This is a simplified account of a complex procedure – read it carefully”

Remember, the genetic code consists of bases on the RNA being “read” in groups of three (triplets). Each group of three bases will code for a single amino acid.

Amino acids will line up in the correct order on the RNA template, according to the genetic code. Organelles called *ribosomes* will hold them in place white *peptide bonds* form between amino acids to make a protein (Fig. 6.64).

Links with other topics

Related topics elsewhere in this book are as follows:

Topic	Chapter and Topic number	Page no
Chromosomes	6.3	214
Mitosis	6.3	217
Meiosis	6.3	219

9 GENETIC ENGINEERING

These are techniques which have been developed quite recently to transfer DNA from one type of organism to another.

A gene is taken from one organism (called the *donor*) and joined to a vector which will transfer it to a different organism (called the *recipient*). The recipient will then be able to make the protein which was coded for on that gene, i.e. it will have acquired a new characteristic. So far, there have been three main uses of genetic engineering:

“Level 10”

- to transfer genes into bacteria, so that they can make useful products e.g. insulin, human growth hormone
- to transfer genes into plants and animals, so that they acquire new characteristics e.g. they are resistant to disease
- to transfer genes into humans, so that they no longer suffer from genetic diseases e.g. cystic fibrosis.

Although there are undoubtedly benefits to be gained from these applications of genetic engineering, it is important to consider the potential dangers.

There are strict legal guidelines which control the *types* of experiment which can be carried out, and which safeguard the *health* of the scientists and the general public.

Key Points

- DNA from a donor is cut into fragments using enzymes
- DNA fragments are transferred to vectors
- vectors carry the donor DNA into recipient cells
- the recipient cell has a new gene and therefore has a new characteristic
- the recipient cell can be cloned to make thousands of identical copies of the gene

Techniques of genetic engineering

We will consider the example of the gene for human growth hormone being transferred to a bacterium (*E. coli*). The bacterium makes the protein (human growth hormone), then it can be collected and used to treat people who do not make enough of this hormone themselves.

Donor organism	:	human
Vector	:	plasmid (small piece of bacterial DNA)
Recipient organism	:	*E. coli* (bacterium)
Product	:	human growth hormone (HGH)

Transferring the gene

There are 4 main stages (Fig. 6.65):

Fig. 6.65 Stages in gene transfer (diagrams not to scale)

(a) Collect donor chromosomes e.g. from human white blood cells.

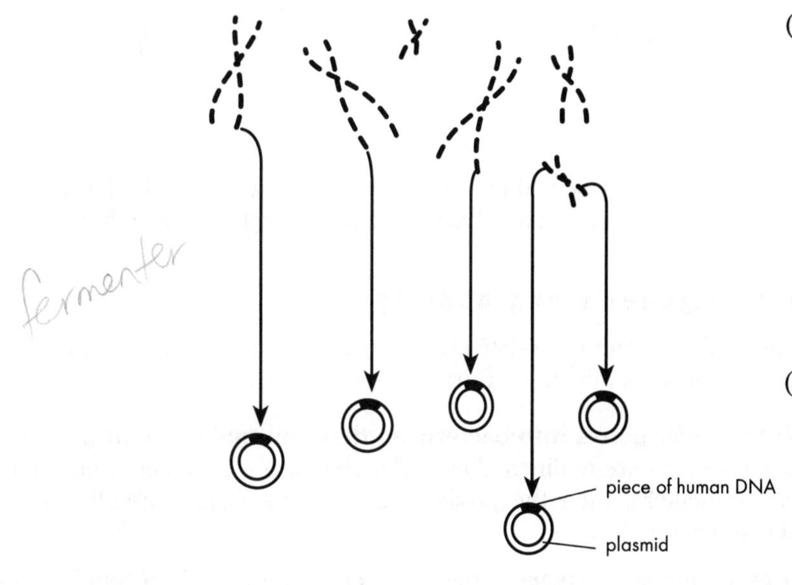

(b) Use an enzyme (restriction enzyme) to cut the chromosomes into fragments.

(c) Transfer each fragment into a vector. The vector is a plasmid – a small circle of DNA from a bacterium. The plasmid and the piece of human DNA are spliced together.

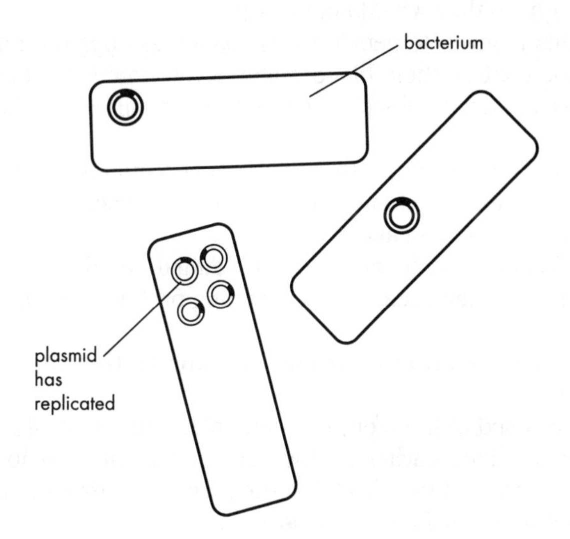

(d) Mix the plasmids and the recipient cells together. The recipient cells (bacteria) will absorb the plasmids. Once they are inside the cells, the plasmids will replicate (make copies of themselves).

One of the recipient cells now has several copies of the gene for human growth hormone, and will be able to make this protein.

Cloning the gene

In order to make large quantities of human growth hormone, we must have large quantities of bacteria carrying the HGH gene. The easiest way to do this is by *cloning* – allowing the bacteria to reproduce asexually, to produce millions of offspring, each containing the gene for HGH.

The cloning procedure is carried out in a large container, called a *fermenter*. The conditions inside the fermenter must be very carefully controlled i.e. correct pH, correct temperature, enough oxygen, suitable nutrients (Fig. 6.66). This gives the bacteria the ideal (optimum) conditions to grow, and to make the protein.

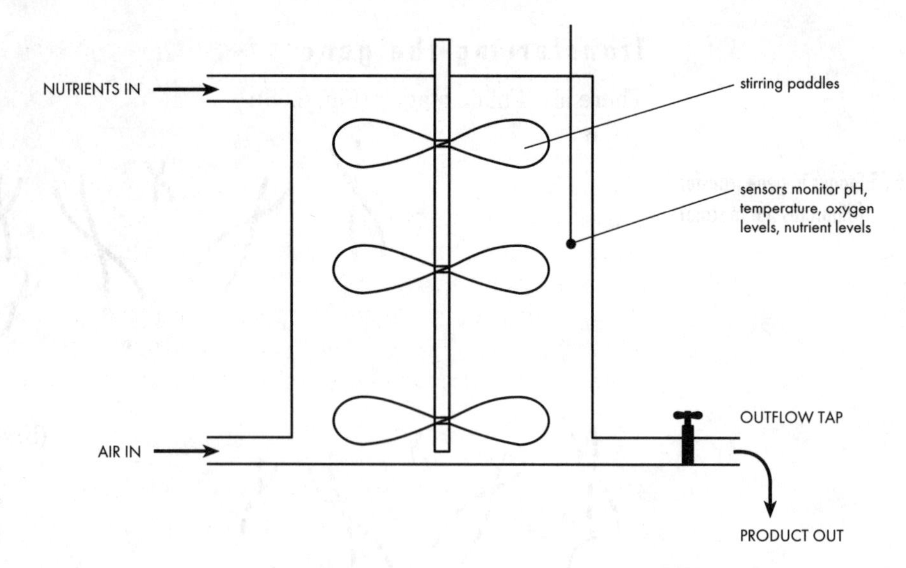

Fig. 6.66 Bacterial Fermenter

The product is collected by opening the outflow tap on the fermenter. However, it must be carefully purified and have any contaminants removed before it can be used.

Uses of genetic engineering

Genetic engineering is a relatively recent invention, and new techniques and uses are being developed all the time. Currently there are 3 main uses:

1 To transfer genes into bacteria so they can make useful products

The techniques are outlined above. The genetically engineered bacteria are grown in large fermenters, then the products are collected and carefully purified. Products made in this way include:

- *Human growth hormone* – this is given to people whose pituitary glands do not make enough, so they would not grow properly.
- *Insulin* – this is given to people whose pancreas does not make enough insulin, so they cannot control their blood sugar level effectively (see Section 4.7). These people are diabetic; previously they were given animal insulin, but this often caused side effects.
- *Rennet* – this is an enzyme used in the cheese-making process to make milk clot. Vegetarian cheeses contain rennet made by bacteria, whereas other cheeses contain rennet from animals.
- *Proteases/lipases* – these enzymes are added to biological washing powders to digest 'biological stains' e.g. food containing proteins and fats.

2 To transfer genes into plants and animals so that they acquire new characteristics

The techniques used to transfer genes into plants and animals are more complex than the procedure outlined earlier in this Section but the principles are the same. This time there is no product to collect, but the recipient organism has a new characteristic which makes it more useful to humans.

e.g. gene for *frost resistance* has been transferred from fish to tomato plants. This means that the plants will be able to grow in colder weather, so farmers will make more profit.

Other organisms which have been modified in this way include:

- **cereal plants** which have been given the gene for disease resistance, so they are more healthy and produce a higher yield of grain.
- **cattle** which have been given the gene for increased growth rate, so they reach their adult size more quickly, and can be sold or used for food sooner.
- **crop plants** which have been given a gene which makes them glow when they are diseased. This means that diseased plants can easily be identified and removed before the disease spreads.
- **trout** which have been given a gene to make them paler, so they can be more easily seen by fishermen.

- **crop plants** which have been given a gene making them resistant to herbicide (weedkiller). This means that fields can be sprayed with herbicides and the crop plants will not die.

One recent, and very controversial, application of this technology is the development of *transgenic pigs.* These pigs have been given a human gene so that all cells in their bodies contain human 'marker proteins' (antigens). Organs from these pigs e.g. kidneys, hearts, could be transplanted into humans, because the 'marker proteins' would stop them being rejected.

Once the required gene has been transferred into the recipient organism, it must be cloned to make many copies, so that many organisms carrying this gene will be formed. Cloning of animals and plants is a more complex process than cloning of bacteria.

Cloning of animals

This involves separating the cells of a developing embryo and implanting the new embryos which are formed into 'surrogate mothers' (Fig. 6.67).

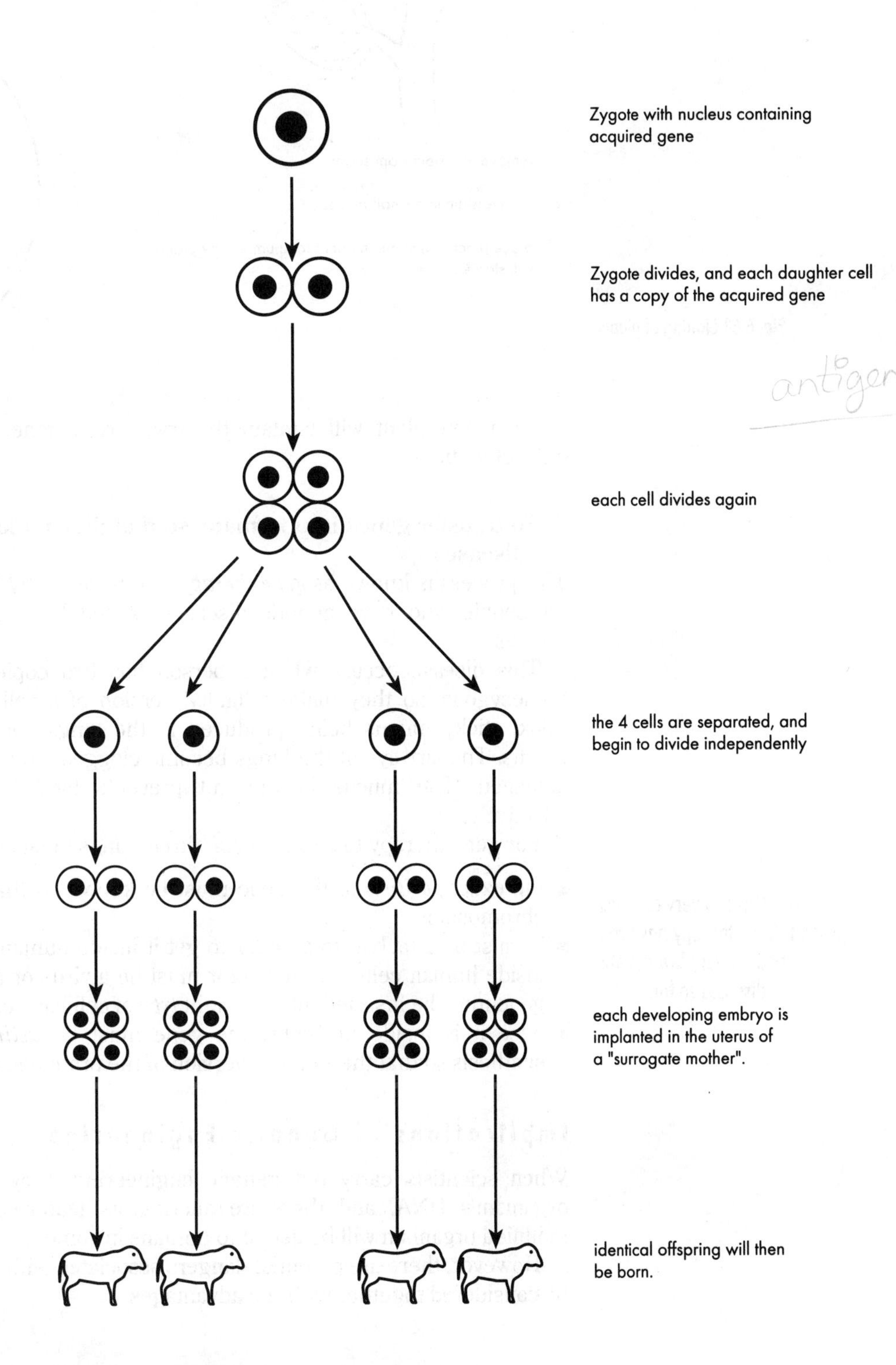

Fig. 6.67 Cloning of animals

Genetically *identical* offspring will be produced. Each will have the transferred gene, so it will have acquired the new characteristic.

Cloning of plants

At the tip of the roots and shoots of plants are areas called *meristems*. Here the cells divide rapidly by mitosis, so the meristems are known as growing points.

If cells are removed from the meristem and placed in suitable conditions, new plants will develop (Fig. 6.68). The new plants will be genetically identical to each other.

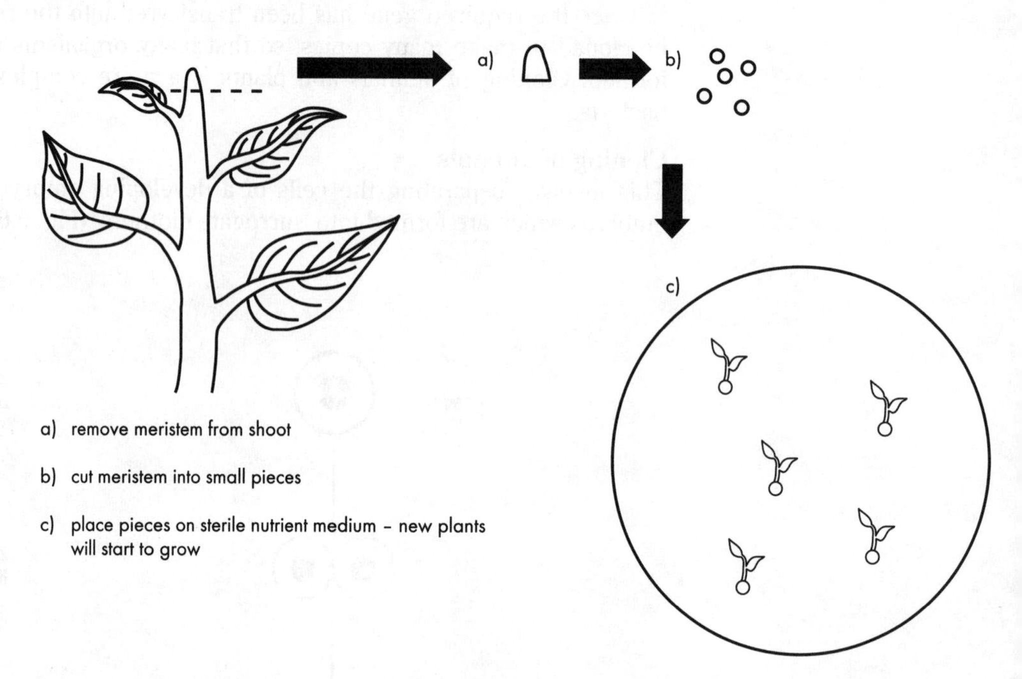

Fig. 6.68 Cloning of plants

Each new plant will contain the transferred gene, so it will have the required characteristic.

3 To transfer genes into humans so that they no longer suffer from genetic diseases

This process is known as *gene therapy*, and is currently being developed. It is suitable for people who have genetic diseases caused by a single faulty gene e.g. cystic fibrosis.

This disease occurs when a person has two copies of the 'faulty gene' i.e. is homozygous, so they make a 'faulty' version of a cell protein. This results in very thick, sticky mucus being produced in the lungs and gut, instead of clear, runny mucus. The airways of the lungs become clogged, and the lungs can be permanently damaged. The mucus in the gut prevents food being digested and absorbed effectively.

For gene therapy to be *successful*, three things must occur

- the exact *position* of the gene must be known so that it can be 'cut out' from the chromosome
- it must be attached to a *vector* to get it inside human cells. Plasmids will not work inside human cells, so the vector must be a virus or an artificial chromosome. The gene must be inserted into the right *type* of cell e.g. cells lining the air passages.
- once it is inside the cells, the gene must be *active* i.e. it must cause protein synthesis so that the 'correct' version of the protein is made.

"This is a very complex process. Gene therapy has been tried for very few genetic diseases so far."

Implications of Genetic Engineering

When scientists carry out genetic engineering, they are permanently altering an organism's DNA, and therefore altering its features. They do this because the modified organism will be useful to humans in some way, or to prevent disease.

However, there are potential dangers associated with altering DNA, and these must be considered together with the advantages.

Example 1 using bacteria to make human growth hormone

Advantages
- large amounts of growth hormone are available to treat humans who need it
- it does not cause side-effects in the people who use it
- manufacturers make a large profit

Disadvantages
- none apparent so far.

Example 2 making crop plants disease resistant

Advantages
- less plants are damaged or killed by disease
- higher yields of crops are produced, so more food is available for human consumption
- farmers make a large profit

Disadvantages
- the genetically-engineered plants might grow much better than other plants i.e. be more successful, so some plants may die out
- animals which depend on those plants i.e. for food, or as a habitat to live on, will be affected.
- food webs may become disturbed as numbers of plants and animals change i.e. the balance of nature is affected
- if some breeds die out there is less variety in the population, so natural selection cannot occur properly.

Example 3 gene therapy to treat genetic diseases

Advantages
- the person no longer has a dangerous disease
- they do not have to use medicines or drugs to treat the symptoms
- they will be healthier and live longer

Disadvantages
- only the body cell genes can be changed, not the genes in eggs or sperm, so the person's children might still have the disease
- if the transfer process goes wrong, then other genes might be affected, causing severe health problems for that person.

A Moral dilemma?

These are some of the questions currently being asked about genetic engineering:
- What would happen if genetically engineered organisms 'escaped' into the environment? how would this affect the balance of nature?
- Is it right to change animals and plants so that farmers can make a bigger profit? does it cause harm to the organisms involved?
- Is it right to alter human genes? Should this only be done to prevent disease, or should 'better humans' be developed?
- Who should decide how plants, animals and humans should be changed?

"Think carefully about the issues raised here. How would you answer?"

Links with other topics

Related topics elsewhere in this book are as follows:

REVIEW QUESTIONS

Q15 It is sometimes necessary for members of a family to prove that they are related. One way in which scientists can help is by 'DNA fingerprinting'.

(a) What is the function of DNA in a cell?

___

(b) In a 'DNA fingerprinting' test cells were taken from the mother, the husband and the child. The cells were broken open and DNA fragments were extracted. The DNA fragments in each sample were separated into a column of bands, called a DNA fingerprint. The diagram below shows the DNA fingerprint of each member of the family. The bands have been numbered to help you answer the questions which follow.

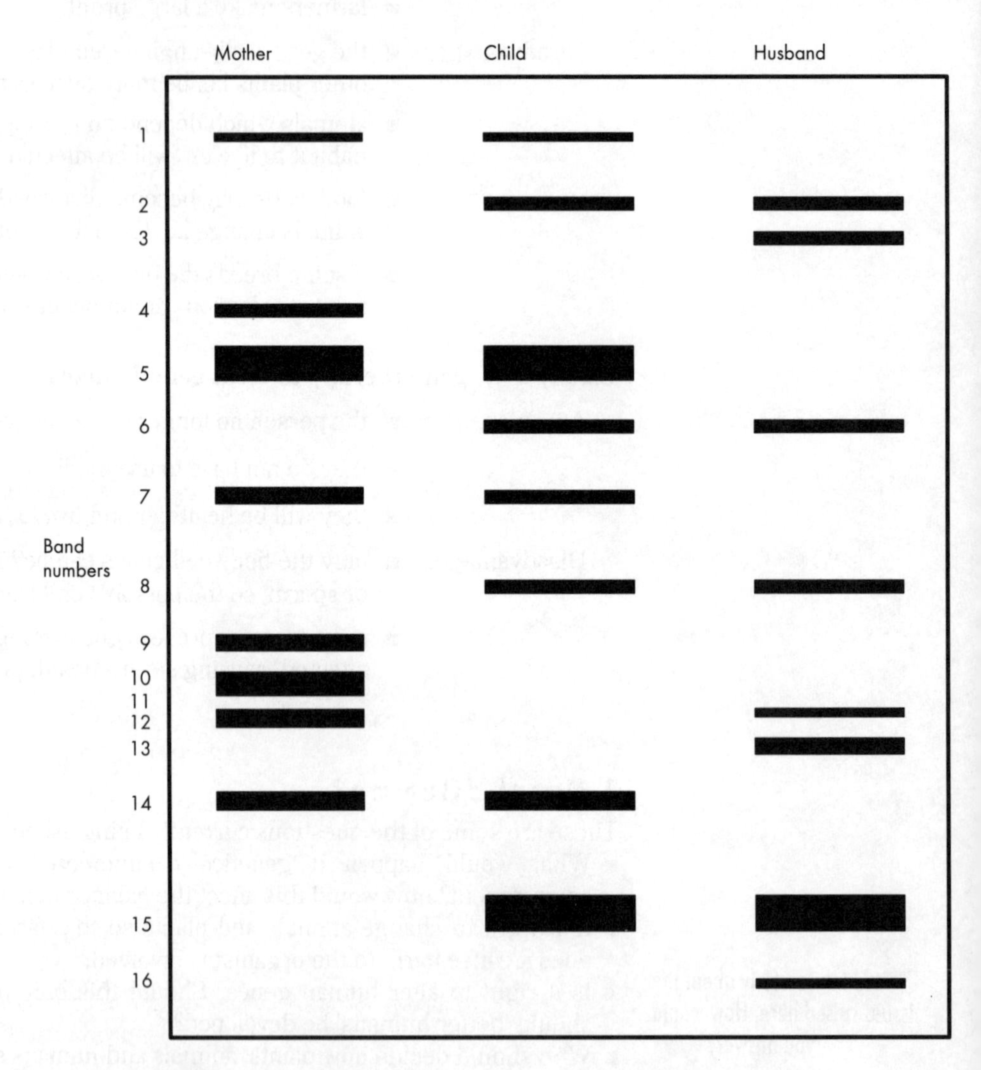

Fig. 6.69

(i) Which of the bands in the child's DNA fingerprint are in exactly the same position as the bands in the mother's DNA fingerprint?

___

(ii) Explain why some of the bands in the mother's DNA fingerprint are absent from the child's DNA fingerprint.

___

(iii) The bands in the child's DNA fingerprint which are absent from the mother's DNA fingerprint must have been inherited from the child's natural father. Does the husband's DNA fingerprint show that he is the child's father? Explain your answer.

___

Q16 (a) Both cystic fibrosis and Down's Syndrome are inherited conditions. Briefly describe the cause of each.

____

(b) Explain how two people, neither of whom shows symptoms of cystic fibrosis, can produce a child with the disease. Illustrate your answer by means of a genetic diagram.

____

____

____

(c) Why might it be advisable for a pregnant woman, one of whose relatives had cystic fibrosis, to obtain genetic counselling?

____

Q17 Genetic engineers are trying to produce a soya bean plant which is resistant to weedkillers.

They have identified an enzyme which makes tomato plants resistant to a common herbicide. They plan to take DNA from the tomato plant and transfer it into the soya bean plant in order to produce a strain of soya bean resistant to the herbicide.

(a) Explain the **advantages** of producing soya bean plants which are resistant to this herbicide.

____

____

____

(b) Explain in detail, how, using an appropriate bacterium, soya bean plants resistant to weedkillers might be produced.

____

____

____

(c) Explain, in detail, how cloning would be used to produce large numbers of herbicide-resistant plants.

____

____

____

ANSWERS TO REVIEW QUESTIONS

A1 (a) passed on from parents to offspring through genes

(b) because she has inherited the gene for glaucoma from her mother

(c) his parents did not have the gene for glaucoma

(d) David's father, because David had the disease but Jean did not

A2 (a) William Aa, Ben aa, Dot aa, Claire Aa

(b) 1 in 2 (50%)

(c) Paul is Aa, Cher is aa so 50% of their children are Aa, and 50% are aa.

A3 (a) 1 = BB, 7 = bb, 9 = Bb.

(b) White fur colour

(c) 1, 2, 7, 8, 11

(d) 3 black : 1 white

Comments

Genetics question like this need very careful attention; don't rush your answer! In this particular question, you could make a guess at the answer, since no explanations are asked for. However, many genetics questions will require explanations, including genetics diagrams, as part of your answer.

Since the genotype of animal number 5 is Bb, we know that the dominant allele is black (B). (a) Since all offspring (animals 3 to 6) in the first generation are black, and one parent (Number 2) has the recessive phenotype (bb), animal number 1 must be BB. If it was Bb, some offspring would show the recessive phenotype (white, bb). This is in fact what has happened in the second generation (animals 8 to 11). Animal number 7 must be bb (i.e. homozygous for the recessive allele), since either BB or Bb would result in a black animal. Since all the first generation of are Bb, and animal 7 is bb, all black animals (including number 9) in the second generation must be Bb (i.e. heterozygous).

(c) These animals are homozygous (i.e. BB or bb), and would normally be true-breeding if only bred with animals with the same genotype.

(d) This 3:1 ratio is the classic 'monohybrid ratio', resulting from a cross between two heterozygous parents (e.g. Bb × Bb).

A4 (a) More groups of tomato fruits. Many large fruits in each group.

(b) Improved taste.

(c) Germinate the seeds to obtain adult plants with flowers. Select any two plants which have desirable characteristics. Cross-pollinate the flowers by transferring pollen from one to the stigma of the other (e.g. using a paint brush). Germinate the resulting seeds to check that a new variety has been produced.

Comments

This question requires an understanding of genetics and plant reproduction, as well as the principles of selective breeding.

(a) Make sure you *only* give characteristics which are visible, and which are different from either parent plant. This requires careful observation.

(b) Many other suitable answers would get you a mark; e.g. improved storage of fruit, improved disease resistance.

(c) Avoid the danger of 'waffling' here! You only have six lines of space, so a concise, clear answer is needed. If necessary, plan your answer in rough first (e.g. as a 'flow diagram', with words and arrows).

A5 (a) coat colour e.g. black and white; coat length e.g. long hair, short hair.

(b) dark coat, short hair are dominant.

(c) F1 generation are hybrids (heterozygous) so only dominant features show. Some of the F2 generation are homozygous, and some are heterozygous, so there is a greater variety of phenotypes.

A6 (a) (i) Females are XX, males are XY. X chromosomes are bigger (contain more DNA) than Y chromosomes.
(ii) 1 in 2 (50%) see Fig. 6.10
(b) (i) 4 (ii) haploid
(c) (i) 22 (ii) 1 (iii) 22

A7 (a) C, B, E, A, D
(b) 46 (c) 20 (d) because it does not have all the genes necessary to make vital proteins, so development is affected.

A8 Topic Group 2.4 Genetics and cells
(a) (The answer is shown in the diagram (Fig. 6.70))
(b) (i) Meiosis (ii) Mitosis
(c) (i) It is being halved, (ii) No change

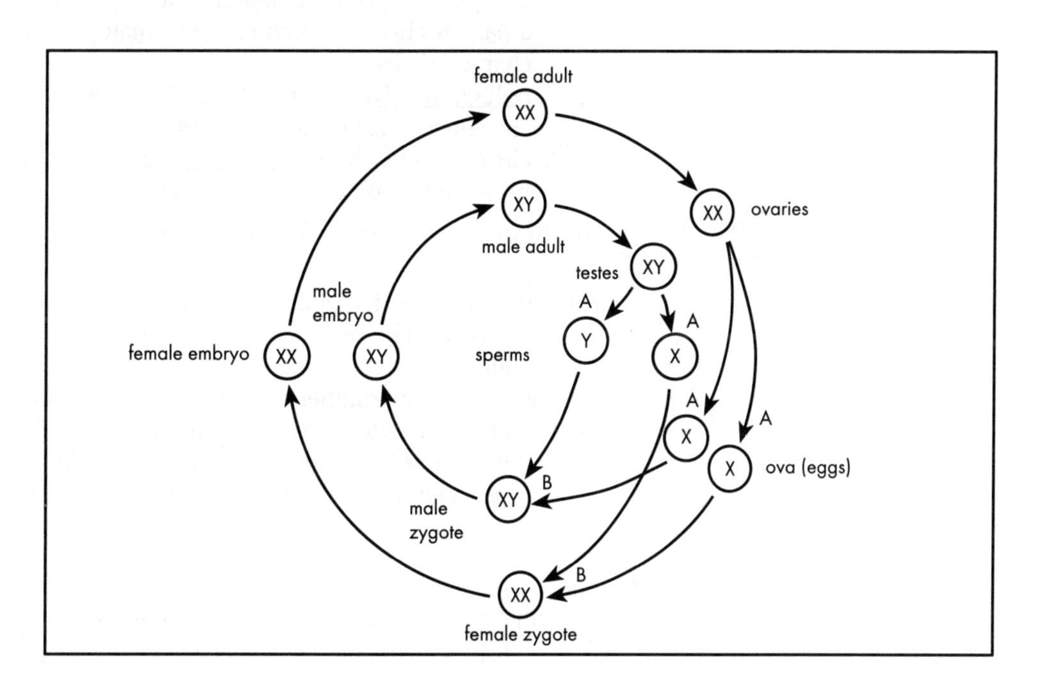

Fig. 6.70

Comments
You may never have seen a diagram like this before! That doesn't matter, since it simply shows a summary of processes which you should already be familiar with. There are several 'clues' in the diagram to help you answer the questions.

(a) To answer this question, you need to know two facts about sex chromosomes. Firstly, that body cells in males are XY, and in females they are XX. Secondly, that these sex chromosomes separate when sex cells are made (i.e. during meiosis). Note that where you put 'X' and 'Y' in the circles at 'A' for the sperms is decided by the rest of the diagram, since males and females are stated.

(b) and (c) Revise the sections on cell division if you got these wrong, or if you're not sure why you got them right!

A9 (a) RR or rr
(b) R and r
(c) (i) 1 : 1 ratio (ii) 3 : 1 ratio
(d) wrinkled seeds
(e)

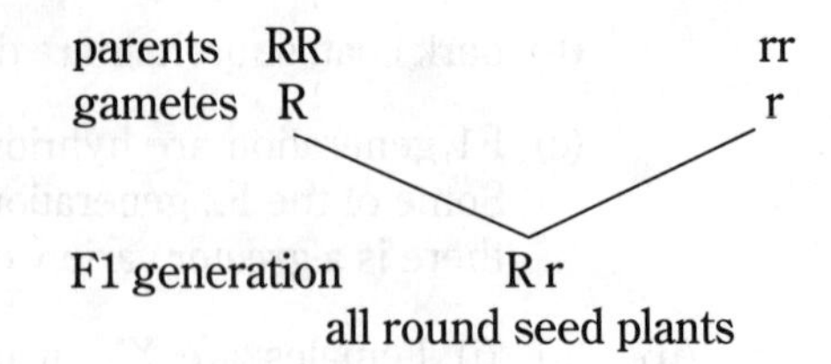

A10 (a) (i) I^AI^B (ii) I^OI^O (iii) I^BI^B and I^BI^O
(b) AB
(c)

Father	I^AI^O		Mother I^BI^O	
Gametes	I^A, I^O		I^BI^O	
Genotypes of offspring	I^AI^B	I^BI^O	I^AI^O	I^OI^O
Phenotypes of offspring	AB	B	A	O

A11 (a) 1. a spontaneous change in a gene or chromosome.
2. chemical or type of radiation which increases the mutation rate.
3. a pair of chromosomes carrying matching genes ie. genes for the same characteristics.
4. reduction division resulting in cells with the haploid number of chromosomes. Occurs in gamete formation.
5. chromosomes break and rejoin during meiosis so material is swapped between homologous chromosomes.

NB Keywords in these definitions are underlined.

(b) (i) genes are changed so that they no longer code properly for a particular protein. The resulting protein may be 'faulty' or may not be produced at all.
(ii) if the environment is changing, a mutation may result in a new characteristic better suited to the new environment.
(iii) sometimes the mutation makes no difference to the organism. This particularly true if it is a point (gene) mutation, resulting in only a tiny change.

A12 (a) (i) Fig. 6.48
(ii) 6.47 – discontinuous; 6.48 – continuous
(b) 6.47 – characteristic controlled by one gene, environment has little effect.
6.48 – characteristic controlled by several genes and/or strongly affected by environment.
(c) (i) mutation
(ii) gene mutation has probably occurred i.e. error in copying when DNA is replicating. This mutation results in the gene for skin pigmentation being affected, so no skin pigment is made. The individual is therefore albino.

A13 (a) B, C, D (b) C
(c) Samples from populations taken further north tend to have longer wings.
(d) Larger birds may be able to store proportionally more food to sustain them during long flights. Also, larger birds will tend to lose heat less readily in the colder northern conditions.

(e) 58 mm. (The graph is shown in Fig. A).

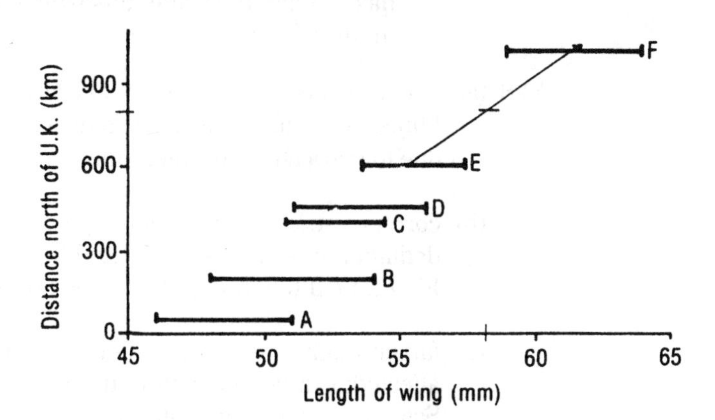

Fig. A

(f) Wing length must be strongly controlled by one or more genes. A mutation (change) in those genes may result in longer wings in some offspring. These birds are likely to be able to fly further or for longer, and some reach more remote islands where there may be more resources available, e.g. food, nesting sites. Birds with longer wings therefore have a competitive advantage over birds with shorter wings. Birds with longer wings may therefore have more breeding success, and the gene(s) for longer wings will increase in frequency, by natural selection.

Comments

This is an nice example of a 'real' problem in biology, one that needs to be solved by careful thought and cool reasoning.

Examine the graph carefully – the answers you need for (a)–(e) are all there!

(f) Is a chance to show your understanding of the process of natural selection. If you had problems, revise the topic again.

A14 (a) The shape and length of a bird's beak is determined by inheritance. Variations occur in the genetic information that determines beak characteristics, e.g. by mutations. Birds are in competition with each other (and other species) for limited amounts of food. New beak characteristics which provide improved adaptations for feeding give the bird a selective advantage. Such birds have an increased chance of survival, including reproduction. The characteristics are therefore more likely to be inherited by future generations.

(b) Populations of a species which become isolated, e.g. by being on different islands. Genetic changes within one isolated group occur independently of other groups, if no 'gene flow' can take place. After prolonged 'genetic isolation', individuals from different populations may be so genetically different that they cannot (or do not) interbreed. This process is species formation. Where several species co-exist, they are likely to have quite different needs for resources from their environment, so will not be in direct competition with each other.

Comments

This is demanding question, designed to test how much you understand the processes of natural selection and species formation in evolution. Note that the answers given here are more detailed than you need to provide to gain full marks, but you must get the same basic ideas across in your answer. In (a), make sure you refer to mutations.

A15 (a) protein synthesis; using the base code.

(b) (i) 1, 5, 7, 14

(ii) When meiosis occurs, only half of the chromosomes are passed on to the egg or sperm – the rest will not be present.

(iii) Yes; some of the child's bands (2, 6, 8, 15) are not from the mother; all have come from the husband; all the bands in the child are from mother/husband.

A16 (a) both are due to abnormal gene/DNA; can therefore be inherited; cystic fibrosis is result of defective gene for abnormal mucus; Down's syndrome due to extra chromosome; (number 21).

(b) correct use of terms 'phenotype' or 'genotype', both parents are carriers; definition of symbols used e.g. F, f; Ff × Ff (parents); F. f × F. f (gametes); FF, Ff, Ff, ff (offspring); ff = cystic fibrosis sufferer.

(c) father might carry cystic fibrosis gene; could be 1:4 chance of offspring affected; can be detected in uterus by antenatal test; CVS (Chorionic Villus Sampling); consider chance of termination.

A17 (a) can use herbicide in soya bean fields without crop damage; improves yield/can grow more plants.

(b) DNA transferred from nucleus of resistant tomato cell into bacterium; tomato DNA becomes joined to bacterial DNA chromosome; bacterial DNA with tomato DNA put into soya bean cell joins to soya bean DNA.

(c) undifferentiated cell taken from resistant soya bean plant divides in suitable conditions to form clones; cloned cells transferred to medium to grow further; each cloned cell can give rise to resistant soya bean plant.

EXAMINATION QUESTIONS

QUESTION 1

Fig. 6.71 shows the offspring of crosses between pure bred Aberdeen Angus bulls, which are black, and pure bred Redpoll cows, which are red. The ratio of the colours of the offspring of the first generation is also shown. Coat colour is controlled by a single gene which has two forms (alleles): one for black and one for red coat.

(a) What letters are suitable to represent the two forms (alleles) of the gene?

black coat ______________________

red coat ______________________ *(1)*

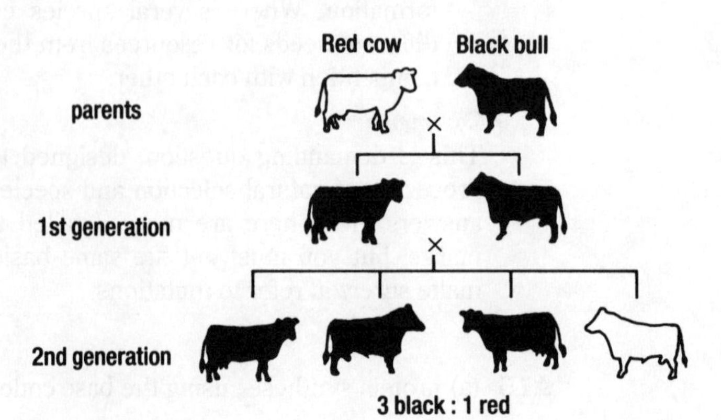

Fig. 6.71

(b) (i) Draw a circle around each animal in the diagram which is definitely homozygous for the gene for coat colour. *(1)*

(ii) Draw a square around each animal in the diagram which is definitely heterozygous for the gene for coat colour. *(1)*

(c) Explain why some of the animals in the diagram could be either homozygous or heterozygous for the gene for coat colour. *(2)*

Total 5 marks (ULEAC)

QUESTION 2

(a) (i) State all the possible genotypes of the four blood groups phenotypes shown in the table.

Blood group	Genotypes
A	__________
B	__________
AB	__________
O	__________

(4)

(ii) Which alleles (forms of the gene) are dominant? *(1)*

(iii) Which alleles (forms of the gene) show incomplete dominance (co-dominance)? *(1)*

Explain your answer. *(1)*

(b) (i) Complete the following diagram to show the possible offspring of the marriage between a man of blood group O and a woman of blood group AB.

Parents	Man (group O)		Woman (group AB)	
genotypes	__________			
gametes	____	____	____	____
F1 genotypes	____	____	____	____
F1 phenotypes	____	____	____	____

(4)

(ii) What is the probability of the first child of the marriage having

1 blood group O? __________

2 blood group B? __________

3 blood group AB? __________

4 blood group A? __________ *(1)*

Total 12 marks (IGCSE)

QUESTION 3

Nucleic acids are long-chain molecules consisting of units called nucleotides each containing one of five different bases denoted by the letters:

A C G T U

The diagram shows a section of a DNA (deoxyribonucleic acid) molecule beginning to 'unzip'. The letters, representing some of the bases, are missing from the bottom end of the drawing.

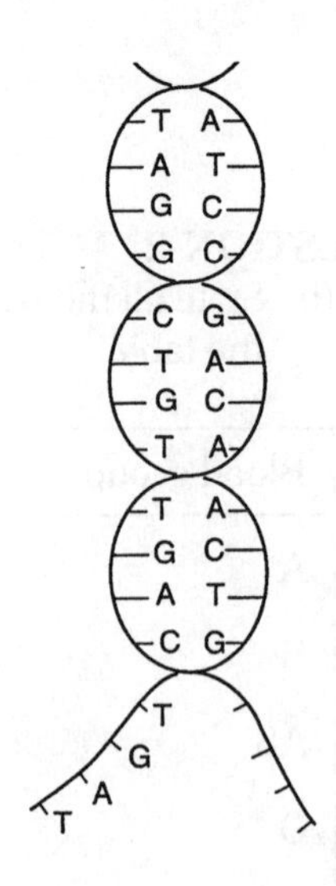

Fig. 6.72 (a)

(a) On the unzipped part of the DNA molecule, write the letters of the missing complementary bases. *(1)*

(b) A DNA molecule unzips and replicates itself just before mitosis. Why does it need to replicate. *(1)*

When unzipped, a strand of DNA can act as a template for building molecules of messenger RNA.

(c) What would be the base sequence of a molecule of messenger RNA built onto part of a strand of DNA with the following base molecule.

Complete the diagram by adding the letters of the missing complementary bases.

part of DNA strand | messenger RNA strand

A
C
T
G
T
G
A
T

Fig. 6.72 (b)

(d) The sequence of bases in a molecule of messenger RNA determines the sequence of amino acids in a polypeptide chain.

(i) How many bases are necessary to code for each amino acid in a polypeptide chain?

____ *(1)*

(ii) Where in the cell are amino acids assembled into polypeptides?

____ *(1)*

QUESTION 4

(i) State the term which is applied to a change in a gene or a chromosome.

(ii) How may such a change affect future generations of a species?

(iii) The appearance of members of that species may change over several generations as a result of the processes named in (i) and (ii). Name this process of change.

QUESTION 5

There is a type of poor vision in humans called 'night blindness' due to a single dominant gene. A man suffering from this eye defect marries a woman whose sight is normal. The couple have five children three of whom have normal vision and two have 'night blindness'. Using the symbol B for the dominant gene and b for the recessive gene answer the following:

(a) Draw a genetic diagram indicating the genotypes and phenotypes for each generation.

(b) One of the sons who suffered from 'night blindness' married a woman who also suffered from the eye defect. They had only one child. Draw the genetic diagrams for this family indicating the possible genotypes and phenotypes for this one child.

QUESTION 6

Fig. 6.73 shows *some* of the stages by which evolution might occur when two different types of animals evolve from a common ancestor over many thousands of years.

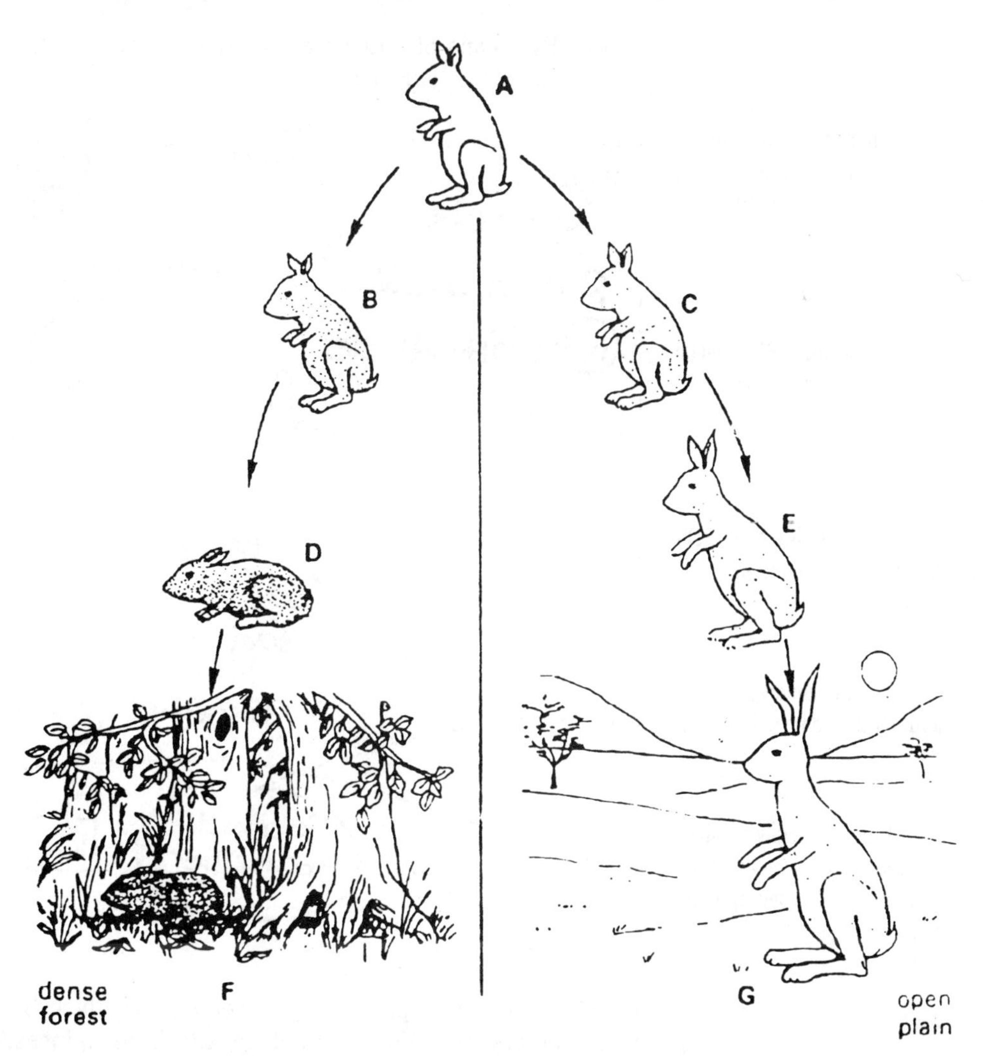

Fig. 6.73

By reference to each of the lettered stages, suggest how the diagram illustrates this process. *(12)*

STUDENT'S ANSWER – EXAMINER'S COMMENTS

QUESTION 7

When the mechanism of inheritance of flower colour in garden peas was investigated, red-flowered plants were crossed with white-flowered plants. The first generation of plants all had red flowers. However, when these red-flowered plants were allowed to self-fertilise, about 25% of the offspring had white flowers, the remainder having red flowers.

In a similar investigation with snapdragon plants, when red-flowered plants were crossed with white-flowered plants, the resulting first generation all had pink flowers. When these pink-flowered plants were self-fertilised, 25% of the offspring had white flowers, 25% had red flowers and 50% had pink flowers.

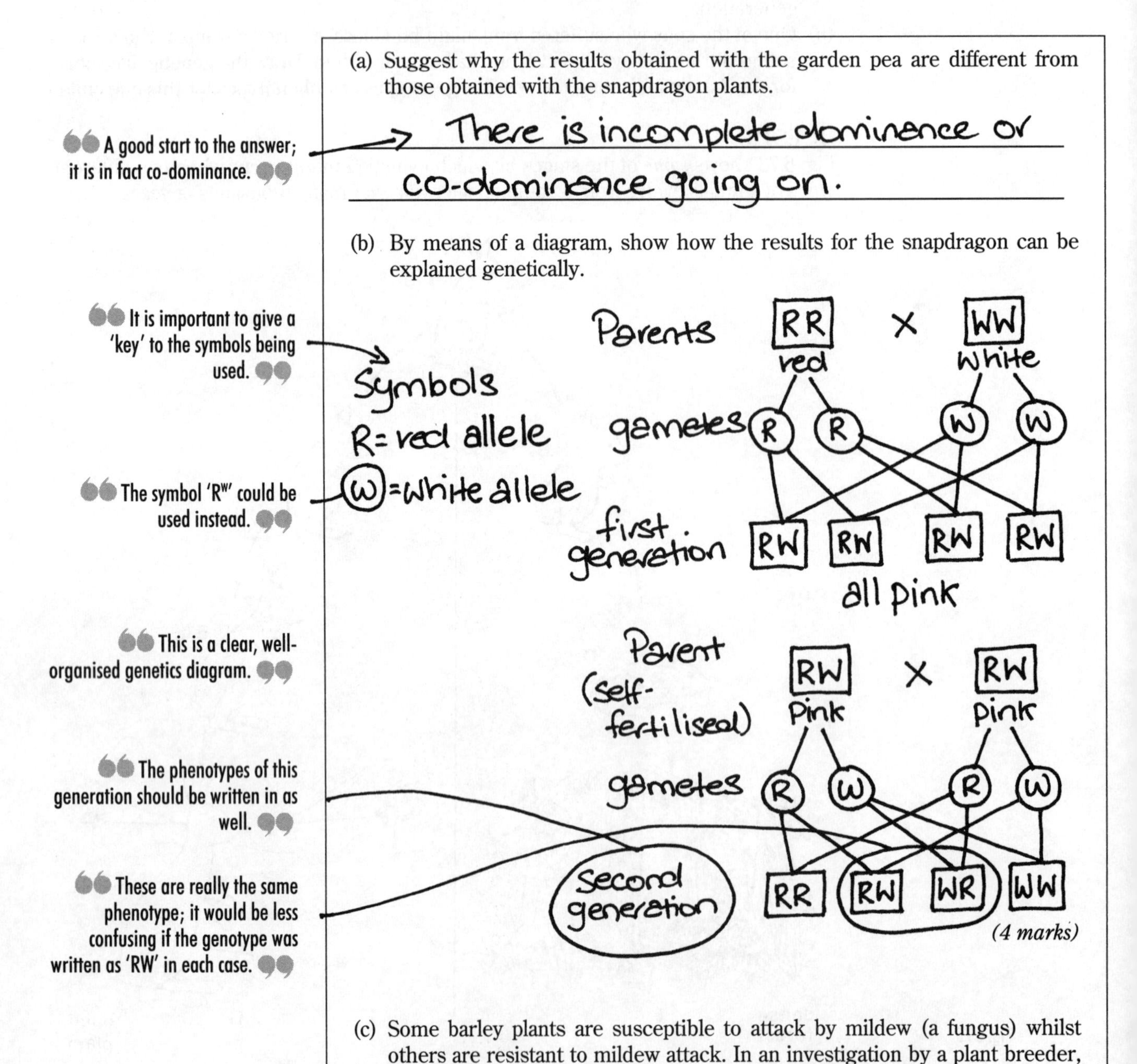

(a) Suggest why the results obtained with the garden pea are different from those obtained with the snapdragon plants.

(b) By means of a diagram, show how the results for the snapdragon can be explained genetically.

(4 marks)

(c) Some barley plants are susceptible to attack by mildew (a fungus) whilst others are resistant to mildew attack. In an investigation by a plant breeder, it was found that susceptible plants produced only susceptible offspring when self-fertilised, but that a resistant plant produced a mixture of resistant and susceptible plants when self-fertilised.

(i) How would the plant breeder obtain a stock of barley plants which were all resistant to mildrew?

find resistant plants which produce other resistant plants all the time.

(2 marks)

This answer is far too vague; it should refer to the selection (by repeated 'crossings') of true-breeding (homozygous dominant) resistant plants.

(ii) Assuming that resistance to mildew is controled by a single gene, what must be the genotype of the resistant stock?

heterozygote (Rr).

(1 mark)

(Total 8 marks)

NEAB

Correct answer, using the available facts.

QUESTION 8

There are two types of clover plant. One type makes cyanide in its leaves, the other type cannot make cyanide. Cyanide is poisonous to most organisms but not to the clover plants which make it. The clover plants growing on a mountain were studied. The chart shows the proportions of two types of clover plants at different heights.

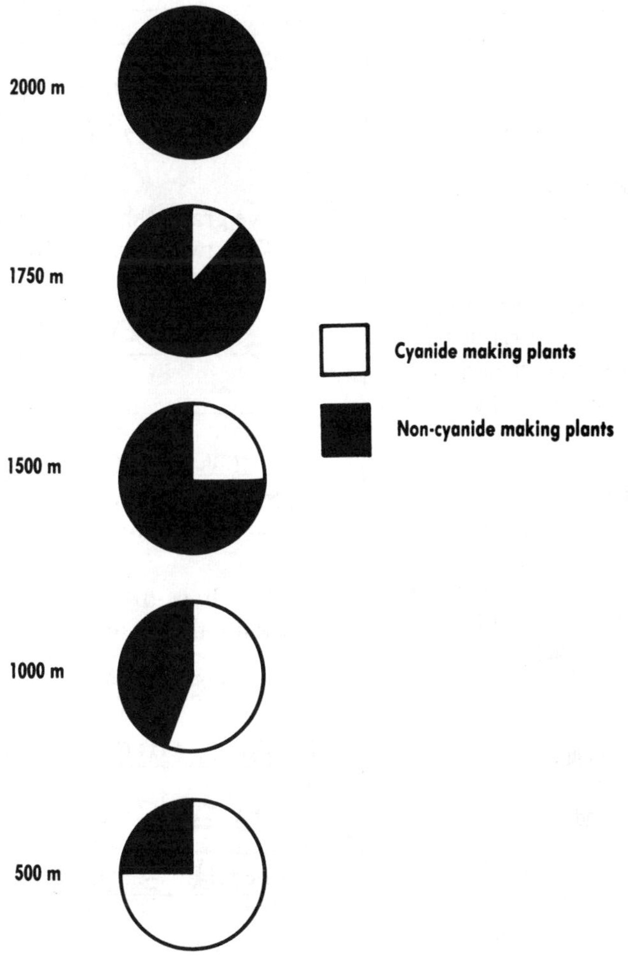

2000 m

1750 m

1500 m

1000 m

500 m

Cyanide making plants

Non-cyanide making plants

(a) Suggest *one* reason why cyanide in the leaves might help clover plants to survive.

it makes them poisonous

❝ This is correct, but is not a very 'complete' answer. They are less likely to be eaten by herbivores. ❞

(b) Suggest why the proportions of the two types of plant change as we go up the mountain.

there are less animals further up the mountain to eat the clover.

❝ Again, this answer needs to be more detailed: there is less 'selective pressure' on plants growing higher up the mountain to produce cyanide. ❞

(2 marks) (Total 3 marks)
(NEAB)

QUESTION 9

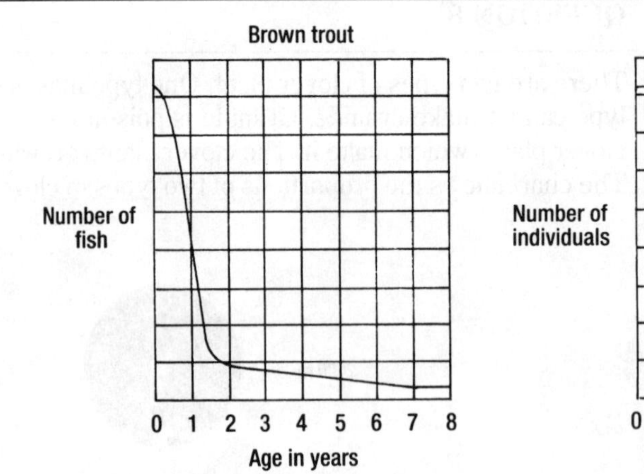

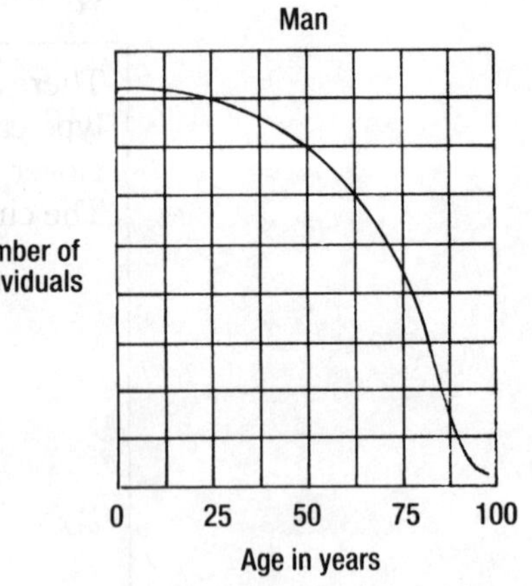

The above graphs show the patterns of survival in the Brown Trout and Man.

(a) Give two reasons why the pattern of survival for Man is different from that for the Brown Trout.

1. Humans live longer than Brown Trout

❝ Correct, though there may be exceptions: mean life span is greater in humans than Brown Trout. ❞

2. Humans are more likely to reach maturity so they can reproduce.

❝ A good answer; this is well-stated. ❞

Explain how the pattern in the graph for the Brown Trout would support Darwin's theory of Natural selection.

most fish do not live for very long and die when they are young. The fish that survive may be 'fitter' and they get a chance to reproduce.

❝ An excellent answer, showing a good understanding of the theory of Natural Selection. ❞

❝ It would be better to use the term 'Have a competitive advantage' here. ❞

(3 marks) (Total 5 marks)
(NEAB)

QUESTION 10

This simplified diagram shows part of a DNA (deoxyribose nucleic acid) molecule.

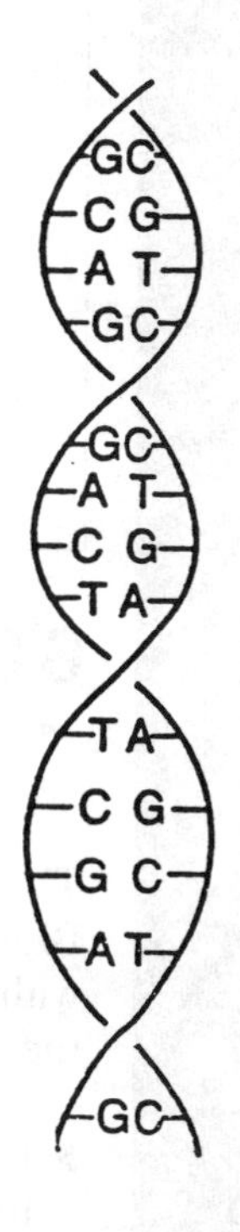

(a) Describe how the DNA molecule makes copies of itself.

The strands 'unzip' so that the bases are exposed and each strand can then be copied using new bases

(2)

'Good', idea of strands acting as a 'template' for copying is the important point.

(b) How is the coded information in the DNA transferred to the part of the cell where protein is made?

It is copied into messenger RNA and this moves to the ribosomes where the cell protein is made.

(2)

Correct but, mention that ribosomes are in the cytoplasm

(c) Explain why a change in the sequence of bases in the DNA is likely to change the structure of the protein made by the cell.

The DNA bases act as a code for a sequence of amino acids. Three bases code for each amino acid. If you alter a base code in a triplet of bases it alters the amino acid sequence.

(2)

Excellent Threebase = triplet. Change this and a different amino acid code is formed.

Note: Answers to these questions can be found at the end of the book

C H A P T E R

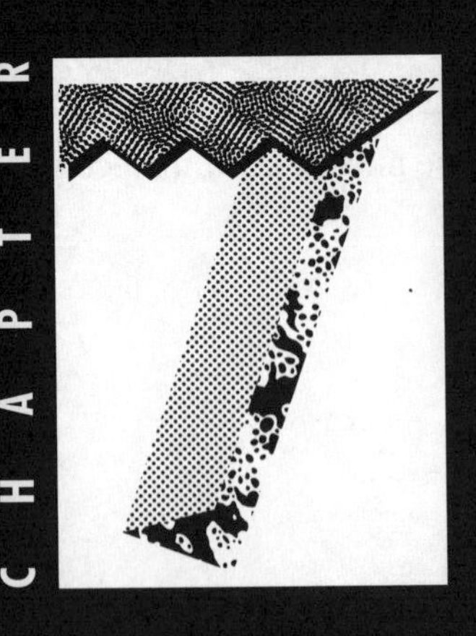

ECOLOGY AND BIOTIC FACTORS

INVESTIGATING ECOSYSTEMS

POLLUTION & SURVIVAL

NATURAL POPULATIONS

HUMAN POPULATIONS AND ENVIRONMENTAL IMPACTS

STRAND (III) POPULATION AND HUMAN INFLUENCES WITHIN ECOSYSTEMS

GETTING STARTED

This chapter covers the materials involved in the core area of the Biology Key Stage 4 Syllabus 'Populations and Human Influences within Ecosystems'. By the end of Chapter 7 you should have met all the Strand (iii) statements of the National Curriculum outlined below.

- Pupils should make a more detailed and quantitative study of a habitat, including the investigation of the abundance and distribution of common species, and ways in which they are adapted to their environment.
- They should explore factors affecting population size, including human populations.
- They should have opportunities, through fieldwork and other investigations, to consider current concerns about human activity leading to pollution and effects on the environment, including the use of fertilisers in agriculture, the exploitation of resources, and the disposal of waste products on the Earth, in its oceans and in the atmosphere. They should relate the environmental impact of human activity to the size of population, economic factors and industrial requirements.
- The work should encourage pupils to use their scientific knowledge, weigh evidence and form balanced judgements about some of the major environmental issues facing society.
- The study of the relationships between living organisms with their environment is called **ecology**. The environment consists of non-living (inorganic, physical) **abiotic factors** and living (organic) **biotic factors**. Abiotic and biotic factors are components of **ecosystems**, which are 'ecological units' within the environment; ecosystems have characteristic abiotic *and* biotic features.

Biotic factors consist of all the organisms living within a **community**, i.e. within a *particular* ecosystem. Each organism is part of another organism's environment and they **interact** in various ways. Although a variety of biotic factors are considered in this chapter, the main focus is on *human* influences on ecosystems. Other biotic interactions are considered in Chapter 8 as well as a range of abiotic factors.

ESSENTIAL PRINCIPLES

Level 4

To get a better picture of how living things survive and interact with each other, we can study them in their natural surroundings.

Fieldwork is an important part of Biology, so you need to plan your work carefully before you go out. Make sure you take a **key** to help you to identify the organisms you find, and take suitable equipment to trap or record the organisms.

Always replace animals where you found them, and take care not to damage living things or their habitat in any way.

Important terms

There are several important terms you should be familiar with in this topic:

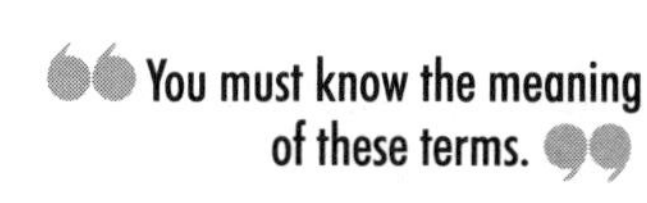

ecology: study of living things in their natural surroundings

habitat: the place where an organism lives e.g. rock pool habitat, pond habitat, woodland habitat

population: all the members of a particular species within a habitat e.g. a rock pool may have a population of limpets

community: all the types of living things within a habitat e.g. a rock pool may have a community which includes seaweeds, limpets, topshells, periwinkles, shrimps and small fish.

ecosystem: all of the living and non-living things in a habitat e.g. plants, animals, rocks and water in a rock pool would make up the rock pool ecosystem.

COLLECTING DATA

Key Points

- you will need keys to identify plants and animals
- record plants or non-moving animals by using transects or quadrats
- record moving animals using nets, beating trays, or traps
- estimate the number of plants by quadrat sampling
- estimate the number of animals by the mark – recapture method

The methods we use to investigate an ecosystem depend on the type of habitat involved. We will consider 3 habitats: rocky shore, pond and woodland.

For each habitat there are *key pieces of information* which should be collected:

- temperature
- pH of soil or water
- variety of plants present i.e. number of species
- type of plants present i.e. names of species
- distribution of plants i.e. where they are growing
- variety of animals present
- type of animals present
- distribution of animals.

Techniques for recording plants

1. Line transect

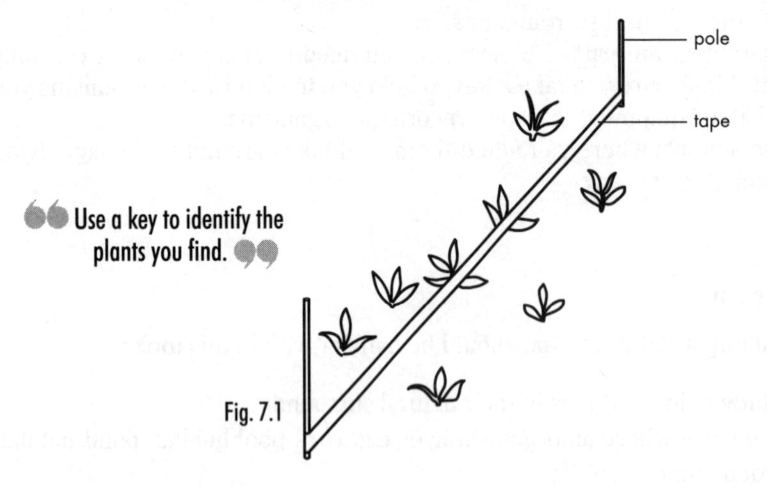

Fig. 7.1

Place a tape or string between 2 poles and record any plant which *touches* the tape: (Fig. 7.1) If the transect is too long, record plants touching the tape at *regular intervals* e.g. every 10 cm.

Use a key to identify the plants you find.

Suitable for:	■ plants which grow close to the ground ■ areas which involve a change from one habitat type to another e.g. edge of a pond, field border adjoining hedgerow.

2. Point transect

Place a tape or string between 2 poles. The tape should be 0.5 m above the ground. Place a piece of dowelling or a metal strip vertically against the tape at *intervals* e.g. every 10 cm and record every plant it *touches*. (Fig. 7.2)

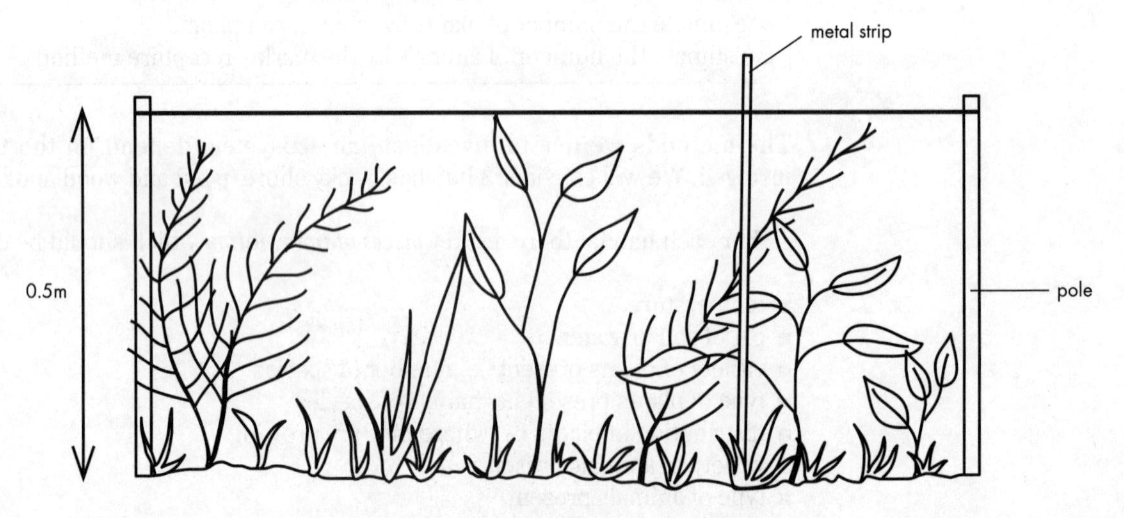

Fig. 7.2

Suitable for:	areas where there are several '*layers*' of plants.

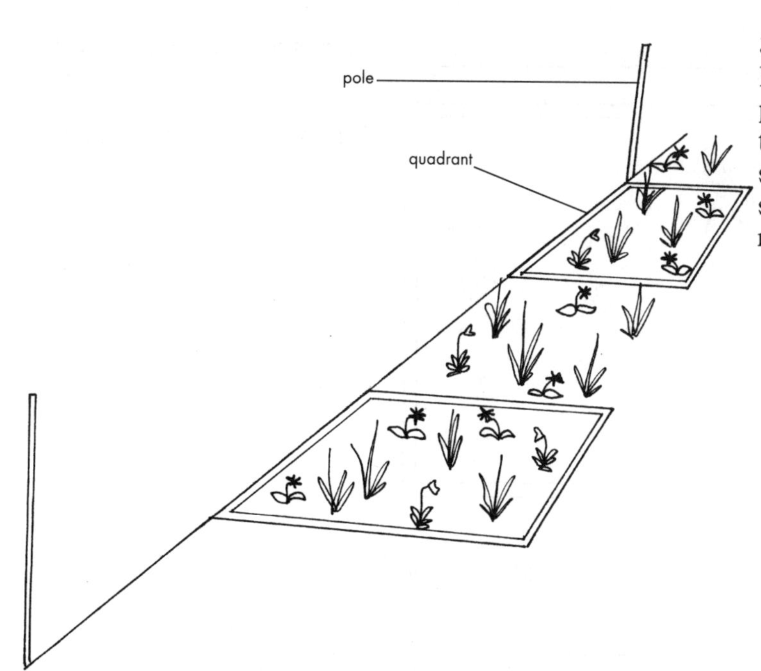

3. Belt transect

Place a tape or string between 2 poles and place a quadrat on it. Record all plants ***within*** the quadrat. (Fig. 7.3) Either move the quadrat so that all plants on the belt are recorded, or so that plants at regular intervals e.g. 1 m, are recorded.

Fig. 7.3

Suitable for: ■ long transects
■ collecting large amounts of data

4. Quadrat

This is a frame (usually 0.5 m × 0.5 m) which is placed on the ground, and plants inside it are identified. It can either be placed *randomly* e.g. by throwing it or using random number tables, or in a *fixed pattern* e.g. a belt transect.

(a) It can be used to assess the *variety* of plants present e.g. in a rocky shore habitat there may be 3 species present in one quadrat, in grassland there may be 25 species present.
(b) It can be used to assess which plant is the *dominant species* in a particular habitat. We can estimate the area inside a quadrat taken up by each species, this is *percentage cover*. In Fig. 7.4, species **A** occupies about 80 per cent of the quadrat, while species **B** occupies the remaining 20 per cent.

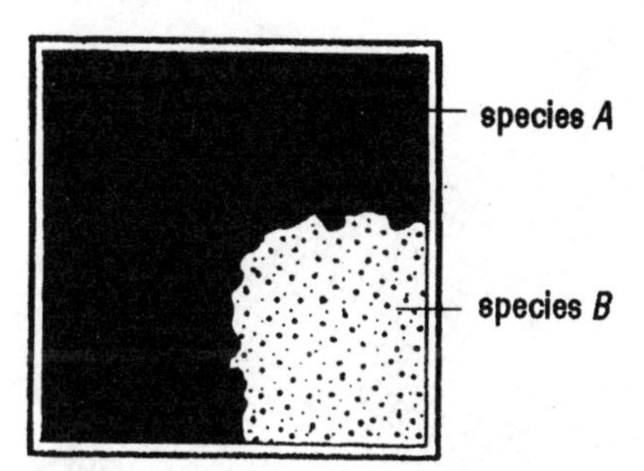

Fig. 7.4

(c) It can be used to estimate *number of plants* present. In a field 100 m × 100 m there are too many daisy plants to count. We can *sample* the field using a quadrat. (Fig. 7.5)

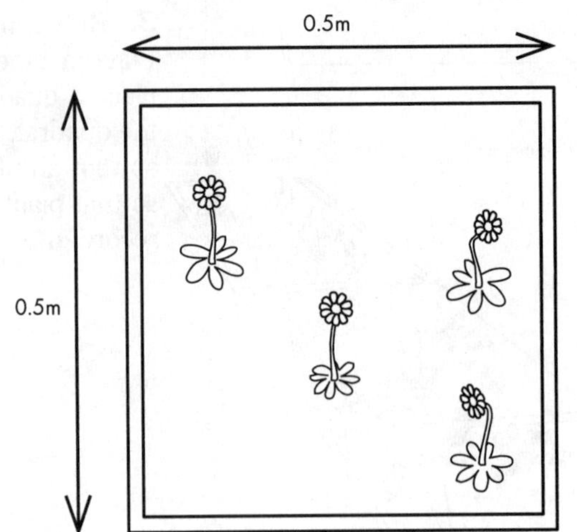

Area of quadrat = 0.5 × 0.5 m
= 0.25 m²

Number of daisy plants present
= 4

Fig. 7.5 Estimation of percentage cover

If we place 9 other quadrats in this field, we record these results.

Quadrat number	1	2	3	4	5	6	7	8	9
Number of daisy plants	0	6	11	3	2	10	0	0	1

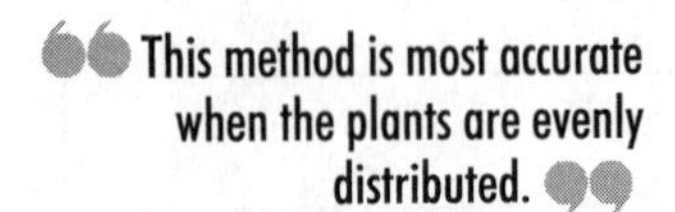

Total number of daisy plants = 37

Mean number per quadrat $= \frac{37}{10} = 3.7$

One quadrat has an area of 0.25 m²

The whole field has an area of 100 × 100 = 10,000 m² so the field is 40,000 times bigger than one quadrat

We can use this information to *estimate* the number of daisy plants in the field.

There are 3.7 daisy plants in one quadrat, so in the whole field there will be

$3.7 \times 40{,}000$ daisy plants
= 148,000 plants.

This gives us a *rough estimate* of the daisy plants in the field.

3 Techniques for recording animals

1. Nets

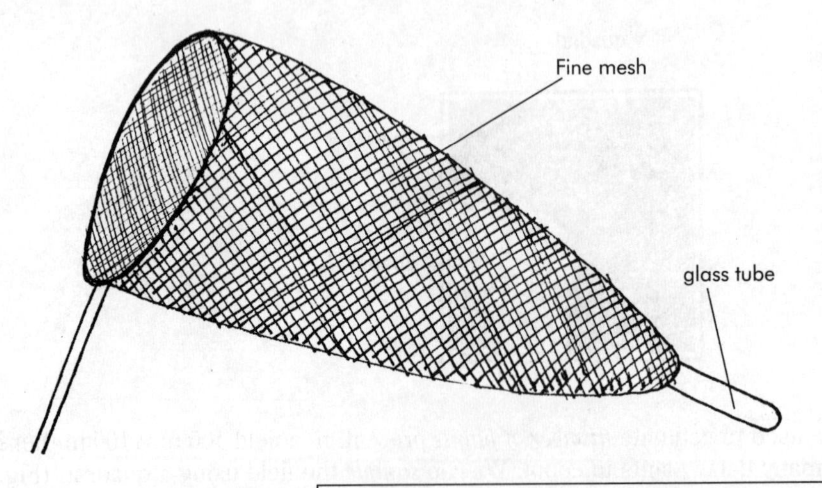

Use a fine mesh net on a wire frame to collect small animals from water (a sieve could also be used for this.) Empty the net into a white enamel dish containing a small amount of water, so that you can see the animals clearly.

Some nets have a glass tube at the base to collect trapped animals (Fig. 7.6). These are called *plankton nets.*

Fig. 7.6 A net

Suitable for: collecting small animals from ponds, streams or rock pools.

2. Beating tray

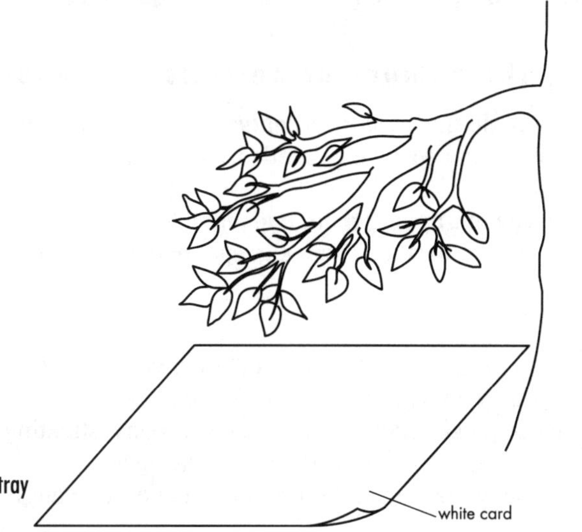

Fig. 7.7 A beating tray

Place a large piece of white card or material (e.g. old sheet) under the branches of a bush or tree. Gently hit the branches to shake off any animals attached to the bark or leaves – take care *NOT* to damage the branches. (Fig. 7.7). The animals will fall onto the card and can be identified.

Suitable for: small animals living in hedgerows or trees e.g. caterpillars, beetles

3. Pitfall trap

"Always use a key to identify the animals you find"

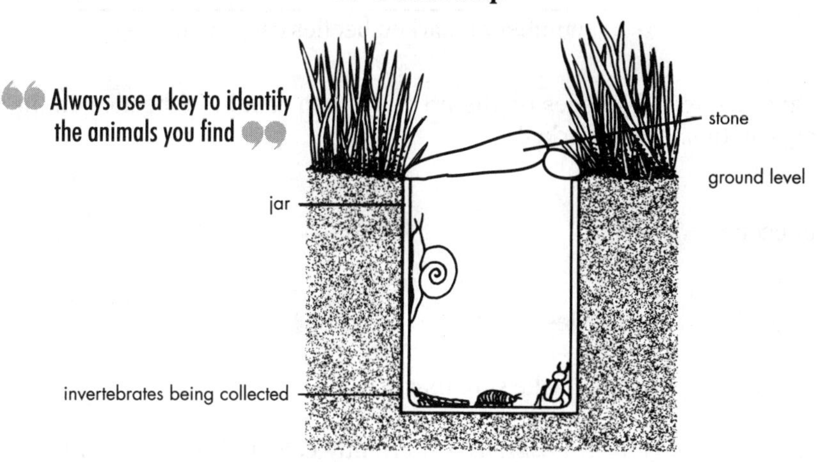

Bury a container e.g. yogurt pot or jam jar, so that the top is level with the ground. Balance a stone over the top so that it will not fill with water if it rains. Leave it for 24 hours, then collect and identify the animals which have crawled into it. (Fig. 7.8)

Suitable for: small animals in grassland or woods e.g. beetles, spiders.

Fig. 7.8 A pitfall trap

4. Tullgren funnel

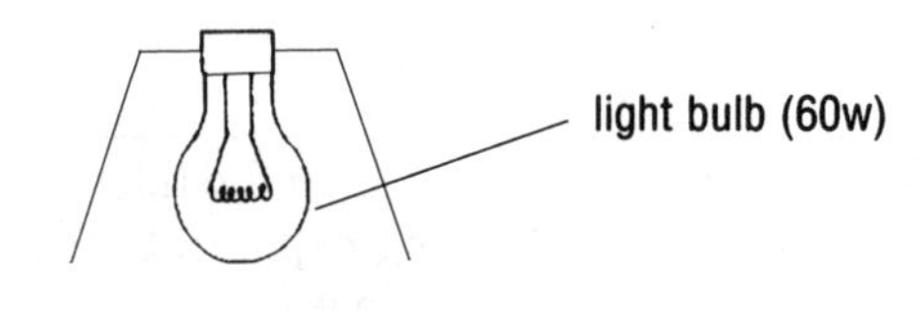

Collect a sample of leaf litter and put it on the gauze platform inside the funnel. Small animals will move away from the heat and light, and will fall into the alcohol, which kills them. (Fig. 7.9)

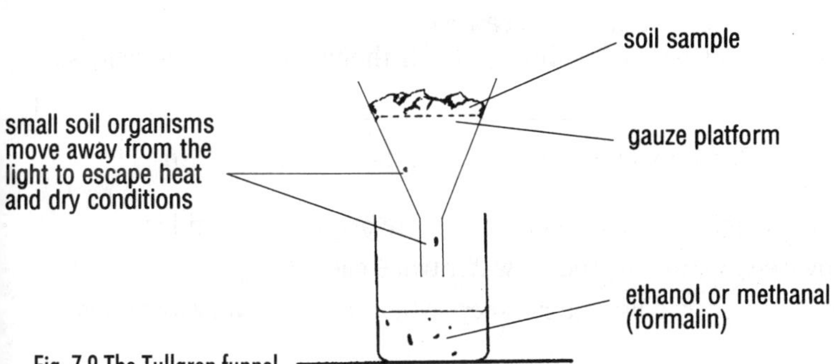

Fig. 7.9 The Tullgren funnel

Suitable for: small animals in leaf litter e.g. beetles, millipedes, worms.

5. Quadrats
Some animals do not move at all, or move only slowly e.g. barnacles, limpets, mussels on rocky shores. These can be recorded using a quadrat.

Estimating the number of animals in a habitat

The easiest way to do this is *mark-recapture technique*. This method works for animals which can move around easily e.g. beetles, as long as:

(a) you have a way of trapping them e.g. pitfall trap
(b) you have a way of marking them e.g. spot of paint on body

It involves 5 stages:

1. Set up the trap to catch the animals you are investigating e.g. beetles.
2. Record how many are caught, and mark each one.
3. Release the marked beetles in the area you are investigating.
4. Some time later, e.g. 24 hours, set up the trap again.
5. Record how many marked and unmarked beetles are caught.

You can use this formula to work out the total number of beetles in the area you are investigating:

$$\text{Total number of beetles} = \frac{\text{number caught 1st day} \times \text{number caught 2nd day}}{\text{number of marked beetles caught 2nd day}}$$

e.g. if I caught and marked 27 beetles on the first day, then caught 30 beetles on the 2nd day, but only 6 of them were marked

$$\text{Total number of beetles} = \frac{27 \times 30}{6}$$

$$= 135 \text{ beetles}$$

When using this method, you must make sure that:

Note these points.

- marked beetles are *equally likely to survive* i.e. brightly coloured paint would make them more noticable to predators, so they would be less likely to survive.
- you leave a *suitable time* between stages 3 and 4 for the beetles to spread out.

ROCKY SHORE ECOSYSTEM

Key Points

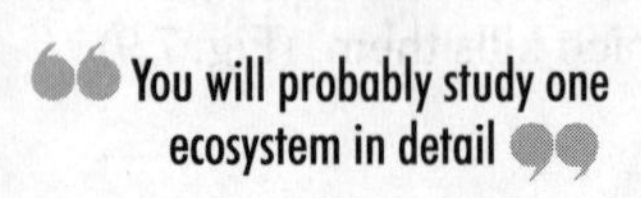

- the shore is divided into 3 regions depending on the tides
- organisms must be able to cope with being exposed to the air when the tide is out, and pounded by the waves when the tide is in
- the only plants found on rocky shores are seaweeds
- many of the animals found on rocky shores attach themselves to the rock, so they can be recorded using a quadrat

The rocky shore can be divided into 3 regions, based on the tides (Fig. 7.10)

Upper shore – is usually uncovered, except when there are very high tides
middle shore – is covered and uncovered by water twice each day
lower shore – is usually covered by water, except when there are very low tides

Within these 3 regions there are many different types of habitat e.g. rock pools, cracks and crevices in rocks, areas under boulders etc.

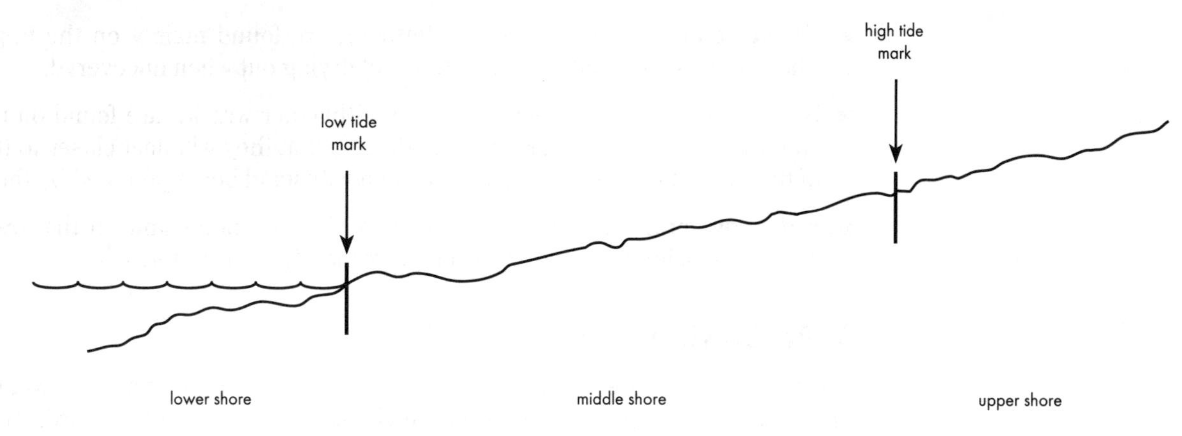

Fig. 7.10 Profile of a rocky shore

1 How do conditions vary?

As a result of the tides, conditions vary significantly:

daily
- amount of *water* covering an organism changes as the tide goes in and out (twice each day)
- *temperature* varies due to the tides. Organisms that are covered in water will have a reasonably constant temperature compared to those that are uncovered
- amount of *light* that reaches an organism varies. When the tide is in, less light can penetrate the water.

seasonally
- the heights of the *tides* vary so that different regions of the shore are covered and exposed. The highest tides are in spring and autumn.

Organisms which live on a rocky shore must be able to cope with the following problems:

- drying out when they are left uncovered
- large changes in temperature
- being moved around and swept off the rocks by the power of the waves
- animals must get enough oxygen from the water to survive
- plants must trap enough light for photosynthesis.

“Zonation is a key concept on rocky shores – make sure you understand *why* it occurs.”

The differing conditions on different parts of the shore leads to *zonation* i.e. animals and plants have a particular region of the shore where they can survive best, and where they are abundant.

2 Rocky shore plants

The only plants found on rocky shores are seaweeds. These are algae which attach themselves to rocks with a holdfast. They have no proper roots, stems or leaves, but they have a region called the thallus which can photosynthesise (Fig. 7.11).

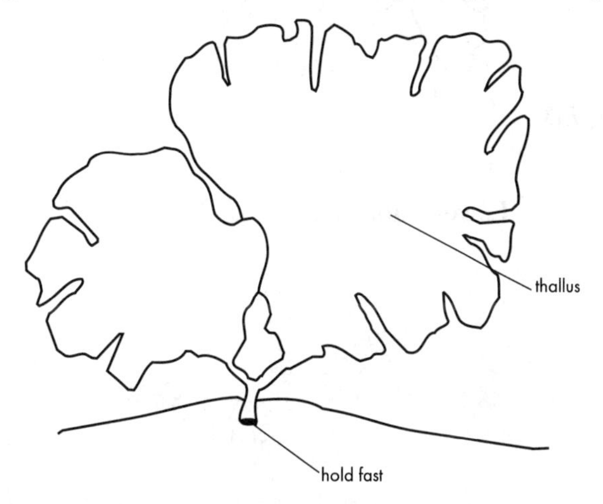

Fig. 7.11 Structure of a seaweed *(Ulva)*

Seaweeds can be divided into 3 groups

- **green seaweeds** e.g. *Ulva* (sea lettuce), are found mainly on the upper shore. They are usually small and can withstand drying out when uncovered.
- **brown seaweeds** e.g. *Fucus vesiculosus* (bladder wrack), are found on the middle and lower shore. Some have air bladders so that they will float closer to the surface of the water. Some are very large and can withstand being pounded by the waves.
- **red seaweeds** e.g. *Corallina,* are found in rock pools and on the lower shore. These vary a lot in size and can live permanently underwater.

3 Rocky shore animals

There is a large variety of animals found on unpolluted rocky shores. Many of them are fixed to the rock, or move very slowly, so they are easy to record. Animals you might see include:

sponges
sea anemones
molluscs e.g. periwinkles, limpets
crustaceans e.g. barnacles, crabs, shrimps
starfish
fish e.g. blenny, wrasse

Their position on the shore will depend on several factors

e.g. how they feed
how they prevent themselves drying out
how they prevent themselves being damaged by waves.

We will consider one animal from each region (diagrams are not to scale).

Upper shore e.g. Periwinkle (*Littorina neritoides*)

Level 4

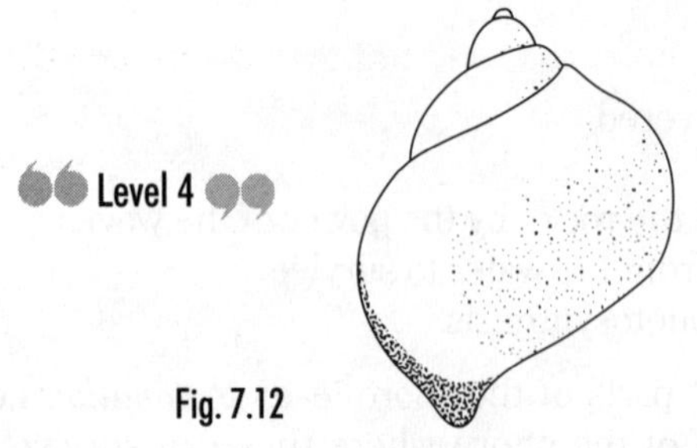

Fig. 7.12

- lives in rock crevices
- clamps itself to rock when tide is out to reduce water loss
- shell protects it from predators
- feeds on algae on rock surfaces of upper shore

Middle shore e.g. Barnacle (*Balanus*)

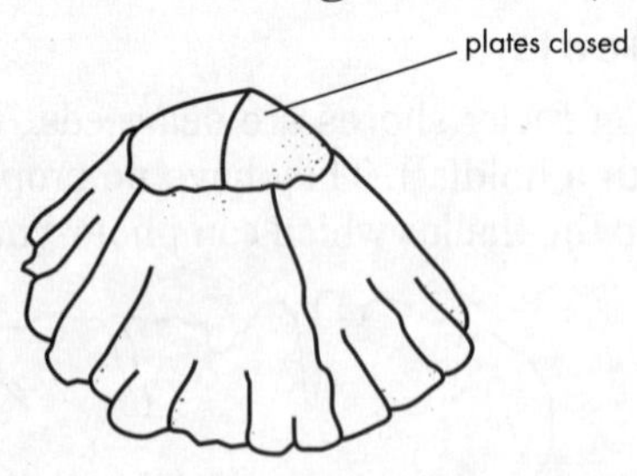

Fig. 7.13

- is firmly attached to rock surface by a cement-like substance
- when the tide is out, the plates move to close the shell, to prevent drying out and attack from predators
- when the tide is in, 6 pairs of legs are exposed, and these filter food particles from the water

Lower shore e.g. Starfish (*Asterias*)

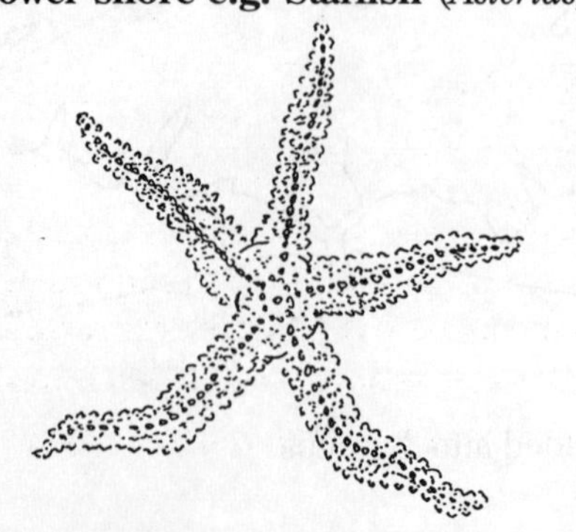

Fig. 7.14

- shelters under weeds or in crevices when the tide is out
- is a carnivore which grips its prey with 'tube feet'
- can grow new arms if damaged by waves

4 Collecting data on rocky shores

Plants – use line or belt transects, quadrats
Animals – use belt transects or quadrats for fixed animals and nets for swimming animals e.g. fish, shrimps.

5 Ideas for fieldwork

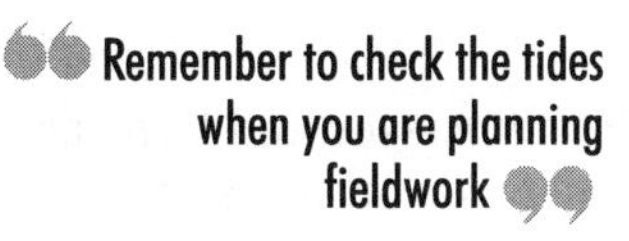
Remember to check the tides when you are planning fieldwork

1. Investigate zonation of plants or animals.
2. Investigate how plants or animals prevent themselves being swept off rocks.
3. Investigate the size and shape of limpets from different regions of the shore.
4. Investigate the rate at which different types of seaweed dry out when exposed to air.

Always avoid disturbing or damaging the habitat.

POND ECOSYSTEM

Key Points

- the pond is divided into 4 regions
- pond plants are an important source of oxygen, food and shelter for animals
- make sure you record which region of the pond animals were found in
- search samples of bottom mud carefully – lots of animals live there

A pond is a small area of quite shallow, stagnant water (Fig. 7.15). It can be divided into 4 regions:

Surface film – small animals and plants live on the surface film of water or hang underneath it.
Vegetation zone – this consists of floating and submerged plants, and the animals that live on them.
Open water – floating plants, and swimming or drifting animals are found in this region.
Muddy bottom – light and oxygen are scarce here, but several types of animal thrive in these conditions.

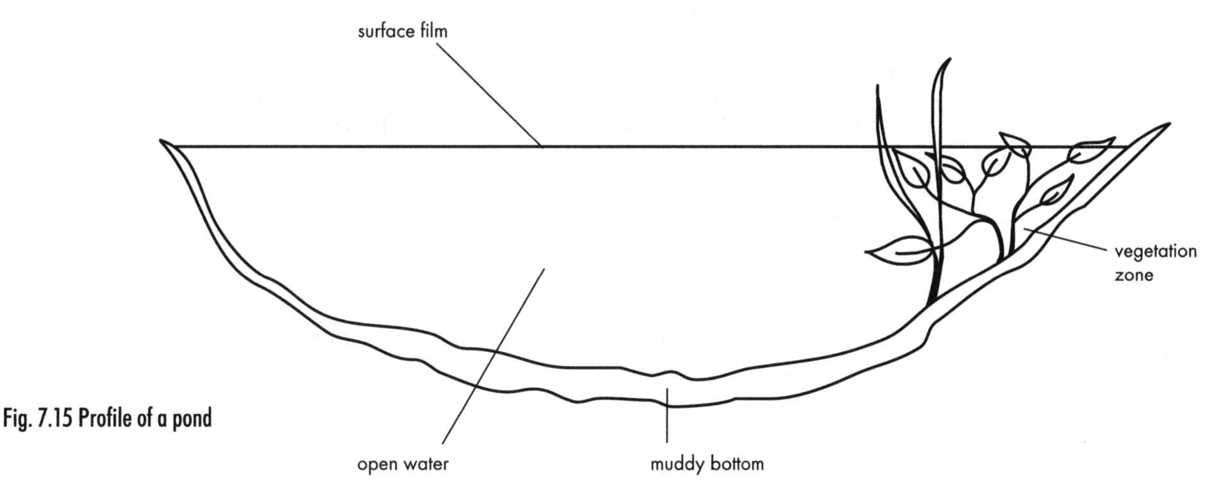

Fig. 7.15 Profile of a pond

1 How do conditions vary?

Conditions will vary much more in small ponds than larger ones.

daily
- amount of *oxygen* varies. At night, plants have no light for photosynthesis so they stop producing oxygen. All pond organisms continue to use oxygen for respiration, so oxygen levels will fall very low. During the day more oxygen is made by photosynthesis
- *temperature* is higher during the day than at night.

seasonally – amount of *light* varies. In the summer there is more light, so plants grow faster
– *temperature* varies. In summer it is warmer so many animals will be more active
– amount of *water* varies. In summer some ponds will become much shallower, or even dry up, as water evaporates in the heat.

2 Pond plants

Most of the plants found in ponds are algae or flowering plants; mosses and ferns are very rare. Algae are mainly microscopic, and live in open water or on the surface of stones or leaves e.g. *Spirogyra, Chlorella*. Flowering plants can be divided into 3 main groups:

Rooted plants with floating leaves e.g. water-lily (*Nymphaea*)
Roots anchor the plant at the bottom of the pond, but are not very important for absorbing water or minerals. The leaves have air spaces so they will float to obtain maximum light for photosynthesis. The stomata are on the upper surface of the leaf, and this surface is usually waxy so water does not collect there.

Floating plants e.g. duckweed (*Lemna*)
Small, circular fronds (leaves) float on the surface of the water, and short roots hang down. These help to balance the plant, and collect minerals from the water. Gas exchange occurs through stomata on the upper leaf surface and through the roots.

Submerged plants e.g. Bladderwort (*Utricularia*)
There are no roots, but horizontal stems with many small leaves absorb water and minerals. Some of the leaves have small sac-like bladders to trap tiny water organisms which provide the plant with minerals. Bladderwort grows best in ponds where the water contains few minerals, because it has an advantage over the other plants there.

Plants are important in ponds in the following ways:

Do not underestimate the importance of plants.

- they produce oxygen when they photosynthesise
- they are an important food source for animals
- they provide an important place for animals to shelter and lay their eggs.

3 Pond Animals

The variety of animals will depend on the size and permanence of the pond. Some of the animals will spend their whole life there, while others have a complicated life-cycle and spend only part of it in the pond e.g. dragonflies, frogs. Animals you might see include:

Level 4

flatworms
annelid worms
molluscs e.g. pond snails
crustaceans e.g. water fleas, waterlice
insects e.g. pondskater, water boatman, midge larva
fish
frogs (or tadpoles)

We will consider one animal from each region of the pond (diagrams not to scale).

Surface film e.g. pondskater (*Gerris*)

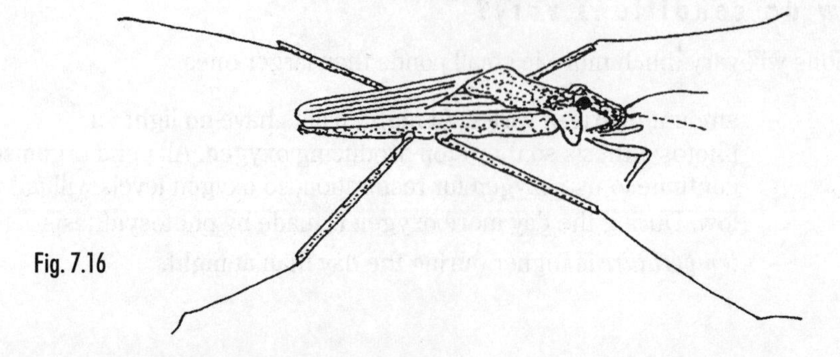

Fig. 7.16

- small animal which walks on surface film of water
- 4 long legs angled away from the body to spread its weight
- very light for its size
- carnivore which feeds on other animals on or near the water surface.

Vegetation zone e.g. pond snail (*Limnaea*)

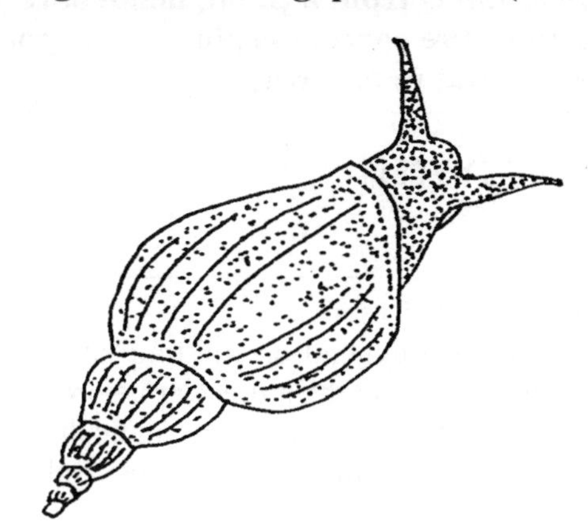

Fig. 7.17

- found on the surface of leaves or stones where they feed on algae
- produce slime to help them move easily
- have tentacles with an eye at the base
- come to the pond surface to take air inside their shell, then use this to breathe underwater.

Open water e.g. water boatman (*Notenecta*)

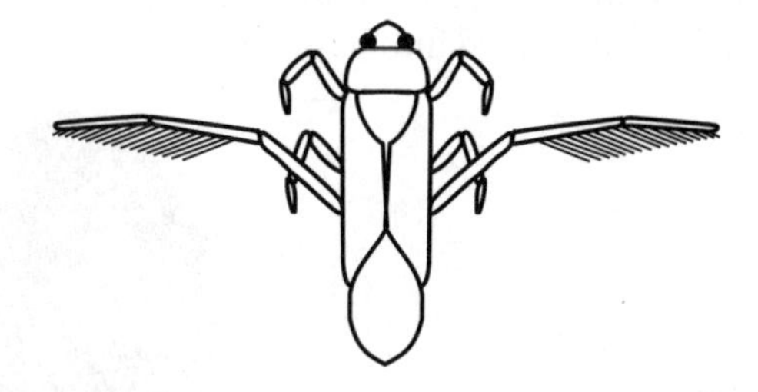

Fig. 7.18

- streamlined shape to move easily in water
- powerful legs to swim fast
- hairs on legs to increase area in contact with water
- wings to fly from pond to pond
- piercing mouth and poisonous saliva to kill prey.

Muddy bottom e.g. midge larva (*Chironomus*)

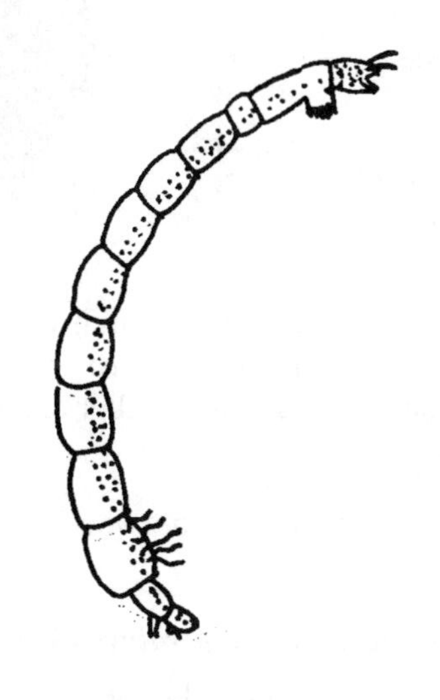

Fig. 7.19

- live in mud tubes at the bottom of pond
- feed on small organic particles in mud
- contain haemoglobin to help them absorb oxygen from the water
- move their bodies to create water currents, to increase absorption of oxygen.

4 Collecting data in ponds

Plants – use nets or jars

Animals – use nets for surface film or open water
– search vegetation by hand
– dig up a small amount e.g. a teaspoonful, of bottom mud and place in a white enamel dish with some water.

5 Ideas for fieldwork

1. Investigate the distribution of plants or animals in a pond.
2. Investigate the growth rate of duckweed with different amounts of light.
3. Investigate the growth rate of duckweed with different amounts of minerals.
4. Investigate which type of substrate pond snails prefer e.g. sandy, muddy, gravel etc.

If you remove animals from a pond, make sure you return them to the same pond, and to the correct region of the pond. Avoid disturbing or damaging the habitat in any way.

WOODLAND ECOSYSTEM

Key Points

- the wood is divided into 5 regions
- the plants you see will depend on the time of year you visit the wood. Some plants flower and die back early in the year before the leaf canopy forms
- many of the animals living in woods are difficult to observe. Use beating trays and set up pitfall traps to catch invertebrates, and take a sample of leaf litter to put in a Tullgren funnel.

Woods can be divided into two types: broad-leaved (deciduous) woods and coniferous woods. In this section we will consider *broad-leaved* woods.

A typical wood can be divided into 5 regions:

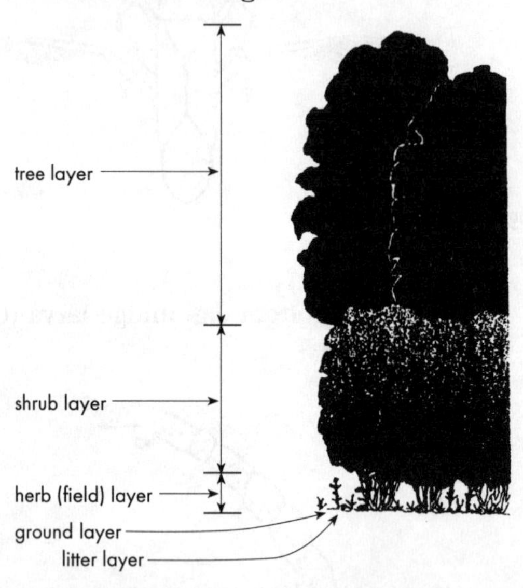

Fig. 7.20 Profile of a wood

In *coniferous* woods, the trees form a much denser canopy layer, so less light penetrates through to the woodland floor. There are far fewer plants in the shrub, field and ground layers, and consequently the number of animals is smaller.

1 How do conditions vary?

Conditions will depend on the size of the wood, the type of soil present and how closely the trees are growing.

daily – amount of *light* and *temperature* will vary between day and night

seasonally – amount of *light* varies as leaves in the canopy layer grow in spring and fall in autumn

– *daylength* varies throughout the year. This is important because it triggers flowering in many plants and beheaviour patterns e.g. mating or hibernation in some animals.

2 Woodland plants

A typical wood contains a large variety of plants, including lichens, mosses, ferns and flowering plants. Many of these plants can cope well with shady conditions.

Tree layer – there is usually one dominant species e.g. oak (*Quercus*) or two species may be equally dominant e.g. oak and ash.

Shrub layer – this includes hawthorn, hazel, elder and young trees.

Field layer – this region includes ferns, bramble, bluebell, mercury, and wild garlic.

Ground layer – mosses, lichens and grasses are found in this layer.

Litter layer – only dead and decaying plants are found here.

Many plants found in the field layer have adaptations for living in the shade e.g.

These are important adaptations.

- early growth and flowering before the leaf canopy forms
- plants grow taller to make the best use of available light
- plants have larger leaves than those of the same species growing in sunny places. This gives a larger surface area for the absorption of light
- leaves contain more chloroplasts to trap sunlight.

3 Woodland animals

This is a huge variety of animals in a typical wood, including:

annelid worms e.g. earthworm
molluscs e.g. snail, slug
crustaceans e.g. woodlice
arachnids e.g. spiders, harvestmen
insects e.g. butterflies, moths, beetles, ants
myriapods e.g. centipedes, millipedes
birds e.g. crows, rooks, woodpeckers, owls
mammals e.g. dormice, woodmice, squirrels.

Many of these animals will be very secretive or nocturnal, and therefore difficult to observe.

We will consider one animal from each layer of the wood (diagrams not to scale).

Tree layer e.g. Tawny owl

Fig. 7.21

- forward-facing eyes so it can accurately see prey
- sharp talons to catch prey
- hunts at night when small mammals are most active
- nests in holes in tree trunks.

Level 4

Shrub layer e.g. squirrel

Fig. 7.22

- sharp teeth to gnaw through nuts
- claws help it to climb up trees
- good eyesight to avoid predators
- hibernates during the coldest part of the winter.

Field layer e.g. speckled wood butterfly

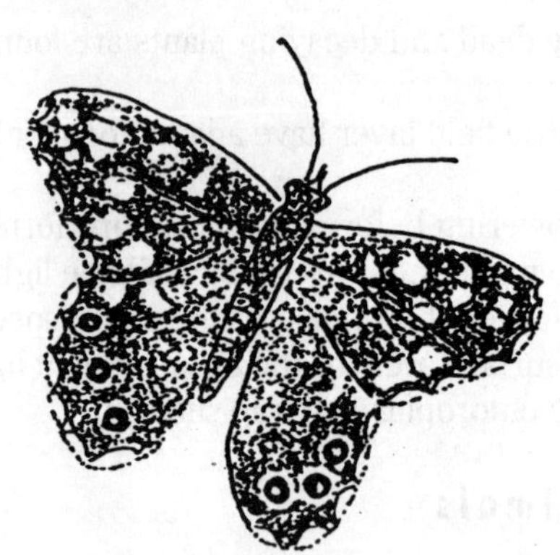

Fig. 7.23

- eggs are laid on underside of grass leaves
- caterpillars are green, so are well camouflaged on grass
- caterpillars have strong jaws to feed on grass
- adult is mottled brown colour (well camouflaged)
- adult is active even at low temperatures e.g. in shady woods.

Ground layer e.g. wood ant

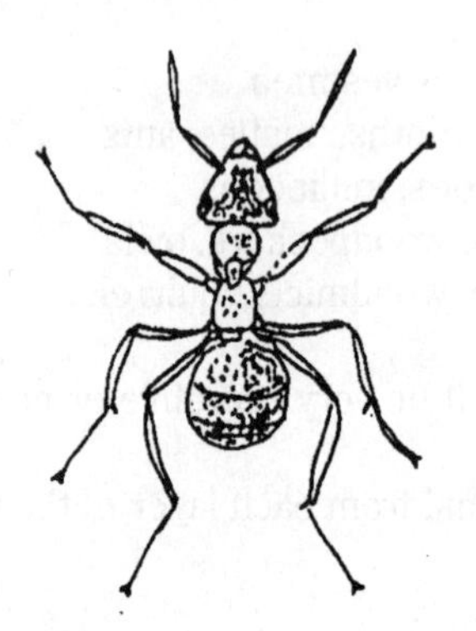

Fig. 7.24

- lives as part of a group (up to 100,000 ants) in a large mound
- produces poison to deter other insects
- has strong jaws to kill prey
- removes vegetation to make 'tracks' to trees where prey are found.

Litter layer e.g. earthworm

Detritus feeders in the litter layer are very important in the process of decay.

Fig. 7.25

- has bristles to anchor itself in soil and help it to move
- mucus covers the body to prevent it from drying out
- feeds on soil, and absorbs the organic materials i.e. a detritus feeder
- pulls leaves underground into its burrow i.e. mixes litter into soil.

4 Collecting data in woods

Plants – use line, point or belt transects, or quadrats

Animals – use beating trays for animals in shrubs or trees
pitfall traps for animals in ground layer
Tullgren funnel for animals in litter layer.

5 Ideas for fieldwork

1. Investigate distribution of plants or animals in a wood.
2. Compare heights of a single species of plant growing in the sun and shade.
3. Compare leaf area of a single species of plant growing in the sun and the shade.
4. Investigate the numbers of a particular animal species using mark-recapture techniques.
5. Compare the animals found in a deciduous and a coniferous wood.

If you remove any animals from the wood, make sure they are returned to the correct region. Avoid damaging or disturbing the habitat in any way.

Links with other topics

Related topics elsewhere in this book are as follows:

Topic	Chapter and Topic number	Page no
Adaptations for survival	4.4	44
Classifying living things	5	180
Populations and resources	7.3	298
The Biological community	8.3	325
Sequence of decay	8.2	318
Role of living things in the process of decay	8.2	323

REVIEW QUESTIONS

Q1 Bill noticed some tiny white insects called springtails in the soil below an oak tree. Springtails feed on fungi which decay the leaves.

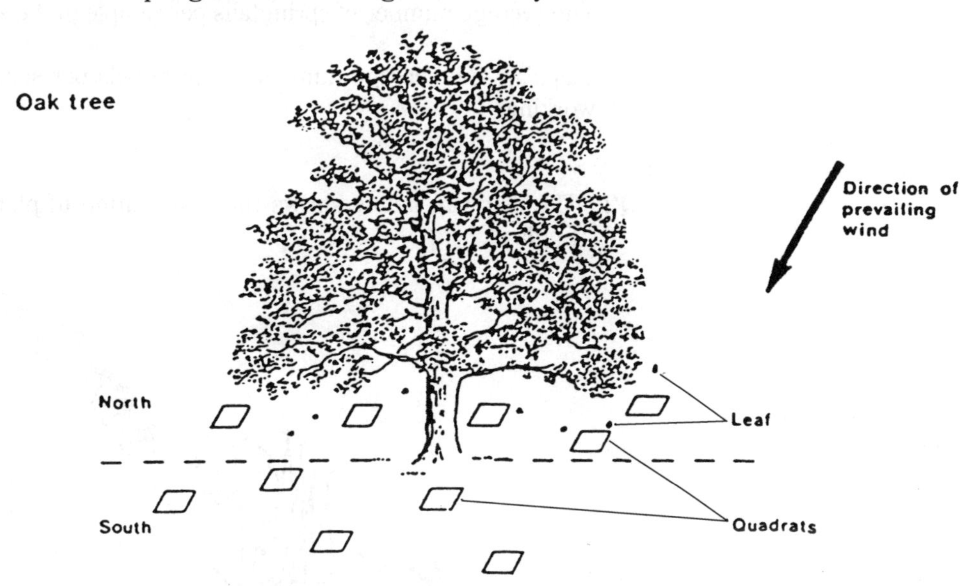

Fig. 7.26

He investigated the distribution of the insects around the tree using five quadrats to the north and five to the south of the tree.

Bill took an equal sample of soil from each quadrat. He counted the number of springtails in each sample using the apparatus below.

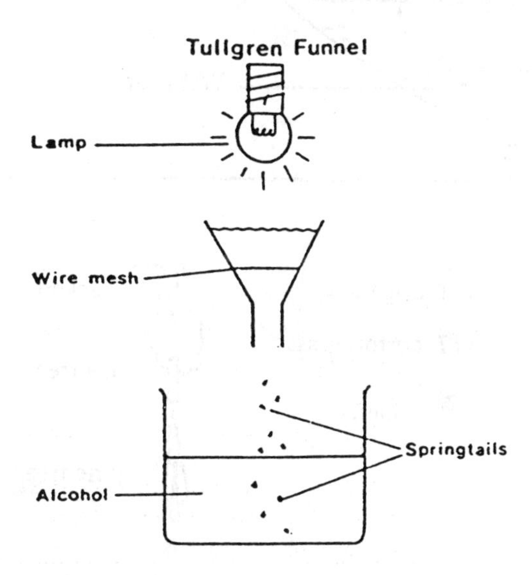

Fig. 7.27

(a) (i) Why did Bill take an equal sample of soil from each quadrat?

____ *(1)*

(ii) Why did he use a lamp?

____ *(1)*

(iii) Why was the wire mesh used?

____ *(1)*

(b) The results of the investigation are shown in the table below.

	North of tree					South of tree				
Number of springtails per soil sample	20	30	45	10	25	10	15	20	5	5

The average number of springtails per sample in the south was 11.

Calculate the average number of springtails per sample in the north. Show your working.

Q2 The diagram below shows the distribution of plants in a habitat in Northern England.

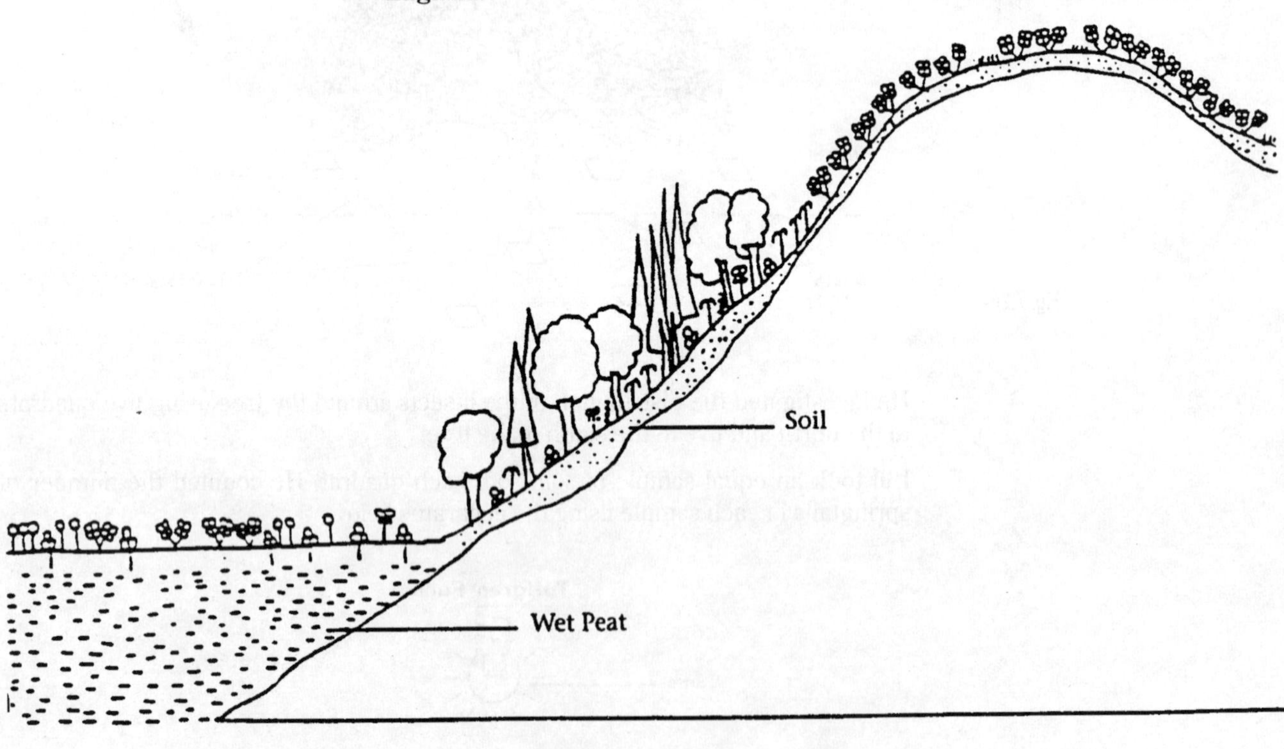

Fig. 7.28

(a) (i) Name two plants which live only amongst the trees.

________________ and ________________ *(2)*

(ii) Suggest ONE reason why these plants live only amongst the trees.

________________________________ *(1)*

(b) Suggest ONE reason why there are no trees on the upper slopes of the hill.

________________________________ *(1)*

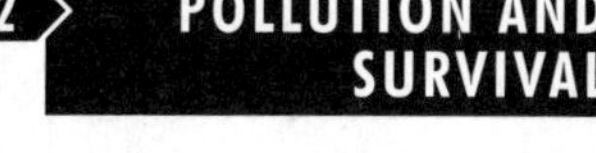

Pollution is the presence in the environment of substances in the wrong amounts, in the wrong place and at the wrong time. Most pollution occurs as a direct result of human activities. It tends to cause most concern when it affects humans, but most organisms in most habitats are likely to be affected by pollution. Each of the main types of habitat – air, land and water – are affected by pollution. In this Topic Group, we look at examples of pollution in the air and in water.

Despite our improved understanding of the causes and effects of pollution, the problem is likely to become more serious as the human population continues to expand. One type of pollution, which has global rather than simply local effects, is carbon dioxide. This is described in more detail in a later section.

AIR POLLUTION

Key Points

- *Air pollution* mostly results from the combustion of fossil fuels (coal and oil) by industry, domestic use and vehicles. The main pollutant gases, produced directly or subsequently, are *sulphur dioxide, nitrogen oxides, ozone, carbon monoxide* and *carbon dioxide.*
- Air pollutants, particularly at high concentrations, can have serious direct and indirect effects on living things. For example, sulphur dioxide is thought to increase the incidence of respiratory diseases in humans, increase 'die back' in forest trees and death in freshwater fish.

Air pollution is derived from many sources, mostly as waste products from combustion of *fossil fuels*, particularly oil and coal. Volcanoes contribute air pollutants to the atmosphere, but most pollutants are produced by human activities. The main sources of pollutants are in industrialised areas, and consist of heavy industry and emissions from vehicles. Domestic burning of various fuels also add pollutants to the atmosphere. The main air pollutants are summarised in Fig. 7.29. *Primary pollutants* are those which are produced directly from processes on the ground. *Secondary pollutants* result from chemical reactions of primary pollutants in the atmosphere, and include *acid rain*, which is a 'cocktail' of acids derived from pollutant gases.

A clear relationship has been shown between human respiratory disease and air pollution. For example, the incidence of deaths due to respiratory disease (bronchitis, pneumonia) and heart failure dramatically increases during episodes of very high pollution (Fig. 7.30). In the UK, the Clean Air Act (1956) has done much to reduce high levels of air pollution, though emissions of certain pollutants (e.g. from vehicles) continues to increase. In some cities in the world (e.g. Mexico City, Los Angeles), this is a major problem.

The effects of air pollutants on natural ecosystems are numerous and complex. For example, the effects of acid rain on lakes and forests depends on how sensitive these habitats are in a given area. The sensitivity of water and soil depends on several factors, including the normal pH (acidity or alkalinity) of the surrounding soil. Acid

Air pollutant	Source	Damaging effects
Primary pollutants		
sulphur dioxide	combustion of fossil fuels (mainly industry)	'dieback' in trees, reduced crop plant growth
nitrogen oxides	combustion of fossil fuels (mainly vehicles)	'dieback' in trees, reduced crop plant growth
carbon monoxide	combustion of fossil fuels (mainly vehicles)	human respiratory disease
carbon dioxide	combustion of fossil fuels	global warming?
chlorofluorocarbons (CFCs)	aerosol cans, refrigerators	destruction of ozone layer
particulate matter	smoke (mainly from industry)	human respiratory disease, cancer
radioactive substances	nuclear tests, nuclear power stations	increased mutation rates
lead	vehicle exhaust, from leaded petrol	mental development in children impaired
Secondary pollutants:		
ozone	formed from chemical reaction with nitrogen oxides and UV light	'dieback' in trees, reduced crop plant growth
acid rain	mixture of sulphuric and nitric acids, derived from sulphur dioxide and nitrogen oxides in rainwater	death of freshwater fish, corrosion of stone buildings

Fig. 7.29 Main pollutants

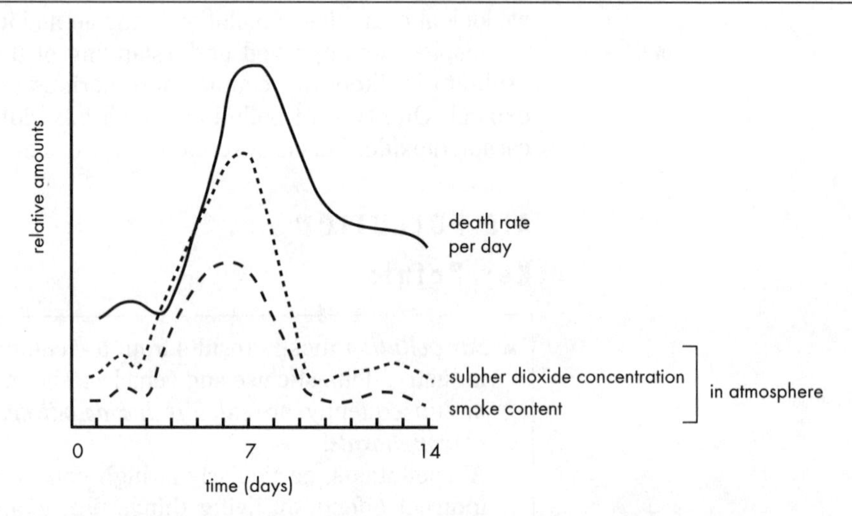

Fig. 7.30 Incidence of human death rates in London during a two week period in Winter, 1952

rain has been widely attributed to be the cause of forest 'dieback' (large-scale death or poor growth of trees) in North America and in Europe. Although acid rain may be the principle cause, the presence of other factors such as disease and drought may increase the vulnerability of the forest.

One striking example of the likely effects of acid rain is on the occurrence of lichen species. Lichens are symbiotic associations of fungi and algae, often found growing on trees and buildings. Some lichen species are very sensitive to acid rain, and it has been shown that the number of species declines at distances closer to sources of air pollution, such as city centres (Fig. 7.31).

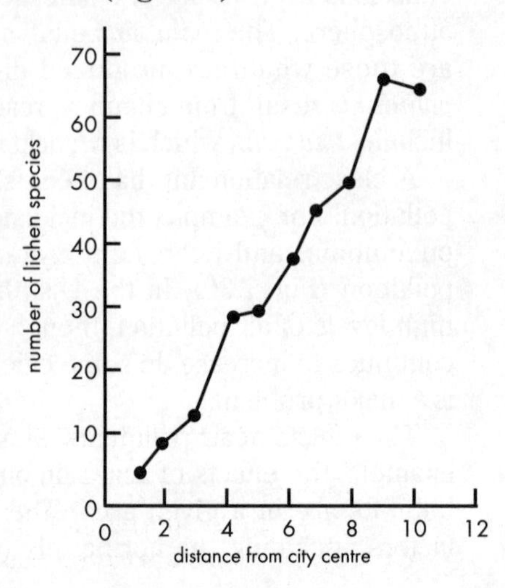

Fig. 7.31 Changes in the number of lichen species with distance from city

REVIEW QUESTIONS

Q3 Acidic gases released during the burning of fossil fuels are seriously affecting the biosphere.

(a) What is meant by the 'biosphere'?

___

In many parts of the world, acid precipitation and the run-off from the land have lowered the pH of lakes.

In Sweden, there are about 83,000 lakes of varying size. A recent survey of the pH value of the water in these lakes produced the results below (Fig. 7.32).

	number of lakes			
size of lakes in km²	**pH 5.0 or lower**	**pH 5.0–5.9**	**pH 6.0–6.9**	**pH 7.0 or higher**
bigger than 100	0	0	9	13
10–100	0	2	260	100
1–10	28	380	3 000	590
0.1– 1	600	4 400	12 700	1 550
0.01–0.1	4 000	24 500	28 500	2 700
total	4 628	29 282	44 469	4 953

Fig. 7.32

The chart (Fig. 7.33) shows the range of pH values in which populations of various organisms cannot survive in lakes.

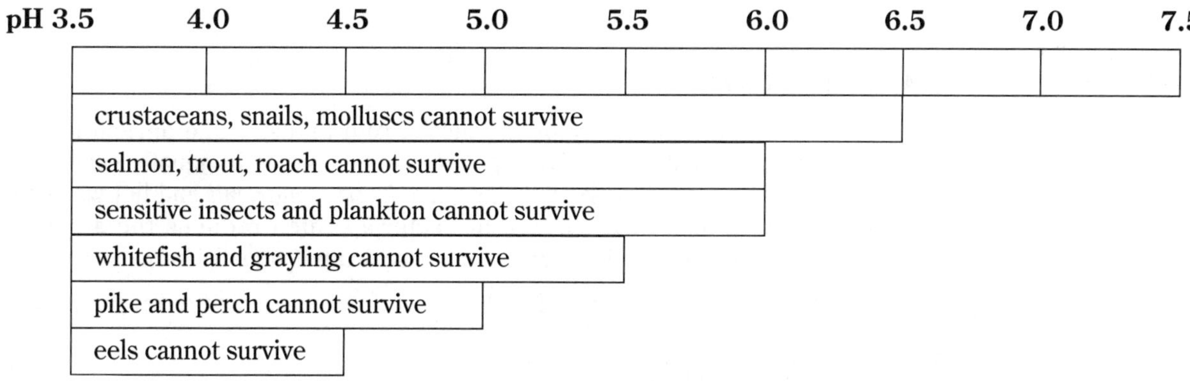

pH 3.5 **4.0** **4.5** **5.0** **5.5** **6.0** **6.5** **7.0** **7.5**

crustaceans, snails, molluscs cannot survive
salmon, trout, roach cannot survive
sensitive insects and plankton cannot survive
whitefish and grayling cannot survive
pike and perch cannot survive
eels cannot survive

Fig. 7.33

(b) How many of the lakes in the survey cannot support fish life, except perhaps eels?

___

(c) What will be the effect on the salmon population in a lake, if the pH value of the water falls from 6.5 to 5.5?

___

___

Aluminium salts in the soil are dissolved by acid rain. They enter the lakes in run-off from the land. When the pH of the lake water falls below 5.8, toxic aluminium salts are also dissolved from the sediment at the bottom.

Aluminium salts in solution are toxic to fish. Their gills become clogged by mucus.

The presence of these aluminium salts in the lakes also reduced the phosphates available to phytoplankton and other aquatic plants.

(d) Explain how **these** effects of aluminium salts cause changes in the fish population.

400 of the lakes have been treated with slaked lime (calcium hydroxide).

(e) Explain **one** benefit of adding slaked lime to these lakes.

(f) Suggest **one** reason why smaller lakes were treated rather than larger lakes.

WATER POLLUTION

Water pollution consists mainly of chemicals from industry and agriculture, and sewage from human habitation. *Toxic pollutants* accumulate in aquatic ecosystems, and become incorporated into food chains. *High levels of nutrients* and hot water from pollution result in excessive growth of bacteria, which reduces the amount of dissolved oxygen in water. This, in turn, can cause death in aquatic animals.

There are two main types of water pollution; *toxic substances* and *excess nutrients*.

1 Toxic substances

Toxic substances include wastes discharged from industrial and mining sites into rivers, lakes and oceans. Examples of these pollutants include cyanide and the heavy metals: lead, mercury, copper, zinc and cadmium. Oil spillage and the dumping of low-level radioactive waste at sea occur on a large scale. Toxins from agriculture include pesticides, herbicides and fungicides which have been washed into streams, rivers and lakes. Examples of these substances include chlorinated hydrocarbons and organophosphates.

“Note that there are both direct and indirect effects of water pollution.”

Effects on aquatic organisms may occur directly, by consumption or absorption, or indirectly by accumulation in food chains. Humans can be affected either by consuming contaminated fish and other aquatic organisms, or by drinking water which contains dissolved toxins. Although many toxic substances occur at fairly low concentrations, many accumulate in the body over time.

2 Excess nutrients

Many waste substances are in fact also nutrients. Organic waste such as sewage (effluent) is a food source for bacteria, which break it down. However, such bacteria are often oxygen-demanding, and their activity reduces the amount of dissolved oxygen available to other organisms. The activity of bacteria is further increased by the presence of hot water, which often accompanies the discharge of effluents. The amount of oxygen consumed by aquatic organisms is called the *biochemical oxygen demand (BOD)*. Polluted water typically has a high BOD value.

Nutrients such as nitrates and phosphates are contained in fertilisers, which are widely used in agriculture to promote crop growth. However, not all these nutrients are absorbed by the crops. Surplus amounts contained in the soil may eventually soak into streams, rivers and lakes. High levels of nitrate from urine are also released from the sewage of human populations and farm animals. These nutrients are then used particularly by aquatic algae, which grow very rapidly (= 'algal bloom'). When the algae die, they are broken down by bacteria, which consume much of the dissolved oxygen in water. This process of excess growth with a lowering of dissolved oxygen is called *eutrophication*.

❝ Make sure you understand the term *eutrophication* – it is very important. ❞

fertilizer

Links with other topics

Other related topics elsewhere in this book are as follows:

Topic	Chapter and Topic number	Page number
Global warming	7.4	305

REVIEW QUESTIONS

Q4 Fig. 7.34 shows the concentration of oxygen, the numbers of bacteria and the numbers of fish in a river over a distance of 50 km, measured from point **P** which is upstream from a source of pollution.

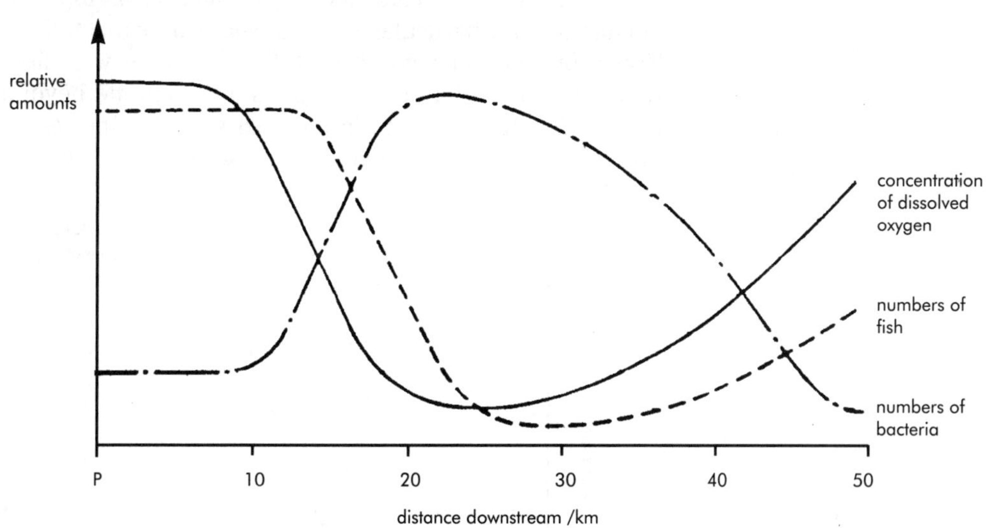

Fig. 7.34

(a) At what distance from point **P** did the river become polluted?

____

(b) (i) With reference to the three curves on the graph, describe the effect of the pollution.

____

____

____

(ii) Suggest a possible cause of the pollution.

____

(c) The oxygen concentration 50 km downstream from point **P** is returning to its original level, yet the numbers of fish are still much reduced. Suggest **two** reasons for this.

____

____

3 NATURAL POPULATIONS

This Topic Group describes the ways in which natural populations are regulated. In general, organisms tend to increase their numbers to the point at which they are prevented. What prevents unlimited population increases includes lack of resources, an accumulation of toxic waste products, and other organisms. Organisms influence each other by competition and predation.

This part of the syllabus provides an important basis for the topics on the human population and on human effects on the environment.

PREDATOR-PREY POPULATIONS

It is a good idea to become familiar with examples of predator–prey relationships

Key Points

Level 6

- Populations of organisms tend to increase to the maximum rate permitted by the availability of resources. In practice, resources needed by many organisms include other organisms. Predators eat prey, and both the populations of predators and prey are affected by each other. In situations where predators eat mainly one type of prey, the population of each regulates and is regulated, by the other.

Populations of organisms in natural habitats are controlled by the resources available to them. Resources include food, space and access to breeding partners. In theory, the population of a particular species would be expected to *increase exponentially* (i.e. '*increasing at an increasing rate*') if no resources were limiting (Fig. 7.35). In practice, this either does not occur, or does not continue beyond a few generations. Some reasons for this are given in the next section. One reason is due to the presence of *predators*, i.e. animals which capture and consume live animals, their *prey*.

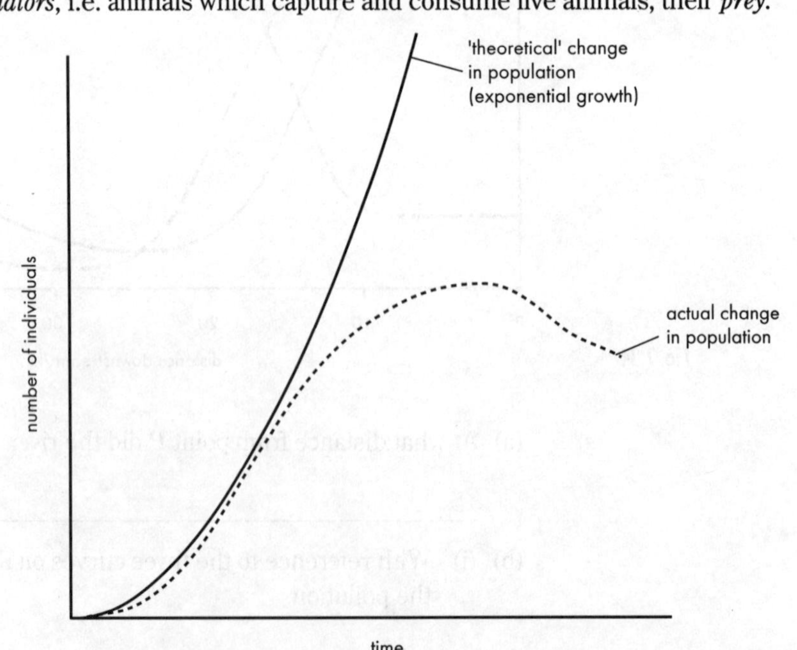

Fig. 7.35 Theoretical and actual changes in the population of an organism

If a predator relies mainly on a particular prey for its food, the populations are likely to be strongly affected by each other. For instance, if the prey population increases, the population of the predator will also increase, at least initially. This is because the predators will tend to survive for longer, and reproduce more successfully, if there is abundant food available. However, an increase in predator numbers will cause a reduction in the prey population. This, in turn, will result in a reduction in predator

This is a very important relationship – make sure you understand it.

numbers. This is an example of ***density-dependent regulation*** of populations, since numbers of both populations affect each other. There are two important features shown in Fig. 7.36. Firstly, the relative numbers of the predator (*Hydra*) are always less than those of the prey (***Daphnia***). The reason for this is explained in a later section on food chains. Secondly, there is a 'lag' after the increase in the prey population, before the predator population increases. This is the time needed for the predator to respond to the increased food resource, and to reproduce.

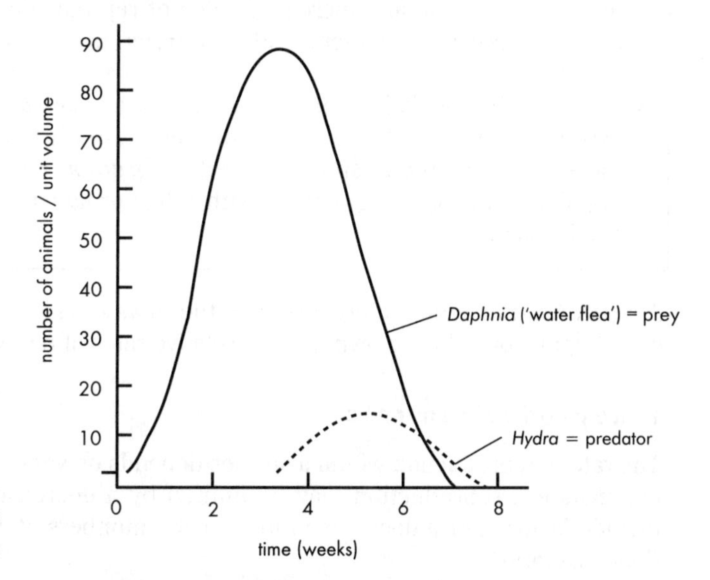

Fig. 7.36 Example of a simple predator-prey relationship

If a decline in predator numbers allows the prey population to recover, then the whole process may be repeated. The result is continuing fluctutations of numbers of prey and predators over time (Fig. 7.37). This is used as a method of controlling pest (= prey) populations by predators, in ***biological control.***

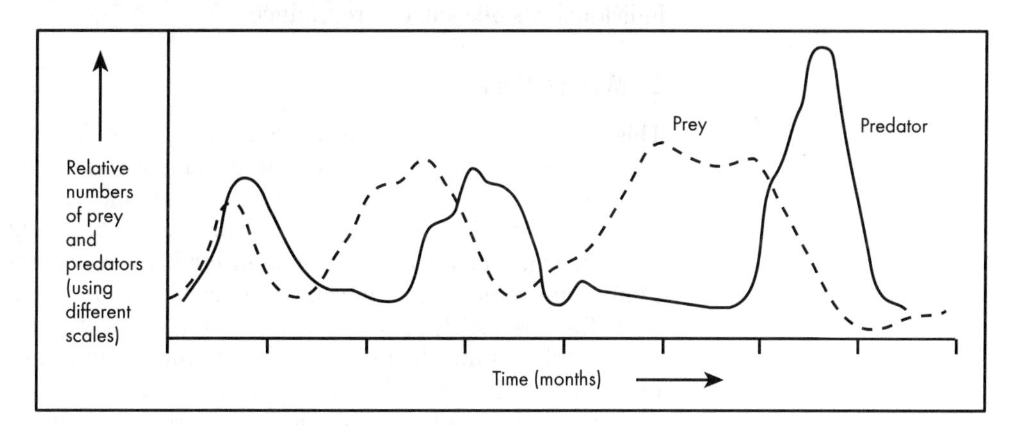

Fig. 7.37 Fluctuations of predator-prey populations over time

Links with other topics

Other related topics elsewhere in this book are as follows:

Topic	Chapter and Topic number	Page number
Competition and survival	4.4	43
Principles of food chains	8.1	316
Pyramids of numbers	8.3	327

REVIEW QUESTIONS

Use the Review Questions following the next topic to check your understanding of Natural Populations.

POPULATIONS AND RESOURCES

Key Points

Level 7

- The number of organisms in a population is determined by many factors, including the availability of necessary resources. Most organisms exist in a constantly changing environment. Favourable conditions often result in increased survival and increased rates of reproduction. However, this may be accompanied by a reduction of resources, preventing continued population growth.
- Increases in population result from increased *reproductive rates* and increased *immigration* into the population. Decreases in population are caused by increases in *mortality* and increased *emigration*. Factors affecting changes in population include interactions with other organisms, e.g. in competition and predation.

A *population* is a group of organisms of the *same species* in the *same place* at the *same time*. Populations change owing to the relative rates of *reproduction, death, migration.*

1 Reproduction rate

The rate of reproduction within a species depends on various abiotic and biotic factors. For instance, reproduction may be limited by a decrease in space or temperature (abiotic factors), or a decrease in food or the numbers of sexually mature individuals (biotic factors).

2 Death rate (mortality)

This depends on the rate at which individuals are removed from the population by death. Death occurs in most organisms and may take place before or after an individual has been able to reproduce.

3 Migration

This is the movement of individuals from one population to another. Such movement, or *dispersion* may involve all of the organism (e.g. a migrating bird) or part of an organism (e.g. a seed). Migration may be temporary or permanent. There are two types of migration, *emigration* and *immigration*. *Emigration* is the *departure* of individuals from a population. *Immigration* is the *entry* of individuals into a population. Migration can be a rapid means of increasing or decreasing populations, for instance when there is a fairly sudden change in conditions within the environment.

A population may be established by immigration from a relatively small number of individuals and, if conditions are suitable and the organisms are adapted there will be an increase in numbers. This process of *colonisation* may be very important in the early stages of *succession*. The increase might be quite rapid initially and in most species then slows down and possibly decreases as conditions become less favourable. For instance, resources such as food and space will become limiting. This increase

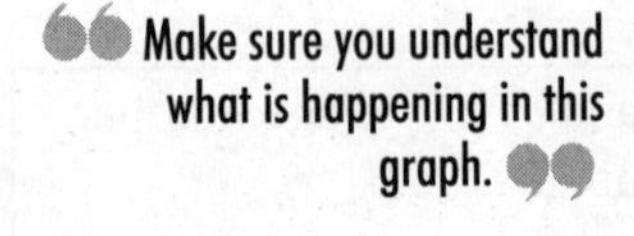

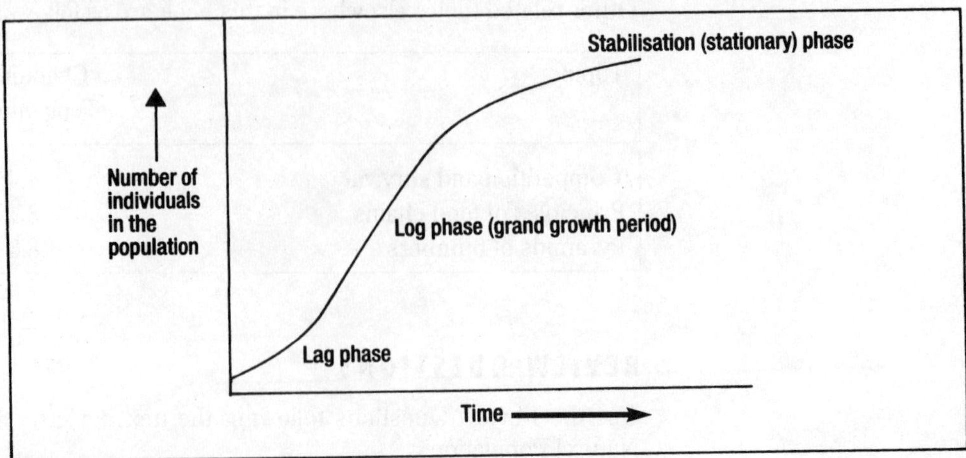

Fig. 7.38 Typical population graph

and subsequent decrease in numbers can be shown in a graph (Fig. 7.38). The shape of this graph is characteristic for most species.

The shape of the graph is known as *sigmoid* ('S'-shaped) and consists of three phases:

- **Lag phase**
This occurs when individuals may be emerging from dormancy or adjusting to their new environment.

- **Log phase**
This is a period of very rapid increase in numbers. The log (logarithmic, exponential) phase continues until resources become limiting.

- **Stabilisation**
This takes place when the limits of the environment are reached; the population is maintained at a level which can just be supported by the environment. This may continue indefinitely in some populations; in others, there may be a decline in numbers if the log phase has exceeded the *carrying capacity* of the environment, or if conditions deteriorate. The carrying capacity represents the maximum population which can be maintained in a given situation without a long-term depletion of resources.

Factors which are particularly significant in determining population growth are biotic factors such as *food supply, predation* and *disease*. In most species, population increase is self-regulated by density-dependent factors, caused by the individuals themselves. These factors may determine the relative rates of birth, death and migration.

4 Competition

Competition occurs between organisms when resources are limiting. There are two main types of competition; interspecific competition, between species, intra-specific competition, within species.

Interspecific competition may result in populations affecting each other directly. This is particularly apparent in cases where there is a close feeding relationship, such as occurs between predator and prey animals. An example of this involves two species of mite; one pest species (*Eotetranychus*) is the prey to another species (*Typhlodromus*) which is its predator, and which can be used as a biological control (refer to the previous section).

“Questions on biological control may ask for you to provide examples of prey and predator animals.”

The artificial introduction of predators into a community can be used as a means of regulating the production of prey species which are pests. This is called ***biological control*** and can reduce the damage caused by pests, for instance to crops, to acceptable levels. Biological control may be used as an alternative to ***chemical control***, which for instance involves the applicationof toxic (poisonous) substances, called pesticides, to the pest population. Biological control has certain advantages over chemical control because it may not need continual application, it regulates prey numbers at an acceptable level, it is often very specific in its action, and it does not result in the possible accumulation of toxic substances in the ecosystem. Examples of biological control include rabbits by the Myxomatosis virus, prickly pear by the moth *Cactoblastis*, and certain caterpillars by bacteria (e.g. *Bacillus thuringensis*).

Intra-specific competition is likely to be particularly intense because members of the *same species* usually require similar resources from the environment. Competition *within* a population can be an important process in evolution by natural selection.

Links with other topics

Other related topics elsewhere in this book are as follows:

Topic	Chapter and Topic number	Page number
Competition and survival	4.4	43
Principles of food chains	8.1	316
Pyramids of numbers	8.3	327

REVIEW QUESTIONS

Q5 The table below shows the number of living yeast cells in a culture solution during a 24 hour period. The temperature of the solution was kept constant throughout the experiment.

time (hours):	2	4	6	8	10	12	14	16	18	20	24
number of yeast cells (millions per ml):	10	60	90	200	400	600	650	700	700	50	0

(a) On the graph below (Fig. 7.39), plot a graph to show the changes in populations of yeast.

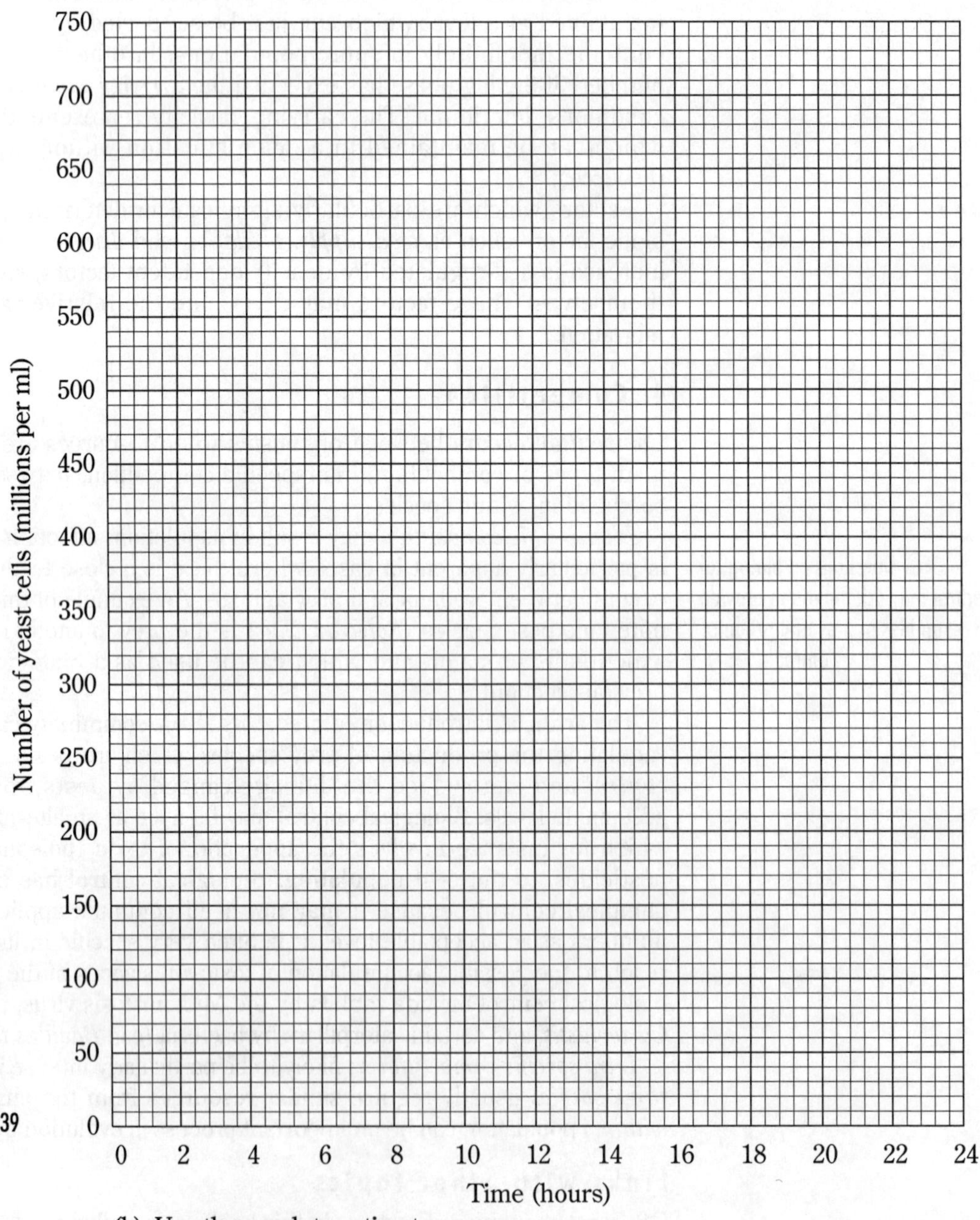

Fig. 7.39

(b) Use the graph to estimate:

(i) the number of yeast 7 hours after the beginning of the experiment

____

(ii) the time taken for the population to reach 500 million cell per ml of culture solution

____

(c) suggest two reasons for the changes in the yeast population during the last 8 hours of the experiment:

____

____

____

____

The human population continues to expand, unlike populations of other species, which in most cases are stable. Humans have both the need and the ability to obtain vast quantities of resources from the environment. Continued rates of increase of human populations and resource utilisation cannot continue at their present rates, because resources are finite.

This Topic Group describes some of the main effects that humans are having on their environment. These need to be understood in relation to ecological principles which control natural populations and their use of resources, which ultimately apply also to humans.

TRENDS IN HUMAN POPULATION

Key Points

- The world's human population is increasing, and this is happening at an increasing rate. In many countries, *birth rate* exceeds *death rate*. Infant mortality is decreasing in many countries and life expectancy is increasing in most countries. The main causes of this trend are improved nutrition, sanitation and health care.
- Population growth increases are higher in developing countries compared with industrialized regions of the world. The world's resources are finite, so some slowing of the human population increase will be necessary in order to avoid a global catastrophe.

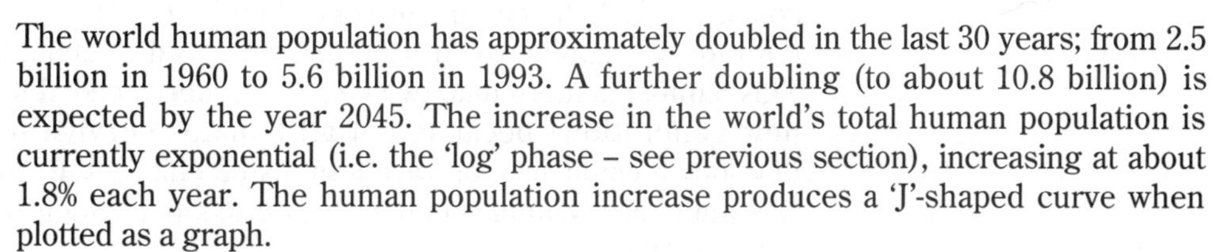

The world human population has approximately doubled in the last 30 years; from 2.5 billion in 1960 to 5.6 billion in 1993. A further doubling (to about 10.8 billion) is expected by the year 2045. The increase in the world's total human population is currently exponential (i.e. the 'log' phase – see previous section), increasing at about 1.8% each year. The human population increase produces a 'J'-shaped curve when plotted as a graph.

The main factors in determining human population rates are *birth rate, death rate,* the *age structure* of the population, *generation time* and *fertility and birth control.*

1 Birth rate and death rate

There has been a slight decline in birth rate throughout the world. However, it is the *relative difference* between birth rate and death rate which determines actual population increase. Birth rate exceeds death rate in many countries, particularly in the developing world.

Factors which tend to increase birth rate (natality) and decrease death rate (mortality) include an increased availability of food, improved sanitation and better medicine and health care. Antibiotics such as penicillin and streptomycin, and insecticides such as DDT, have made a major impact on disease control. Selective breeding of crops and domestic animals, and increased mechanisation in agriculture, have had a dramatic effect on food production. However, two-thirds of the world's population is still without sufficient food.

2 Age structure of the population

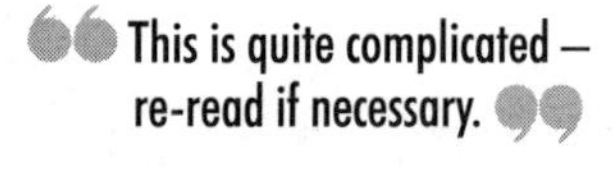

The relative number of individuals in each age group in the population is very important in determining both current and future increases in population. Population increase is strongly influenced by the relative number of people of 'reproductive age' (i.e. who can have children). However, even if the number of children born to parents is reduced, the total population will still increase in many countries, because a high proportion of individuals belong to younger age groups, and many will be future parents (Fig. 7.40). In developing countries, the proportion of individuals under 15 years old is approximately 40 per cent, compared with 28 per cent in developed countries. This, along with the increasing proportion of old people in the population who need to be supported, has been called a 'demographic time-bomb'.

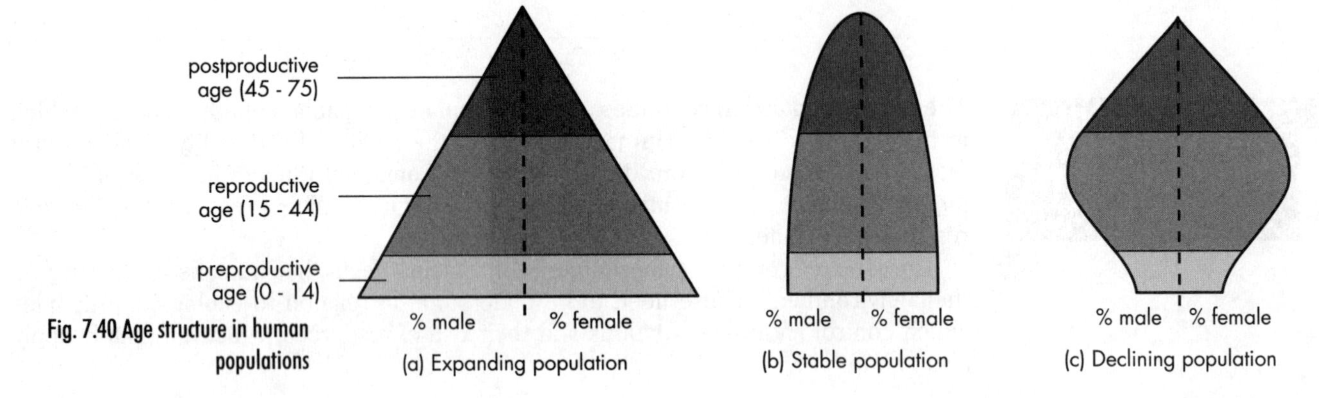

Fig. 7.40 Age structure in human populations

3 Generation time

The age when women begin to have children is an important factor for human population growth for two reasons. Firstly, having children later in life means there is 'less time' to have children. Secondly, an older woman may be more aware of the consequences of having more children than she can cope with, and may also have taken on other responsibilities. In some countries, the minimum age allowing marriage is relatively high, which 'delays' the onset of the next generation.

4 Fertility and birth control

'Fertility' is an expression of the number of children born to each family, in each generation. Currently, the number of children in each family throughout the world is on average 4.7. In developing countries, the average is higher (5.7) than in developed countries (2.6). If the human population is to stop growing, the number of children born to each family would need to be reduced to about 2.5; the *replacement number*. This number is more than two (i.e. the number needed to 'replace' the parents in each new generation) because not all females survive to reproduce (especially in developing countries), or have less than 2 children. Even if this replacement number was immediately implemented, the world's population would continue growing until the year 2100.

Birth control, for example by *contraception* is currently used by many adults to limit the size of their family. Contraception in some communities is either not available, or not understood, or not acceptable. This is a complex issue because it involves personal choices, but the effects of continued population growth affect society in general, for instance because of the world's limited resources. Rates of food production and distribution are being increased, but this is not a long term solution to the problem of an ever-increasing human population.

Refer to the topic on human reproduction if necessary.

Links with other topics

Other related topics elsewhere in this book are as follows:

Topic	Chapter and Topic number	Page number
Competition for survival	4.4	43
Reproduction (animals)	4.5	65
Populations and resources	7.3	298
Global warming	7.4	305

HUMAN EFFECTS ON THE ENVIRONMENT

1 Urbanisation and industrialisation

"Level 8"

Urban and industrial development often have a profound influence on the natural environment. Urbanisation and industrialisation include the construction of buildings and rail and road links. This requires land clearance, and may be accompanied by an increase in **pollution**; the release of substances into the environment in potentially harmful quantities. Pollution can be described as the presence in the environment of substances in the wrong amount, at the wrong place or at the wrong time. Pollution occurs in each of the three main types of habitat; air, land and water.

■ Air pollution

This results from the burning (combustion) of fossil fuels such as coal, oil and natural gas. Combustion occurs in power stations, heavy industrial plants, vehicle engines and in domestic use. The products of combustion include lead, sulphur dioxide (SO_2) and nitrogen oxides (NO_x); sulphur and nitrogen oxides are often known collectively as acid rain. This is currently causing serious damage to crops, forests and aquatic ecosystems. One indirect effect of **acid rain** is to cause aluminium to be released in increased amounts into the environment, and this is affecting various organisms, such as birds and fish. Air pollution also includes smoke particles, which may reduce photosynthesis in plants, and nuclear fallout, which may cause mutations (Chapter 6) in organisms.

■ Land pollution

This results from the accumulation of industrial and urban waste, such as scrap metal and plastics.

■ Water pollution

"Look back at topic group 7.2 for more details of pollution."

This is caused by the addition of **effluents** (wastes) from three main sources; domestic, industrial and agricultural. Domestic pollution includes sewage; although this is often treated (by being filtered and oxidised), organic pollutants may be released in significant amounts. Industrial pollution includes toxic wastes which have not been adequately treated before release. Agricultural wastes include excess inorganic or organic fertilisers and pesticides. These may be washed off the land by leaching and irrigation. Pesticides accumulate in the tissues of organisms, especially those towards the 'top' of food chains.

2 Agriculture

A major purpose of agriculture is to produce food for human consumption. Both the efficiency and the intensity of food production are being continually increased to meet the demands of the human population. There are several environmental implications of this:

■ Land clearance

Land is cleared for cultivation and for grazing; this reduces the number of potential habitats available. Tropical forests are cleared for timber and land use on a massive scale; this destroys important habitats and makes the soil unstable. Destruction of a habitat reduces variation and hence the so-called gene pool. This decreases the diversity of species and makes land more exposed and vulnerable to wind, which can blow away topsoil; this is called *erosion*.

■ Monoculture

Monoculture is the cultivation of a single species of crop on a particular area of land, for instance wheat and barley. Monoculture is an 'artificial' situation because there will normally be a *succession* leading to a greater diversity of species. This is resisted in monoculture by the use of selective *herbicides* to prevent the growth of weeds, and *pesticides* to remove insect and other pests. Herbicides are used to control or eliminate weeds. Several herbicides contain an impurity called *dioxin*, which is highly poisonous to wild life and to humans. Insecticides are used to kill insect pests. Organochlorine insecticides, including DDT, do not break down very rapidly in the environment (i.e. they are not biodegradable), which can result in pollution problems. DDT and similar pesticides accumulate in the tissues of animals, and become concentrated, particularly in those occupying higher trophic levels (see 8.1). The pollutant therefore becomes concentrated along the food chain. One result of DDT has been to reduce the thickness of egg shells in birds of prey, causing a decrease in successful reproduction. A further problem with DDT is that many insect pests have now become resistant to it. In addition to this, mechanical processes require a substantial amount of fuel.

"Exam questions on the effects of humans on the environment are almost guaranteed!"

■ Over-production

Maximum use is made of available agricultural land by intensive cultivation, including the use of nitrate fertilisers. One possible effect of this over-exploitation of soil is that it becomes more susceptible to erosion by wind and water. This may lead to desertification. Over-use of fertilisers may cause minerals such as nitrates to be leached away; nitrates can accumulate in aquatic ecosystems, resulting in *eutrophication*. Eutrophication involves rapid, excessive growth of aquatic plants which then die and decay; their decomposition lowers oxygen levels in the aquatic environment (see 7.2).

3 Fishing

Fishing is important in many human communities as a source of protein and oil, and also to provide feed for domestic animals and ingredients for fertilisers used in agriculture. In many coastal and island countries, fish can provide up to 90 per cent of the proteins required by humans. Most (about 90 per cent) of the fish and shellfish that are consumed are caught by fishing boats.

There have been dramatic increases in the intensity and efficiency of commercial fishing methods. This has resulted in *overfishing* in many areas of the world. Overfishing results in a depletion of younger fish, so that the 'breeding stock' is unable to maintain previous population levels. A *commercial extinction* level is reached, beyond which it is unprofitable to continue fishing for a particular species. Some countries have introduced laws to control net mesh sizes to reduce this problem; larger-mesh nets allow juvenile fish to escape, and so go on to reproduce. Other control methods include restrictions on where or when fishing is permitted, to allow fish populations to recover. *Aquaculture* and *fish farming*, in which captive fish and shellfish are bred specifically for food, is being used increasingly to supply human needs (see topic group 8.4).

4 Mining

Mining is used to obtain a wide range of minerals, as well as coal. Minerals are both metallic (e.g. iron, copper, aluminium) and non-metallic (e.g. phosphates). Various techniques are used, mainly with heavy machinery, to obtain minerals from the surface and underground. Mining is essential to meet human needs for energy and many raw materials. If mining is done carefully, the mined area can be allowed to recover to a large extent after mining operations have been completed.

However, mining can result in serious environmental degradation, for example by land disturbance, erosion, air pollution and water pollution. Underground mining produces proportionally much less land disturbance than surface mining, but is more expensive and dangerous.

For any given mineral which is being mined, there are finite amounts available for extraction. The exact amounts are uncertain in many cases, because reserves have not all been identified. At each site, the amount of mineral that can be extracted is also limited by *economic depletion* levels, beyond which it is not economic to continue mining. The rate at which minerals are used up also depends on how efficiently they are mined and used (Fig. 7.41).

5 Deforestation

Forests cover about 34 per cent of the world's land surface. However, about half the world's forests have been removed (*deforestation*) during the last 30 years, often to provide additional land for cultivation and occupation. Forests are an important source of timber for construction, as a fuel and for paper. Forests also provide rubber, foods and natural medicines. Forests have a much wider ecological role in *stabilising the*

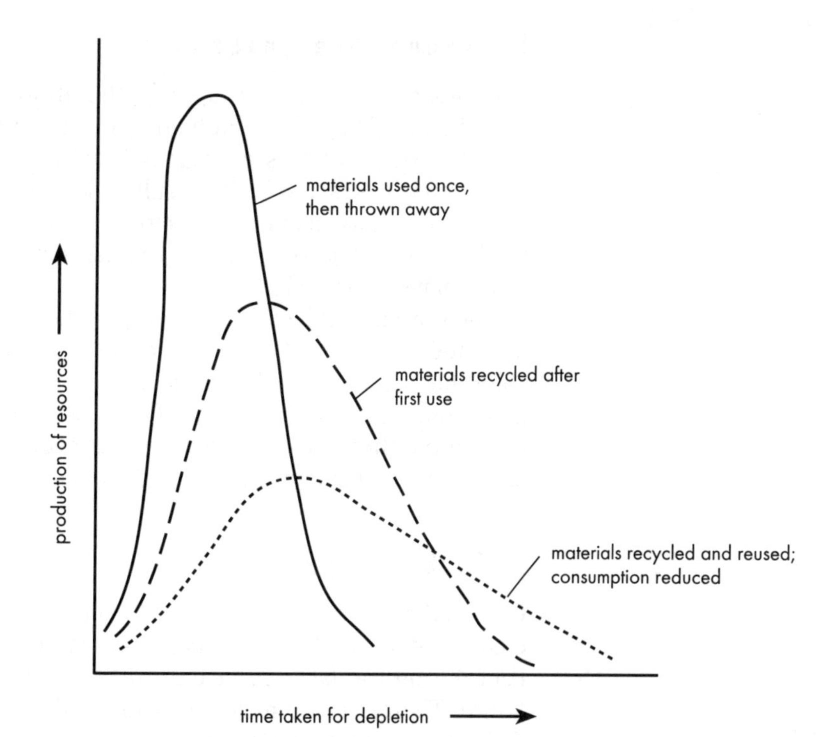

Fig. 7.41 The rate at which non-renewable resources are used up depends on the way in which they are used

Earth's climate, especially in the circulation of carbon dioxide, oxygen and water, and are important in preventing soil erosion. In tropical rain forests in particular, forests, are *centres of genetic diversity*, since many of the world's species exist in these ecosystems. It is estimated that at least 50 per cent of the world's species live in tropical rain forests, even though they only occupy about 7 per cent of the world's land area.

Forests are a potentially renewable resource, providing timber extraction does not exceed the rate of natural forest regeneration. In tropical areas, local people cause some deforestation by shifting cultivation and the collection of fuelwood. However, timber extraction to meet the demand of developing countries often occurs at much higher rates, which are not *sustainable*. It also opens up the forest for further exploitation. Intensive deforestation has many direct and indirect effects, including increased soil erosion and flooding, and displacement of indigenous people. It is also predicted that, with less trees to absorb carbon dioxide, deforestation will contribute to global warming.

Links with other topics

Other related topics elsewhere in this book are as follows:

Topic	Topic number	Page number
Competition for survival	4.4	44
Populations and resources	7.3	298
Trends in human population	7.4	301
Air pollution	7.2	291
Water pollution	7.2	294

GLOBAL WARMING

Key Points

- Global warming is expected to result from increasing concentrations of '*greenhouse gases*' in the atmosphere. Greenhouse gases (e.g. carbon dioxide, methane) and water vapour trap heat in the atmosphere, in what is known as the *greenhouse effect*.
- An increase in carbon dioxide concentrations in the atmosphere mainly result from two types of human activity: *burning of fossil fuels* and *deforestation*.

decomposition

1 Greenhouse gases

Greenhouse gases form a layer in the atmosphere which acts like the glass in a greenhouse. They allow high-energy solar radiation to pass through to the earth's surface. Much of this energy is 'bounced back' towards space as heat, which greenhouse gases 'trap'. The result, at least in theory, is *global warming*. There is still some uncertainty about whether recent increases in the world's temperature are due to global warming, or part of a regular series of long-term temperature fluctuations which have occurred in the past.

The presence of increasing amounts of greenhouses gases in the atmosphere is not disputed, however. Greenhouse gases include carbon dioxide (CO_2), methane, chlorofluorocarbons (CFCs) and nitrous oxide. Many of these gases result from air pollution. Although methane, CFCs and nitrous oxides are not present in high quantities, they have much more absorbing power than CO_2. Water vapour and particulate matter (e.g. smoke) in the atmosphere also act as greenhouse gases.

2 Carbon dioxide

Carbon dioxide represents about 49 per cent of all greenhouse gases. Carbon dioxide concentrations in the atmosphere have increased from 0.027 per cent to over 0.033 per cent during the last 100 years. Further, more rapid, increases are expected in the future. There are two major reasons for the increase in CO_2 in the atmosphere; *fossil fuel burning* and *deforestation*.

Look at the section on the carbon cycle if necessary.

The burning of fossil fuels accounts for about 80 per cent of the increases of CO_2 in the atmosphere. Most (76 per cent) comes from industrialised countries. The burning of fossil fuels (oil and coal) is likely to increase well into the future.

Deforestation accounts for about 20 per cent of the increased CO_2 levels in the atmosphere. There are three reasons for this. Firstly, timber removal means there are less trees to absorb CO_2 by photosynthesis. Secondly, deforestation is often accompanied by burning of remaining vegetation, to prepare the land for cultivation. Thirdly, any unburnt vegetation may quickly die and decompose, releasing CO_2. The regrowth of vegetation after deforestation is often insufficient to offset these effects.

3 Possible effects of global warming

Various predictions have been made concerning global warming. Possible temperature increases during the next 50 years have been suggested in the range 1.5°C to 5.5°C, if concentrations of greenhouse gases rise at current rates. One problem is that, by the time there is indisputable evidence for a link between greenhouse gas concentrations and global warming, the ecological problem will be too big to solve.

Possible effects of global warming might include increased crop yields (which will also be able to utilise higher CO_2 concentrations), but insect pest populations might also increase and there might be less soil water available.

In tropical areas of the world, decreased availability of water might lead to the formation of deserts. In coastal areas throughout the world, some melting of polar icecaps would result in flooding. Other predicted effects include the spread of tropical disease over a wider area, increased frequency of droughts, hurricanes and cyclones, and also forest fires.

Links with other topics

Other related topics elsewhere in this book are as follows:

Topic	Chapter and Topic number	Page number
Photosynthesis	4.6	105
Air pollution	7.2	291
Populations and resources	7.3	298

REVIEW QUESTIONS

Q6 The table below shows some of the pollutants released by aluminium production and from car exhausts.

	Pollution released			
Air pollutant	during extraction from aluminium ore in g per tonne	during recycling of aluminium cans in g per tonne	from car exhausts without a catalytic converter in g per kilometre at average 40 km per h	from car exhausts with a catalytic converter in g per kilometre at average 40 km per h
Carbon monoxide	35 000	2500	10.0	1.0
Sulphur dioxide	88 600	886	0.6	0.05

(a) Using information in the table, name TWO ways of reducing air pollution.

1. ______________________________

2. ______________________________

(2)

(b) A car without a catalytic converter travels 20 000 kilometres in a year at an average speed of 40 km per h. How many kg of carbon monoxide does it produce during this period?
Show your working.

____

____

____

(2)

(c) Name ONE other source of sulphur dioxide pollution.

____

(2)

(d) (i) Carbon dioxide and sulphur dioxide contribute to the effects represented on the postage stamps in the table below.
Write in the name of the gas which makes the **major** contribution in each case.

Postage stamp	Gas

(2)

(ii) Suggest ONE advantage of showing examples of pollution on a postage stamp.

____ *(1)*

(iii) Give TWO possible effects when acid rain kills plants in a lake.

1. ____

2. ____ *(2)*

(Total 10 marks)
(ULEAC, 1994)

Q7 Insecticides are chemicals which kill insects. These chemicals may be passed on in a food chain. The diagram below shows the diet and the levels of an insecticide in some birds living near a pond.

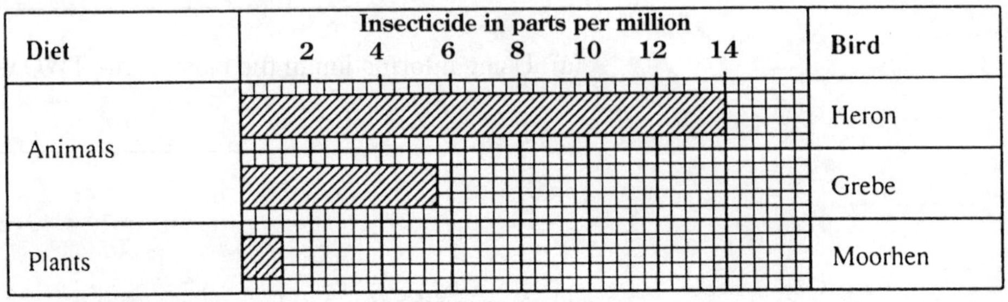

Diet	Insecticide in parts per million 2 4 6 8 10 12 14	Bird
Animals		Heron
		Grebe
Plants		Moorhen

(a) Explain the difference in the insecticide levels of the plant and animal feeders.

____

____

____ *(2)*

(b) (i) Suggest a feature which you might include, if you were to develop a new insecticide, to help protect the birds.

____ *(2)*

(ii) Suggest ONE way other than by the use of chemicals, by which the population of an insect might be controlled.

____ *(1)*

(c) DDT is a chemical insecticide which has been very effective in killing insects. However, there are now many populations of insects resistant to DDT. Suggest how these DDT resistant insects first appeared.

____ *(1)*

(Total 5 marks)
(ULEAC, 1993)

ANSWERS TO REVIEW QUESTIONS

A1 (a) (i) to be a fair test/to make sure results are truly representative

(ii) drive out insects

(iii) to hold back bigger insects/to hold the soil in place

(b) 130 divided by 5 equals 26

A2 (a) (i) fern, sorrel

(ii) prefer shade/less exposed to wind

(b) e.g. too cold

A3 (a) The total global system of habitats within which living organisms exist.

(b) 4 628.

(c) It will die (population will be reduced).

(d) There are direct and indirect effects. Fish are killed directly by being unable to obtain sufficient oxygen through their gills. They are killed indirectly because food chains on which they depend are affected, and they are unable to obtain suffcient nutrition.

(e) Slaked lime raises the pH of the lake. This will reduce the amount of aluminium salts released from the sediment.

(f) There is smaller amount of water, so less slaked lime is needed to raise the pH above the 'threshold' where aquatic organisms are affected.

Comments

A good example of a question in which you have to make careful use of the information provided. Don't allow yourself to be 'swamped' by all the information, though. Read the question through more than once, to see what the examiner is 'driving at'.

(a) There are other acceptable definitions.

(b) This is the total number of lakes at pH 5.0 or lower, only eels can survive below pH 5.0 (but above pH 4.5).

(d) Make sure you use the information given; there are two parts to the answer.

(e) An answer referring to raising the pH to allow more organisms to survive, would also be acceptable. You must give a reason, not just mentioning that pH is raised.

(f) You could also mention the expense of adding slaked lime to larger lakes. Smaller lakes are not necessarily more acidic than larger lakes (see table), so its best to avoid referring to this.

A4 (a) 10 km.

(b) (i) The pollution caused an increase in numbers of bacteria, resulting in decreases in the concentration of dissolved oxygen, and hence numbers of fish. This effect was reduced at increasing distances downstream from the source of pollution.

(ii) Wastes discharged from sewage works.

(c) The numbers of other organisms, needed by the fish for food, are still reduced. Toxic substances may be still be present in the water, or in the food chain.

Comments

(b) (i) needs answering carefully, so that you give yourself a chance to show your understanding of the processes involved.

(ii) Several answers are possible here.

(c) Another possibility is that the fish have not had time to reproduce, following a 'pulse' of pollution.

A5 (a) You must plot a *line graph*. Make sure the points are small and neat, and are joined with a smooth curve.

(b) (i) 125 million per ml.

(ii) 11 hours.

(c) Cells may be running out of food, or running out of oxygen, or being poisoned by build-up of toxic waste products.

A6 (a) 1. Recycle aluminium products.

2. fit catalytic converters to cars.

(b) 10 g of carbon monoxide produced per km.
$10 \times 20\,000$ of co produced in 1 yr
$= 200\,000$ g
$= 200$ kg

(c) Factories, power stations etc.

(d) (i) Acid rain = sulphur dioxide
Greenhouse effect = carbon dioxide

(ii) informs the public about environment issues

(iii) animals die due to lack of food
animals die due to lack of oxygen in water

A7 (a) Herbivores, e.g. moorhen, eat small amounts of plants therefore contain small amounts of insecticides. Carnivores eat animals and therefore accumulate the insecticide from their bodies — it builds up as you go further along the food chain.

(b) (i) Biodegradable insecticides (breaks down in the environment).

(ii) Biological control, ie. use a natural predator to kill the insects.

(c) Mutation.

Comment

Questions like this about accumulation of toxins along a food chain are very common.

EXAMINATION QUESTIONS

Question 1

Fig. 7.43 illustrates some possible sources of pollution.

By reference to the activities which occur in the lettered regions on the diagram explain how pollution is being caused. *(12)*

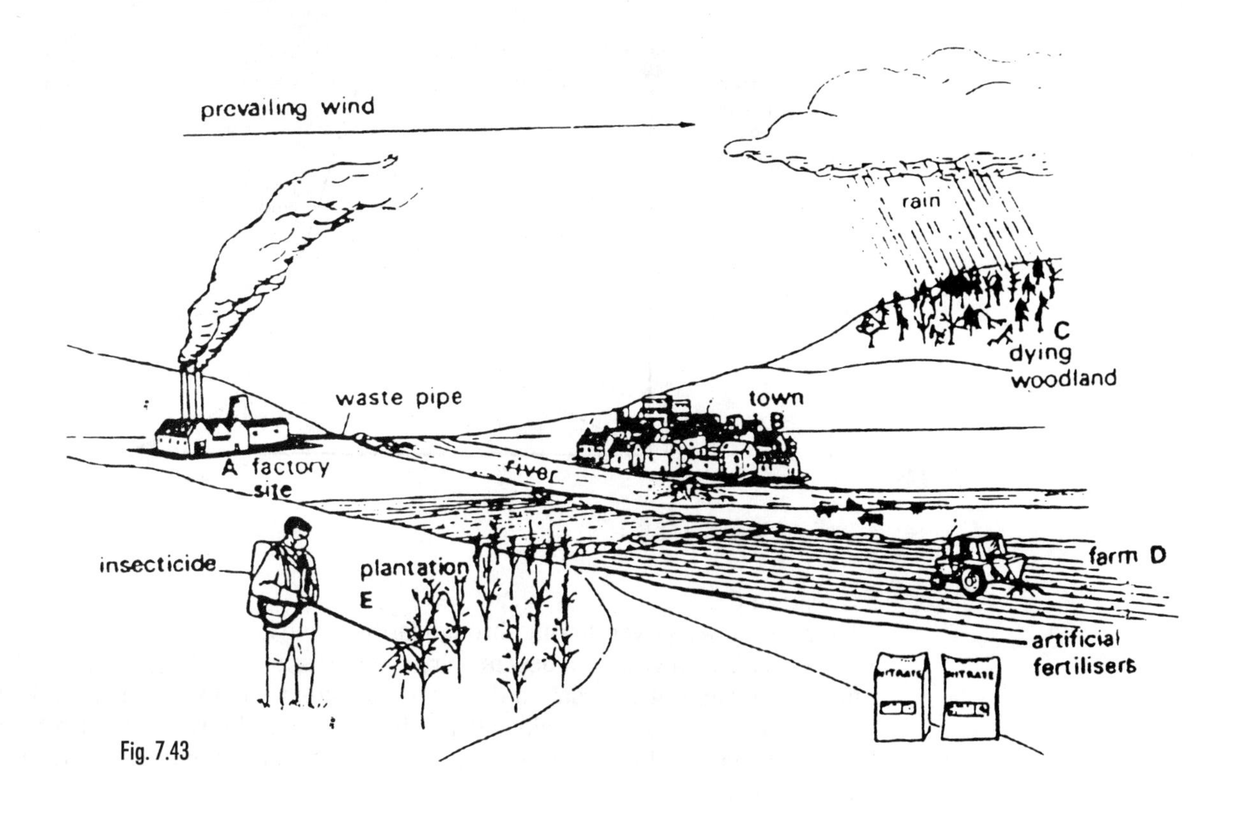

Fig. 7.43

Question 2
A stickleback is a small, freshwater fish.
The graph (Fig. 7.44) shows a growth curve for a population of sticklebacks.

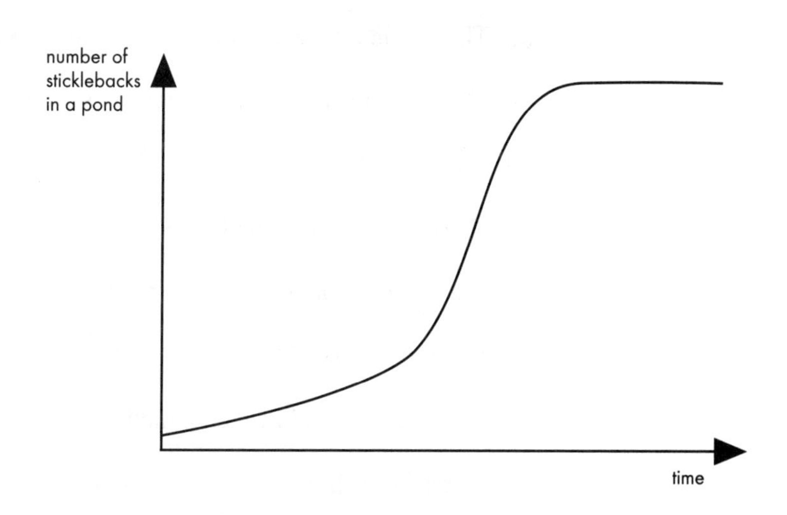

Fig. 7.44

(a) Give **two** limiting factors which could have caused the population growth of sticklebacks to slow down. *(2)*

Sometimes a growth curve for a population of sticklebacks may look like this (Fig. 7.45):

(b) Give **two** possible reasons for a population crash. *(2)*
(4 marks)

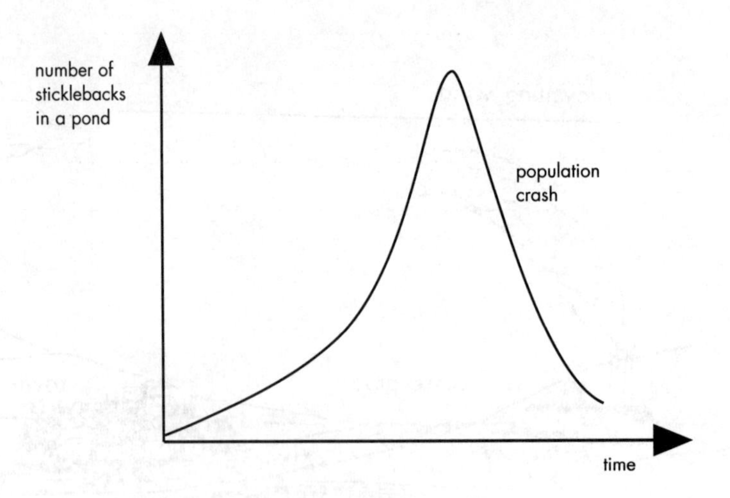

Fig. 7.45

Question 3

The map shows a river flowing through farms **Y** and **Z**.

Farmer Y changed from beef to dairy cattle. His herd no longer stayed all day in the fields, but came twice a day to the farm buildings to be milked. Farmer **Z** had a fish farm and kept his fish in tanks filled with water from the river. His fish began to die. Scientists analysed the river water at **A**, **B**, **C**, **D** and **E** along the river. Their results are shown in the table.

(a) (i) What *three* factors shown by the table could have caused the death of Farmer **Z**'s fish? *(1)*

(ii) Which factor do you think the most likely to have caused the death of Farmer **Z**'s fish? Why would this factor have caused the death of his fish? *(2)*

(b) Suggest *one* explanation each for:

(i) The rise in nitrogen compounds between **A** and **C**. *(2)*

(ii) The fall in nitrogen compounds between **C** and **D**. *(2)*

(iii) The low pH at **C**? *(2)*

(iv) The low oxygen content at **D**. *(2)*

(c) Farmer **Y**'s cows had polluted the river.

(i) What is pollution? *(1)*

(ii) Suggest *one* way in which pollution by the cows could be reduced. *(2)*

(iii) Instead of fish, Farmer **Z** wanted to grow watercress in his tanks, but the local health authority would not let him sell it. Why not? *(1)*

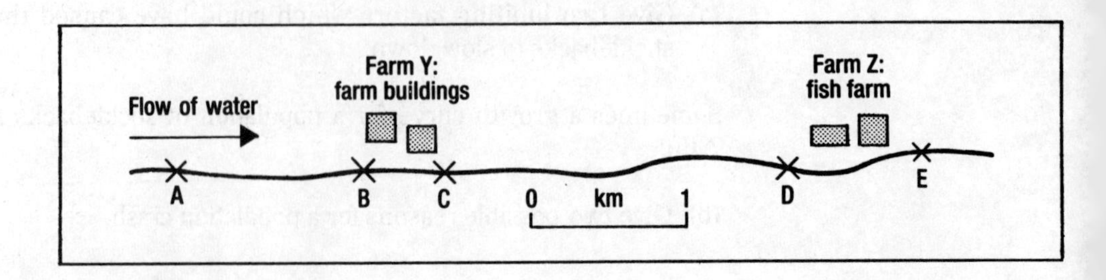

	River-water analysis		
	Total nitrogen in chemical compounds (parts per million)	pH	Dissolved oxygen (parts per million)
A	0.40	8.5	10.0
B	2.60	6.8	3.6
C	2300.00	4.0	10.0
D	0.76	7.8	1.2
E	0.66	7.8	4.0

Fig. 7.46

Total 15 marks (SEG)

Question 4

(a) A mature deciduous tree is cut down and then removed from a grassed area. What changes might occur in the ecosystem due to the tree's absence? (*17 marks*)

(b) Describe the effects of *two* other activities of humans which may harm and *two* activities which may improve the conditions of life for other organisms in the ecosystem. (*8 marks*)

Question 5

The diagram below shows the position of a sewage outflow pipe at a local beach.

The whole beach is covered in different types of seaweeds, growing on rocks and in rock pools. The main species of animals are snails, crabs, mussels, barnacles, limpets and fish. The animals live in the rock pools.

There is concern that the animals and plants are being affected by the sewage from the pipe.

You and a group of friends decide to investigate the situation by collecting some information about the different types of animals and plants.

(a) Describe how you would measure the size of the population of one of the types of animals found in the study area on the beach. *(4)*

(b) Describe how you would compare the seaweeds growing in the study area on this beach with those growing on a beach where there was no sewage pipe. *(4)*

(c) Suggest three factors, other than the presence of the sewage pipe, which could affect the types of plants and animals found on the two beaches. *(3)*

Fig. 7.47

Question 6

The table below shows a number of features of the population of six countries in 1980.

COUNTRY	POPULATION (Millions)	BIRTH RATE (per 1000)	DEATH RATE (per 1000)	INFANT MORTALITY RATE (per 1000)	LIFE EXPECTANCY (years)	POPULATION UNDER 15 years (%)	ESTIMATED POPULATION BY YEAR 2000 (millions)
Canada	24.1	16	6	16	74	30	32.6
France	54.1	16	9	12	73	25	62.9
Egypt	40.2	41	17	100	54	43	65.6
Angola	7.2	51	28	203	41	42	12.9
Turkey	42.3	40	14	119	57	44	73.0
Japan	115.4	20	7	11	75	26	134.7

Fig. 7.48

(a) The world mean life expectancy in 1980 was 61 years.
How many of these countries had a life expectancy greater than this?

________ *(1)*

(b) (i) Which country is estimated to have the largest increase in population by the year 2000?

________ *(1)*

(ii) Suggest **two** reasons why the population of this country is estimated to rise so much.

1. ____________________

2. ____________________ *(2)*

(c) Suggest **two** possible reasons for the high infant mortality rate in Angola.

1. ____________________

2. ____________________

____________________ *(2)*

Note: Answers to the exam questions can be found at the end of the book

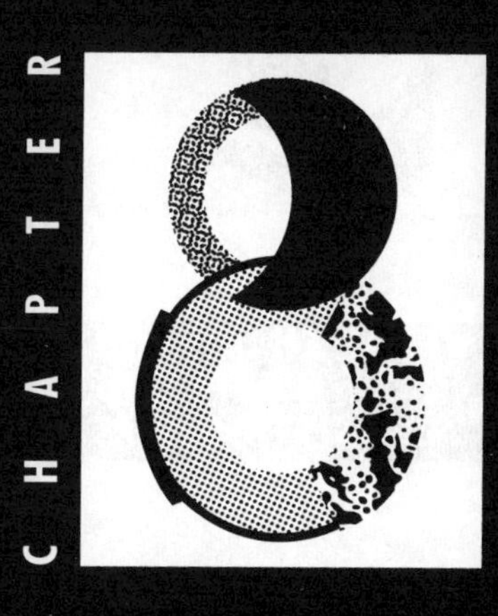

ENERGY TRANSFER AND CYCLING OF MATERIALS

FOOD CHAINS AND FEEDING RELATIONSHIPS

DECAY AND CYCLING

THE BIOLOGICAL COMMUNITY

AGRICULTURE AND FOOD PRODUCTION

CONSERVATION

STRAND (IV) ENERGY FLOWS AND CYCLES OF MATTER WITHIN ECOSYSTEMS

GETTING STARTED

There is some overlap between the content of this Chapter and Chapter 7. Human (biotic) impact on the environment and ecosystem is again considered, although in this chapter there is greater focus on non-living (abiotic) factors than in the previous chapter.

By the end of this chapter you should have met all the strand (iv) statements of the National Curriculum outlined below.

- Pupils should consider energy transfer through an ecosystem and how photosynthesis initiates this process.
- They should consider how food production involves the management of ecosystems to improve the efficiency of energy transfer, and how such management imposes a duty of care.
- They should explore cycling of the elements and biological materials in specific ecosystems, for example, *seas, farms and market gardens*, including the role of microbes and other living organisms in the cycling of carbon and nitrogen.
- They should relate their scientific knowledge to the impact of human activity on these cycles and ecosystems, and to the disposal of waste materials.

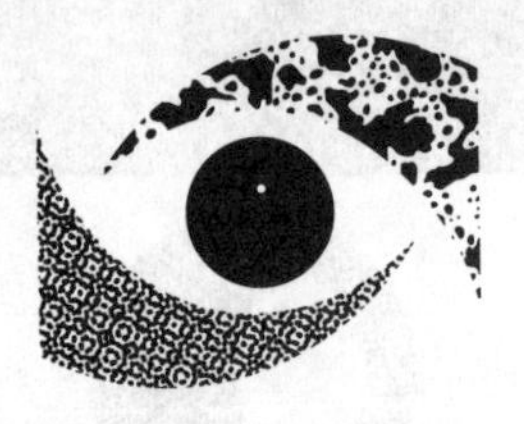

ESSENTIAL PRINCIPLES

This topic is fundamental in the study of ecology. Food chains (and food webs) consist of feeding relationships between different types of organism. When one organism is consumed by another, the nutrients it contains are 'passed on'. This allows nutrients to be recycled within living communities. Each type of organism is adapted to the way in which it obtains its food. When you revise this topic, it is a good idea to also revise the topics on survival in a natural habitat (4.4) and natural populations (7.3).

PRINCIPLES OF FOOD CHAINS AND WEBS

Key Points

- Food chains are the series of *feeding relationships* which exist between organisms in the environment. Each food chain 'begins' with a *producer*, which is always a green plant. The other organisms in the food chain are *consumers*, which are usually animals. Consumers eating producers (plants) directly are called *primary consumers*, and they are *herbivores*. Consumers eating other consumers are called *secondary consumers*, and they are *carnivores*.
- Food chains are usually linked to form *food webs*, since each type of organism normally has a feeding relationship with several other organisms.
- The energy passed along food chains comes from the sun.

1 Food chains

A food chain is a sequence of *feeding relationships* between organisms living within the same community. Feeding relationships involve the transfer of *energy* and *nutrients* (as food) from one organism to another, down the food chain. In natural communities, normally several food chains are interconnected to form *food webs*.
Different organisms obtain their food in different ways. In other words, they occupy a different 'feeding level', or *trophic level*, within the food chain. An example of a simple food chain is shown in Fig. 8.1, with four of the main trophic levels.

It is a good idea to be familiar with other examples of food chains and webs.

Producer		**Primary consumer**		**Secondary consumer**		**Tertiary consumer**
Microscopic algae (plant)	→	Tadpole (herbivore)	→	Water scorpion (carnivore)	→	Perch (carnivore)

Fig. 8.1 Example of a food chain. Arrows represent the direction of energy and nutrient flow

2 Food webs

Food webs arise when organisms consume, or are consumed by, two or more other organisms. In other words, food webs consist of two or more interacting food chains (Fig. 8.2; how many food chains can you spot in the diagram?).

In fact, it is unusual for a consumer to restrict its *diet* (i.e. food intake) to a single species. One reason for this is changes in the availability of food, for instance during different seasons. Another reason is that the full range of nutrients needed by an organism will not necessarily be available from a single source. Some animals (the **omnivores**) even obtain their food from more than one trophic (feeding) level.

A more complete (and also more complex) food web would include *decomposers* and *detritus feeders* (which feed on 'scraps'), since these play an essential role in the circulation of minerals within the environment. In this sense, the transfer of nutrients from one organism to another should really be shown as a *food cycle* because it is a continuing and repeating process for nutrients. However, energy does not re-circulate in this way; it *flows* through the system.

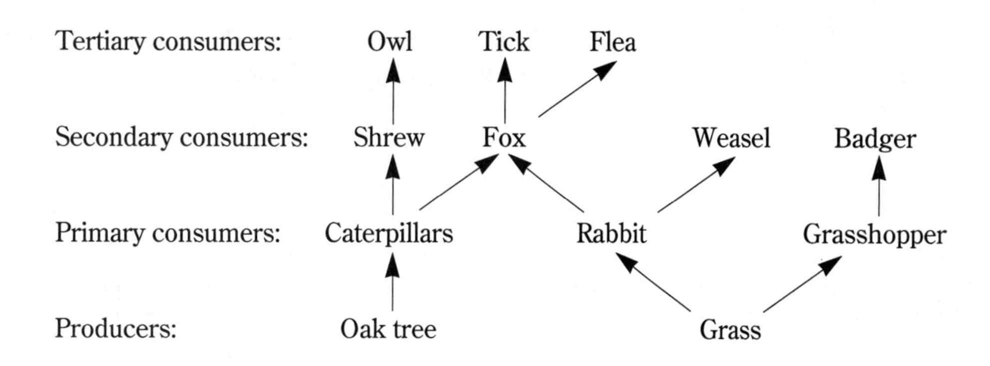

Fig. 8.2 Example of a food web

Links with other topics

Other related topics elsewhere in this book are as follows:

Topic	Chapter and Topic number	Page number
Nutrition	4.5	78
Photosynthesis	4.6	105
Predator-prey populations	7.3	296
Energy and food chains	8.3	327
Pyramids of numbers and biomass	8.3	328
Transfer of materials in the biological community	8.3	331

REVIEW QUESTIONS

Q1 Fig. 8.3 below shows a simple food chain.

(a) Select from the figure

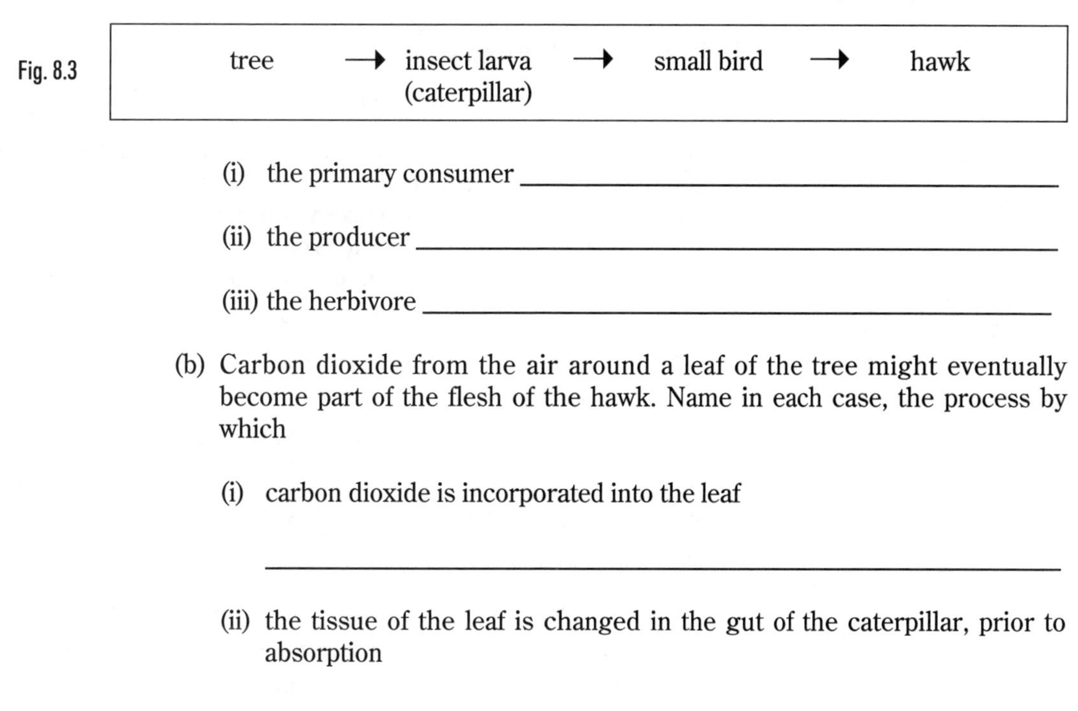

Fig. 8.3

(i) the primary consumer ______________________________

(ii) the producer ______________________________

(iii) the herbivore ______________________________

(b) Carbon dioxide from the air around a leaf of the tree might eventually become part of the flesh of the hawk. Name in each case, the process by which

(i) carbon dioxide is incorporated into the leaf

____

(ii) the tissue of the leaf is changed in the gut of the caterpillar, prior to absorption

____

(iii) carbon dioxide is eventually returned to the atmosphere from the body of the living hawk

____

(c) What else, besides chemical substances, is passed along such a food chain and where does it originate?

___

___

2 DECAY AND CYCLING

The materials which make up organisms are being constantly transferred between organisms, and also between organisms and their non-living environment. The processes which are involved in these transfers include *nutrition, excretion, egestion,* and *death and decay.*

The topics in this section should be revised with related topics on natural populations and the biological community.

All living things will eventually die, and useful chemicals will be released when they decay. In addition to this, living organisms produce organic waste products which can decay e.g. leaves fallen from plants, animal faeces, insect exoskeletons which are shed during growth.

The process of decay is important for two reasons:

- it causes the breakdown of waste products and the remains of dead organisms, so these do not 'pile up'
- it releases valuable substances into the environment to be re-used by other organisms.

Key Points

- All organic matter will eventually decay (break down).
- Decay involves *enzymes* within the organic matter, and living organisms: *scavengers, detritus feeders* and *saprophytes* (decomposers).
- Rate of decay depends on several factors eg. temperature, amount of air etc.
- Decay is very important in the *cycling of nutrients.*

1 Process of decay

There are 4 main stages in the process of decay:

1. **Autolysis**: enzymes inside the organism start to break down body tissues e.g. protease enzymes break down proteins to make amino acids.
2. **Scavengers and detritus feeders**: these are animals which feed on the remains of dead organisms or their waste products.

   Scavengers are fairly large animals which feed on animal remains. Each habitat will contain animals which feed in this way

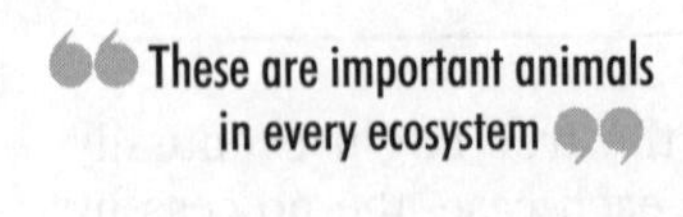

   e.g. woodland – raven, fox
   rocky shore – crab
   pond – water louse, flatworm

   Detritus feeders are usually smaller animals which will feed on small fragments from the dead organism (detritus). Detritus tends to collect at the lower levels of a habitat, so that is where detritus feeders are found

   e.g. woodland – beetles, earthworms (in the litter layer)
   rocky shore – barnacles, mussels (filter detritus from water)
   pond – gnat larvae (filter detritus from water)

3. **Fungi**: these will grow on the dead organism and produce enzymes to digest the cells and their contents. They absorb the nutrients from the cells into their hyphae – this is called *external digestion* (Fig. 8.4). Fungi are particularly important in the decay of plant material (including wood), because they produce cellulase enzymes.

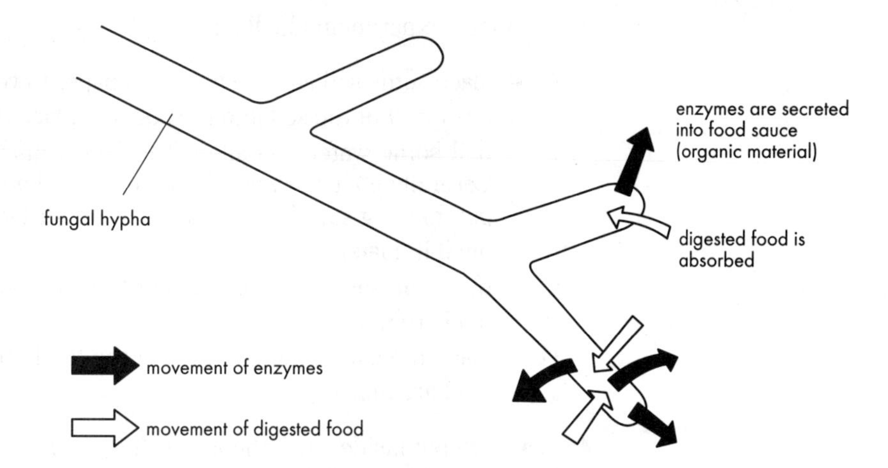

Fig. 8.4 External digestion in fungi

4. **Bacteria**: these microscopic, single-celled organisms complete the process of decay. They secrete enzymes which digest organic material, releasing nutrients and minerals – this is *external digestion*. They absorb the nutrients and use them to grow and reproduce. Eventually the whole organism will decompose.

Make sure you understand what saprophytes do.

Fungi and bacteria which feed in this way are called *SAPROPHYTES*, or decomposers. Saprophytes are an important part of any food chain, because they release the nutrients 'locked in' dead organisms, so they can be re-used. Large numbers of saprophytes are found in soil.

2 Key factors in the process of decay

Remains of organisms do not always decay at the same rate; in some conditions they can be quite well preserved for long periods of time.

The rate of decay is affected by:

1. **Temperature**: saprophytes work faster at warm temperatures than cool temperatures.
2. **Moisture**: saprophytes work faster when there is a suitable amount of water.
3. **Air**: most saprophytes use the oxygen in air to cause decay.

This information is useful if we want to prevent saprophytes from working e.g. to keep food fresh. Think about some of the ways that humans treat food to prevent it 'going off'.

Exam questions are often set on this topic.

e.g. freezing it – used for meat, fish, vegetables
drying it – used for milk, soup, fruits (raisins, currants)
vacuum-packing it – used for cold meats, cheese.

3 Investigating the process of decay

You could do an experiment to investigate the rate of decay in different conditions.

Apparatus: yogurt pots or similar containers (4)
soil (contains saprophytes)
pieces of food e.g. apple, orange, bread
polythene bags (must be transparent)
labels

1. Choose one type of food to test, and cut it into 4 even-sized pieces.

The control is a very important part of the experiment - make sure you know why you have 4 pots.

2. Label the pots like this:

A – no water C – cold conditions
B – no air D – control

3. Set up the experiment like this:

   A – place some soil and the food on a paper towel over a radiator and leave it until it is dry. Put the soil into the pot, and place the food on top.
   B – boil some water and allow it to cool (this removes all the oxygen from it). Carefully pour the water into the pot and add some soil. Put the food into the pot, making sure it is below the water surface (weight it down with a drawing pin if it floats).
   C – place the soil in the pot and put the food on top. Make sure that the soil and food are moist.
   D – place the soil in the pot and put the food on top. Make sure that the soil and food are moist.

4. Seal each pot inside a polythene bag (Fig. 8.5).
5. Put each pot in a suitable place

   e.g. A, B, D – warm place e.g. windowsill, near heater
   C – cold place e.g. fridge.

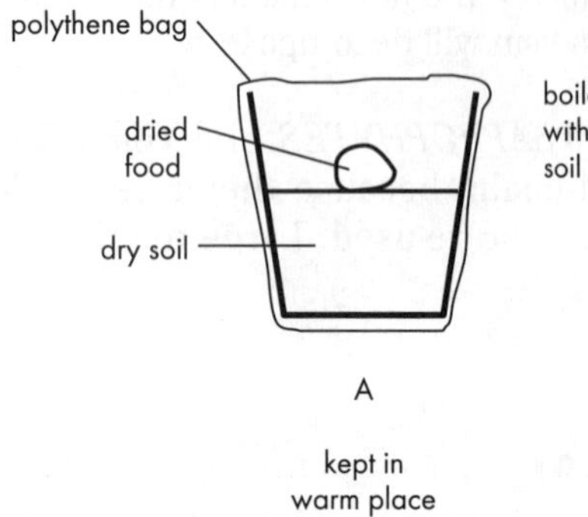

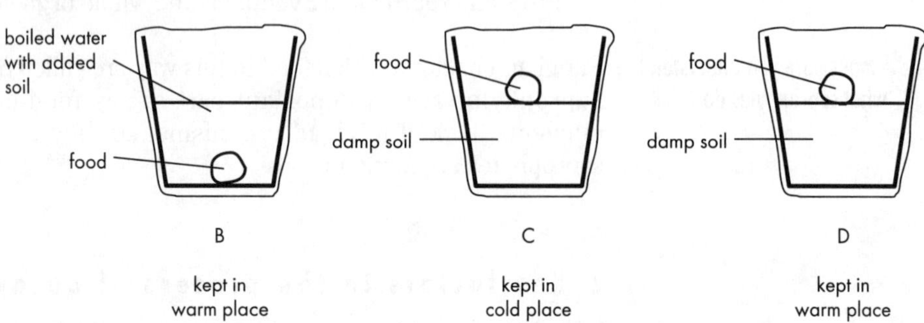

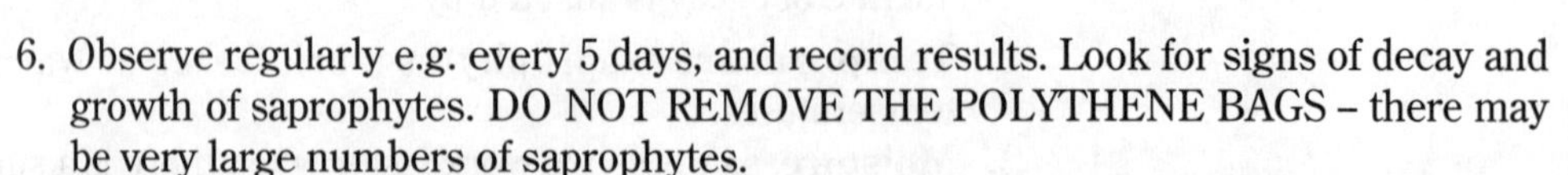

Fig. 8.5 Experiment to investigate rate of decay

6. Observe regularly e.g. every 5 days, and record results. Look for signs of decay and growth of saprophytes. DO NOT REMOVE THE POLYTHENE BAGS – there may be very large numbers of saprophytes.
7. At the end of the experiment e.g. after 20–30 days you must dispose of the pots carefully. Do not open the bags. It is best to burn them, or put them into bin-bags which should be sealed.

REVIEW QUESTIONS

Q2 Susan wanted to find out more about the decay process in soil. She placed 30 leaf discs in two mesh bags and buried them 8 cm below the soil surface.

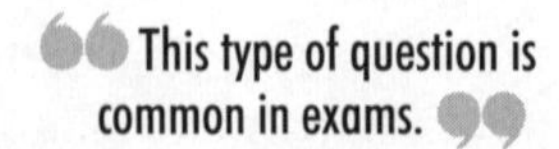

The bags were examined at intervals. The amount of leaf disc lost was recorded. Susan's results are shown below.

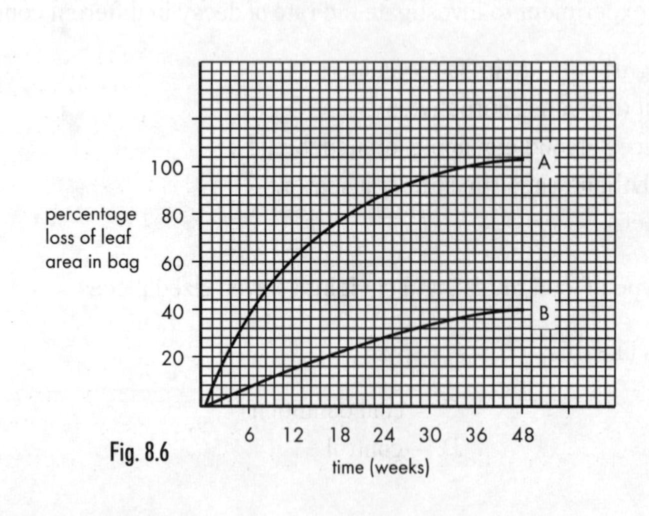

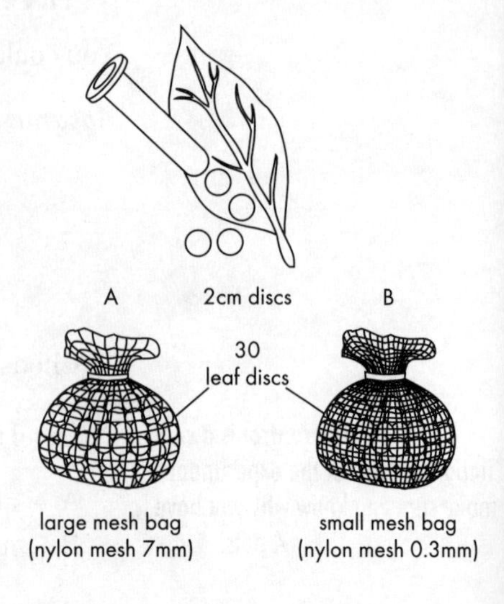

Fig. 8.6

(i) In which bag did the leaves break down at the faster rate?

____

(ii) Name **two** groups of organisms which may have caused the disappearance of some of the material from bag **A**, but not from bag **B**.

1. ____

2. ____

(iii) Name **one** other type of organism which could have caused the decay process in bag **B**.

____

(iv) If Susan buries the bags at a depth of 1 m, the decay is slower. State **two** reasons why.

1. ____

2. ____

DECAY AND THE CYCLING OF MATTER

Key Points

Level 6

- The circulation of matter in the natural environment is vital in making materials available to organisms. The circulation occurs as a result of processes including *nutrition, exceretion, egestion* and *decomposition.*
- Two of the most important substances to be circulated in natural ecosystems are *carbon* and *nitrogen*. The circulation of both substances involves important *decomposer organisms*, such as bacteria and fungi.

The circulation of minerals provides an important link between the abiotic and biotic environment. Minerals from the environment are required directly by plants, and indirectly by all other organisms. Minerals are returned to the environment from organisms by *excretion, egestion* and by *death and decay* of tissues. The circulation of mineral elements, called 'biogeochemical cycling', involves chemical, physical and biological processes. Examples include *carbon, nitrogen* and *water* cycles.

The carbon cycle

Carbon is a component of all organic molecules and is essential for all life. The carbon cycle mainly involves the conversion of the inorganic molecule carbon dioxide to or from organic molecules which are formed within the tissues of organisms. The carbon cycle is summarised in Fig. 8.7.

Refer to the topic on global warming also.

The concentration of carbon dioxide in air is fairly constant; however, there is a possibility of levels gradually increasing because of the increased burning of fossil fuels or because vegetation is being reduced, for instance by the clearing of tropical rain forests, which decreases *carbon fixation* by photosynthesis. An increase in atmospheric carbon dioxide would 'trap' retransmitted solar energy which might otherwise be lost into space. This is called the *'greenhouse effect'* and would raise the earth's average surface temperature.

The nitrogen cycle

Nitrogen is essential for the formation of proteins. Nitrogen gas represents a very high proportion (78 per cent) of the atmosphere but nitrogen cannot be absorbed directly by most organisms. The nitrogen cycle involves the conversion of nitrogen gas by various biological, as well as chemical and physical, processes (Fig. 8.8).

Make a particular effort to learn and understand the carbon and nitrogen cycles. You may be asked to identify particular processes occurring within them.

Carbon dioxide in atmosphere
Autotrophic nutrition (photosynthesis)
Organic (carbon-containing) compounds in plants
Respiration
Heterotrophic nutrition
Respiration
Organic (carbon-containing) compounds in microbes
Death and decay
Death and decay
Respiration
Organic (carbon-containing) compounds in animals
Combustion
Fossil fuels
Fossilisation

Fig. 8.7 The carbon cycle

The *biological process* involves *nitrogen fixation*, either by free-living microbes or by microbes living in a *symbiotic* relationship with certain plants, especially *legumes*. Legumes, for example pea and clover, develop *root nodules* within which *nitrogen-fixing bacteria* live. The resulting *ammonia* and *nitrate* can be absorbed by plants which use them to make amino acids. The organic form of nitrogen is then available directly or indirectly for animals.

Nitrogen also enters the abiotic environment by the *decomposition* of organic material by fungi and bacteria in soil. Decomposition results in the release of ammonia (*ammonification*) which is converted to nitrate (*nitrification*) by bacteria. Nitrate may then be absorbed by plants. Some nitrate is lost by soil leaching or by bacteria which convert it back to nitrogen gas (*denitrification*).

To summarise; there are 4 types of micro-organisms involved in the nitrogen cycle:

You must know about these microbes, and what they do, to achieve level 8.

- nitrogen fixing bacteria – change nitrogen into nitrate
- nitrate/nitrate bacteria – change ammonia into nitrate
- saprophytes – break down dead organic material to ammonia
- denitrifying bacteria – change nitrate into nitrogen gas

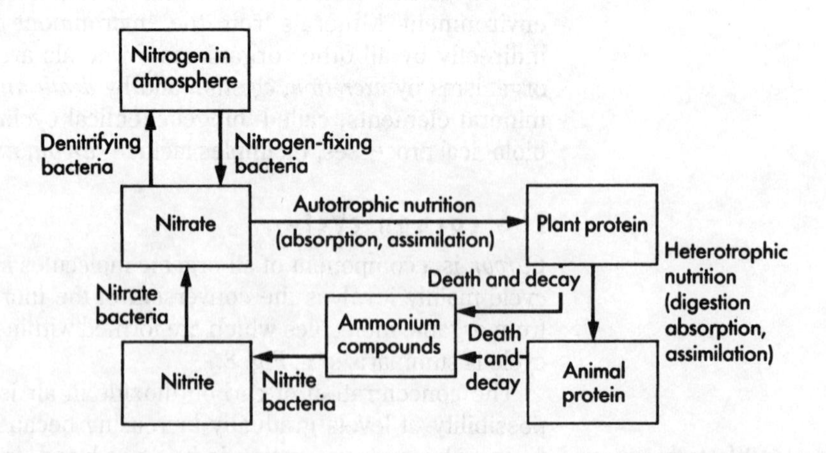

Fig. 8.8 The nitrogen cycle

Chemical and physical processes which converts nitrogen gas into nitrate include the action of *lightning* and the production of nitrate fertiliser. The amount of nitrate added to the soil as fertiliser is about the same as that added by nitrogen fixation. The production of nitrate fertiliser is energy-intensive in production and application. Fertilisers dramatically increase crop yield, and this is important as a means of providing enough food for a growing population (see topic group 7.3). However, increased growth rates of crops remove more minerals from the soil and these may not all be replaced. Excess nitrate may be washed off agricultural land and result in an over-growth of aquatic plants (eutrophication).

THE ROLE OF LIVING THINGS IN DECAY AND THE CYCLING OF NUTRIENTS

Decay

There are 4 types of organism involved in the process of decay:

1. **Scavengers** – these are usually quite large animals which feed on animal remains. They will tear off pieces which they digest internally. Examples of scavengers include ravens and foxes (woodland habitat), crab (rocky shore habitat), water lice (pond habitat).
2. **Detritus feeders** – these are usually smaller animals which feed on fragments of organic material (detritus). They may filter this from water, or extract it from soil, and digest it internally. Examples of detritus feeders include beetles and earthworms (woodland habitat), barnacles and mussels (rocky shore habitat) and gnat larvae (pond habitat).
3. **Fungi** – these grow over the surface of the organic material and secrete enzymes which begin to digest it (this is external digestion). Once digested, the nutrients can be absorbed into the hyphae. Examples of fungi which feed in this way include *Mucor* (pin mould).
4. **Bacteria** – these grow on the remaining organic material and secrete enzymes to digest it externally. This will continue until the organism is fully decomposed. *Pseudomonas* is a bacterium which feeds in this way.

Scavengers and detritus feeders are important because they speed up the rate of decay, by breaking the organic material into small pieces i.e. increasing the surface area. If they are not present, decay will still occur, but it will take much longer.

Bacteria and fungi (saprophytes) are important because they release nutrients from dead organisms or their waste.

Links with other topics

Other related topics elsewhere in this book are as follows:

Topic	Chapter and Topic number	Page number
Air pollution	7.2	291
Water pollution	7.2	294
Global warming	7.4	305
Transfer of materials in the biological community	8.3	331

REVIEW QUESTIONS

Q3 Fig. 8.9(a) shown below, represents the nitrogen cycle.

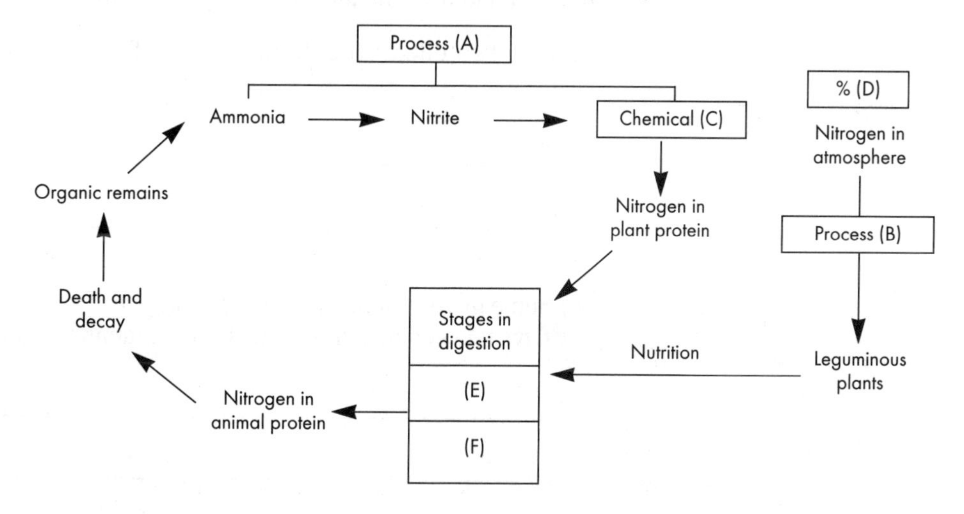

Fig. 8.9 (a)

Give, in the spaces below, the information shown by **A–F** in Fig 8.9 (a).

Name the process (**A**) ______________________

Name the process (**B**) ______________________

Name chemical (**C**) ______________________

State the percentage of nitrogen in the atmosphere (**D**) ______________________

Name an intermediate product of protein digestion (**E**) ______________________

Name the end product of protein digestion (**F**) ______________________

Q4 The diagram (Fig. 8.9 (b)) is one way of showing the carbon cycle.

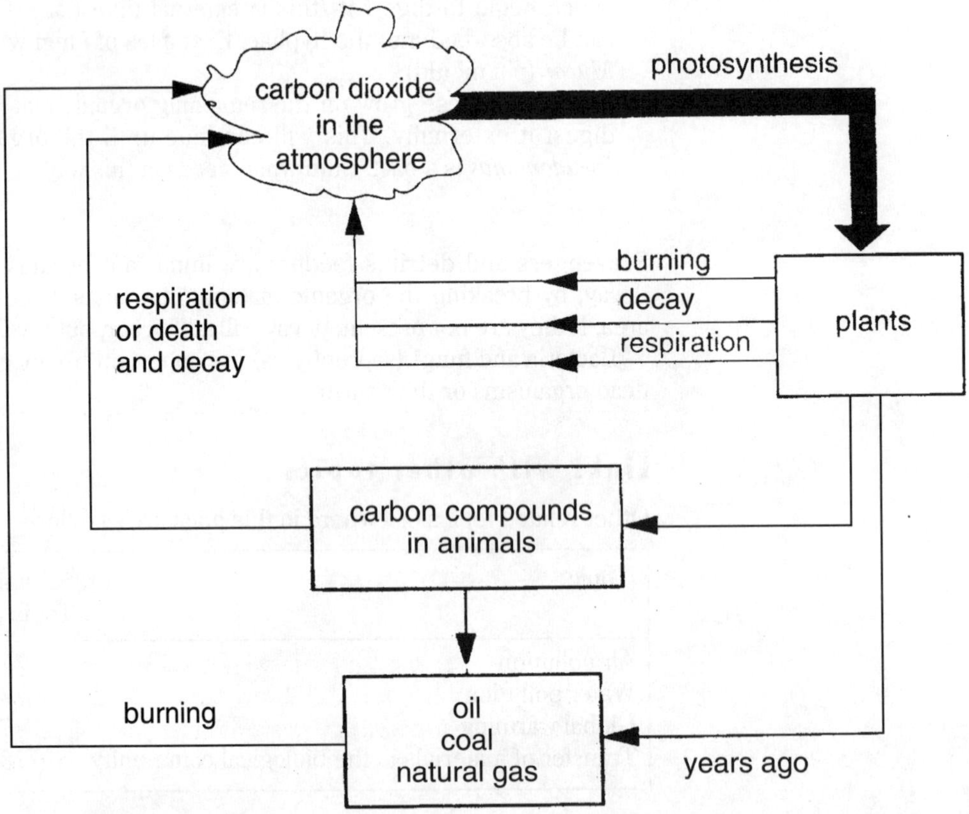

carbon dioxide in the atmosphere
photosynthesis
plants
burning
decay
respiration
respiration or death and decay
carbon compounds in animals
oil coal natural gas
burning
years ago

Fig. 8.9 (b)

The increase in world population has led to an increase in the demand for food. This had led to more land being cleared for crops.

It is estimated that 30,000 km^2 of rain forest is burnt or cut down every year.

(a) Give two effects that burning down the rain forest will have on the carbon cycle.

(i) ______________________

(ii) ______________________

(b) Some governments are trying to stop the rain forests being destroyed. Give two ecological reasons for maintaining rain forests.

(i) ______________________

(ii) ______________________

3 THE BIOLOGICAL COMMUNITY

This Topic Group describes the structure of typical biological communities, and the feeding relationships which exist between organisms. Feeding relationships involve the transfer of both nutrients and energy between organisms. In general, nutrients are *re-cycled* continually within biological communities, and energy *flows* through them. Energy is not efficiently transferred between organisms by feeding, since only energy which is 'locked up' in tissues is transferred. This represents a small proportion (about 10 per cent) of the original energy obtained by each organism during nutrition.

This part of the syllabus deserves careful revision, partly because it presents difficulties for some students. Another reason is that the full understanding of several other subject areas is dependent on a familiarity with concepts of food chains and the transfer of nutrients and energy. Check the sections in the 'Links with Other Topics' tables to see which other parts of the syllabus you may need to revise at this stage.

ECOSYSTEMS AND BIOLOGICAL COMMUNITIES

Key Points

- *Ecosystems* consist of both non-living (*abiotic*) and living (*biotic*) components. Organisms exist within ecosystems in *communities*. Organisms interact with each other in various ways, including *feeding relationships* (mostly between species) and *reproduction* (mostly within species).
- The particular 'role' an organism has within the ecosystem is known as its *niche*. Organisms can co-exist within the same biological community because they have different niches; i.e. they have different requirements for resources.
- The number of species present within an ecosystem depends on many factors, including how 'old' the community is; this determines how many different species have had a chance to enter the ecosystem, during *succession*.

1 Ecosystems

An *ecosystem* is an 'ecological unit' which has characteristic features, determined by *abiotic* (non-living) and *biotic* (living factors). Ecosystems often have fairly obvious limits which separate them from neighbouring ecosystems. Examples of ecosystems include grassland, woodland, ponds, sand dunes, heathland, rocky shorelines and compost heaps. Ecosystems show variations in size. However, generally they are fairly convenient to study because they are often associated with particular physical conditions and characteristic types of organisms.

“Make sure you understand these 'ecological' terms”

Organisms exist on a global scale within a continuous *biosphere*, consisting of *terrestrial* (land) and *aquatic* (water) environments, together with the atmosphere. However, organisms tend to be *adapted* to particular ecosystems within the biosphere, and therefore relatively little movement occurs from one type of ecosystem to another.

There are two main components of all ecosystems:

- **Habitat.** The abiotic (physical) environment within which organisms live; that is, their 'address'.
- **Community.** The collection of organisms within a *particular ecosystem*. The collection of organisms within a *particular species* is called a *population*.

Organisms are dependent on each other in various ways for survival. Examples of interactions between organisms include *feeding relationships* (mostly between species) and *sexual reproduction* (mostly within species). These interactions often involve *competition* between organisms living in the same community.

2 The niche

“If 2 organisms try to occupy the same niche, they will compete, and one may not survive.”

Organisms are adapted through evolution to perform a particular range of activities within a certain type of ecosystem. This *niche* or 'occupation' of organisms means that, in general, they can exist in the same community with other organisms having a different niche. This is because organisms from different niches do not require exactly the same range of resources from the environment (Fig. 8.10).

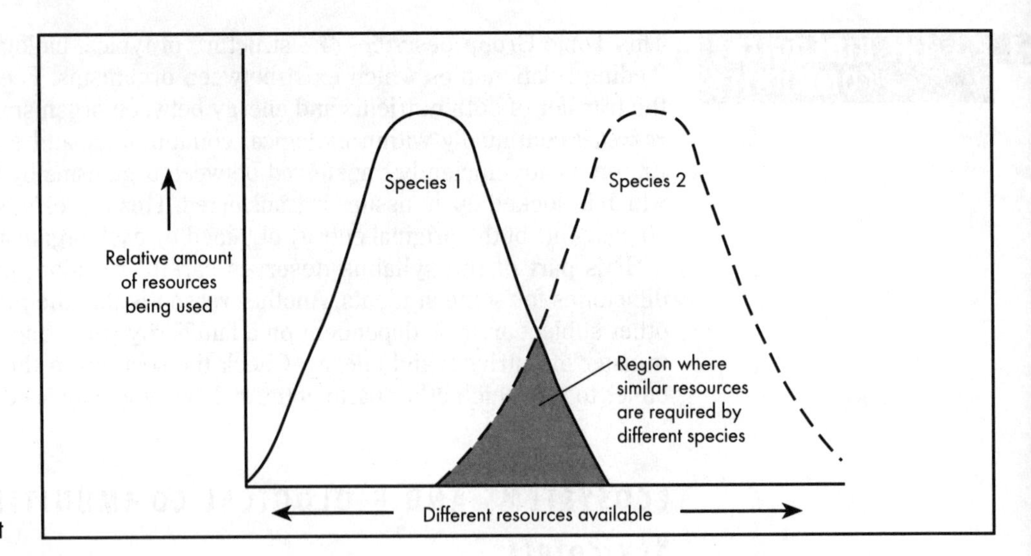

Fig. 8.10 The niche concept

3 Succession

Refer to the section on agriculture (7.4) which describes how succession can be stopped.

Communities become established within a habitat by a gradual process called *succession*. Succession occurs when different species enter and occupy a potential habitat; many organisms have a means of *dispersal* which allows them to move into a new habitat. For example, newly cleared land tends to be *colonised* by certain plants, e.g. rosebay willowherb, which have very good powers of dispersal, and whose seeds can rapidly germinate on newly available land. These plants stabilise the soil, release nutrients through nutrient cycling and provide food and shelter for small animals. Each organism in some way modifies its environment, resulting in a gradually changing habitat; species which are adapted to the changed conditions can then occupy vacant niches.

Succession results in a gradual increase in the numbers of organisms and also in the number of different species; this adds to the *diversity* of the ecosystem. A *maximum* diversity is reached when all available niches are occupied. This is called a *climax community* (Fig. 8.11) and is likely to be quite a stable ecosystem.

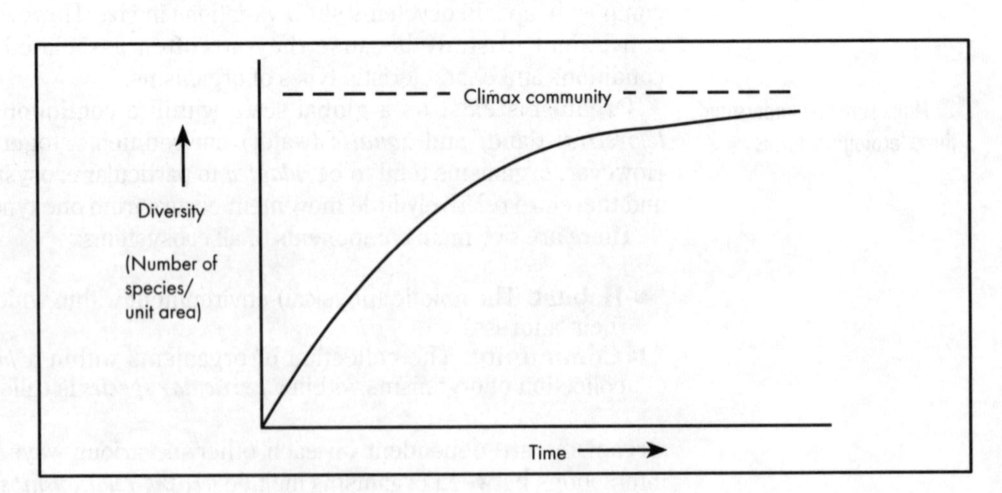

Fig. 8.11 The effect of succession on species diversity

For *terrestrial* (land) ecosystems, the climax community is often a forest or woodland. This sort of ecosystem can often support more species (several thousands) than any other. The climax community can only develop if abiotic and biotic conditions are suitable. For example, lack of sufficient water (an abiotic condition) is an important factor in preventing a desert gradually becoming a rainforest. Grazing (a biotic condition) may prevent grassland eventually becoming a deciduous woodland.

Links with other topics

Other related topics elsewhere in this book are as follows:

Topic	Chapter and Topic number	Page number
Competition and survival	4.4	43
Predator-prey populations	7.3	296
Principles of food chains and webs	8.1	316
Food chains and ecological pyramids	8.3	327

Note: Complete the Review Questions at the end of this Topic Group to test your understanding of ecosystems and communities.

FOOD CHAINS AND ECOLOGICAL PYRAMIDS

Key Points

Level 7

- *Food chains* consist of different 'feeding levels' (*trophic levels*). The main feeding levels are *producers* (green plants), *primary consumers* (herbivores), *secondary consumers* (carnivores) and *tertiary consumers* ('top' carnivores). Organisms tend to be specialised for the feeding level that they occupy. When particular organisms feed on, or are fed on, by more than one organism, *food webs* are established. Food webs consist of two or more connected food chains.
- *Nutrients* and *energy* are transferred through food chains. However energy is 'lost' at each feeding level, and is not available for organisms in the next feeding level. This means that a reduced number or biomass ('living mass') of organisms can be supported in each feeding level 'down' the food chain. This can be shown in 'ecological pyramids', such as *pyramids of numbers* and *pyramids of biomass.*

1 Food chains

Feeding relationships incorporate a transfer of *energy* and *nutrients* from one organism to another through a *food chain*. Each organism occupies a particular *trophic (feeding) level* within a food chain. The main trophic levels are:

■ Producers

Producers are green plants. These can produce their own nutrients using the sun's energy, by photosynthesis. Photosynthesis basically involves the conversion, by the green pigment chlorophyll, of light energy from the sun into chemical energy (see Section 4.6). All organisms depend, directly or indirectly, on this process by which solar energy is made available. The sun is the ultimate source of energy for all food chains.

■ Primary (first level) consumers

Primary consumers are *herbivores.* These can eat plants directly. Plants are not a particularly concentrated source of protein and they contain a large proportion of cellulose which is difficult to digest. For these reasons herbivores may spend a relatively large proportion of their time eating.

■ Secondary (second level) consumers

You should be familiar with more than one example of a food chain, and know to which trophic level each organism belongs within the chain.

Secondary consumers are *carnivores which mostly eat hervicores.* Meat is a fairly concentrated source of protein, so carnivores may eat fairly infrequently. (Other adaptations to carnivorous feeding are described in Section 4.5, page 84).

■ Tertiary (third level) consumers

Tertiary consumers are *carnivores* (sometimes called '*top carnivores*') *which mostly eat other carnivores.*

An additional *quarternary* (fourth level) consumer group may be present in some food chains.

"Into which trophic level(s) would you put an insect-eating plant?"

Omnivores consume a mixed diet of plants and animals and so are both primary and secondary consumers; in other words, they occupy more than one trophic level. *Parasites* may be primary, secondary or tertiary consumers, depending on the sort of host that they derive their nutrition from.

Fig. 8.12 Example of a food chain (arrows represent the direction of energy and nutrient flow)

Producer		**Primary consumer**		**Secondary consumer**		**Tertiary consumer**
Oak (leaves)	→	Caterpillar	→	Shrew	→	Owl

Organisms from different trophic levels are linked together in food chains (Fig. 8.12). When one organism is consumed by another, there is a transfer of energy; energy *'flows'* down a food chain. The only energy which can be transferred from an organism is chemical energy contained within the organism's tissues. This is always a small proportion (perhaps 10 per cent) of the energy originally consumed by the organism. Energy is lost from organisms by processes such as respiration, excretion, egestion and heat loss. This energy loss occurs at each trophic level (Fig. 8.13) and is the reason why food chains do not usually contain more than five trophic levels.

"This is a very important point. Energy is 'lost' at each stage of a food chain, so they are usually 4 or 5 stages long. Energy is lost when organisms move, give out heat, etc."

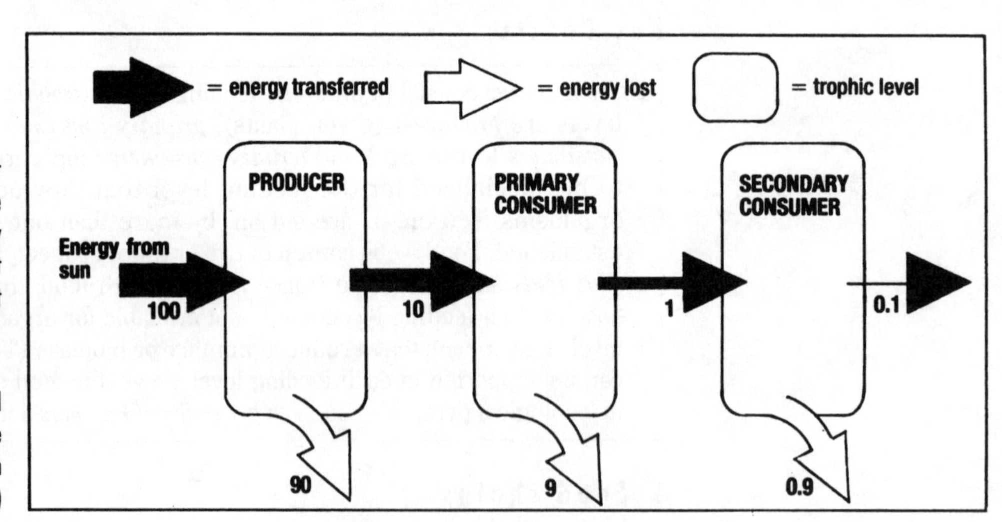

Fig. 8.13 Energy flow in a food chain (The figures indicate the relative amounts of energy at each stage in the food chain)

2 Ecological pyramids

The loss of energy from a food chain means that there is progressively *less energy* available for organisms further 'down' the food chain. For this reason, there is often a *decrease* in numbers of organisms at each successive trophic level. This can be shown in a *pyramid of numbers* (Fig. 8.14), which shows the relative number of organisms in each trophic level. There is often an increase in *size* of organisms in each successive trophic level. However, pyramids of numbers can be misleading in some cases; for instance, where one relatively large organism supports many other organisms. For example, one oak tree is likely to provide food for many thousands of caterpillars; this sort of situation causes a 'partial inversion' of the pyramid of numbers (Fig. 8.15 a).

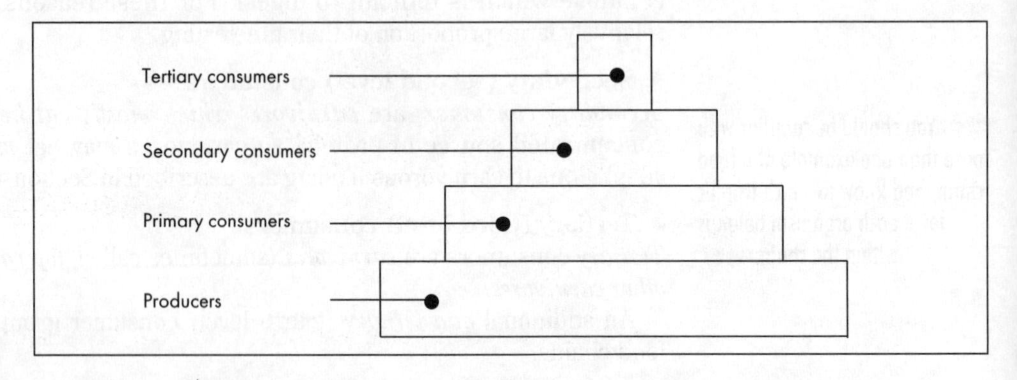

Fig. 8.14 Pyramid of numbers

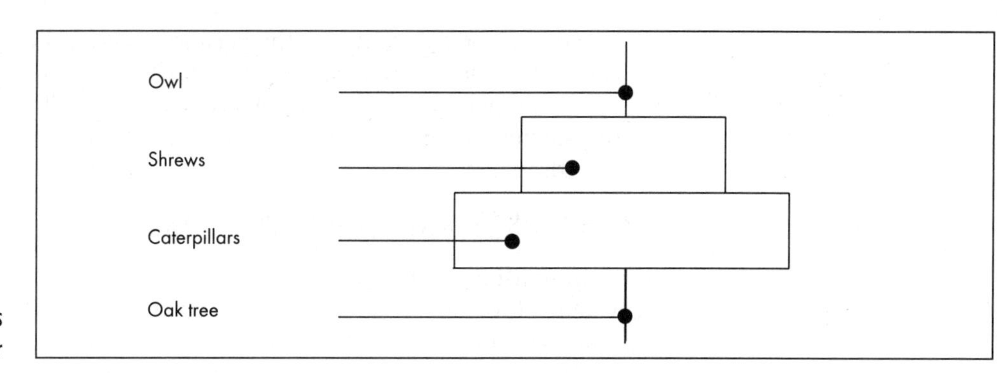

Fig. 8.15 (a) Pyramid of numbers involving one large producer

Also, if parasites are feeding on an animal, e.g. fleas on an owl, this can alter the classic pyramid shape (Fig. 8.15 b).

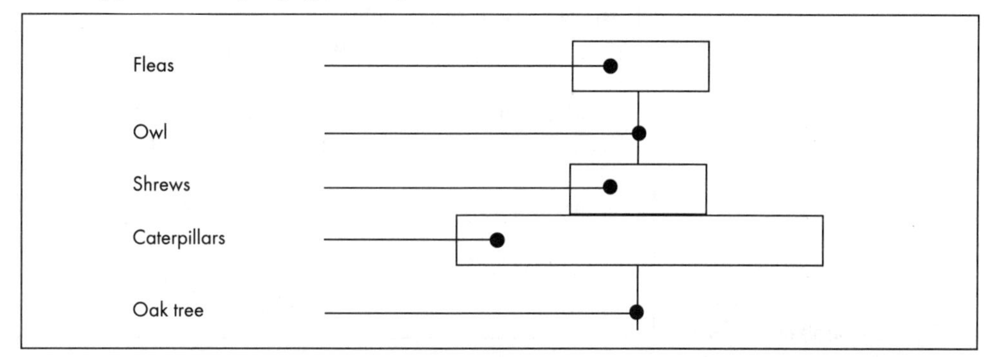

Fig. 8.15 (b) Pyramid of numbers including parasites

The *biomass*, or total mass of organisms in a population, decreases in a progressive way 'down' the food chain. This can be shown in a *pyramid of biomass* (Fig. 8.16); this gives a rather clearer idea than the pyramid of numbers of the 'relative amount' of living matter supported at each trophic level within a community. For instance, there are likely to be more herbivores than carnivores within a particular area. The numbers of carnivores can only increase if there are sufficient herbivores for them to feed on.

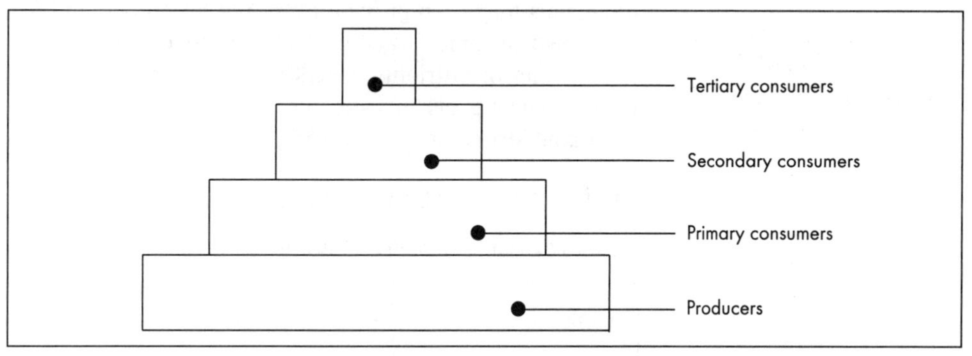

Fig. 8.16 Pyramid of biomass

Similarly, herbivore populations depend on the amount of vegetation available to feed on. This applies to humans, too. It has been estimated that the same area of land can provide about ten times more plant protein than animal protein (Fig. 8.17); many herbivores, including beef cattle, are rather inefficient at converting plant protein in their diet into their own animal protein. This is one reason why some humans consume a vegetarian (non-meat) diet, although humans are adapted as omnivores. As the human population continues to rise, more food will need to be made available by increased production, and possibly by a reduction in meat consumption; in other words by shortening the food chain.

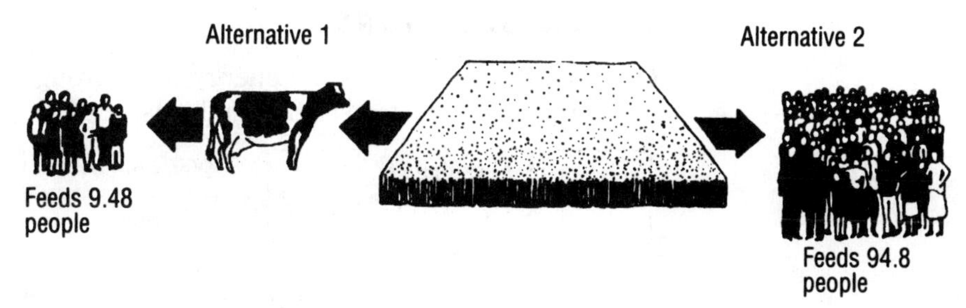

Fig. 8.17 Diagram showing two alternative ways in which the corn produced by one hectare of land could be used (AEB, 1985)

The original examination question to which Fig. 8.17 related asked:

(a) Name *two* food substances in the corn which could be used by man.
(b) alternative 2 supports more people than alternative 1.
  (i) how much more efficient is alternative 2 than alternative 1?
  (ii) suggest *one* reason why alternative 1 represents a less efficient use of the cornfields.

3 Food webs and nutrient cycles

Many individual food chains can be identified within the communities in an ecosystem. However, most consumers obtain their food from more than one type of organism, and many (such as the omnivores) obtain their food from more than one trophic level. For this reason, feeding relationships within communities can be quite complex, consisting of many interlinking food chains, making *food webs* (Fig. 8.18).

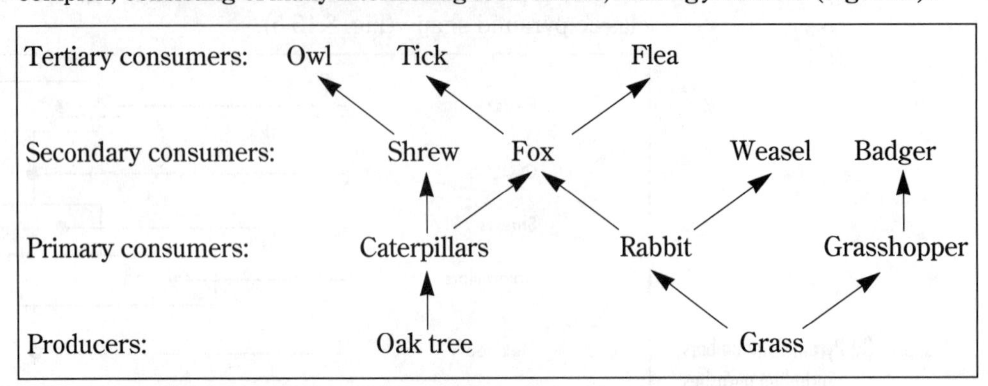

Fig. 8.18 Example of a food web

Energy passes from the sun and through food chains and webs in one direction (shown by the arrows in Fig. 8.1, Fig. 8.2 and Fig. 8.13). Nutrients, unlike energy, can be *cycled* between organisms as excreta (urine) and egesta (faeces), and when organisms die and decay. Organic molecules are broken down by *decomposers* into inorganic molecules, such as carbon dioxide and various minerals. These are then available to green plants which can use them to form organic molecules. The organic molecules may then provide nutrients for consumers to use directly, or to re-form into their own organic molecules. Decomposers, which are *saprophytes,* are essential to this cycling of nutrients, resulting in ***nutrient cycles***. Examples of food cycles include those involving carbon and nitrogen; these elements alternate between organic and inorganic forms (see Section 8.2).

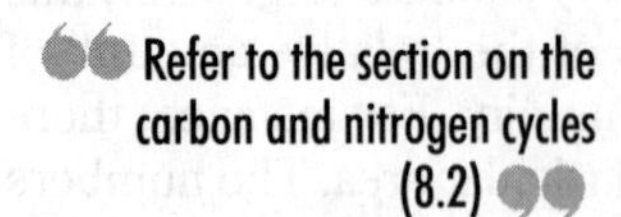

Links with other topics

Other related topics elsewhere in this book are as follows:

Topic	Chapter and Topic number	Page number
Competition for resources	4.4	43
Predator-prey populations	7.3	296
Populations and resources	7.3	298
Principles of food chains and webs	8.1	316
Decay and the cycling of matter	8.2	318

REVIEW QUESTIONS

Q5 Fig. 8.19 represents the energy flow through a food chain consisting of four organisms. The total energy in the tissue of the green plant is shown in the box.

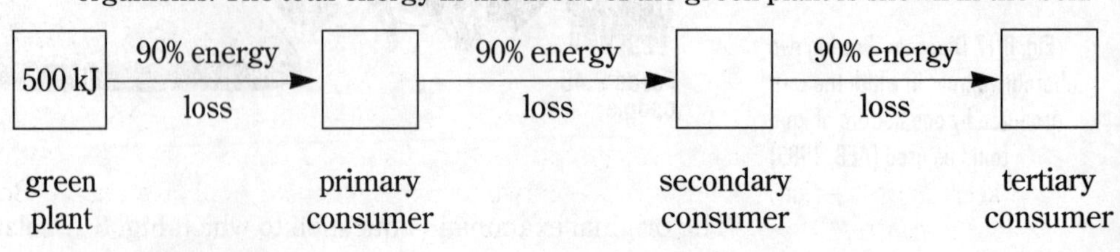

Fig. 8.19

(a) Complete the other boxes to show the total energy in the tissues of each of the consumers, assuming a 90% loss of energy at each of the stages in the food chain.

(b) State **one** process which results in a loss of energy between one step and the next.

____

(c) (i) What is meant by the term **producer** in the food chain?

____

(ii) State the source from which the producer obtains its energy.

____

(d) Explain why there are rarely more than four or five links (or stages) in a food chain.

____

____

(e) How is a food web different from a food chain?

____

(f) Green plants → caterpillars → small birds → eagle

From this example of a food chain, explain how the practice of spraying crops with insecticides can lead to birds of prey like the eagle being harmed.

____

____

____

TRANSFER OF ENERGY AND MATERIALS IN THE BIOLOGICAL COMMUNITY

Key Points

- Solar energy enters each food chain via producers, by *photosynthesis*. This converts the sun's energy into chemical energy, which is available for *consumers* in the food chain. Most energy is 'lost' within each feeding level; only a small proportion is incorporated into biomass that can be passed on to the next feeding level.
- The efficiency of energy transfer is particularly important at the start of each food chain. The efficiency with which producers convert solar energy to 'trapped' chemical energy is known as *productivity*. Productivity and the efficiency of energy transfer 'down' the food chain depends on several factors, including the type of *ecosystem*. Agricultural land, from which most humans derive their food, is intermediate in productivity.

1 Energy flow through the biological chain

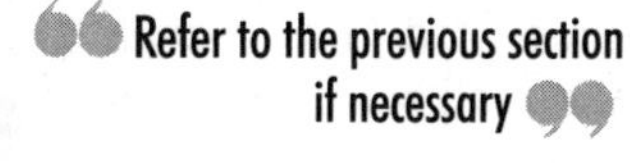

The idea of *food chains* and *food webs* has been described in the previous section. An important aspect of food chains and webs is *energy transfer*. It is important to realise that energy can be transformed from one form to another. Energy enters biological communities in two forms; *light energy* and *heat energy*. Both types of energy originally came from the *sun*. Light energy mostly enters food chains via *producers* (green plants). It gets converted to chemical energy by ***photosynthesis***. Heat energy enters all organisms directly (from their surroundings) and indirectly, through nutrients obtained via the food chain.

Less than half of the sun's energy penetrates the atmosphere and becomes available to organisms. Organisms only absorb between 5 to 10 per cent of the solar energy which reaches the earth's surface. Most of the energy that is eventually used by the biological community is intercepted by green plants. Leaves of plants are only able to use less than half of the solar energy that hits the leaf (Fig. 8.20). The rest cannot be used because it is the 'wrong' wavelength, or because it is reflected from the leaf, or transmitted through the leaf. Most of the rest is 'lost' in respiration and evaporation. Only about 5 per cent of the energy which was intercepted by the leaf is used in making new plant material. This 'trapped energy' is all that is available to the first consumer (a herbivore) in the food chain!

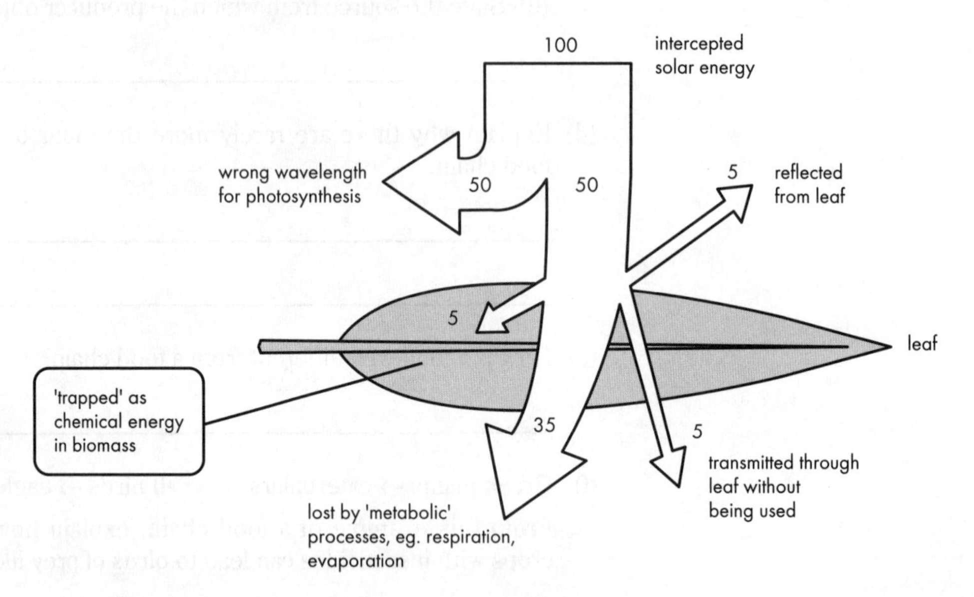

Fig. 8.20 The 'destination' of the sun's energy which is intercepted by a plant leaf. The figures are percentages of the original amount of energy

Energy is also 'lost' from consumers (Fig. 8.21). Cattle can only eat certain plants, and also only parts of plants, so not all the energy 'locked' inside the producer is actually consumed. Of the energy-containing material that is eaten, only about 4 per cent of the energy is 'trapped' in the herbivore's tissues. This is all that is available for the next consumer. A similar 'loss' of energy occurs at each feeding level of the food chain.

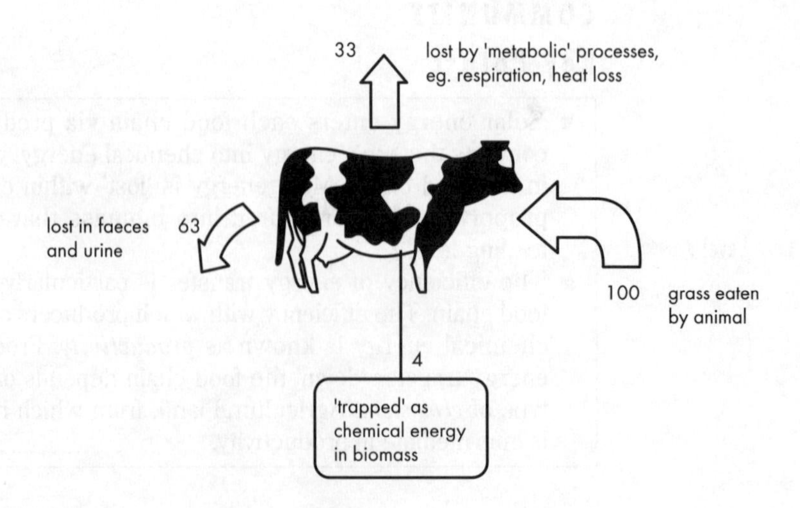

Fig. 8.21 The 'destination' of chemical energy from plant leaves eaten by an animal. The figures are percentages of the original amount of energy

The proportion of energy transferred between feeding levels is usually between 2 to 30 per cent, depending types of organisms and ecosystems. The average energy loss between feeding levels is about 10 per cent. This can be represented in a ***pyramid of energy*** similar in shape to the ***pyramid of biomass*** in the previous section.

2 Ecosystems and productivity

The efficiency of the process by which plants convert solar energy into 'trapped' forms (chemical energy and biomass) is known as *net primary productivity.* Ecosystems with the highest productivity are tropical forests, swamps, marshes and estuaries. The lowest productivity occurs in desert, Arctic tundra and open oceans. Agricultural land is intermediate in productivity, although output is usually increased by management techniques, such as by the use of fertilisers, herbicides and pesticides. The energy contained in the high-productivity ecosystems is important for the animals which live there. However, it is mostly not energy which can easily be used by humans.

In tropical forests, most of the nutrients are 'locked up' in the trees, and the soils are fairly nutrient-poor. Some tropical forests are being cut down to create land for grazing. However, the soils are unable to support high crop production over several years. Combined with the loss of energy in food chains described below, grazing cattle on land created by removal of tropical rain forests is not efficient. This is why an area of tropical rain forest in Central America about the size of a small kitchen (about 5 m^2) is lost to provide grazing to make one quarter-pounder hamburger!

Links with other topics

Other related topics elsewhere in this book are as follows:

Topic	Chapter and Topic number	Page number
Human populations	7.4	301
Principles of food chains and webs	8.1	316
Ecosystems and biological communities	8.3	325
Food chains and ecological pyramids	8.3	327

REVIEW QUESTIONS

Q6 (a) Describe how energy flows through a food chain.

____

____

____

____

(b) With reference to energy transfers, describe the advantages and disadvantages of being a herbivore (plant-eating animal)

____

____

(c) Using a simple example of each, explain how food chains differ from food webs.

____

____

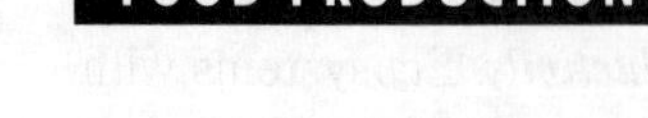

Production of food, either from crops or animals, is more efficient if conditions are carefully controlled. The main principles of this management are:

- to provide optimum conditions for growth e.g. temperature, nutrients
- to improve efficiency of energy transfer, so that more energy is used for growth and less is 'wasted'
- to start with 'high yield' organisms which may have been obtained by selective breeding or genetic engineering
- to eliminate diseases and predators which will reduce yield.

Level 10

As well as maximising the profit to be made, the farmer must consider the costs of the process and the responsibilities to the environment and to the organisms involved.

ANIMAL HUSBANDRY

Key Points

- animals are kept for the products they make e.g. milk, eggs, or their meat
- movement, food intake and temperature of their surroundings are carefully controlled
- farmers must dispose of waste products e.g. urine and faeces carefully.

Farm animals are kept for the *products* they make e.g. milk, eggs, or to be killed and used as *meat*. In both of these cases the animal is fed on a crop, and uses the energy in this food to make the product or to grow. This results in the following type of food chain:

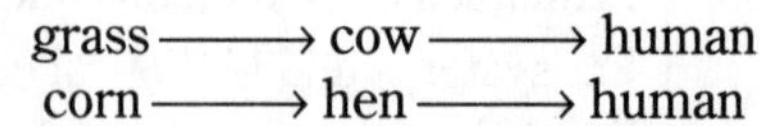

We have already seen that energy transfer through food chains is not very efficient, with only 10 per cent of the animal's food intake being used for growth. Around 90 per cent is 'wasted' on movement, keeping warm and various vital body processes.

1 Principles of animal husbandry

Farmers reduce this energy loss by keeping animals in carefully controlled conditions. In particular they:

restrict movement – animals may be kept in confined spaces so that they do not use energy unnecessarily.

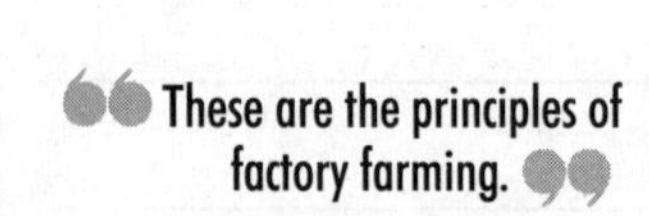

control food intake – animals are fed the optimum (ideal) amount of food for rapid growth, and have this food delivered to them, i.e. they do not have to move around to find it.

control temperature – birds and mammals are endothermic: this means that they maintain a constant body temperature. Many farm animals are kept in heated buildings so that they do not use energy to keep warm.

Farmers must also ensure that animals are not at risk from **predators** or **disease.**

If large numbers of animals are confined in warm buildings, disease can spread rapidly. Many farmers routinely give animals *antibiotics* with their food, to control diseases. However, this can cause problems for two reasons

- bacteria can develop *resistance* to the antibiotics, so the drugs can no longer be used to treat sick animals
- it is important that there are no traces of the drug present in meat or other animal products, as some people are *allergic* to antibiotics.

Farmers try to maximise the amount of product they obtain by using **high yield animals.**

Some of these have been developed over many years through *selective breeding*

e.g. cows with high milk yield
pigs with low back-fat ratio

Some have been developed more recently through *genetic engineering*

e.g. animals with genes for disease resistance
pigs with genes for increased growth rate

Some animals are given injections of **hormones** which will increase their growth rate or milk yield.

The farmer must balance the demands of maximum food production (and therefore high profits) against the cost and responsibilities of the production process.

Costs. These include energy to heat buildings, drugs to prevent or treat disease, and food. Farmers must also pay high prices for good breeding animals.

2 Responsibilities to animals

It is important that the welfare of animals is considered at all stages of food production. There is some evidence that animals which are used to make large amounts of food products, e.g. milk, eggs, are more likely to suffer from some *health problems* than other animals. For example:

- dairy cows have an increased risk of *mastitis* (infection of the udder) than other cows, due to the large volumes of milk produced.
- dairy cows have an increased risk of *lameness*, due to walking difficulties caused by an enlarged udder.
- hens kept for egg production have an increased risk of *broken bones*, due to repeated use of body calcium (from their bones) to develop egg shells. Many people feel that this harm to animals is unacceptable.

3 Responsibilities to the environment

When large numbers of animals are kept in confined spaces for food production, there is often a problem with disposal of *waste products*.

For example in dairy farms, cows will produce large amounts of urine and faeces. In the past, when farmers were involved in mixed farming i.e. they kept animals and grew crops, these waste products would have been used as *fertilisers*. Now, most dairy farmers keep large numbers of cows (so they have large amounts of waste) and do not grow crops, so they do not need fertiliser.

“Eutrophication is a major problem when waste gets into fresh water”

If this waste is allowed to drain into streams or ponds, it will cause *eutrophication* i.e. minerals, especially nitrate, in the waste will make algae in the water grow faster. They will form a mat on the water surface, blocking out the light for other plants. When the algae die, saprophytes will make them decompose and oxygen is removed from the water. This results in deoxygenated streams and ponds which cannot support animal life.

FISH FARMING (AQUACULTURE)

Key Points

- some fish are not suitable for fish farming because they grow slowly, or have a complex life-cycle or expensive diet.
- fish are reared in tanks then kept in outdoor cages until they are big enough to be killed.
- movement and food intake are carefully controlled
- fish farming still accounts for less than 10% of the fish we eat.

This involves rearing fish in an enclosed area until they can be used for food.

Currently, less than 10 per cent of the fish we eat are obtained in this way, whilst 90 per cent are caught using nets or hooks and lines.

Some fish are unsuitable for fish farming, for the following reasons:

- they have a very *slow* growth rate, and may take around 5 years to reach their adult size e.g. cod, plaice
- many popular fish e.g. cod, haddock, sole, are carnivores, found at the top of food chains. It is therefore *expensive* to provide them with suitable food.

The fish farmer must consider public demand for the product.

The best profit returns come from fast-growing, fresh water herbivores e.g. carp, but these are not popular with the public.

1 Principles of fish farming

Eggs and sperm are collected from captive adult fish, and fertilisation occurs in tanks. The eggs are kept in controlled conditions until they hatch, then the larvae are provided with suitable high-energy food for rapid growth.

When the larvae have grown enough, they are transferred to outdoor *cages*. For freshwater fish e.g. trout, these cages are usually in lakes. For marine fish e.g. salmon, the cages may be in sea lochs, or in the open sea.

The cages keep the fish in one place, and allow them to be fed easily, while preventing losses from predators. Fish can be kept in the cages for 1–2 years, before being killed for food.

Within a food chain, energy is lost at each trophic level e.g. in moving around and carrying out vital body processes. Fish farmers try to maximise energy transfer by:

restricting movement – larvae are kept in small tanks and adults are kept in cages, so that they do not use energy unnecessarily.

controlling food intake – for maximum growth rate, fish must be given large quantities of nutrient-rich food. This is normally processed into pellets made from molluscs and small fish (of varieties humans do not choose to eat e.g. sand eels), with added vitamins and minerals. Such a diet is very expensive.

controlling temperature – fish are ectothermic i.e. they do not control their body temperature, so they do not use energy keeping warm. However, they will grow faster if they are kept in warm conditions. It is easy to control the temperature of the indoor stages i.e. eggs and larvae, but once the fish have been transferred to outdoor cages, it is virtually impossible.

Large numbers of fish kept in tanks or cages are at risk of **disease** e.g. from parasites such as lice, and from fungi and bacteria. This is avoided in two ways:

- spraying the water in the cage area with *pesticides* to kill parasites. This is only effective if the cage is in a relatively enclosed area e.g. small lake
- adding antibiotic and anti-fungal *drugs* to the fish diet.

Farmers try to maximise the amount of product they obtain by using **high-yield fish**. Some of these have been developed by selective breeding, and others by genetic engineering.

Farmers must balance the demands of maximum food production (and therefore high profits), against the costs and responsibilities of the production process.

Costs. These include energy to heat indoor stages, drugs and pesticides to prevent disease, and very high food costs.

2 Responsibilities to animals

Fish farming represents a major change in the lifestyle of many fish, e.g.

- they can no longer migrate

- they can no longer range over large areas of water to feed
- they are kept at very high density i.e. large numbers of fish per cage.

Although little work has been done on the effects of these changes, it is likely that they are detrimental to the fish involved.

3 Responsibilities to the environment

Problems arise as a result of two of the management procedures:

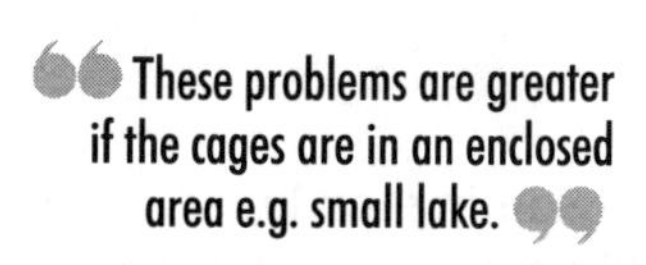

- when caged fish are fed on processed, pelleted food, some of it will fall through the bottom of the cage and settle on the lake or sea bed. Here the pellets will disintegrate, releasing their nutrients. This has a damaging effect on the natural food web in this region.
- when the cage area is sprayed with pesticides to kill fish parasites e.g. sea lice, the pesticide will spread through the water and affect other invertebrates, disrupting food webs.

GLASSHOUSE HORTICULTURE

Key Points

- conditions are carefully controlled inside the glasshouse to give maximum growth rate
- high yield plants should be used
- disease is a problem in enclosed areas, but farmers can use chemical or biological methods to combat this.

Plants use light energy from the sun, and convert it into chemical energy by the process of photosynthesis. This is a relatively inefficient process, with only 1–2 per cent of the light energy which falls onto a plant being converted to carbohydrate.

Horticulturalists aim to maximise growth rate by

- providing optimum conditions for growth of plants
- reducing loss from pests and diseases.

1 Providing optimum conditions for growth

When plants are grown in glasshouses, they are in a totally artificial environment i.e. every aspect of their environment is controlled, e.g.

Amount of light – daylength can be increased, or light intensity boosted by using artificial light. In the middle of the day, shades can be used to protect plants.

Amount of water – plants must have enough water to prevent wilting (and therefore closure of stomata), but root cells must have access to air, so soil must not be waterlogged. *Transpiration rate* will determine the amount of water needed, and this is affected by temperature.

Temperature – the temperature will determine the metabolic rate of the plant i.e. rate of photosynthesis and respiration, as these processes are *enzyme-controlled.* It follows that a warm temperature i.e. 20–30°C, will result in rapid photosynthesis and rapid growth. In addition to this, many plants need a period of time at a specific temperature before they will flower, or before seeds will germinate.

Amount of carbon dioxide – the air around us contains 0.04 per cent carbon dioxide. If plants are given optimum conditions for rapid growth, it may be necessary to add extra carbon dioxide to glasshouses. This can be done by pumping it in, or burning fossil fuels e.g. paraffin, to achieve a concentration of 1 per cent.

virus

Amount of minerals – plants make carbohydrate by photosynthesis, but they need minerals to convert it to protein. Plants also need minerals to make nucleic acids e.g. DNA, hormones e.g. auxin, and chlorophyll, and to control active transport and translocation. Three minerals are needed in relatively large amounts, *–nitrogen, phosphorus* and *potassium*, but many others are vital in smaller amounts. They can be added to the soil or to the watering solution.

"This is a very important point; read the section on limiting factors again."

It is vital that these requirements are considered as a group, otherwise one will act as a *limiting factor* (see Topic 4.6).

In addition to this, **weeds** i.e. non-crop plants, should be removed. Weeds will limit crop growth in two ways,

- they remove water and minerals from the soil
- they may shade the crop plant, reducing its light.

pesticide

Weeds may be removed by hand, which is very time-consuming, or killed using *herbicides*. Herbicides are chemicals which selectively kill some plants, but do not affect the crop.

2 Reducing loss from pests and diseases

Pests are animals which damage the crop

e.g. caterpillars eat crop leaves
aphids (greenfly) pierce phloem tubes and feed on sap
wireworms attack roots, reducing water and mineral uptake.

fungi

disease

They can be controlled in two main ways:

Chemical pesticides sprayed onto the plant. The chemicals kill the pest, but do not harm the crop. Initially pesticides tend to be very effective, but problems may arise later because

bacteria

- the pesticide may kill other animals e.g. bees, birds, small mammals. This is not a major problem if it is used in glasshouses
- it may contaminate the crop, causing harm to humans who eat it
- pests may develop *resistance* to the pesticide by natural selection (see Topic 6.7), so that it is no longer effective.

Biological control methods. This involves using a natural predator to control the pest

e.g. introducing ladybirds to eat aphids
spraying a solution containing bacteria onto plant leaves to kill caterpillars.

These methods are very *specific* i.e. only the pest is affected, while the crop and other animals are unharmed. It is not as rapid as using chemical pesticides, but is a better long-term solution.

Diseases
Diseases are caused by fungi, viruses and bacteria. They affect plant metabolism, causing large losses in crop yield.

They are an important problem in glasshouses, where large numbers of plants are confined in an enclosed environment.

Fungi can be killed using *fungicides* sprayed onto seeds or plants.

Viruses and *bacteria* are spread by insects as they feed on plants, so *controlling insect pests* within the glasshouse will limit the spread of disease.

5 CONSERVATION

Key Points

- conservation involves *action* by individuals as well as governments and industry
- habitats are being destroyed by human activities
- organisms living in those habitats will decline and may be at risk of extinction
- conservation works towards maintaining a *natural ecological balance* in our surroundings

Whenever humans interfere with a natural environment, they will change the ecological balance e.g.

- when land is cleared to build houses and roads, large areas of habitat are destroyed
- when water flow is changed e.g. marshy land is drained, or estuaries are dammed up to make marina areas, natural habitats are destroyed
- when crop plants are grown, rather than native wild plants, food webs in that area will be affected
- when animals are hunted for sport, or for products which can be used by humans e.g. furs, ivory, whale oils, their numbers will fall and they may become extinct.
- when natural resources are taken from the land e.g. in mining, quarrying, logging, habitats are destroyed. Deforestation causes particular damage as it can speed up the process of erosion.

- when chemicals e.g. pesticides, herbicides, are released into the environment, plants and animals may be killed
- some pollutants cause major changes which affect all living things e.g. excess carbon dioxide gas is linked to the greenhouse effect, and global warming; CFCs damage the ozone layer allowing harmful radiation to reach the earth.

Conservation involves doing things to *limit the impact* of humans on the environment, and to maintain a *natural ecological balance* e.g.

- controlling building in green-belt areas, so that local habitats are saved
- preventing the destruction of important habitats by giving them legal protection e.g. as national parks, nature reserves or SSSI (sites of special scientific interest)
- allowing wild plants to live alongside crop plants, by retaining hedgerows and avoiding the use of herbicides
- protecting endangered species by law
  e.g. in Britain it is illegal to disturb bats, or to pick or damage fen orchids.
- maintaining the diversity of habitats by responsible management
  e.g. coppicing, dredging ditches
- recycling natural resources wherever possible to limit environmental damage. Glass, metal, paper and plastics can all be recycled, saving energy and valuable natural resources, and preventing damage to habitats
- encouraging organic farming so that living things are not harmed by agricultural chemicals
  e.g. pesticides, herbicides, artificial fertilisers
- limiting the release of pollutants by enforcing or improving existing laws, and by reducing energy use by individuals e.g. walking or cycling instead of using a car.

Although some of these conservation measures depend on laws and decisions made by governments, there are lots of things that all of us can do in our everyday lives to help to conserve our environment.

Links with other topics

Related topics elsewhere in this book are as follows:

Topic	Chapter and Topic number	Page no
Photosynthesis	4.6	105
Transport of water and minerals in plants	4.7	131
Selective breeding	6.2	209
Air pollution	7.2	291
Water pollution	7.2	294
Populations and resources	7.3	298
Changes in the biosphere	7.4	301
Energy and food chains	8.3	327

ANSWERS TO REVIEW QUESTIONS

A1 (a) (i) insect larvae (ii) tree (iii) insect larvae
(b) (i) photosynthesis (ii) digestion (iii) respiration
(c) energy, from the sun

A2 (i) A

(ii) large soil organisms –insects/millipedes/worms/woodlice

(iii) micro-organisms; bacteria, fungi (saprophytes/decomposers)

(iv) 1. less air for action of aerobic microbes
2. fewer soil organisms at this depth.

A3

A – nitrification | D – 78%
B – nitrogen fixation | E – polypeptides
C – nitrate | F – amino acids

Comments

This is a sort of 'either-you-know-or-you-don't' type question. If you don't, you should revise the relevant section. Note that the question is testing your familiarity with protein digestion as well as the nitrogen cycle.

A4 (a) More carbon dioxide released from burning the forest. (i) Less carbon dioxide absorbed by the forest.
(b) (i) They are centres of very high biological diversity; important as a 'genetic resource' for food, medicines. (ii) They remove carbon dioxide from the atmosphere.

Comments

A straight forward question on a topical subject.
(a) Also, more carbon dioxide is released by decomposition of 'debris' left after burning or cutting.
(b) There are several possible answers here. Give *ecological* rather than economic, environmental or aesthetic reasons.

A5 (a) 50 kJ; 5kJ; 0.5kJ; respectively.
(b) Respiration.
(c) (i) Green plant, capable of photosynthesis, which produces organic (food) material from inorganic raw materials. (ii) The sun.
(d) Since energy transfer is so inefficient within the food chain, animals 'further down' the food chain need to consume a relatively large amount of prey to obtain sufficient energy to stay alive. After 4 or 5 links in the food chain, this is no longer energetically worthwhile.
(e) A food web is more complex, and consists of more than one food chain.
(f) There are two types of effect. One is to reduce (indirectly) the amount of food available to the eagle, since caterpillars die, and less small birds can be supported on them. The other effect is by a progressive 'magnification' of concentrations of toxins down the food chain, which may accumulate in large concentrations in birds of prey.

Comments

(a) In each case, only 10 per cent remains. Calculate 10% of each 'box' to find the value in the next box. All you have to do is move the decimal point (i.e. after the last digit) one place to the left in each case (revise this part of your maths syllabus if you're not confident about this).
(b) Other answers would include loss of body heat, loss of heat in egestion, excretion.
(d) A difficult concept to get across! Use your own way of explaining this, but make sure you refer to 'energy' and 'inefficient transfers', or equivalent.
(e) You could mention that individual organisms often feed, or are fed on, by more than one type of organism.
(f) Again, there are several approaches to this question. Make sure you refer to the actual example given.

A6 Answer to Question 6 see page 350

EXAMINATION QUESTIONS

Question 1

The diagram below shows the nitrogen cycle.

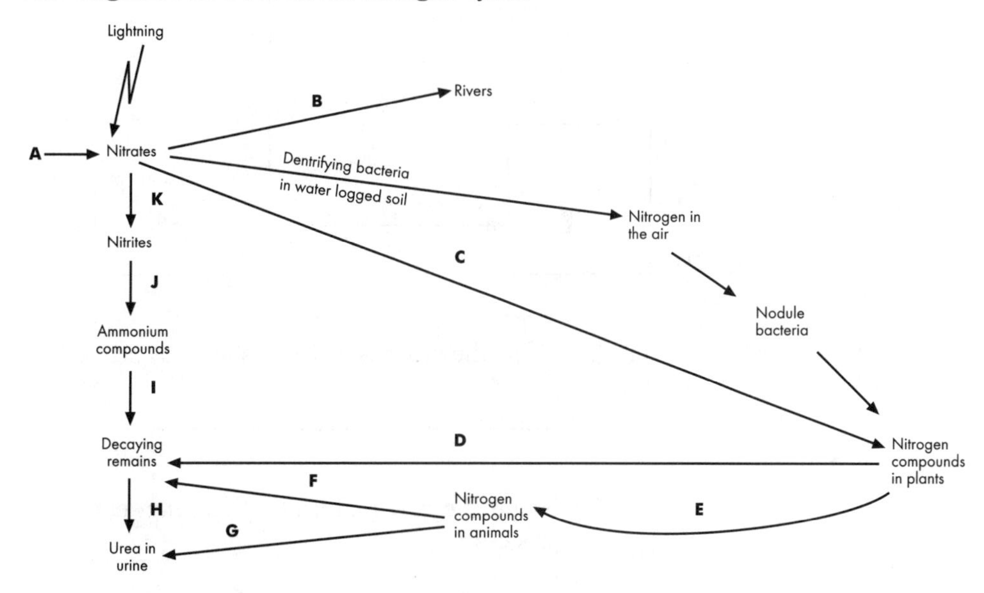

Fig. 8.22

(a) The table below lists some of the processes in the nitrogen cycle.

Complete the table by writing the letter or letters from the diagram showing the position of the process in the nitrogen cycle.

The first box has been done for you.

Process	*Letter(s)*
Nitrates absorbed by plant roots	C
Animals feeding	
A farmer adding fertiliser	
Death of living things	

(3)

(b) Name **two** types of organism which bring about process **I**.

1. ______________________________

2. ______________________________

(2)

(c) Organisms which bring about process **I**, **J** and **K** need oxygen. Use this information, your own knowledge and the facing diagram to suggest two reasons why nitrate levels fall in waterlogged soil.

1. ______________________________

2. ______________________________

(2)

Question 2
The diagram below shows the carbon cycle.

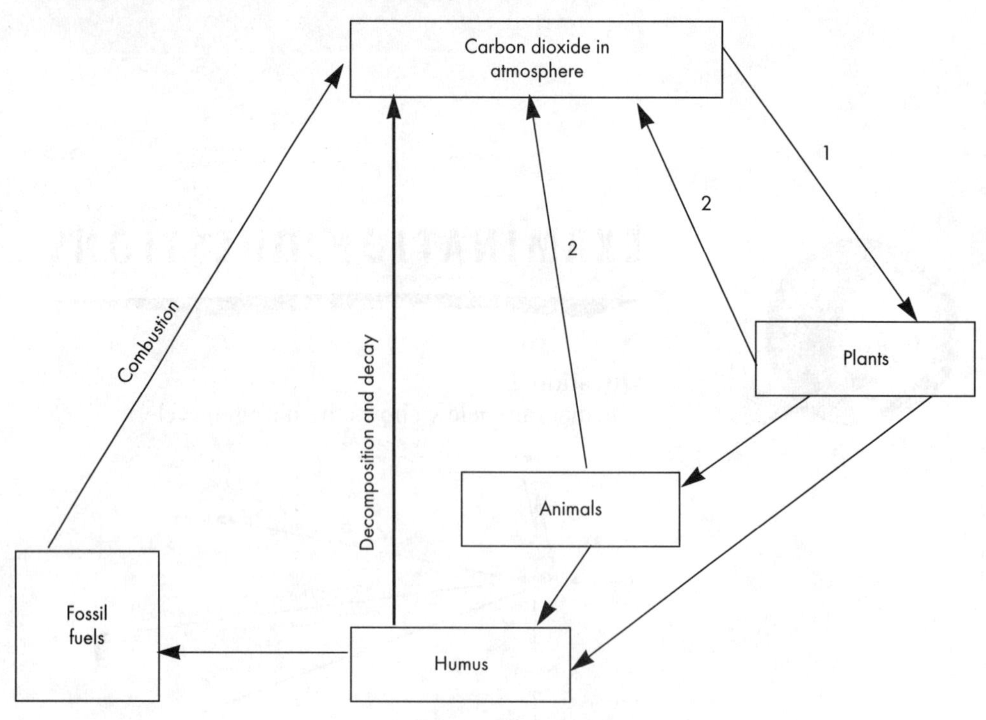

Fig. 8.23

(a) (i) Name the process which is shown by arrow 1.

____ *(1)*

(ii) Name the raw material, other than carbon dioxide, which is needed for process 1.

____ *(1)*

(b) Name the process shown by the arrows labelled 2.

____ *(1)*

(c) Name **two** groups of organisms which bring about decomposition and decay.

1. ____

2. ____ *(2)*

(d) Name **two** fossil fuels in common use.

1. ____

2. ____ *(2)*

Question 3
(a) Many bacteria which are saprophytes take part in the breakdown of plant material in nature. Name another type of organism which is a saprophyte.

____ *(1)*

(b) Some saprophytic bacteria break down starch into ethanol (a type of alcohol). The diagram shows a system in which saprophytic bacteria can be grown to produce ethanol. Suggest the role of each of the following in this system:

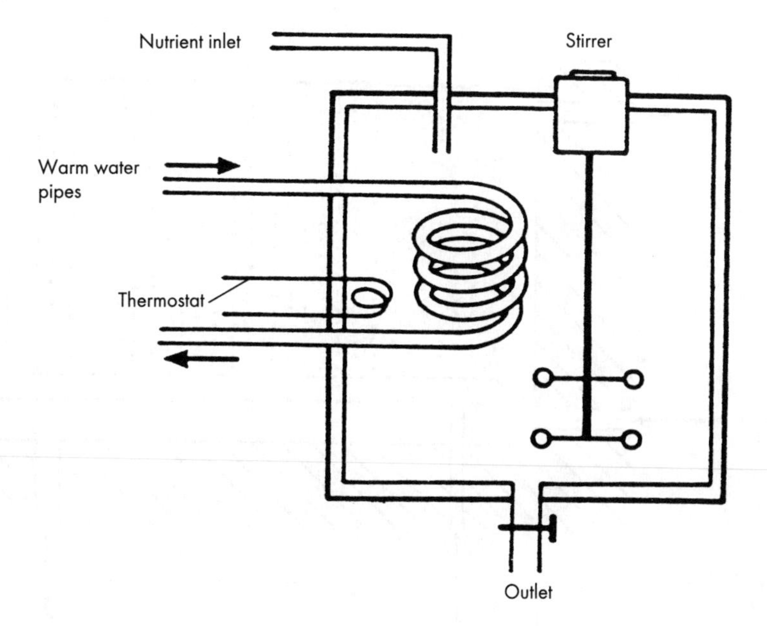

Fig. 8.24

(i) The warm water supply

______________________________ *(1)*

(ii) The stirrer

______________________________ *(1)*

(iii) The thermostat

______________________________ *(1)*

(c) Because these bacteria are easily damaged by heat, the system is run at about 28°C. A new variety of these bacteria is being developed which can remain active at higher temperatures. Suggest why this might be an advantage.

____

____

______________________________ *(2)*

(d) Suggest **one** advantage of using ethanol rather than petrol as a fuel in cars.

____

______________________________ *(1)*

Question 4

The diagram shows the main stage of the 'activated sludge' method of sewage treatment. Activated sludge is rich in microorganisms and is added to the sewage in large open tanks. Air is bubbled through the sewage/sludge mixture continuously to supply the microorganisms with oxygen.

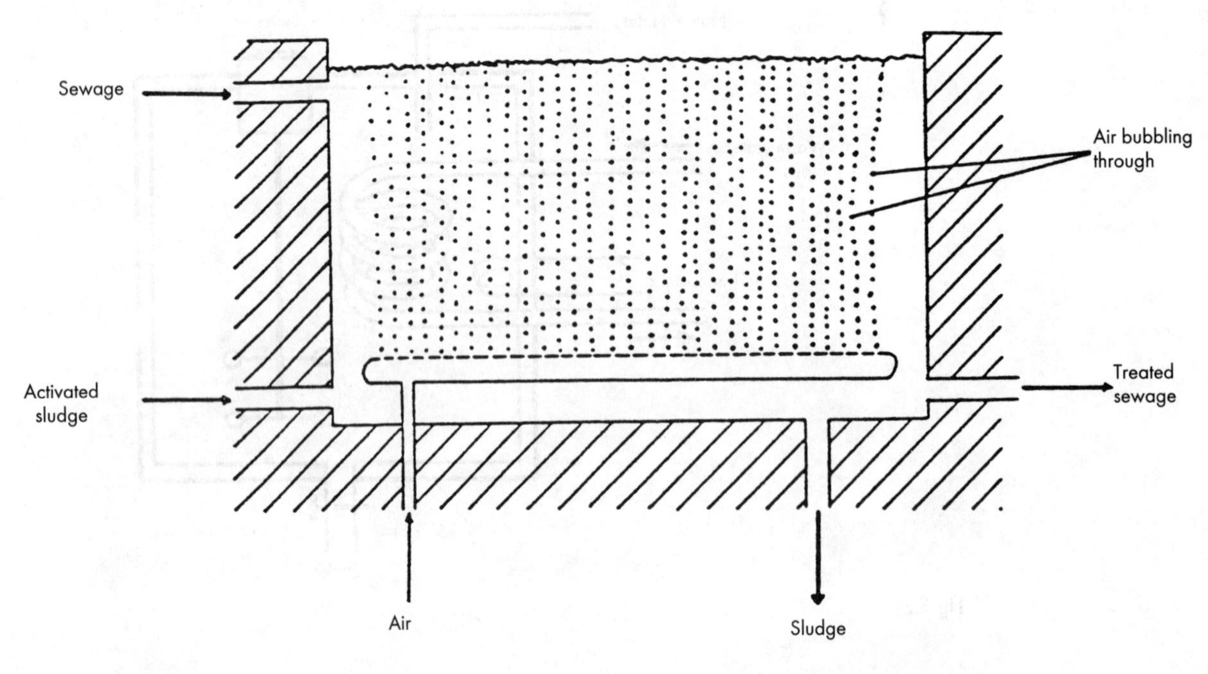

Fig 8.25

(a) For what biological process do the micro-organisms need oxygen?

____

____ *(2)*

(b) Describe how micro-organisms in the activated sludge make the sewage harmless.

____

____

____ *(4)*

Question 5

Fig. 8.26 shows the food web of an oak tree.

(a) Give the **names** of each trophic (feeding) level (1 to 4) of the food web.

(b) Select **three** organisms in the food web which would be affected if the flowers of the oak tree were not pollinated and fertilised. State a reason for each of your choices. *(3)*

(c) The oak tree is shown in winter, after leaf fall. Write down the names of the materials **A**, **B** and **C** in the following cycle, showing how the oak tree receives some of its required mineral salts for the next year's growth. *(3)*

(d) Construct a simple carbon cycle based on the leaves of the oak tree. *(2)*

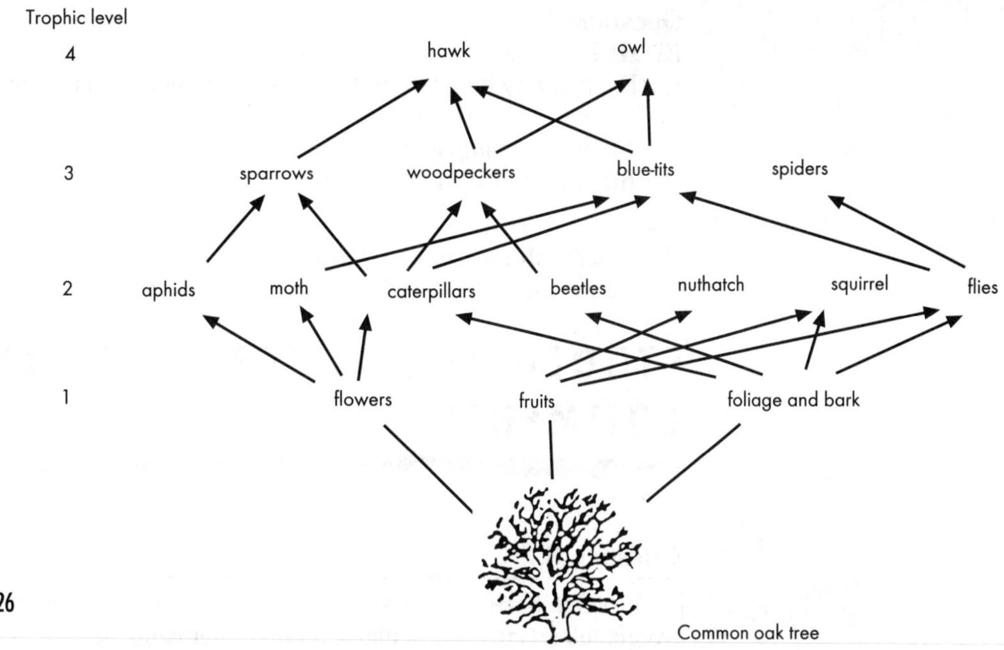

Fig. 8.26

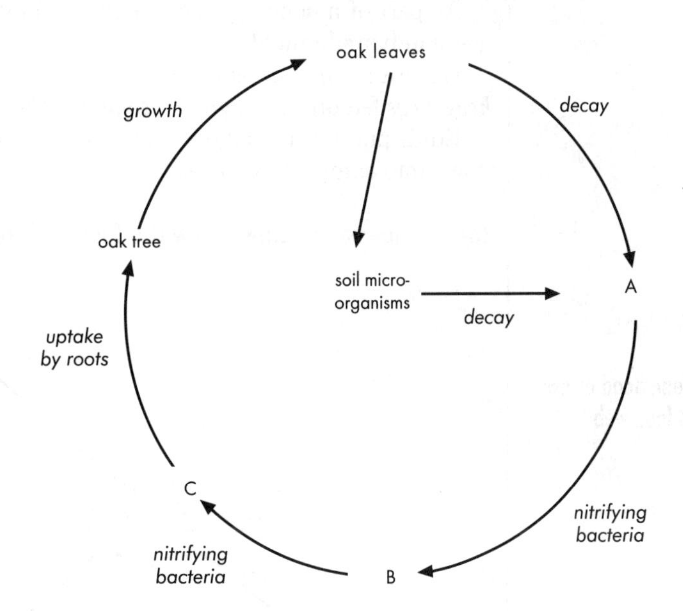

Fig 8.27

Question 6

The Figure (Fig. 8.28) indicates the major stages in the Carbon Cycle.

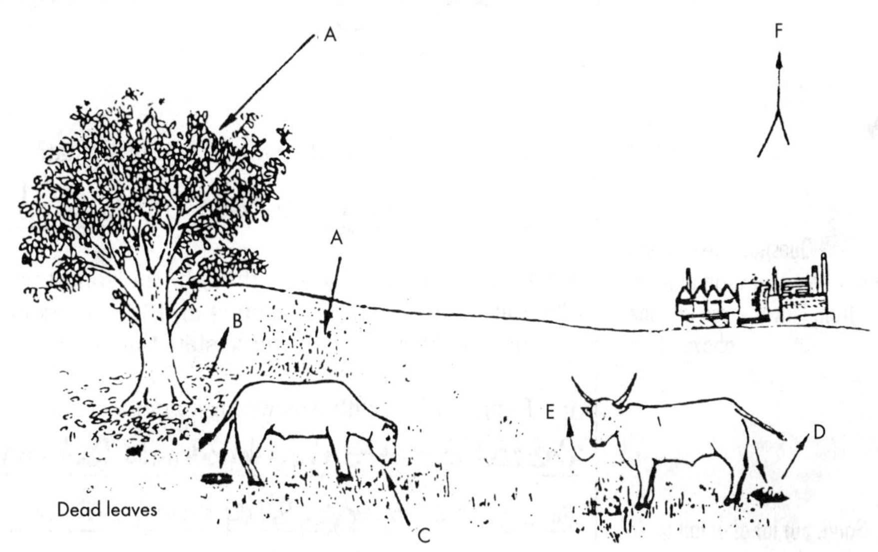

Fig. 8.28 With reference to the letters **A** to **F**, explain the process involved. *(12)*

Question 7
EITHER
(a) Explain why it is important to conserve living species for the benefit of

(i) the human population, and
(ii) the natural environment.

OR
(b) Discuss the reasons for recycling manufactured materials, using named examples. *(6)*

STUDENT'S ANSWERS WITH EXAMINER'S COMMENTS

Question 8

Edith Sidebotham is a keen gardener. She has an allotment where she grows vegetables. She is particularly proud of her cabbages.

As part of a school project, her son Roger investigated the plants and animals living in her allotment.

He noted that *slugs* could eat the cabbages, as could *caterpillars*. A pair of *hedgehogs* fed on the slugs and *song thrushes* ate both slugs and caterpillars.

Edith put any damaged leaves on the compost heap where they were broken down into compost by *worms*.

(a) In the space below, draw the food web based on Edith's cabbages. *(3)*

Excellent, clear, good arrows and correct food web

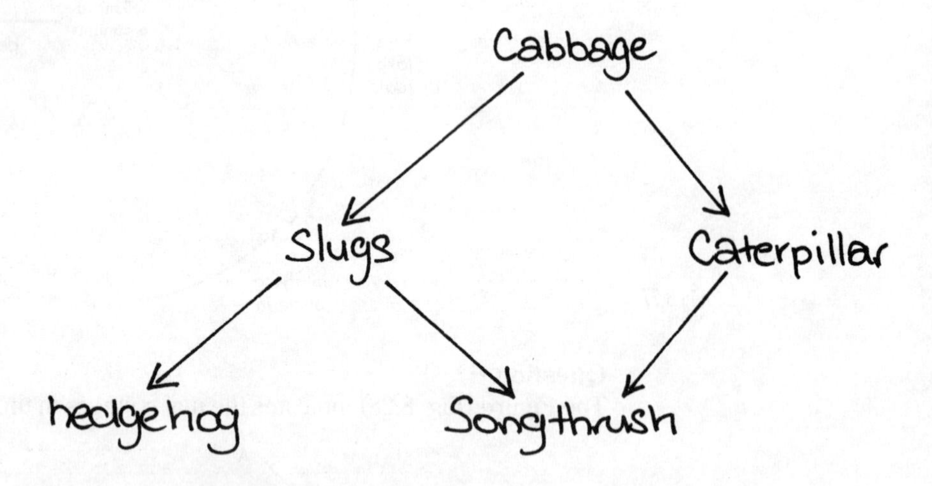

Fig. 8.29

(b) What is the role of the worms in this habitat?

They aerate the soil *(1)*

Question refers to *this* habitat where they form compost from leaves – see last line above

Edith got a cat because she thought the song thrushes were eating her cabbages. The cat kept the song thrushes and the hedgehogs away. Edith found that her cabbages were in an even worse state than before.

(c) Explain why Edith was wrong to get a cat for her allotment

The cat ate the thrush which fed on slugs and caterpillars and not cabbage. Slug numbers and caterpillars increased because no birds left to keep their numbers down. *(1)*

Good, but lot of detail given for one mark!

Question 9

(a) Soil from a deciduous wood in the British Isles contains an abundance of micro-organisms, annelids and wingless insects.

(i) Name the two main groups of micro-organism that you would expect to find

Bacteria

Fungi

(2 marks)

❝ Correct. ❞

(ii) Name one annelid and one wingless insect that you would expect to find

Annelid: Earthworm

Wingless insect: Caterpillar

(2 marks)

❝ This is not a suitable example; springtails (Collembola) would be preferable. ❞

(b) 10 mm discs cut from fallen oak leaves were placed in nylon bags of various mesh sizes and buried about 3 cm down in the soil in a deciduous wood. The bags were dug up every month, the area of leaf discs remaining was measured and the bags returned to the soil. It was found that the total area of leaf material decreased most rapidly in the bags of 7 mm mesh, about seven times more slowly in the bags of 0.5 mm mesh and that there was hardly any decrease at all in the bags of 0.003 mm mesh, even after six months. The experiments assumed that all leaf material that disappeared was broken down, and concluded that the rate of breakdown was determined only by the size of the mesh.

(i) List the organisms named in your answer to (a) in the appropriate part of the following table.

Size of mesh	Organisms which could NOT pass through the mesh
7 mm	Caterpillar (if big and fat)
0.5 mm	Earthworm . Caterpillar
0.0003 mm	Earthworm . Caterpillar

(4 marks)

❝ This is a very muddled answer. An improved answer would be: 7mm – earthworms; 0.5mm – springtails; 0.0003mm – fungi. ❞

(ii) Comment on the assumption and conclusion of the experimenters stating clearly what criticisms may be made of them.

The rate of breakdown was determined by the amount of animals small enough to enter. Not all the leaf material will have been brokendown. It will have been eaten washed away (soluble bits) by water. The rate of breakdown will also be determined by the season and the temperature and other food levels to determine the number of micro-organisms.

(4 marks)

❝ This would be part of the breakdown process. ❞

❝ This is a good, complete answer, showing a fairly thorough understanding. ❞

(c) Examination of the faeces of millipedes (small herbivorous arthropods) feeding on fallen oak leaves revealed that their faeces were very similar to the leaf litter except that the oak leaves were now in very small fragments. In view of this finding an investigation was carried out on the rate of carbon dioxide release from entire oak leaf litter, ground up oak leaf litter and millipede faeces. The following results were obtained.

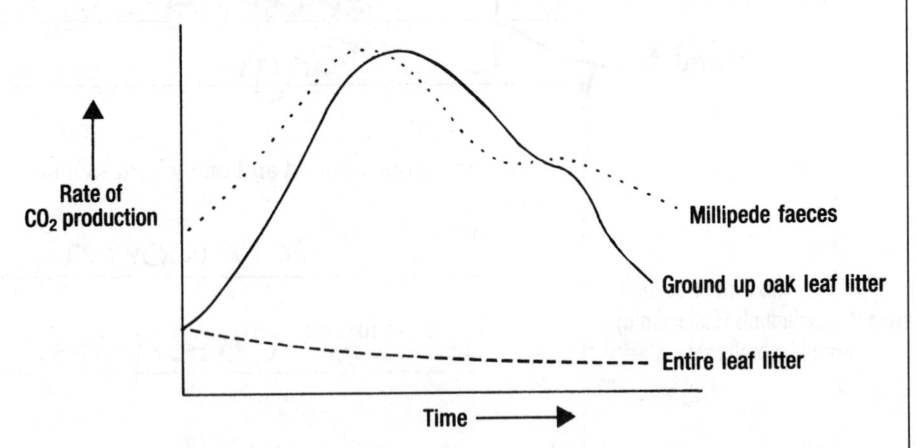

(i) The rate of carbon dioxide release was taken as a measure of the extent of microbial activity. Explain the reasoning behind this.

Respiration provides the energy for the microbial activity and it produces Carbon dioxide

Correct.

(ii) In view of the information provided in this and in earlier sections of the question, put forward a hypothesis to explain the part played by soil invertebrates in the breakdown of leaf litter. State clearly what support your hypothesis is given by the results described earlier.

Soil organisms breakdown leaves and the bigger the organisms allowed to enter the faster the material is broken down as shown by the mesh experiment.

This answer is correct, though the hypothesis is rather limited, and insufficient reference is made to the available information.

(5 marks)

(Total 16 marks)

Question 10

(a) Describe how energy enters, passes through and is lost from a woodland ecosystem *(10 marks)*

Energy enters a wood in the form of solar energy. It is converted to chemical energy by producers (green plants) during photosynthesis, and is then stored as the organic molecules that make up the structure of the plant. When a producer

This is an excellent description of energy flow.

is eaten by a primary consumer (a herbivore eg rabbit), this energy is passed on again, in this case it will mostly be stored as glycogen. The chain continues as a secondary consumer (a carnivore eg Stoat) eats the primary consumer. The secondary consumer might then have to give up this energy as it eaten by a tertiary consumer (eg fox). In this way energy passes through a woodland ecosystem.

> **This is just one of many ways in which energy is available to a consumer; fats and even proteins are important, too.**

Solar -> producer -> primary consumer
energy eg grass eg rabbit

> ***Flow diagrams* like this can be a very effective part of an answer.**

Some energy however is lost as heat through respiration (ie for bodily processes eg digestion, movement in obtaining more food). Because of this there can never be more energy passed on as there was gained and so energy is slowly lost.

> **Good; this is a very important point.**

> **Other processes such as excretion, egestion, death and decomposition should also be referred to.**

(b) Why are woodland ecosystems being lost? What measures can be taken to conserve them? *(4 marks)*

Woodland ecosystems are being lost through damage caused by pollution erosion and by too many trees being cut

> **Air pollution is a factor in the decline of some woodlands.**

It is important to refer to clearance of trees for *urban development.*

down for things like paper and fuel.
In order to conserve them we must try to rearrange our lives in a more (thoughtful) way to the world we live in. By using unleaded petrol, by trying to clean the gasses produced as waste industrially and by using water as power (eg for generating electricity). We can cut down drastically on the poisons we allow to fill the atmosphere. Simple things like recycling paper or by using methane as a fuel for Third world countries. We can stop the number of trees that are unnecessarily being cut down. As well as doing these things it would help if we replaced what we take away by planting more trees and also by using biological methods of controlling pests (eg by introducing predators) instead of using chemicals that are hard to breakdown and produce poisonous oxides that kill off many other things as well as the pests.

Conservation *does* require an understanding and awareness of ecosystems.

There are some good general points here, but the answer seems to refer more to large rain forests than woodlands.

This is a good point to make.

The student has not explained the particular effect of this on *woodland* ecosystems.

(Total 14 marks)
ULEAC

A6 Answer to Question 6, page 333

(a) All energy comes from the *sun*. Light and heat energy is used by *producers* when they photosynthesise. They use it to make organic compounds e.g. sugars, proteins. When the producers are eaten by *herbivores*, the energy in the organic compounds is passed on to the herbivores. When herbivores are eaten by *carnivores*, the energy is passed on to the carnivores. There are *energy losses* eg. due to movement, loss of heat, at *each trophic level.*

(b) Advantage: herbivores are close to the start of a food chain, so relatively little energy has been lost. A piece of land can support about 10 times as many herbivores as carnivores.
Disadvantage: plant material is difficult to digest, so not all the energy is extracted from it. Herbivores spend a lot of time eating.

(c) food chain = simple feeding relationships
e.g. algae → tadpole → water scorpion → perch
food web = several interconnected food chains
e.g. grass → grasshopper → badger
grasshopper → kestrel
grass → rabbit → weasel
rabbit → fox

ANSWERS TO EXAMINATION QUESTIONS

CHAPTER 4

A1

(a) leaf
(b) any *one* of: absorbs water; absorbs minerals; anchors plant in soil (increases surface area).
(c) 1. holds plant more firmly in soil.
2. allows it to collect water (or minerals) from larger area.

An oblique line (/) means *alternative marking points;* either point is worthy of credit. A semi-colon (;) means *separate marking points;* both points are worthy of credit

A2

(a) 1. roots of A are much shorter than roots of B
2. A has no main root of tap root, unlike B
(b) (i) 1. Collect water from soil 2. Collect minerals from soil 3. anchor plant
(ii) Storage of food substances, e.g. starch

A3 (a) A. Cytoplasm B. Nucleus

(b) Large surface area

(c) Chloroplast absorbs sunlight for photosynthesis

A4 (a) Cytoplasm

(b) (i) Animal cell (ii) It contains only a cell membrane and does not have a cell wall.

(c) Measure the width of the drawing; divide this measurement by 0.03 mm.

A5 (a) A – Cell Wall B – Cell Membrane C – Vacuole
D – Nucleus E – Cytoplasm.

(b) Cell wall, membrane, vacuole, nucleus drawn and labelled as shown.

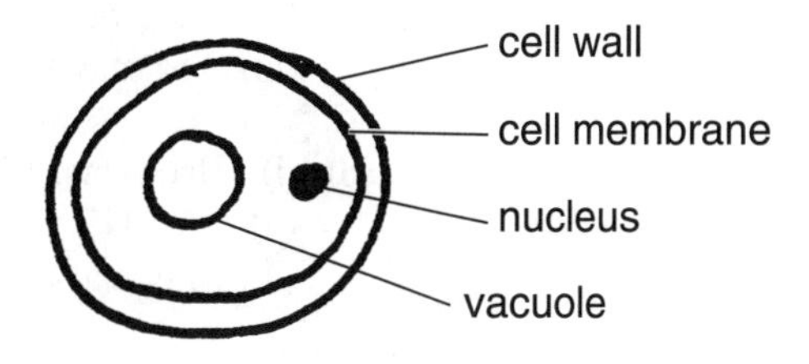

A6 (a) From the top: 2, 3, 6, 5.

(b) Contraction of the biceps muscle raises the arm, which bends at the elbow joint. The triceps muscle is relaxed during this process.

(c) Tendon.

A7 (a) 1 = cortex; 2 = medulla; 3 = ureter

(b) (i) 10 per cent (ii) region 1
(iii) It is reabsorbed from the nephron into the blood.

(c) (i) Less water is lost through sweating on a cold day; the water is lost instead in urine.
(ii) The salt content of the body is maintained at a constant level to avoid problems of osmosis; if a constant amount of salt is present in the diet, a constant total amount will be lost.

(iii) Urine is a solution of the waste urea and other toxins, which must be removed at intervals from the body.
(iv) Consume enough water and salts to maintain minimum amounts in the body.

A8 (a) Homeostasis is the process by which a fairly constant internal environment is maintained within an organism. This is achieved by detecting and correcting any change in the internal environment.

Normal body temperature is 37°C. The body may be subjected to temperatures very different from this, e.g. hot or cold external temperatures, hot or cold food or drink, fever. However, any fluctuations in body temperature are quickly responded to by temperature regulation. Negative feedback is the means by which the body is constantly informed of any difference between normal temperature and actual temperature; the body then responds appropriately.

(b) The water content of the body is controlled by a series of linked processes. The water content of the blood is monitored in terms of changes in blood volume (or pressure). An increased amount of water in the blood causes an increase in blood volume. This is detected by the brain, which releases less ADH (anti-diuretic hormone) into the blood. Reduced amounts of ADH reaching the kidneys cause more water to be lost in urine during excretion; i.e. less water is reabsorbed by the kidneys. This reduces blood volume.

If the water content of the body is too low, blood volume will be decreased and the processes described above will operate in the opposite way; i.e. ADH levels will increase, and less water will be lost in urine.

(c) The sugar content of the body is controlled by regulating the relative amounts of blood sugar and stored sugar. Blood sugar exists as glucose. If levels of glucose exceed the normal concentration, excess glucose is stored as glycogen, for instance in the liver. The conversion of glucose→glycogen is under the influence of the hormone insulin, which is secreted by the pancreas (islets of langerhans). If blood glucose levels are too low, glycogen is converted back to glucose, under the influence of the hormones glucagon (from the pancreas) or adrenalin (from the adrenal glands).

A9 (a) (i) 4060 kJ (i.e. 13460–9400)
(ii) Gain weight (fat)
(iii) She may be pregnant, or be very active.

(b) (i) Breakfast (70:14)
(ii) 2:1 (347:175)
(iii) May cause blocked arteries, coronary thrombosis (heart attack).

(c) Ice cream; 1.3 mg/g

(d) Iron, anaemia; or vitamin C, scurvy

(e) Excess protein in the diet (presence of urea indicates that protein has been broken down).

A10 (a) A = Bronchiole   B = Cells of the alveolus wall
C = Red blood cell.

(b) Pulmonary artery

(c) (i) Diffusion
(ii) 1 Air in the alveolus is constantly changed.
2 Blood in capillary is constantly being circulated.

(d) Thin (single cell thickness) alveolus wall allows short diffusion distances; Moist lining of alveolus allows oxygen to dissolve before entering blood;

Large surface area of alveolus in contact with capillary allows more sites for diffusion to take place

(e) (i) Plasma; Bicarbonate. (ii) Red blood cells; Oxyhaemoglobin.

Comments

Note that this question is really testing your understanding of more than one topic; i.e. breathing, blood circulation, diffusion.

(a) Don't worry if you haven't 'learnt' the 'official' names for structures in questions like this! You can still get marks by *accurately describing* what you can see (or what you know is there).

(b) Remember it's an artery because it carries blood *from* the heart, but unlike other arteries it carries de-oxygenated blood.

(c) (ii) and (d) Use any convenient wording to get your answer across! In (d) make sure you refer only to adaptations which are visible in the diagram. Another possible answer would be that diffusion distances between the lining of alveolus and blood are short.

A11 (a) (i) Pulmonary vein, left atrium, left ventricle, aorta

(ii) The valve is opened by the flow of blood when the left ventricle contracts; the blood pressure in the aorta causes blood to start returning to the heart when the left ventricle relaxes; returning blood fills the valve flaps and closes the valve, so preventing back flow.

(b) (i) 0.8 seconds

(ii) 0.6 seconds; this is the time during each cycle when the pressure in the aorta is greater than that in the left ventricle. This pressure difference is maintained by the closed valve.

(iii) 7.77 kilopascals

(c) Contraction of the muscles in the wall reduces the internal volume of the left ventricle; closure of the bicuspid valve prevents blood returning to the left atrium.

A12 (a) Adrenaline is referred to as the 'fight, fright or flight' hormone, because it is secreted when a person is exposed to potential threats, or the need for rapid, vigorous activity. The general effect of adrenaline is to prepare the animal for muscular exertion.

There are several ways in which adrenaline has its effects on the body. The hormone stimulates the breathing centre in the brain to cause the rate and depth of breathing to increase. This delivers oxygen more rapidly to the lungs, and removes waste carbon dioxide quickly. This effect allows increased rates of respiration in the muscles, and releases more energy. Adrenaline causes an increase in sweating rate, which 'dumps' excess heat generated by muscular activity. Adrenaline stimulates the heart pacemaker to make the heart beat faster. This causes essential substances in the blood to be circulated more rapidly, e.g. oxygen, glucose to the muscles, carbon dioxide from the muscles. Blood supply to the skin and gut are decreased by the effects of adrenaline. This diverts blood away from non-urgent activities to the muscles. Adrenaline also stimulates increased conversion of glycogen to glucose in the liver, and fats to fatty acids in areas of fat deposits. Both these effects liberate more substances which can be used to release energy in muscle respiration. Pupils of the person's eyes widen, allow more visual information to be collected.

Comments

This is an 'unstructured', essay-type question, and needs to be answered carefully! It is important to make your answer logical, clear and complete. Note that the main part of the answer here is really only a list! If you had revised the main points as a sort of 'mental list', you could then 'expand' the list to include more detailed descriptions and explanations. Use the number of marks or space available to tell you how much information to give in your answer. In fact, the answer above is more than enough to earn full marks.

CHAPTER 5

A1 A – Horse Chestnut B – Holly C – Beech
D – Ash E –Sycamore F – Oak

A2 (a) 1 mark for showing the differences in the form of a table headed CENTIPEDE : MILLIPEDE

3 marks, one each for 3 of the following:

1. Centipede legs are longer than the millipede's legs
2. Longer antennae in centipede, shorter antennae in millipede
3. Poison claws present in centipede but absent in millipede
4. Fewer body segments in centipede/more body segments in millipede

(b) (i) Arthropoda
(ii) 1. Jointed legs 2. Segmented body 3. Antennae present

A3 Lycosa – B Scutigerella – C
Geophilus – A Enchytraeid – D

A4 (a) (i) A – 17mm B – 25mm C – 13mm D – 25mm
(ii) A – 5.7mm B – 25mm C – 4.3mm D – 12.5mm
(iii) shortest C, A, D, B longest

(b) (i) 1. Muscular foot 2. Tentacles (or shell) (ii) One (A).

A5 (a) A – 4 B – 1 C – 2 D – 3.
(b) A – 4 B – 3 C – 2 D – 1.

A6 (a) (i) Dolphin (ii) Streamlined shape/fins

(b) Seaweed

(c) (i) Wings (ii) Winged aphid

(d) Birds, Rose, Lizard, Kangaroo/Dolphin, Aphid

A7 C, D, E, B, A, respectively

A8 (a) Slug (Agrolimax)

(b) A = adult thrip, B = harvestman

(c) jointed exoskeleton; more than three pairs of jointed walking legs; body divided into many distinct segments.

A9 (a) Antennae as long as body *Chrysopa septempunctata* B
One pair of wings *Stratiomyia potamida* D
Body more than three times longer than wide *Lepisma saccharina* A
Legs shorter than body *Pediculus humanus* C

(b) Size is a variable feature within the same species for different conditions, for instance age and food availability.

(c) (i) A parasite feeds directly from its living host, with which it is in close relationship.
(ii) Hooks and bristles on its legs allow it to hold on to the host's skin.

CHAPTER 6

A1 (a) Suitable symbols would be:

B (= black) and b (= red)

(b) (i) Circle both parents and red bull in F2.
(ii) Square around both F1 animals.

(c) Homozygous dominant (black) and heterozygous have the same phenotype.

A2 (a) (i)

Blood group	**Genotypes**
A	AA or AO
B	BB or BO
AB	AB
O	OO

(ii) A and B are both dominant to O.

(iii) A and B. Alleles A and B are equally dominant. If they appear together (i.e. AB) both are expressed.

(b) (i)

Parents	**Man (group O)**		**Woman (group AB)**	
genotypes	OO		AB	
gametes	O	O	A	B
F1 genotypes	AO	BO	AO	BO
F1 phenotypes	A	B	A	B

(ii)

1 blood group O?	nil
2 blood group B?	1 in 2
3 blood group AB?	nil
4 blood group A?	1 in 2

A3 (a)

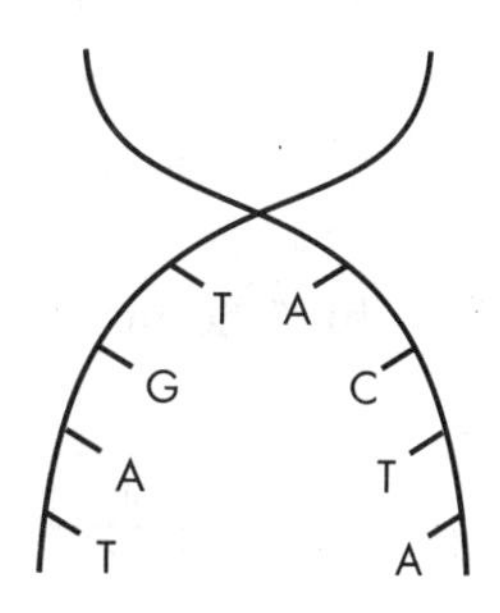

(b) so that each daughter cell gets a copy of the DNA.

(c)

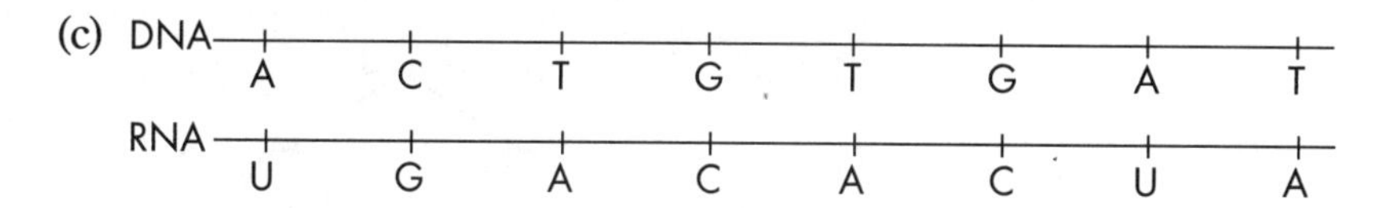

(d) (i) 3 (ii) cytoplasm.

A4 (i) Mutation
(ii) Provided they are inherited by subsequent generations, and if they are not lethal, mutations add to the genetic variation in future generations. Mutations will only affect subsequent generations when they are expressed in the phenotype. When they are expressed, mutations can have three main types of effect: harmful, harmless, beneficial.
(iii) Evolution.

Comments

This is rather a tricky question, and links the topics of genetic variation (mutation), inheritance and natural selection. Like all questions, it is worth reading the entire question to see what the examiner is 'driving at'. You can see from part (iii) that *names* of processes are expected. That should get you off to a good start with your answers to (i) and (iii).

(ii) In your answer, you should focus on how mutations affect future generations. (There is no need to describe how mutations actually occur!) You should show that you understand the importance of *inherited genetic change.* Remember that mutations are passed on to future generations in two ways. In organisms which reproduce asexually, genetic changes are passed onto future generations if the cells containing the mutation form part of the offspring. In organisms which reproduce sexually, mutations are only passed onto the next generation if they involve the sex cells. If mutations are not lethal, they will continue to be carried in the cells of future offspring until a further mutation occurs.

(iii) One way of describing the term 'evolution' is an 'inheritance of genetic change over successive generations'. The answer 'natural selection' would not be correct, since this is simply a mechanism of evolution.

A5 (a) (The answer is shown in the diagram (Fig A.1))

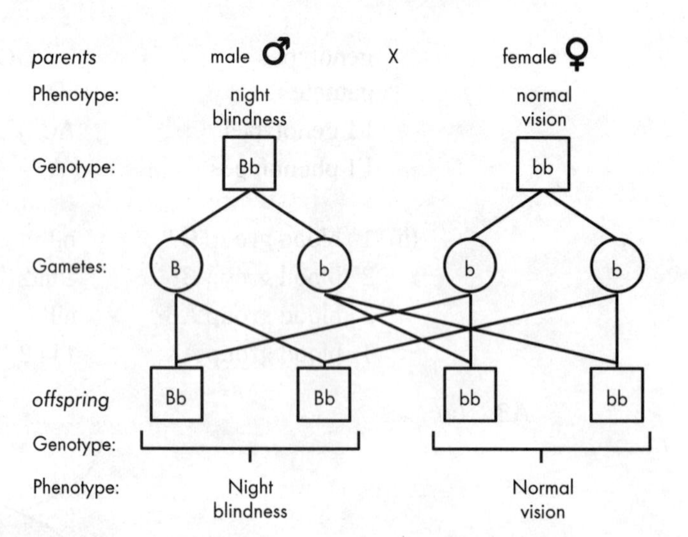

Fig. A1

(b) (The answer is shown in the diagram (Fig. A2))

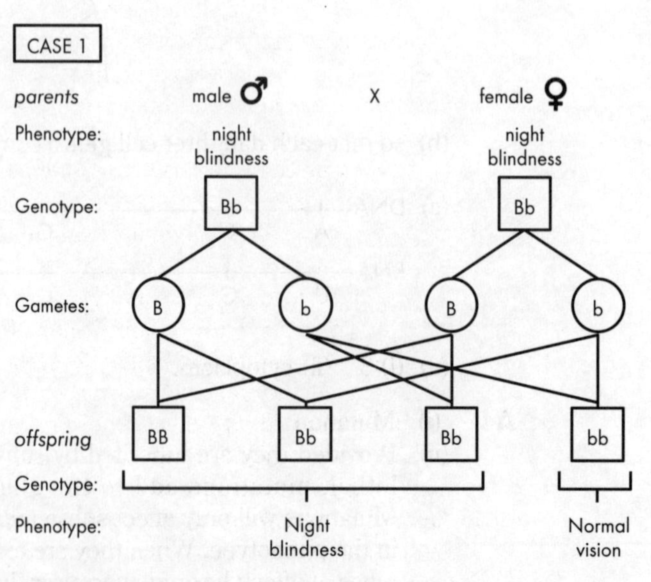

The probability of having a child with night blindness is 75% (3 in 4) if the mother is heterozygous (Bb)

Fig. A2

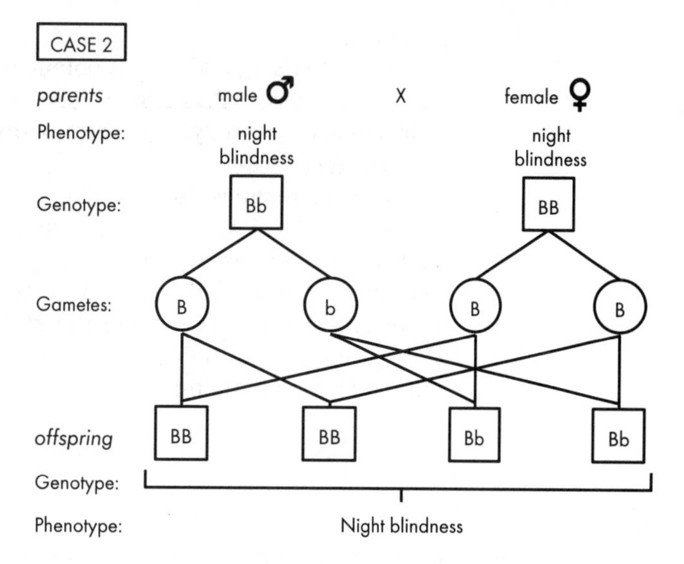

Fig. A2 (*cont.*)

The probability of having a child with night blindness is 100% if the mother is homozygous dominant (BB)

Comments

You need to be familiar with the basic principles of genetics to answer this question. The answers are given as diagrams; use the style of genetics diagram that you are most familiar with.

(a) The mother must be homozygous recessive (bb), since her phenotype (normal vision) indicates that the dominant allele is not present in her genotype. The father must be heterozygous (Bb), since BB would result in all the children having night blindness. The genotypes and phenotypes of the children can easily be worked out, once the parents' genotypes have been determined.

(b) Two genetics diagrams are needed, since you can't be sure whether this mother was BB or Bb. Note that genetics diagrams also include probabilities of night blindness occurring in each case.

A6 Stages B, D and F are each darker in appearance compared with the stage before, and with stage A. These alleles will have resulted from mutations, i.e. inheritable changes in the organism's genotype. The progressive darkening through many generations, including A, B, D and F, could be explained by the natural selection of alleles for darker fur. A darker appearance would provide greater camouflage for animals that were spending an increasing proportion of their time in the dark environment of the dense forest. This would allow darker animals to be better hidden from predators compared with lighter animals. Alleles for lighter fur would therefore be selected against.

There was a slight darkening of appearance in animals from generations C, E and G, but a generally light coloration is retained in these animals. This provides a good camouflage for animals living in an open environment. In this case, alleles for dark fur would be selected against.

Animals from generations B, D and F show a progressive shortening in stature. This would presumably allow them to move more easily through dense undergrowth, and to hide in small recesses in trees.

Animals living in the open environment become taller over many generations (including C, E and G), and develop longer limbs – especially hindlegs. This would presumably allow them to move faster over large distances in search of food, and to escape predators, compared with animals from the dense forest. Animals from the open environment also have longer ears. This is presumably an adaptation to more easily detect the presence of predators. All these changes would have resulted from natural selection of 'beneficial' alleles, which would have arisen by mutations.

Comments

This essay-style question is a potential minefield for candidates who have not properly prepared! Preparation takes two forms. Firstly, you must really know your subject. Secondly, you must be experienced in organising your thoughts clearly and concisely.

There are many ways in which you could answer this question which would earn full marks. Use a style that suits you. However, it is important to refer to such as processes as mutation, genetic variation, natural selection and adaptation, and to show that you understand what is meant by them. It is also important to refer directly to the stages in the diagram (see the question). Remember that individual animals don't change; the changes occur from generation to generation.

CHAPTER 7

A1 The factory at A is releasing both air pollution and water pollution into the environment. Smoke from the chimneys contains sulphur dioxide, smoke particles and carbon dioxide. Sulphur dioxide can be harmful to the growth of crops and other plants, can result in respiratory disease in humans, and can cause corrosion of stonework. Carbon dioxide is a 'greenhouse gas', which is thought to be a major contributor to possible global warming.

Water pollution of the river by the factory at A and the town at B might have several harmful effects on aquatic life. Toxic substances might kill aquatic organisms directly, and may also make the water unfit to drink by humans living downstream.

Rain at the dying woodland C is likely to be 'acid rain', polluted by sulphur dioxide from the factory, and also nitrogen oxides from other sources. Acid rain is though to be responsible for forest 'dieback' and death, particularly in areas where the soil is sensitive to pollution.

At farm D, water pollution may result from excreta from farm animals and from excess fertilisers washed into the river. This may result in excessive growth of algae (eutrophication), which then die and decompose. The activity of decomposer bacteria may then lower the oxygen content of the river, resulting in death of aquatic organisms.

Insecticide used at the plantation E is a possible pollutant, since excess amounts disturb the balance of natural communities. For example, populations of animals not directly affected by the insecticide may increase or decrease in an uncontrolled way. Numbers of insects which are essential for pollinating flowers may be severely reduced.

Comments

The answer to this question needs to be carefully organised, to avoid omissions and repetitions. If necessary, plan your answer in rough first. Use all the information in the diagram. Note that the answer given above is just one of many ways of tackling this question, and doesn't include all possible pollutants.

A2 (a) Insufficient food to support more fish. Gradual increase in the population of a predator.

(b) Food resources used up. Rapid increase in a predator population.

Comments

This is a fairly straightforward question. There are several alternative answers to both parts of this question. Note that the answers for (a) and (b) look very similar, but there are important differences between them. In the first graph, the population is stable and in equilibrium. This indicates that limiting factors are only partially limiting. In the second graph, the population cannot be supported at all, and there is a massive reduction.

Other possible answers for (b) include disease, temperature change (e.g. seasonal change) and pollution.

A3 (a) (i) Nitrogen, pH, dissolved oxygen. High levels of nitrogen, low pH, low levels of dissolved oxygen.
(ii) Low amounts of dissolved oxygen; oxygen is needed for respiration.

(b) (i) Excretion, egestion from cattle contains nitrogenous waste.
(ii) Nitrogen is absorbed by various aquatic organisms.
(iii) Acids from excretion, decomposition of waste produces CO_2.
(iv) Oxygen is used by the increased activity of aquatic organisms, including bacteria.

(c) (i) Pollution is the addition of an unnatural, and potentially harmful, concentration of a substance in the environment.
(ii) Products of excretion and egestion could be diverted away from the river or collected, then perhaps used as organic fertiliser on the land.
(iii) Watercress could become contaminated with bacteria if grown in these conditions.

A4 Essay plan:

(a) Describe dominant role of tree in its environment. Wide range of functions/effects:

1. Habitat, e.g. root system, bark, leaves, fruit for animals, including insects, birds, fungi.
2. Nutrient input, e.g. from leaf fall.
3. Food source, e.g. fruits, leaves.
4. Shade for certain ground layer plant species.
5. Interception of rain water.
6. Protection for animals living on, in or around, e.g. against predators, wind, excess temperature.
7. Competition, e.g. with grassland, which may undergo gradual succession as a result.
8. Stabilisation of soil, which might otherwise become eroded.

(b) Suitable examples:

harmful – pesticides, e.g. DDT; effects on birds of prey
– air pollution, possible effects on forest decline

useful – recycling, e.g. paper – reduce deforestation
– conservation, e.g. national parks, protect threatened species.

A5 (a) 1. Measure the total area of the beach to be studied.
2. Using a $\frac{1}{4}$ m^2 quadrat, place the quadrat at random and count the numbers of individuals in that small area.
3. Record results and repeat until 10 quadrats have been sampled.
4. Add all results together; divide by 10 to find average result.
5. Multiply by total area of beach to find total population.

(b) Using random quadrats, sample the proportion of a particular species of seaweed on the two different areas of beach. Repeat for four different species. Observe differences in colour, size, etc., of seaweeds of the same species on the two different beaches.

(c) 1. exposure of the beach to strong waves and wind.
2. amount of human interference on the beach, e.g. tourism, boating, etc.
3. different types of pollution such as oil and other chemicals in the seawater.

A6 (a) Look for the countries where life expectancy is above 61 years
Answer = 3 (1)

(b) (i) Look for the country where the figure for estimated population and actual population show the greatest difference. You can roughly work out that the biggest differences are for Turkey and Egypt, and then work out the accurate increase. Turkey has an increase of 30.7 millions whereas Egypt is only 25.4 millions.
Answer = Turkey (1)

(ii) The word of instruction asks you to 'suggest' so any reasonable scientific answer will obtain the mark. For example: immunisations have reduced infant mortality (1) birth control may not be widely used (1)

(c) Again you are only asked to suggest reasons, you do not have to know the correct answer, and there may not be one correct answer. Possible answers include: lack of immunisation / lack of basic health care facilities (1) poor nutrition (1)

CHAPTER 8

A1 (a) E; A; D & F.

(b) 1. bacteria 2. fungi

(c) 1. waterlogged soils have low oxygen level; prevents I, J & K working because they cannot respire.

2. denitrifiers do not require oxygen; denitrifiers break down nitrates.

A2 (a) (i) photosynthesis (ii) water

(b) respiration

(c) 1. bacteria
2. fungi

(d) coal, oil

(e) increases carbon dioxide level of the air.

(f) reduces oxygen level of the air; increases carbon dioxide levels.

A3 (a) fungus

(b) (i) so that the starch remains in solution/to increase the rate of decay
(ii) thoroughly mix the starch and bacterial solution
(iii) maintain a constant temperature in the system

(c) fewer bacteria killed; more starch converted to ethanol; quicker.

(d) less pollution/can be obtained from plant waste

A4 (a) Aerobic; respiration

(b) use of enzymes; digestion to form small molecules; the enzymes are secreted; and the products of digestion absorbed.

A5 (a) 1 = Producers 2 = Primary Consumers
3 = Secondary Consumers 4 = Tertiary Consumers.

(b) Nuthatch, because they feed only on the fruits, which would not be formed if no pollination or fertilisation had occurred. Squirrel, because their diet includes fruits, which would not be formed. Flies, because their diet includes fruits, which would not be formed.

(c) A = Ammonium compounds, B = Nitrites, C = Nitrates.

(d) A carbon cycle is shown in Fig. A3 below

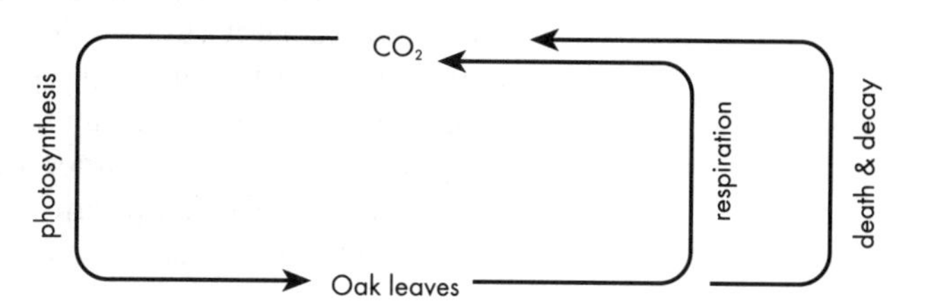

Fig. A3 Carbon cycle

Comments
This question requires an understanding of several topics, including food chains, plant reproduction, and the nitrogen and carbon cycles.

(b) Note that these animals are all directly dependent on fruits. Other animals in the food web would also be affected indirectly (e.g. spiders, blue-tits), but you would need to explain exactly how, if you were using such animals for your answer.

(c) and (d) are straightforward questions about the nitrogen and carbon cycles, respectively.

A6 Green plants, such as trees and grass at A, absorb carbon dioxide for photosynthesis. Carbon is used to build carbohydrates in the plant. When leaves die and decay, they become food for decomposer organisms (saprophytes), which break down the carbohydrates, which releases carbon dioxide (shown at B). Living leaves are consumed by herbivores (C). Undigested material, mainly cellulose, is egested and is then broken down by decomposers, releasing carbon dioxide (D). Carbohydrates which are digested and absorbed by animals may eventually be used in respiration, which produces carbon dioxide as a waste product (E). Carbon is also released into the atmosphere as carbon dioxide and also carbon monoxide from the burning of fossil fuels (originally from plant material) by domestic and industrial use (F).

Comments
If you are familiar with the carbon cycle, this question should present no serious difficulties. Make sure you refer to each of the processes A to F shown in the diagram. Since there are six processes shown, and twelve marks available, expect to gain up to two marks for correctly describing what is happening in each case.

A7 (a) (i) Humans are totally dependent on other living species, for a variety of reasons. For example, humans are dependent on other animals as a source of food, materials for clothing, and for pulling farm machinery. Humans are dependent on plants also for food, building and clothing materials, and as a source of medicines. The genetic diversity of other living species is important as source material for selective breeding programmes. The potential value of many living species has not yet been exploited in many cases. This is because species themselves have not yet been identified, or because the human need for them has not yet become apparent. The conservation of living species in their natural environment is also important for recreation.

(ii) The balance of living species in natural communities has often become established gradually, and is usually self-regulating and stable. For example, forests are important in regulating the world's climate and for stabilising the soil. However, natural environments can be disturbed by intensive and/or ignorant exploitation by humans. This may result in a degradation in the environment, which in some cases is irreversible. The diversity of natural environments may be reduced as a result.

(b) Manufactured materials are often made from non-renewable resources (which cannot be replaced) or potentially-renewable resources (which can only be replaced if not used too quickly). Examples of materials derived from non-renewable resources include plastics (made from oil) and aluminium (made from mined minerals). An example of a material from a potentially-renewable resource is paper (made from trees). The rate at which these resources are used up can be reduced by recycling the manufactured product. Re-cycling also conserves energy, and saves money.

Comments

Note that *either* (a) or (b) only need be answered. With this sort of question, it's important not to get carried away by your own 'pet obsession'! Take care to present cool, factual answers which are relevant to the question. There may be much more that you could say on a given topic, but remember that you cannot gain more than six marks.

INDEX